CREEP AND FRACTURE OF ICE

Ice creeps when loaded slowly and fractures when loaded rapidly. This book is the first complete account of the physics of the creep and fracture of ice, and their interconnectivity. It investigates the deformation of low-pressure ice, which serves as the basis for the study of a variety of natural phenomena – glaciers, polar ice sheets, floating ice covers and the uppermost region of icy moons of the outer Solar System.

The central argument of the book focuses on the structure of ice and its defects, and proceeds to describe the relationship between structure and mechanical properties, including elasticity. It reviews observations and measurements, and then interprets them in terms of physical mechanisms. A section on creep concludes with a discussion on modeling the ductile behavior of isotropic and anisotropic ice in relation to the flow of glaciers and polar ice sheets. A section on fracture assesses ice forces on engineered structures and analyses the deformation of the ice cover on the Arctic Ocean from the perspective of physical mechanisms.

The book provides a road-map to future studies of the mechanics of ice by exploring its new contexts – the behavior of glaciers and ice sheets in relation to climate change, the dating of deep ice cores, and the projected behavior of sea ice. It also highlights how this knowledge of ice can be transferred into an understanding of other materials, especially the mechanical behavior of rocks, minerals, metals and alloys. Written by two experts in the field (Duval on creep and Schulson on fracture), the book is ideal for graduate students, engineers and scientists in Earth and planetary science, and materials science.

ERLAND M. SCHULSON completed his Ph.D. at the University of British Columbia, and is the George Austin Colligan Distinguished Professor of Engineering at Dartmouth College, USA. He has worked on physical metallurgy, materials science and ice mechanics in laboratories and universities in Canada, France, the USA and the UK. He has served as a Fellow of the Minerals, Metals and Materials Society and of ASM International, and been listed in ISI as a highly cited researcher in materials science. He holds four patents in the field, has published over 250 papers in international technical literature, and been invited to lecture at over 100 conferences, workshops and research centers.

PAUL DUVAL completed his Ph.D. at the University of Grenoble, and has been a CNRS Research Scientist in the Laboratory of Glaciology and Geophysics of the Environment since 1969. His areas of research include ice physics and mechanics, glaciology and materials science. His studies on densification of snow have been developed with applications in paleoclimatology, and he has recently worked on the viscous behavior of the ice on Europa, one of Jupiter's moons. He has published over 100 refereed papers in international journals.

CREEP AND FRACTURE OF ICE

ERLAND M. SCHULSON
Thayer School of Engineering at Dartmouth College, USA

PAUL DUVAL
Laboratoire de Glaciologie et Géophysique de l'Environnement, CNRS, France

CAMBRIDGE
UNIVERSITY PRESS

CAMBRIDGE UNIVERSITY PRESS
Cambridge, New York, Melbourne, Madrid, Cape Town, Singapore, São Paulo, Delhi

Cambridge University Press
The Edinburgh Building, Cambridge CB2 8RU, UK

Published in the United States of America by Cambridge University Press, New York

www.cambridge.org
Information on this title: www.cambridge.org/9780521806206

First published 2009

Printed in the United Kingdom at the University Press, Cambridge

A catalog record for this publication is available from the British Library

Library of Congress Cataloging in Publication data
Schulson, E. M.
Creep and fracture of ice / by Erland M. Schulson and Paul Duval.
p. cm.
Includes bibliographical references and index.
ISBN 978-0-521-80620-6
1. Ice mechanics. 2. Ice crystals. 3. Ice – Creep. 4. Ice – Fracture.
I. Duval, Paul. II. Title.
TA714.5.S38 2009
551.31–dc22
2009004024

ISBN 978-0-521-80620-6 hardback

To Sandra and Marie-Thérèse

Contents

Preface

The purpose of this book is to discuss the creep and fracture of ice and the transition from ductile to brittle behavior in terms of underlying physical mechanisms. Accordingly, the book focuses on ice as a material, particularly on low-pressure or ordinary ice, and proceeds to describe the relationship between structure and mechanical properties, including elasticity. It reviews observations and measurements and then interprets them, generally in terms of concepts that have been developed to explain the deformation of other crystalline materials. Occasionally, new physical mechanisms are introduced, which appear not to be limited to ice. The creep behavior of both the ice crystal and the polycrystal is described. A brief description is given of some polycrystal models developed by the materials science community for the computation of the mechanical behavior of the ice polycrystal. A short analysis of the creep behavior of high-pressure ices is also made. The understanding of creep is applied to the deformation of glaciers and ice sheets, and of fracture, to ice forces on engineered structures and to the break-up of the sea ice cover on the Arctic Ocean. Passing reference is made to certain satellites within the outer Solar System and their icy crusts. Unanswered questions are noted, as is controversy.

Although some knowledge of materials science would be helpful, none is assumed. The book begins with a discussion of structure – crystal structure, defect structure and microstructure – and describes the elements that are fundamental to understanding creep and fracture. Thus, both dislocation theory and fracture mechanics are introduced. Where appropriate, attention is given to experimental methods of measurement.

In reviewing the literature, emphasis is placed upon peer-reviewed papers. Most of the references, therefore, are to articles in scientific and technical journals. Occasionally, conference proceedings are cited. Attempts have been made to be as thorough as possible, and as up to date, although inadvertent omissions almost certainly occur.

Acknowledgements

Eland Schulson acknowledges his former students, whose work forms the basis of much of this book: Rachel Batto, Stephanie Buck, Neil Cannon, Michelle Couture, John Currier, Douglas Fifolt, Andrew Fortt, Mark Gies, Narayana Gulding, Johan Grape, Eric Gratz, Rebecca Haerle, Steven Hoxie, Daniel Iliescu, Douglas Jones, Gary Kuehn, Gregory Lasonde, James Laughlin, Russell Lee, Pooi Lim, Michael Manley, Jeffrey Melton, Oleg Nickolayev, Suogen Qi, Erik Russell, Bradley Smith, Timothy Smith, Katy Taylor and Luke Wachter.

He also acknowledges his colleagues, whose thoughtful comments and constructive criticism helped to clarify things over the years: Ali Argon, Ian Baker, Zdenek Bazant, David Cole, Max Coon, Tom Curtin, John Dempsey, James Dorris, William Durham, Ed Earle, Robert Frederking, Harold Frost, Robert Gagnon, Lorne Gold, Richard Greenberg, Chris Heuer, Bill Hibler, Geir Horrigmoe, Roy Johnson, Ian Jordaan, Mark Kachanov, Francis Kennedy, Ron Kwok, Francois Louchet, Dan Masterson, Jacques Meyssonnier, Bernard Michel, Maurine Montagnat, Don Nevel, Wilfrid Nixon, James Overland, Robert Pappalardo, Victor Petrenko, Robert Pritchard, Terry Ralston, Charles Ravaris, Carl Renshaw, Jackie Richter-Menge, Kaj Riska, Martin Rist, Quincy Robe, Peter Sammonds, Charles Smith, Devinder Sodhi, Walt Spring, Harry Stern, Garry Timco, Ron Weaver, Willie Weeks, Johannes Weertman, Jerome Weiss, Rudy Wenk, Robert Whitworth and Jinlun Zhang.

Thanks go to his assistant Mary Moul, whose tireless reprocessing and good humor got the book ready to go; his editor Matt Lloyd, for advice along the way; and his family, for support and encouragement, and for putting up with it for so long.

Paul Duval would like to acknowledge Ph.D. students whose work also forms the basis of much of this book: H. Le Gac, P. Pimienta, P. Kalifa, O. Castelnau, L. Arnaud, M. Montagnat and J. Chevy. He will remember encouraging smiles from friends worrying about the progress of the manuscript. He especially thanks

Sophie de la Chapelle for her contribution and friendship. During the writing of the book, he has had particularly useful discussions with O. Castelnau, F. Louchet, J. Meyssonnier, M. Montagnat, O. Gagliardini, G. Durand and J. Weiss. He also wishes to stress how much this book owes to discussions and correspondence with M. F. Ashby, P. Bastie, Y. Bréchet, W. B. Durham, C. Fressengeas, T. Hondoh, R. Lebensohn and V. Lipenkov. He would like to thank the encouragement of several colleagues and friends in LGGE.

Finally, he acknowledges the great patience and understanding shown by Marie-Thérèse. He hopes to have kept some time for grandchildren, Victor, Etienne, Camille and Paul.

The authors acknowledge the permission given by several editors and journals (AGU, IGS, CRREL, Elsevier, Springer, *Nature*, *Science*, *Comptes Rendus de L'Académie des Sciences*, Hokkaido University Press, *Canadian Journal of Physics* and Taylor & Francis Ltd) to reproduce some of their figures.

1
Introduction

Creep and fracture of ice are significant phenomena with applications in climatology, glaciology, planetology, engineering and materials science. For instance, the flow of glaciers and polar ice sheets is relevant to the global climate system and to the prediction of sea-level change. The ice stored within the Greenland and the Antarctic ice sheets, should it flow into the sea, would raise the level by $\sim$7 m and $\sim$60 m, respectively. The creep of ice sheets, which can impart strains that exceed unity in several thousands of years, is also relevant to paleoclimatology. The determination of the age of ice and the age of greenhouse gases (Chapters 2, 3) entrapped within deep ice cores depends in part upon constitutive laws (Chapters 5, 6) that describe the deformation of the bodies – laws that must incorporate plastic anisotropy (Chapters 5 and 7) and the presence of water near the bottom. Fracture, too, plays a role in ice sheet mechanics. When the bed steepens, the ice flows more rapidly and creep can no longer accommodate the deformation (Chapters 9, 10). Fracture ensues, leading to the formation of glacier icefalls and crevasses. Fracture is of paramount importance in the catastrophic failure of icefalls, and it is a key step in the calving of icebergs and thus in the equilibrium between accumulation and loss of mass from polar ice sheets. Fracture is also fundamental to the integrity of the ice cover that forms on the Arctic Ocean (Chapter 15) and to the attendant release through open cracks or leads of heat and moisture to the atmosphere.

Water ice is abundant in the outer Solar System. Within the icy moons of the outer planets, pressure can reach as high as several GPa and temperature can reach as low as 40 K. As a result, several high-pressure phases exist (Chapter 8). Their behavior, whether ductile or brittle, influences the thermal evolution and tectonic history of the moons. Jupiter's satellite Europa and Neptune's satellite Triton, for instance, are encrusted in ice that, in both cases, overlies a putative ocean. The thickness of the crust is difficult to model owing to uncertainties in the thermal state of the satellite and to incomplete understanding of the creep and fracture of crustal material. The other significant point is that frictional sliding across strike-slip-like faults may generate sufficient heat to locally raise the temperature of the crust tens of degrees above background.

In engineering, creep limits the service life of a floating ice cover when used as a landing for aircraft or as a site for stationary objects. Also, it sets the load produced on an offshore structure by a floating ice cover moving slowly against it. When the cover moves more rapidly, it fractures (Chapter 14), under stresses that exceed the creep strength. In other applications: the fracture of ice is key to polar marine transportation; it plays a role in the removal of atmospheric ice from ships, power-lines and other structures; and it is a factor in the integrity of engineered freeze walls.

Creep and fracture of ice are also relevant to the deformation of other materials. Ice is elastically relatively isotropic (Chapter 4), but is plastically highly anisotropic (Chapter 5). Owing to the latter quality, which arises from the hexagonal crystal structure that the material adopts under terrestrial conditions (Chapter 2), ice is a useful analogue for validating micro-macro polycrystalline models that are used to simulate the behavior of anisotropic materials (Chapter 7). It is proving to be useful also in elucidating the nature of the intermittent and spatially variable plastic flow within materials that possess intrinsically low lattice resistance to slip (Chapters 5, 6). In such cases, deformation is better characterized by power-law distributions and fractal concepts and by the cooperative movement of dislocations than by simple averages and the movement of individual defects. The localization of deformation in slip lines and slip bands with scale-invariant patterns, as shown by X-ray topography in ice crystals (Chapter 5), appears to be an intrinsic feature of crystal plasticity. Through detailed microstructural analysis, ice provides evidence for the role of grain boundaries as sources of dislocations (Chapter 2) and for the importance of dynamic recrystallization in long-term creep (Chapter 6). On fracture, owing to its optical transparency and its relatively coarse grain size (1–10 mm) ice is instructive in revealing the character of primary and secondary cracks and their role in the process of brittle compressive failure (Chapters 11,12). Wing cracks, long postulated as fundamental elements, are important features. So, too, are other secondary features termed comb cracks and their attendant fixed-free microcolumns: under frictional drag across the free ends, the columns appear to bend and then break, thereby triggering a Coulombic fault (Chapter 12). The comb-crack mechanism appears to operate in rocks and minerals as well. Ice has led also to a better understanding of the ductile-to-brittle transition in rocks and minerals, in terms of a competition between creep and fracture (Chapter 13), and it has helped to elucidate the difference between Coulombic or frictional faults and plastic or non-frictional faults and thus between brittle and brittle-like compressive failure (Chapter 12). Ice has also made clear the governing role of friction in post-terminal failure under confinement (Chapter 12). In other words, while the understanding of ice benefits greatly from the understanding of other materials, ice offers something in return.

Ice offers some insight into the scale effect in brittle failure. It is now clear that the failure envelope of the arctic sea ice cover has the same shape as the envelope for test specimens harvested from the cover (Chapter 15). Similar, too, is the appearance within the cover of deformation features that resemble the wing cracks, comb cracks and Coulombic shear faults of test specimens. The implication is that, on both spatial scales, brittle compressive failure is governed by similar physical processes.

In discussing the creep and fracture of ice, we focus not on applications per se. Rather, our primary objective in this book is to elucidate physical mechanisms that govern the deformation. That said, we do address the flow of ice of polar ice sheets (Chapter 7) through micro-macro modeling, the nature of ice forces on offshore structures (Chapter 14) and the processes underlying the fracture of the arctic sea ice cover (Chapter 15).

For a discussion of topics beyond the scope of this book, we recommend: Hobbs (1974) and Petrenko and Whitworth (1999) on ice physics; Hutter (1983), Hooke (1998), Lliboutry (2000) and Paterson (2000) on glaciology; Bamber and Payne (2003) on the overall mass balance of the cryosphere and on ice dynamics; Schmitt *et al.* (1998) on extra-terrestrial ice; Lock (1990) on the growth and decay of ice; Wadhams (2000) on ice in the ocean; and Weeks (2009) on sea ice. For a discussion on atmospheric icing, we recommend a series of papers edited by Poots (2000).

References

Bamber, J. L. and A. J. Payne, Eds. (2003). *Mass Balance of the Cryosphere: Observations and Modeling of Contemporary and Future Changes*. Cambridge, New York: Cambridge University Press.

Hobbs, P. V. (1974). *Ice Physics*. Oxford: Clarendon Press.

Hooke, R. L. (1998). *Principles of Glacier Mechanics*. Upper Saddle River, N.J.: Prentice Hall.

Hutter, K. (1983). *Theoretical Glaciology: Material Science of Ice and the Mechanics of Glaciers and Ice Sheets*. Dordrecht: Reidel/Terra Scientific Publishing Company.

Lliboutry, L. (2000). *Quantitative Geophysics and Geology*. London, New York: Springer.

Lock, G. S. H. (1990). *The Growth and Decay of Ice*. Cambridge: Cambridge University Press.

Paterson, W. S. (2000). *Physics of Glaciers*. 3rd edn., Oxford: Butterworth Heinemann.

Petrenko, V. F. and R. W. Whitworth (1999). *Physics of Ice*. New York: Oxford University Press.

Poots, G. (2000). Ice and snow accretion on structures: Introductory remarks. *Phil. Trans. R. Soc. Lond., Ser. A*, **358**, 2803–2810.

Schmitt, B., C. deBergh and M. Festou, Eds. (1998). *Solar System Ices: Based on reviews presented at the International Symposium "Solar System Ices" held in Toulouse, France on March 27–30, 1995*. Dordrecht: Kluwer Academic Publishers.

Wadhams, P. (2000). *Ice in the Ocean*. Amsterdam: Gordon and Breach Science Publishers.

Weeks, W. F. (2009). *Sea Ice*. Fairbanks: University of Alaska Press.

2

Structure of ice

2.1 Introduction

There are 12 crystalline forms of ice.[1] At ordinary pressures the stable phase is termed ice I, terminology that followed Tammann's (1900) discovery of high-pressure phases. There are two closely related low-pressure variants: hexagonal ice, denoted Ih, and cubic ice, Ic. Ice Ih is termed *ordinary ice* whose hexagonal crystal symmetry is reflected in the shape of snow-flakes. Ice Ic is made by depositing water vapor at temperatures lower than about −130 °C. High-pressure ices are of little interest in relation to geophysical processes on Earth, but constitute the primary materials from which many extra-terrestrial bodies are made. We describe their structure and creep properties in Chapter 8.

In addition to the 12 crystalline forms, there are two amorphous forms. One is termed *low-density ice* (940 kg m^{-3} at −196 °C at 1 atmosphere) and the other, *high-density ice* (1170 kg m^{-3}, same conditions). The density of ice Ih is 933 kg m^{-3} at the same temperature and pressure (Hobbs, 1974). Amorphous ices can be made at low temperatures in five ways (see review by Mishima and Stanley, 1998): by condensing vapor below −160 °C; by quenching liquid; by compressing ice Ih at −196 °C; by electron irradiation; and through transformation upon warming from one amorphous state to another. Although once thought to be a nanocrystalline material, amorphous ice is now considered to be truly glassy water. Its mechanical behavior remains to be explored.

In this chapter we address the structure of ice Ih. In relation to creep and fracture, the important elements include crystal structure and defect structure where the latter includes point, line, planar and volumetric features. Dislocations play an essential

[1] Ice is generally understood to be frozen water. Astrophysicists use the term *water ice* to distinguish frozen water from ices of NH_3, SO_2, CO_2 and CH_4–N_2 when they exist in the solid state either as molecular solids, mixtures of solids, hydrates or clathrate hydrates. Molecular ices represent a large fraction of the mass of the outer Solar System (Schmitt *et al.*, 1998). In this book, ice means frozen water.

role in creep and in the formation of micro-cracks and also in the evolution of the microstructure. We consider the elements in turn.

2.2 Crystal structure of ice Ih

2.2.1 Unit cell and molecular stacking

The crystal structure of ice Ih is well established (Pauling, 1935). It is illustrated by the ball and stick model shown in Figure 2.1. The oxygen and hydrogen atoms are represented by large and small spheres, respectively, and the oxygens are numbered for reference. Hexagonal symmetry can be seen by connecting oxygen atoms. The unit cell, also shown in the image and defined in the caption, contains four molecules – one for each of the eight corner sites (occupied by atoms numbered 1, 2, 3, 7, 21, 22, 23, 27), which are shared eight times, one for each of the four edge sites (14, 15, 16, 20), which are shared four times, plus two for the two body sites (8, 11). Each oxygen atom has four oxygens as nearest neighbors that are situated near the vertices of a regular tetrahedron, which is centered about the atom of interest. The tetrahedron around oxygen 8, for instance, is made from oxygens 1, 2, 7 and 11; and the tetrahedron around oxygen 11 is made from oxygens 8, 14, 15 and 20. All other crystalline forms of ice are also built from tetrahedral units of this kind, a consequence of the fact that the free H_2O molecule is bent through an H–O–H angle of 104.52°. The actual tetrahedral O–O–O angle of ice Ih is 109.47°.

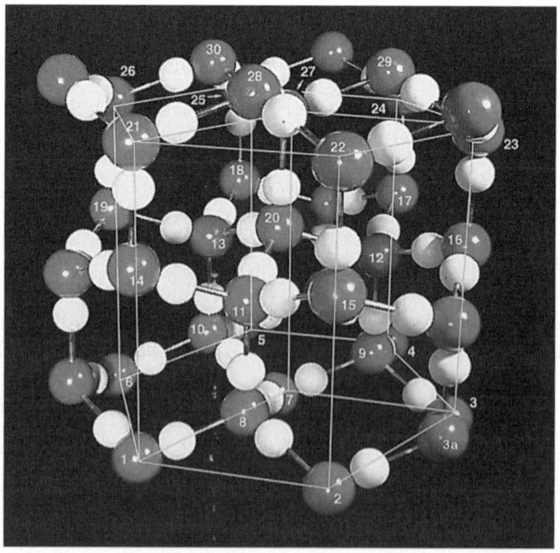

Figure 2.1. Photograph of a ball and stick model of the crystal structure of ice Ih. The larger balls represent oxygen and the smaller balls, hydrogen. The sticks represent hydrogen bonds between H_2O molecules. The corners of the Ih unit cell are delineated by oxygen atoms 1, 2, 3, 7, 21, 22, 23, 27.

The O–O distance is 0.276 nm and the O–H distance is 0.0985 nm at temperatures of terrestrial interest (Hobbs, 1974). Strong covalent bonds join the oxygen atom of each molecule to two hydrogen atoms, while weak hydrogen bonds link the molecules to each other. The weak bonds account for the relatively low melting temperature of ice.

The sequence of molecular stacking, when projected onto the basal plane (which in Figure 2.1 is defined by the hexagonal ring of atoms 1, 2, 3, 4, 5, 6, 7) is ... ABBAABBA ... Six layers are shown: atoms 1 through 7 define layer-A; 8–10 layer-B; 11–13 another layer-B; 14–20 another layer-A; 21–27 layer-A; and 28–30 layer-B. Alternatively, the sequence may be viewed as ... A′B′A′B′A′B′ ... where the prime denotes pairs of molecules joined in a dumb-bell manner along the *c*-axis. In the latter scheme dumb-bells from atoms 8 and 11, from 9 and 12, and from 10 and 13 constitute a B′ layer; and from 14 and 21, 15 and 22, 16 and 23, 17 and 24, 18 and 25, 19 and 26, and 20 and 27 constitute an A′ layer. The alternate scheme is reminiscent of atomic stacking within closely packed hexagonal metals. However, contrary to comments one occasionally reads, ice Ih does *not* possess the closely packed crystal structure of hexagonal metals. Instead, its structure is rather open (more below).

In crystallography, the lattice of oxygen atoms in the Ih structure is described by the space group $P6_3/mmc$. The letter P means that the space lattice is primitive; 6_3 means that the principal axis of symmetry (i.e., the *c*-axis) is a hexagonal screw axis that passes through the center of horizontally oriented, hexagonal puckered rings of atoms (e.g., atoms 2, 3a, 3, 9, 7, 8 of Figure 2.1). When a ring group is translated along the *c*-axis and simultaneously rotated about that axis, the atoms in the group coincide with atomic positions in the crystal when the group has translated a distance of $3c/6$ and rotated by $2\pi/6$. Of the letters mmc, the first means that there are planes normal to the principal axis of symmetry that mirror atomic positions. The remaining m and c mean that there are two sets of symmetry planes parallel to the *c*-axis and that one of these defines another set of mirror planes and the other defines a set of glide planes. Within this context a "glide plane" is not one on which dislocations slip (Section 2.4), but one in which, upon displacement of $c/2$ along the *c*-axis, a mirror image of an oxygen atom corresponds to an atomic position in the crystal.

2.2.2 Lattice parameters, thermal expansion and density

The lattice parameters of ice Ih vs. temperature under atmospheric pressure were measured by Röttger *et al.* (1994) through diffraction measurements using synchrotron radiation, Table 2.1. At −20 °C the values commonly taken are $a = 0.4510$ nm and $c = 0.7357$ nm. The c/a ratio shows no significant variation with temperature

Table 2.1. *Lattice parameters of ice Ih vs. temperature at atmospheric pressure*

T (°C)	−263	−233	−203	−173	−143	−113	−83	−53	−23	−8
a (nm)	0.449 69	0.449 67	0.449 59	0.449 66	0.449 88	0.450 21	0.450 63	0.451 17	0.451 81	0.452 14
c (nm)	0.732 11	0.732 05	0.731 98	0.732 04	0.732 40	0.732 96	0.733 72	0.734 47	0.735 61	0.736 16

From Röttger *et al.* (1994)

Table 2.2. *Density of ice Ih vs. temperature under atmospheric pressure*

T (°C)	−260	−220	−200	−160	−138	−80	−60	0
Density (kg m^{-3})	933.3	933.0	932.8	931.9	929.6	924.5	922.7	916.4

From Hobbs (1974)

and has the value $c/a = 1.628$. This is very near the ideal value 1.633 for hard spheres that are closely packed, even though, as already noted, the crystal structure of ice Ih is not closely packed.

The fact that the c/a ratio is essentially independent of temperature implies that thermal expansion occurs isotropically. The coefficient of linear expansion has the value $\alpha = 5.3 \times 10^{-5}$/°C at −20 °C. At lower temperatures the coefficient is lower, reaching 1.0×10^{-5}/°C at −173 °C. At still lower temperatures, the coefficient reaches zero (at ~−200 °C) and then becomes negative before approaching again the value zero at absolute zero. Petrenko and Whitworth (1999) discuss the nature of this property.

Incidentally, the thermal expansion coefficient near the melting point is relatively high. When coupled with a low Young's modulus of $E \sim 10$ GPa (Chapter 4) and with a relatively low tensile strength of $\sigma_t \sim 1$ MPa (Chapter 10), this translates to a *low thermal shock resistance* of $\Delta T \sim \sigma_t/E\alpha \sim 2$ °C. For concrete and diamond $\Delta T \sim 10$ °C and ~1000 °C, respectively (Ashby, 1989).

The density of ice can be deduced from the lattice parameters and from the fact that there are four molecules per unit cell. Table 2.2 lists values under atmospheric pressure. At the melting point the expected density is 916.4 kg m^{-3}, in agreement with the value obtained using the hydrostatic weighing method on glacier ice single crystals (Bader, 1964). The density ρ is higher under higher pressure. At −35 °C, for instance, $\rho = 942.6$ kg m^{-3} under a pressure of 200 MPa (Gagnon *et al.*, 1988). At this pressure and temperature, ice Ih and ice Ic are almost in thermodynamic equilibrium with each other.

2.2.3 Bernal–Fowler rules

Pauling (1935) recognized that under ordinary conditions the orientation of the H_2O molecules does not exhibit long-range ordering. That is, of the six possible configurations that a pair of hydrogen atoms may adopt within the tetrahedrally coordinated arrangement of oxygen atoms, none dominates throughout the crystal. The hydrogen atoms, in other words, impose a limited degree of disorder within the crystallographically ordered arrangement of the oxygen atoms. Disordered though it is, the arrangement is constrained by two rules, known as the Bernal–Fowler ice rules (1933):

- two hydrogen atoms must be located near each oxygen atom; and
- only one hydrogen must lie on each O–O bond.

Disobedience leads to point defects (described below), at least at higher temperatures where the atoms are dynamic. At lower temperatures, the hydrogen atoms effectively freeze into a disordered configuration, owing to the extraordinarily slow reorientation of the H_2O molecule in pure ice (estimated to be more than 100 years at liquid nitrogen temperature). This metastable state leads to a large amount ($3.41\ \mathrm{J\ °C^{-1}\ mol^{-1}}$) of zero-point entropy (Pauling, 1935).

2.2.4 Physical characteristics

The Ih crystal structure accounts for several physical characteristics of ordinary ice. The spacing between the oxygen atoms is large relative to the size of the atoms themselves (0.276 nm vs. 0.12 nm in diameter). Consequently, even when the hydrogen atoms are taken into account there is a large amount of open space within the lattice. This translates to an atomic packing factor of 0.34, compared with a packing factor of 0.74 for closely packed hexagonal metals of ideal c/a ratio. This openness accounts for ordinary ice being less dense than water. It accounts also for the pressure-induced reduction of the melting point of 0.074 °C per MPa at elevated temperatures (Chapter 12).

Another characteristic is radial isotropy. The Ih lattice has only one major axis of symmetry; namely, the *c*-axis. This accounts for the fact that the physical properties (thermal conductivity, elastic stiffness and atomic diffusivity) are isotropic in all directions perpendicular to the *c*-axis. There is some anisotropy, however, between properties measured in directions parallel to and perpendicular to the *c*-axis (see Chapter 4 for a detailed discussion of elastic moduli).

The other characteristic stemming from the Ih structure is that the H_2O molecules are concentrated close to the basal planes (defined in Figure 2.1 by atoms 1 through 7, for instance, or by atoms 21 through 27). This, plus the fact that crystal dislocations (described below) are highly mobile on basal planes, allows for easy slip (Chapter 5).

2.3 Point defects

Point defects are atomic-sized features that form within the ice lattice: vacancies, interstitials, solutes, ionic and Bjerrum defects. The first three are similar to point defects within crystals of metals and compounds. The other two are unique to ice and are often termed *protonic defects*. Each plays a role in inelastic behavior, particularly in creep.

2.3.1 Vacancies

A vacancy is defined as an empty molecular site. The defect exists in thermal equilibrium at all temperatures above absolute zero and is stabilized by an increase in entropy that more than compensates for an increase in internal energy. The equilibrium vacancy concentration C_v is given by the Boltzmann relationship (e.g., see Swalin, 1962, or any other book on defects in solids):

$$C_v = \exp(S_v/k_B)\exp(-E_v^f/k_BT) \tag{2.1}$$

where S_v denotes the extra entropy associated with each vacancy, E_v^f the internal energy that must be added to remove a molecule from the interior of a crystal and place it on the surface, k_B Boltzmann's constant and T absolute temperature. Fletcher (1970) estimated that $E_v^f \sim 0.5$ eV, based upon the sublimation energy of ice. This implies an equilibrium molecular fraction at the melting point of around 10^{-10}, assuming $S_v \sim 0$. Although relatively low compared with melting-point concentrations of 10^{-3} to 10^{-4} for vacancies in metals and alloys, the concentration is large enough to account for the formation of vacancy-type prismatic dislocation loops (as well as interstitial loops, more below) upon rapid cooling from −20 °C to −60 °C (Liu and Baker, 1995).

2.3.2 Interstitials

An interstitial is formed when a molecule becomes situated somewhere within the open space of the Ih crystal lattice. It, too, is thermodynamically stable at all temperatures above absolute zero, for the same reason. Interstitials are created at free surfaces and interfaces, but independently from vacancies that are also created there. The equilibrium interstitial concentration C_i may be approximated by the expression:

$$C_i = \exp(S_i/k_B)\exp(-E_i^f/k_BT) \tag{2.2}$$

where S_i denotes the extra entropy per interstitial molecule and E_i^f the energy to form an interstitial. Goto *et al.* (1986) deduced that $S_i = 4.9k_B$ and that $E_i^f = 0.40$ eV. The melting-point interstitial concentration is thus estimated to be $\sim 10^{-6}$. This is several orders of magnitude higher than that for metals and alloys, owing to the more open unit cell of ice Ih. It is also several orders of magnitude greater than the equilibrium vacancy concentration in ice Ih and probably contributes to a greater propensity for the formation of interstial dislocation loops vs. vacancy loops upon rapid cooling from near the melting point.

Vacancies and interstitials are mentioned again in connection with self-diffusion in Chapter 4. There it is noted that the H_2O molecule diffuses as a unit, mainly through an interstitial-assisted mechanism, at least at temperatures above −50 °C (Goto *et al.*, 1986).

2.3.3 Solutes

The solubility of foreign species within the Ih crystal lattice is very low (Gross and Svec, 1997). As a result, when water freezes it almost completely rejects them. In some cases, the impurities may be incorporated within the solid matrix as liquid inclusions, such as brine pockets within sea ice (Chapter 3). An exception is ammonium fluoride, which can form substitutional solid solutions of up to 10 wt. % (Gross and Svec, 1997), an effect that is attributed to isomorphism between the NH_4F and the ice Ih crystal lattices. Also notable is an apparent synergistic effect of the cation NH_4^+ on the solubility of CO_3^-, NO_3^- and other anions (Gross and Svec, 1997).

2.3.4 Ionic and Bjerrum defects

Violations of the Bernal–Fowler ice rules create *protonic defects*. If an atom of hydrogen moves from one equilibrium position to another along an O–O bond, then the first rule is violated and ionic defects are created: one proton near an oxygen atom creates an OH^- ion; three create an H_3O^+ ion. The ionization reaction is similar to that in water, except that in ice the ions do not migrate as complete units because the oxygen is tied to a given lattice site. If a proton moves around the oxygen atom instead of moving along an O–O bond, then the second rule is violated: no hydrogen atom on an O–O bond creates L-type Bjerrum defects (from the German leere, meaning empty); two protons create a D-type Bjerrum defect (from doppeltbesetzte, meaning doubly occupied). Figure 2.2 illustrates these defects. Both kinds are accompanied by an energy penalty, $E_\pm$ for the ionic defect and E_{LD} for the Bjerrum defect. The probability of the defect existing at any non-zero temperature

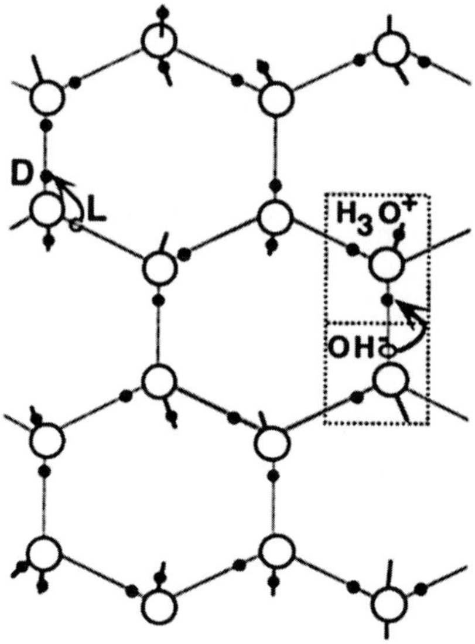

Figure 2.2. Schematic sketch of ionic (H_3O^+ and OH^-) and Bjerrum (L and D) point defects.

is again given by a Boltzmann function. In pure ice the concentrations C_- and C_+ of the two ionic defects are equal and may be estimated from the expression:

$$C_- = C_+ = \exp(S_\pm/k_B)\exp(-E_\pm/2k_BT). \tag{2.3}$$

The concentrations C_L and C_D of the two Bjerrum defects in pure ice are also equal and may be estimated from the expression:

$$C_L = C_D = \exp(S_{LD}/k_B)\exp(-E_{LD}/k_BT). \tag{2.4}$$

The parameters $S_\pm$ and S_{LD} again denote an increase in entropy. Petrenko and Whitworth (1999) estimate with caution pair formation energies of $E_\pm > 1.4$ eV and $E_{LD} = 0.66$–0.79 eV at −20 °C, implying (for $S \sim 0$) intrinsic ionic and Bjerrum concentrations, respectively, of $<10^{-14}$ and $\sim 10^{-7}$. Both kinds of defect contribute to the electrical conductivity of ice: migration of ions allows protons to move from one end of a bond to the other, and the movement of Bjerrum defects allows protons to move from one side of an oxygen atom to another. Without the migration of both kinds of defects, long-range conduction could not occur. Protonic defects also affect dislocation mobility and hence creep.

2.4 Line defects: dislocations

2.4.1 Basic concepts

Dislocations are line defects within the crystal lattice. They are created during the thermal-mechanical history of the material and are fundamental to plasticity and

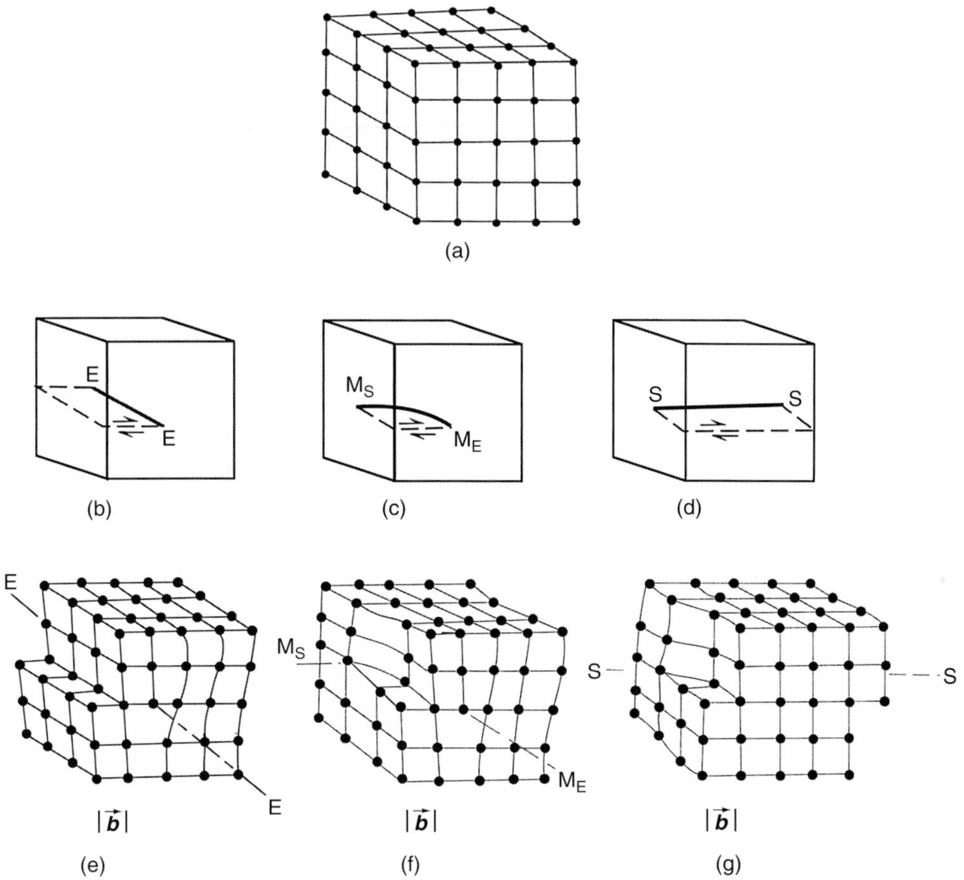

Figure 2.3. Schematic sketches of dislocations in a hypothetical simple cubic crystal. ***b*** denotes the Burgers' vector. (a) The perfect crystal. Shearing across a horizontal slice creates an edge dislocation EE, (b) and (e); a mixed dislocation M_S and M_E, (c) and (f); and a screw dislocation SS, (d) and (g).

strength. Here we describe only basic concepts. Readers interested in fuller treatments should consult excellent books by Cottrell (1953), Read (1953), Friedel (1964), Hull and Bacon (1984), Hirth and Lothe (1992), Weertman and Weertman (1992) and by Nabarro (2003).

Crystals commonly contain three types of dislocation lines, or dislocations for short. Figure 2.3 illustrates them schematically, with reference to a hypothetical simple cubic lattice. One type is termed the *edge dislocation*, shown again in Figure 2.4. Imagine slicing the crystal all the way through its thickness and part-way through its width, displacing or slipping by a unit distance (set by the lattice dimensions) the atoms above the slice over those below it in the direction shown, and then rejoining the atomic bonds across the cut. Terminating at the inside boundary of the re-bonded slice is an atomic half-plane that is imperfectly bonded

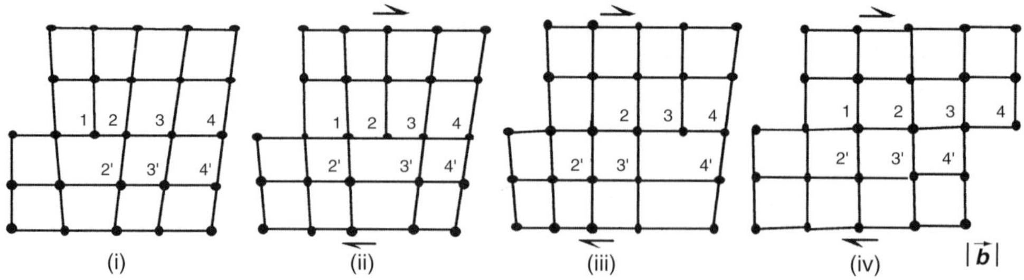

Figure 2.4. Schematic sketch showing four stages in the consecutive glide of an edge dislocation of Burgers' vector ***b*** within a hypothetical simple cubic crystal loaded under a shear stress.

at its edge. The edge EE marks the dislocation line. Another type is the *screw dislocation*, Figure 2.3g. It can be "created" in a similar manner, by partially slicing the crystal as shown in Figure 2.3d, by slipping atoms above the slice a unit distance over those below it, and then by rejoining the bonds across the cut. In this case, the inside boundary of the re-bonded slice is defined by line SS, which marks the axis of an atomic-scale spiral ramp. Line SS marks the screw dislocation line. The third and most common type is the *mixed dislocation*, Figure 2.3f. It is a mixture of edge and screw types and can be made by slicing the crystal only part way through its width and thickness, by displacing the upper over the lower part by the same amount and in the same direction as in the first two cases, and then by reforming the bonds across the cut. The degree of edge and screw character varies along the dislocation line, from pure edge M_E where it emerges on the front face to pure screw M_S where it emerges on the side face. In real crystals mixed dislocations often appear as closed loops whose Burgers' vector (defined below) lies within the plane of the loop. Whether edge, screw or mixed, a dislocation may be viewed as the boundary between the slipped and the un-slipped part of a crystal.

A fourth type of dislocation, less common than the other three, is the *prismatic loop*. Again we imagine cutting atomic bonds, this time ones that thread a disc-shaped planar area internal to the body. Instead of displacing atoms on either side of the cut in a direction parallel to it, we separate the two surfaces on either side and fill the space with more atoms. The periphery of the filled-in space then defines a loop of pure edge dislocation, termed an *interstitial prismatic loop*. Should a layer of atoms be removed instead of added, a *vacancy prismatic loop* of pure edge character would be created. Interstitial prismatic loops form in ice via the condensation of point defects in excess of the equilibrium concentration (Oguro, 1988).

Dislocations are fundamental to crystal plasticity because they allow slip to occur under an applied (shear) stress that is several orders of magnitude lower than the theoretical shear strength. This happens because dislocations generate an internal shear strain (evident from the distortion at their core, Figure 2.3) that partially bends the atomic bonds within the vicinity of the core. To bend the bonds all the way to the point of failure then requires a much smaller applied shear stress than would be required were the defect absent. The dislocation advances incrementally and new bonds form in its wake. As sketched in Figure 2.4, slip across the plane occurs in a consecutive versus a simultaneous manner: movement to the right occurs first by breaking bond 2–2′ and forming bond 1–2′, then by breaking 3–3′ and forming 2–3′, and so on. Eventually, the dislocation moves all the way through the crystal and emerges at the surface, creating a step that is termed a *slip line* (Chapter 5).

It would be a mistake to conclude that the crystal becomes free from line defects once the dislocation emerges. Slip is more complicated than that. Suffice it to say that dislocations multiply as they glide, leading in the end to a greater dislocation density than in the beginning. Dislocation density is defined as the total length of dislocation/unit volume and in well-annealed crystals of ice Ih can be as low as 10^6 m/m^3. (Normally, the density is much higher than this; more below.) That dislocations multiply is evident from the fact that a few dislocations gliding across the slip plane and then emerging at the free surface would not lead to the macroscopic plasticity that crystals of ice Ih can exhibit (Chapter 5). For that to occur, large numbers of dislocations must participate.

In the wake of a dislocation, perfect atomic registry is usually restored. However, within ice Ih this is not strictly true. Owing to the random arrangement of hydrogen atoms, the translation of part of the crystal relative to the rest by a unit distance prescribed by the Burgers' vector (defined below) will not always exactly reproduce the atomic arrangement (Glen, 1968). Instead, the translation can introduce Bjerrum defects. For instance, if the dislocation shown in Figure 2.5 glides one unit to the

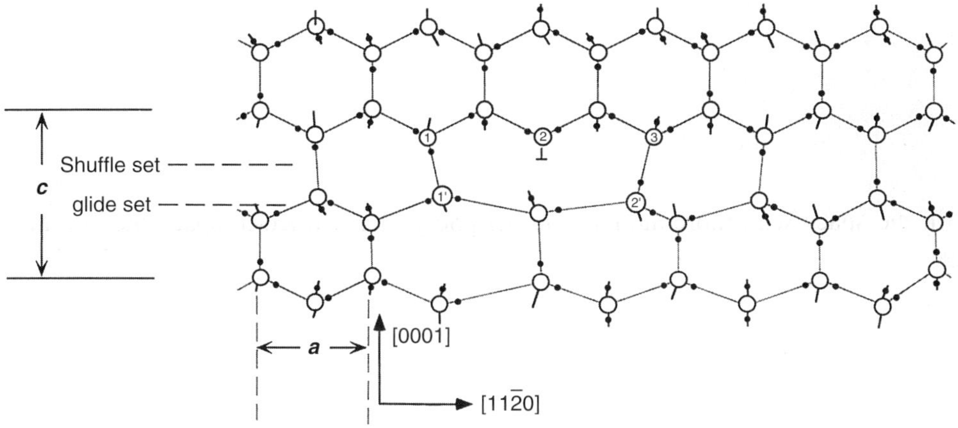

Figure 2.5. Schematic sketch of a section of the Ih crystal structure projected onto the $(1\bar{1}00)$ plane showing a dislocation of $\boldsymbol{b} = \frac{a}{3}[11\bar{2}0]$ on the shuffle set of (0001) basal planes.

right, then bond 3–2′ is broken and 2–2′ is formed. This is not a problem. However, if the dislocation glides to the left, then bond 1–1′ is broken and 2–1′ is formed. This movement requires an L-defect to be created. Glen (1968) showed that the stress needed to create such defects is orders of magnitude greater than can be accounted for by the actual flow stress, implying that some kind of *protonic rearrangement* must occur. The precise way in which dislocations overcome this obstacle is not yet understood.

An important characteristic of a dislocation is its *Burgers' vector* ***b***, Figure 2.3. Its sense and magnitude are determined by constructing an atomic-scale circuit around the dislocation line such that the number of steps up and down are the same and the number of steps to the left are the same as the number to the right: ***b*** is defined by the failure to close the circuit. The construction shows that ***b*** is perpendicular to edge dislocations and parallel to screw dislocations. For mixed dislocations, ***b*** is inclined to the dislocation line as one moves along it. The Burgers' vector is constant along a dislocation and defines the increment of slip. It also defines the slip direction. Detailed dislocation analysis requires that the direction of the dislocation as well as the sense of the circuit around it (left-handed vs. right-handed) be specified. For the purposes in this book, such details need not be given.

The combination of the dislocation line, denoted by the vector ***l***, and the Burgers' vector define the normal, ***n***, to the slip plane; i.e., $\boldsymbol{n} = \boldsymbol{l} \otimes \boldsymbol{b}$. The implication is that edge dislocations are confined to slip on one set of crystallographic planes. Screw dislocations, on the other hand, can *cross-slip* from one set of planes to another in the zone of the Burgers' vector, provided that they are not too widely dissociated (more below). Although edges cannot cross-slip, they can *climb* either up or down from the slip plane. It is important to recognize that dislocation slip is driven by shear stresses, specifically by the component of the stress tensor that acts on the slip plane in the slip direction. Climb, on the other hand, is assisted by normal stresses, specifically by the normal component that acts perpendicular to the extra half-plane.

Dislocations introduce *strain energy* or *line energy* to the system. The energy is equal to the work done in creating the defect and is given approximately by the expression:

$$E_\mathrm{D} = \frac{Gb^2}{A} \ln\left(\frac{R}{R_\mathrm{o}}\right) \tag{2.5}$$

where E_D denotes the energy per unit length of a dislocation within an isotropic crystal, G the shear modulus and A is a dimensionless parameter that depends on the character of the dislocation (edge, screw or mixed); R is the outer cutoff radius taken to be the separation of dislocations and R_o is the radius of the dislocation core within which linear elastic theory cannot be applied; R_o is a little greater than the Burgers' vector. For edge dislocations $A_e = 4\pi\,(1-\nu)$ and for screws $A_s = 4\pi$ where ν denotes Poisson's ratio (Chapter 4). Edge dislocations possess about 30% more energy than screw dislocations. The function $\ln(R/R_o)$ is relatively insensitive to R and so a fair approximation of the line energy is $E_D \sim Gb^2/2$ (Humphreys and Hatherly, 1996).

How large is the line energy? Taking $R/R_o = 10^2$ and $|b| = a = 0.452$ nm, the line energy of a basal edge dislocation in ice Ih is about $E_D \sim 5 \times 10^{-10}\,\mathrm{J\,m^{-1}}$. Consequently, for a crystal of ice Ih that contains 10^{11} meters of dislocation line per cubic meter, which is typical of crystals within deformed ice sheets (de La Chapelle *et al.*, 1998), dislocations raise the internal energy by $\sim 50\,\mathrm{J\,m^{-3}}$. Although small in comparison with the energy of fusion, which is about $3 \times 10^8\,\mathrm{J\,m^{-3}}$, the energy stored by dislocations is more than an order of magnitude greater than the elastic energy, which is $\sim 3\,\mathrm{J\,m^{-3}}$ under an applied stress of ~ 0.1 MPa (Chapters 6, 7).

Unlike point defects, dislocations do not increase thermodynamic stability. The reason is that for line defects the increase in strain energy within the crystal is *not* more than compensated for by an increase in entropy (Cottrell, 1953). Given time at a temperature sufficiently high to allow molecular diffusion (Chapter 4), dislocations rearrange themselves. They tend to anneal out of the crystal. This behavior is fundamental to creep.

2.4.2 Dislocations and slip systems in ice Ih

When loaded slowly (defined later), crystals of ice deform plastically. As discussed in Chapter 5, this occurs primarily through the glide of dislocations on the basal plane (0001) in the $\langle 11\bar{2}0\rangle$ direction. The combination of plane and direction defines the slip system. Table 2.3 lists this one and several other systems according to the self-energy of the dislocation line. (The crystallography is described in Appendix A.) Owing to the hexagonal symmetry of the crystal structure, dislocations lying at 60° to the Burgers' vector are said to be *60° dislocations*. The three

Table 2.3. *Dislocations and slip systems in ice*

Type of dislocations	Burgers' vector	Self-energy[a]	Slip plane
Perfect	$1/3\langle 11\bar{2}0\rangle$	1.0	(0001), $(10\bar{1}0)$, $(10\bar{1}1)$
	$\langle 0001\rangle$	2.7	$(10\bar{1}0)$, $(11\bar{2}0)$
	$1/3\langle 11\bar{2}3\rangle$	3.6	$(10\bar{1}0)$, $(10\bar{1}1)$, $(11\bar{2}2)$

The four-indicial notation for planes and directions is described in Appendix A.
[a] Relative to the perfect dislocation.
From Hondoh (2000)

$\langle 11\bar{2}0\rangle$ directions are the most common slip directions because they define dislocations with the lowest relative line energy. A combination allows slip to occur in any direction in the basal plane (Kamb, 1961). (The magnitude of these vectors is defined by the shortest distance between oxygen atoms in the same basal plane and thus by the distance between next-nearest neighbors, Figure 2.1). Also, prismatic loops with the $\langle 0001\rangle$ Burgers' vector can form in both pure and doped ice freshly grown, as the result of the condensation of excess interstitials (Oguro, 1988; Oguro *et al.*, 1988). Prismatic loops are essentially immobile and so do not contribute to macroscopic plasticity.

2.4.3 Dissociation of basal dislocations into partials and stacking faults

In principle, basal slip in ice can take place on two sets of planes (Whitworth, 1980). One set is more widely spaced and is termed the *shuffle set*. It is defined by the plane between atoms 8 and 11 in Figure 2.1. The dislocation illustrated in Figure 2.5 lies in this plane. The other set is less widely spaced and is termed the *glide set*, defined by the plane between atoms 1 and 8 in Figure 2.1. The distinction is significant. Adjacent planes of oxygen atoms of the glide set relate to each other in a manner similar to the relationship between adjacent planes of atoms within face-centered-cubic and hexagonal-close-packed metals in which dislocations dissociate into partials (more below). This relationship raises the possibility of *dislocation dissociation*. Dislocations in the shuffle set cannot dissociate. Figure 2.6 shows sketches of possible partial dislocations, proposed by Hondoh (2000).

It is not known with certainty on which set of planes dislocations glide. The possibility of dislocation dissociation, however, seems to point to the glide set. This may be seen as follows: Dissociation is driven by a reduction in self-energy. Because line energy is proportional to the square of the Burgers' vector, as evident from Equation (2.5), it is energetically favorable for any dislocation with a total Burgers' vector $\boldsymbol{b}$ to dissociate into two partial dislocations with Burgers' vectors $\boldsymbol{b}_1$ and $\boldsymbol{b}_2$ such that $b^2 > b_1^2 + b_2^2$, provided that the crystal lattice permits the dissociation. The partials are connected by a stacking fault, Figure 2.7, the energy of which balances the reduction in elastic self-energy. For example, the possible dissociation $\boldsymbol{b} = a/3[\bar{1}2\bar{1}0] \rightarrow \boldsymbol{b}_1 = a/3[\bar{1}100] + \boldsymbol{b}_2 = a/3[01\bar{1}0]$ would be driven by a reduction in line energy of about 30% and would be accompanied by the formation of a stacking fault whose width is expected to be ~ 24 nm for a dissociated screw dislocation and ~ 47 nm for a dissociated 60° dislocation, Table 2.4. Faults this wide would be expected to suppress cross-slip (Louchet, 2004). However, constrictions should pre-exist along dissociated screw dislocations making cross-slip possible (Chapter 5).

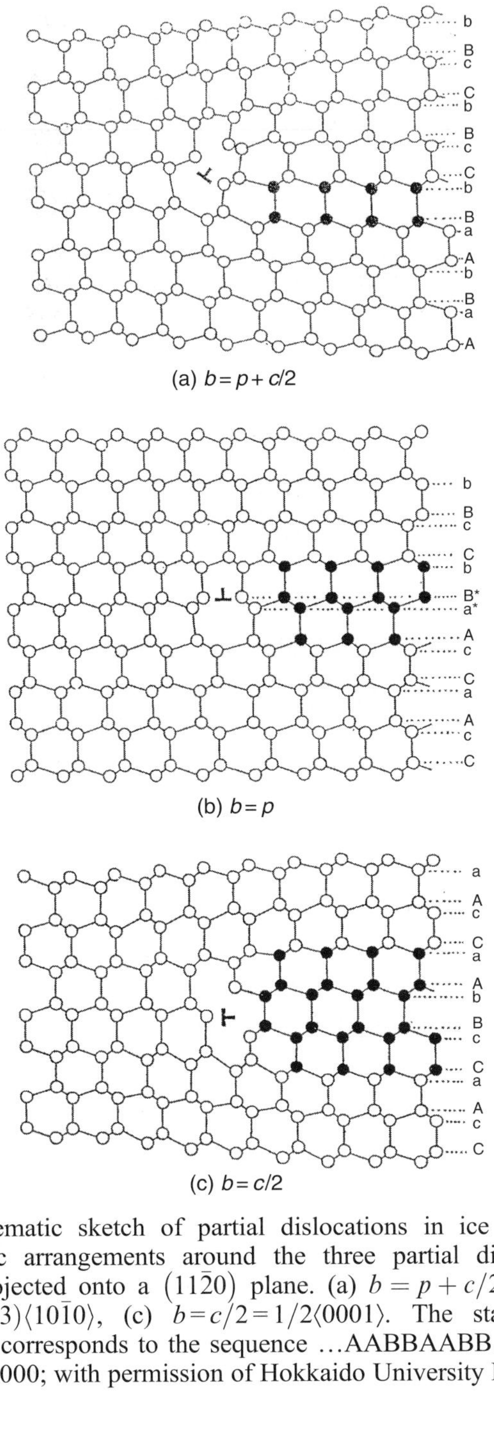

Figure 2.6. Schematic sketch of partial dislocations in ice Ih. The figures show the atomic arrangements around the three partial dislocations given in Table 2.4, projected onto a $(11\bar{2}0)$ plane. (a) $b = p + c/2 = (1/6)\langle 20\bar{2}3\rangle$, (b) $b = p = (1/3)\langle 10\bar{1}0\rangle$, (c) $b = c/2 = 1/2\langle 0001\rangle$. The stacking sequence …AaBbAaBb… corresponds to the sequence …AABBAABB… of Figure 2.1. (From Hondoh, 2000; with permission of Hokkaido University Press.)

Table 2.4. *Partial dislocations in ice*

Perfect dislocations	Burgers' vector	Self-energy[a]	Fault energy (mJ m^{-2})	Extended width (nm)
$1/3\langle 11\bar{2}0\rangle$	$1/3\langle 10\bar{1}0\rangle$	0.33	~0.6	24 (screw), 47 (60°), 55 (edge)
$\langle 0001\rangle$	$1/2\langle 0001\rangle$	0.66	~0.9	>100
$1/3\langle 11\bar{2}3\rangle$	$1/6\langle 20\bar{2}3\rangle$	1.0	~0.3	>400

[a] Relative to the perfect dislocation.
From Hondoh (2000)

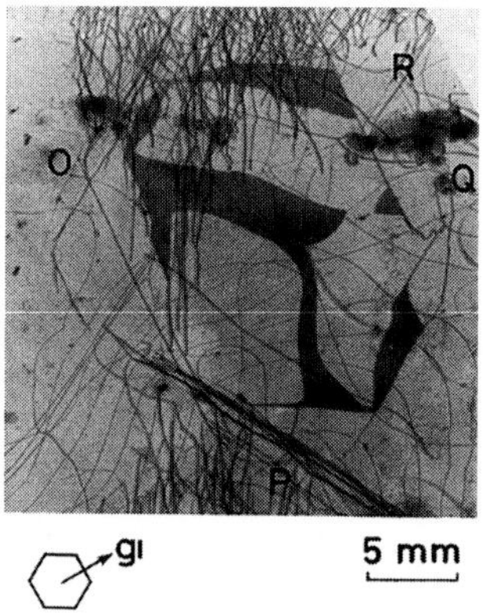

Figure 2.7. X-ray topograph taken with the $(10\bar{1}0)$ diffracting plane of wide stacking faults (dark areas) formed by slow cooling. Dislocations appear as dark lines. OPQR represents a large interstitial loop. (From Higashi, 1988; with permission of Hokkaido University Press.)

We caution against drawing a firm conclusion. Further experiments are needed. Unfortunately, limitations in current techniques described in Section 2.4.5 make that a difficult task.

2.4.4 Non-basal dislocations

A general feature of the plastic deformation of ice is rapid motion of short segments of edge dislocation on the $(10\bar{1}0)$ prismatic plane (Higashi *et al.*, 1985; Fukuda *et al.*, 1987; Ahmad and Whitworth, 1988; Fukuda and Higashi, 1988; Shearwood and Whitworth, 1989, 1991). The glide of such short segments on non-basal planes imparts little macroscopic plasticity. However, as discussed in Chapter 5, non-basal glide provides an efficient mechanism for the generation of basal dislocations (Ahmad and Whitworth, 1988; Ahmad *et al.*, 1992). Non-basal edge dislocations can be produced through the climb of 60° basal dislocations

(Ahmad and Whitworth, 1988) and/or through the cross-slip of basal screw dislocations (Taupin *et al.*, 2007).

2.4.5 Observation of dislocations

Direct observations are difficult, but not impossible, to make. The method of transmission electron microscopy, which has been used so successfully to examine dislocations in other crystalline materials, has led to little success with ice. The problem is that thin specimens suitable for examination are difficult to prepare. Moreover, once prepared, they sublime quickly within the vacuum of the microscope. Another technique, which has also been useful with other materials and which has been employed on ice (Muguruma, 1961; Krausz and Gold, 1967; Sinha, 1977, 1978), is etch pitting. The problem with this method is that it suffers from a questionable relationship between surface pits and bulk dislocations (Jones and Gilra, 1973; Baker, 2003). The technique is also sensitive to surface preparation (Sinha, 1978; Liu and Baker, 1995).

X-ray topography offers the best approach (Baker, 2003). It is based upon Bragg diffraction. The absorption of X-rays by ice is low, allowing thick samples (1–2 mm) to be examined. Ice does not need to be subjected to vacuum and so is less susceptible to sublimation. Burgers' vectors can be determined and through the use of high-brightness synchrotron sources individual dislocations can be seen to glide, provided the dislocation density is sufficiently low. Indeed, the first observations of dislocations in ice were made using this method (Hayes and Webb, 1965) and most of the useful observations since then (Ahmad and Whitworth, 1988; Higashi, 1988; Shearwood and Whitworth, 1989, 1991; Hondoh *et al.*, 1990; Liu and Baker, 1995; Liu *et al.*, 1995) have been made using the technique. Transmission X–ray diffraction, for instance, generates a two-dimensional projection of the three-dimensional structure by scanning a crystal with the X-ray beam. Figure 2.8 shows an example,

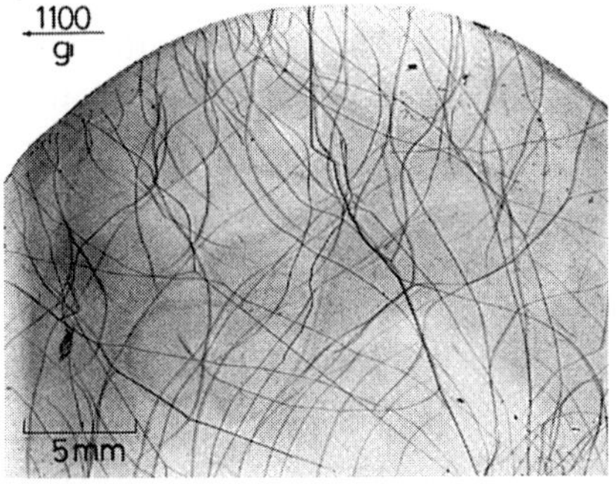

Figure 2.8. X-ray topograph of an ice single crystal grown in the [0001] direction by the modified Bridgman method. The image was obtained with the $(\bar{1}100)$ diffracting plane and scanned over the basal plane. The density of basal dislocations is about $10^6/m^2$. (From Higashi, 1988; with permission of Hokkaido University Press.)

in this case a topograph of a high-quality single crystal grown parallel to the c-axis that contained basal dislocation loops of density $\sim 10^6\,m^{-2}$. Higashi (1988) offers a good introduction to the technique.

The main limitation of X-ray topography is spatial resolution, which is a little less than one micrometer. This means that partial dislocations cannot be detected and that individual dislocations cannot be resolved when their density exceeds $\sim 10^9\,m^{-2}$. However, this technique is well suited to the observation of stacking faults (Higashi, 1988), slip lines (Chapter 5) and any lattice distortion induced by several dislocations (Montagnat *et al.*, 2003).

2.5 Planar defects

In addition to possible stacking faults linking partial dislocations, the other types of planar defects relevant to creep and fracture are grain boundaries and free surfaces. Deformation twins, which are crystals that have two parts symmetrically related to each other across a crystallographic plane, have not been observed in either well-annealed or plastically deformed crystals. Growth twins have been reported (Kobayashi and Furukawa, 1975) and could lead to 12-sided snow-flakes.

2.5.1 Grain boundaries

Grain boundaries are nanometer-wide regions within polycrystals that separate individual grains or crystals. The majority are characterized by a difference in orientation of >10°, which is taken to define a high-angle boundary. Others are termed small-angle or *sub-boundaries* and develop from intragranular arrays of either edge dislocations (*tilt boundary*) or screw dislocations (*twist boundary*). High-angle boundaries generally contain ledges and facets, some up to 1 mm in height (Liu *et al.*, 1995). Near the melting point, grain boundaries contain liquid water in sub-millimeter-sized veins that lie along lines of intersection (Nye, 1992).

The structure of large-angle boundaries has been studied using X-ray topography and interpreted in terms of the coincident site lattice (CSL) (Higashi, 1978, 1988). Accordingly, lattice points of the two crystals are thought to match each other in a periodic manner along the boundary. For instance, grain boundaries that resemble growth twins and satisfy CSL dictates can be created (Kobayashi and Furukawa, 1975) by rotating 47° about the axis $[10\bar{1}0]$ and by 21.8° about [0001]. A problem in applying the CSL model to ice is proton disorder. This makes it impossible to form a perfect periodic lattice, implying that the boundary must be a region of considerable molecular disorder. Grain boundaries increase Gibbs free energy and so, given sufficient time at elevated temperatures where molecular diffusion occurs, they migrate. As a result, some grains grow at the expense of others and the average grain size of an aggregate increases. Grain growth is a process of particular significance in relation to the microstructure of glaciers and polar ice sheets (Chapters 3 and 6).

When a grain boundary intersects a free surface it exerts tension and this causes a groove to form. The included angle of the groove φ is related to the grain boundary energy γ_{gb} and to the surface energy γ_{sv} through the relationship $\gamma_{gb} = 2\gamma_{sv}\cos(\varphi/2)$. Ketcham and Hobbs (1969) found that $\varphi = 145° \pm 2°$ at the melting point, which gives the value $\gamma_{gb} = 41$ mJ m^{-2} for $\gamma_{sv} = 69$ mJ m^{-2} (van Oss *et al.*, 1992). A value of $\gamma_{gb} = 65$ mJ m^{-2} was given by Hobbs (1974) who took $\gamma_{sv} = 109$ mJ m^{-2}. For comparison, the ice–water interfacial energy has a value $\gamma_{sl} \sim 29$ mJ m^{-2} (Hardy, 1977) and the water–vapor interfacial energy has the value $\gamma_{lv} \sim 76$ mJ m^{-2} at 0 °C (Petrenko and Whitworth, 1999). All these energies result from unpaired molecular bonds. Grain boundary energy is important in relation to energy stored within the solid following deformation-induced recrystallization (Chapter 6). Polycrystals that are composed of grains around 1 mm in diameter, for instance, store about as much energy in their grain boundaries as they store in dislocations of density around 10^{11} m^{-2}. The ratio $\gamma_{gb}/\gamma_{sv} \sim 0.6$ and this is about a factor of two greater than that for elemental metals near their melting points. The reason for the higher ratio in ice is not clear.

In addition to storing energy, grain boundaries perform several other functions. They serve as sinks for vacancies and interstitials, as evident from the absence upon rapid cooling of dislocation loops within their vicinity (Hondoh *et al.*, 1981). They also serve as sources of point defects. Grain boundaries serve as well as preferential sites for impurities (Baker and Cullen, 2003; Baker *et al.*, 2003), through either segregation or sweeping up as boundaries migrate. For example, H_2SO_4 appears to be swept up (Iliescu *et al.*, 2003). High-angle grain boundaries impede slip, evident from the piling up there of dislocations (Liu *et al.*, 1993). Also, they as act as sites from which dislocations are generated (Liu *et al.*, 1993). Under appropriate conditions of loading, grains slide past each other along the boundary between them (Nickolayev and Schulson, 1995; Picu and Gupta, 1995a, b; Weiss and Schulson, 2000) and this mode of deformation leads to characteristic peaks in low-frequency *internal friction* (Perez *et al.*, 1979; Tatibouet *et al.*, 1987; Cole, 1995; Cole and Durell, 1995). Grain boundary sliding can also lead to the initiation of cracks (Picu and Gupta, 1995a,b; Frost, 2001). Finally, grain boundaries serve as sites of crack nucleation (Gold, 1997, 1999) and thus play an important role in fracture. (Chapters 9–14).

2.5.2 Free surfaces

The nature of surfaces in ice is a complex subject and an unsettled one. We refer the reader to a review by Dash *et al.* (1995) and to a discussion by Petrenko and Whitworth (1999).

Perhaps the most often cited characteristic is the *quasi-liquid layer*. First proposed by Faraday (1850, 1860), this pre-melting feature is now recognized as an important characteristic of warm ice. Through the application of ellipsometry, X-ray scattering, proton channeling, surface conductivity and atomic force microscopy, its thickness has been found to decrease with decreasing temperature, by an amount that unfortunately seems to depend upon the method of measurement. Recent measurements using atomic force microscopy indicate a thickness of about 30 nm at −1 °C and about 11 nm at −10 °C. The temperature at which surface melting starts is not well defined, although estimates range from −13 °C (Makkonen, 1997) to −34 °C (Döppenschmidt and Butt, 2000) to even lower. Petrenko and Whitworth (1999) review the evidence. The properties of the liquid-like layer are different from those of bulk water and appear to change with the depth from the surface (Furukawa and Nada, 1997). Interestingly, pre-melting appears not to be unique to ice, but to be a characteristic of other materials as well (van der Veen *et al.*, 1988).

In terms of creep and fracture, free surfaces appear either to be a source of dislocations or to affect the rate of dislocation multiplication. Hence, their condition affects the creep strength of single crystals (Chapter 5). Surfaces might also act as crack nucleation sites, and thus affect the fracture strength of crystals (Chapter 10). The creation of new surface is a major impediment to fast crack propagation (Chapter 9) and the free surface is of special interest to friction which, in turn, is fundamental to brittle compressive failure (Chapters 11–15). Thus, although not well understood, the free surface is a significant element in the deformation of ice.

2.6 Volumetric defects

Pores are the most common volumetric defect. When ice is made through solidification, pores form as the result of rejecting oxygen and nitrogen from water and from the ice lattice. Within fresh-water ice they contain air only. Within salt-water ice, they also contain brine (Chapter 3). When ice is made through the sintering and densification of snow (Chapter 3), pores are again air-filled. They may be either connected as in firn or isolated as in ice. The character of the shape and distribution of porosity depends upon the thermal-mechanical history of the product. In polar ice sheets, air bubbles transform into clathrate hydrates from a depth of about 500 m (Chapter 3).

Hard particles constitute the other kind of volumetric defect. During the formation of glaciers, atmospheric particulates can become entrapped within the matrix via either dry deposition on the snow or directly in the atmosphere during the nucleation of ice crystals. During the growth of a sea ice cover, precipitates of sodium chloride and other salts form within entrapped brine pockets once the temperature of the ice falls below the appropriate water–salt eutectic temperature

(Chapter 3). Pores tend to lower the resistance to creep deformation (Chapter 6), while hard particles can both increase and decrease the creep resistance, depending on their size and number density and on the temperature of the ice. Pores also lower resistance to crack propagation (Chapter 9).

Appendix A Miller–Bravais indices for hexagonal crystals

The hexagonal unit cell is defined by two coplanar vectors $\boldsymbol{a}_1$ and $\boldsymbol{a}_2$ of equal length at 120° to each other, plus the perpendicular vector $\boldsymbol{c}$, Figure A.1a. Owing to the

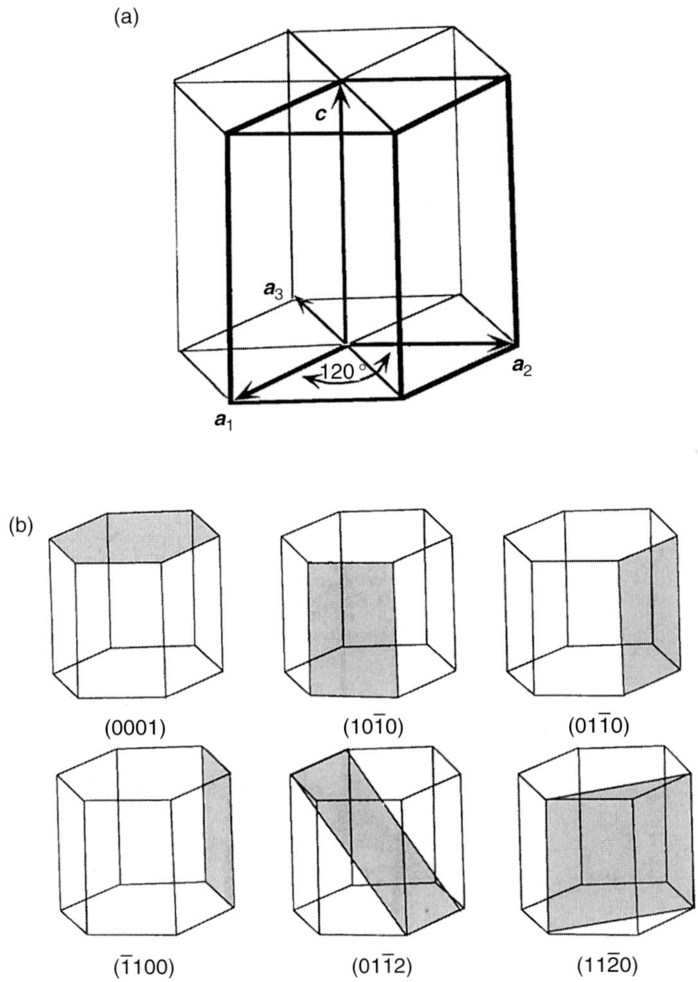

Figure A.1. (a) Sketch showing the hexagonal prism (light lines) and the hexagonal unit cell (dark lines). (b) Sketch showing specific (*hkil*) planes in the hexagonal prism.

six-fold symmetry about the c-axis, a third vector $\boldsymbol{a}_3$ lying in the basal plane is used when defining planes and directions.

Planes are defined in terms four Miller–Bravais indices (*hkil*). The index h is the reciprocal of the fractional intercept the plane makes in cutting the a_1 axis, k is the reciprocal of the fractional intercept on the a_2 axis, i the reciprocal of the fractional intercept on the a_3 axis, and l the reciprocal of the fractional intercept on the c-axis. Because the intercepts on the a_1 and the a_2 axes determine the intercept on a_3 the relationship follows: $-i=h+k$. For instance, with reference to Figure A.1b, the uppermost surface corresponds to the basal plane of the crystal. It is parallel to $\boldsymbol{a}_1$, $\boldsymbol{a}_2$ and $\boldsymbol{a}_3$ and so intersects the three axes at infinity whose reciprocals are zero. However, it intersects the c-axis at unity. Hence, the basal plane is described by the indices (0001). This also designates all parallel planes in this set whose spacing is the lattice parameter c. The side planes of the hexagonal prism in Figure A.1b are all similar and their indices are $(10\bar{1}0)$, $(01\bar{1}0)$ and $(\bar{1}100)$. Round brackets denote specific members of a family of planes, which in the previous example is denoted by curly brackets $\{10\bar{1}0\}$. The inclined plane shown in the figure is designated $(01\bar{1}2)$ and the other vertical plane shown there is $(11\bar{2}0)$.

Directions are also described in terms of four indices. Again, the third index equals the negative of the sum of the first two. In this case, reciprocals are not used. Consider the a_1-axis. This axis has the same direction as the sum of three vectors $2\boldsymbol{a}_1-\boldsymbol{a}_2-\boldsymbol{a}_3$ with no component along the c-axis. Its direction is thus specified as $[2\bar{1}\bar{1}0]$. Similarly, the direction of the a_2-axis is given as $[\bar{1}2\bar{1}0]$ and the direction of a_3 is $[\bar{1}\bar{1}20]$ These three directions, specified in square brackets, are members of the family of directions which is specified in pointed brackets $\langle 11\bar{2}0\rangle$. The direction of the c-axis is [0001].

References

Ahmad, S. and R. W. Whitworth (1988). Dislocation-motion in ice – a study by synchrotron x-ray topography. *Phil. Mag. A*, **57**, 749–766.

Ahmad, S., C. Sherwood and R. W. Whitworth (1992). Dislocation multiplication mechanisms in ice. In *Physics and Chemistry of Ice*, eds. N. Maeno and T. Hondoh. Sapporo: Hokkaido University Press, pp. 492–496.

Ashby, M. F. (1989). Materials selection in conceptual design. *Mater. Sci. Technol.*, **5**(6), 517–525.

Bader, H. (1964). Density of ice as a function of temperature and stress. *U.S. Army Cold Regions Research and Engineering Laboratory (USA CRREL), Special Report*, 64.

Baker, I. (2003). Imaging dislocations in ice. *Microsc. Res. Tech.*, **62**, 70–82.

Baker, I. and D. Cullen (2003). SEM/EDS observations of impurities in polar ice: Artifacts or not? *J. Glaciol.*, **49**, 184–190.

Baker, I., D. Cullen and D. Iliescu (2003). The microstructural location of impurities in ice. *Can. J. Phys.*, **81**, 1–9.

Cole, D. M. (1995). A model for the anelastic straining of saline ice subjected to cyclic loading. *Phil. Mag. A*, **72**, 231–248.

Cole, D. M. and G. D. Durell (1995). The cyclic loading of saline ice. *Phil. Mag. A*, **72**, 209–229.

Cottrell, A. H. (1953). *Dislocations and Plastic Flow in Crystals*. London: Oxford University Press.

Dash, J. G., F. Haiying and J. S. Wettlaufer (1995). The premelting of ice and its environmental consequences. *Rep. Prog. Phys.*, **58**, 115–167.

de La Chapelle, S., O. Castelnau, V. Lipenkov and P. Duval (1998). Dynamic recrystallization and texture development in ice as revealed by the study of deep ice cores in Antarctica and Greenland. *J. Geophys. Res.*, **103** (B3), 5091–5105.

Döppenschmidt, A. and H. J. Butt (2000). Measuring the thickness of the liquid-like layer on ice surfaces with atomic force microscopy. *Langmuir*, **16**, 6709–6714.

Faraday, M. (1850). On certain conditions of freezing water. *The Anthenaeum*, **1181**, 640–641.

Faraday, M. (1860). Note on regelation. *Proc. Roy. Soc. Lond.*, **10**, 440–450.

Fletcher, N. H. (1970). *The Chemical Physics of Ice*. Cambridge: Cambridge University Press.

Friedel, J. (1964). *Dislocations*. Addison-Wesley Series in Metallurgy and Materials. New York: Pergamon Press.

Frost, H. J. (2001). Crack nucleation in ice. *Eng. Fracture Mech.*, **68**, 1823–1837.

Fukuda, A. and A. Higashi (1988). Generation, multiplication and motion of dislocations in ice crystals. In *Lattice Defects in Ice Crystals*, ed. A. Higashi. Sapporo: Hokkaido University Press.

Fukuda, A., T. Hondoh and A. Higashi (1987). Dislocation mechanisms of plastic deformation of ice. *J. Physique*, **38**, 163–173.

Furukawa, Y. and H. Nada (1997). Anisotropic surface melting of an ice crystal and its relationship to growth forms. *J. Phys. Chem. B*, **101**, 6167–6170.

Gagnon, R. E., H. Kiefte, M. J. Clouter and E. Whalley (1988). Pressure dependence of the elastic constants of ice Ih to 2.8 kbar by Brillouin spectroscopy. *J. Chem. Phys.*, **89**, 4522–4528.

Glen, J. W. (1968). The effect of hydrogen disorder on dislocation movement and plastic deformation in ice. *Phys. Kondens. Mater.*, **7**, 43–51.

Gold, L. W. (1997). Statistical characteristics for the type and length of deformation-induced cracks in columnar-grain ice. *J. Glaciol.*, **43**, 311–320.

Gold, L. (1999). Statistical characteristics for the strain-dependent density and the spatial postion for deformation-induced cracks in columnar-grain ice. *J. Glaciol.*, **45**, 264–272.

Goto, K., T. Hondoh and A. Higashi (1986). Determination of diffusion coefficients of self-interstitials in ice with a new method of observing climb of dislocations by X-ray topography. *Jpn. J. Appl. Phys.*, **25**, 351–357.

Gross, G. W. and R. K. Svec (1997). Effect of ammonium on anion uptake and dielectric relaxation in laboratory-grown ice columns. *J. Phys. Chem. B*, **101**, 6282–6284.

Hardy, S. C. (1977). Grain-boundary groove measurement of surface-tension between ice and water. *Phil. Mag.*, **35**, 471–484.

Hayes, C. E. and W. W. Webb (1965). Dislocations in ice. *Science*, **147**, 44–45.

Higashi, A. (1978). Structure and behaviour of grain boundaries in polycrystalline ice. *J. Glaciol.*, **21**, 589–605.

Higashi, A., Ed. (1988). *Lattice Defects in Ice Crystals*. Sapporo: Hokkaido University Press.

Higashi, A., A. Fukuda, T. Hondoh, K. Goto and S. Amakai (1985). Dynamical dislocation processes in ice crystal. In *Dislocations in Solids*. Tokyo: Yamada Science Foundation, University of Tokyo Press, pp. 511–515.

Hirth, J. P. and J. Lothe (1992). *Theory of Dislocations*, 2nd edn. Malabar: Krieger Publishing Company.

Hobbs, P. V. (1974). *Ice Physics*. Oxford: Clarendon Press.

Hondoh, T. (2000). Nature and behavior of dislocations in ice. In *Physics of Ice Core Records*, ed. T. Hondoh. Sapporo: Hokkaido University Press, pp. 3–24.

Hondoh, T., T. Itoh and A. Higashi (1981). Formation of stacking-faults in pure ice single-crystals by cooling. *Jpn. J. Appl. Phys.*, **20**, L737–L740.

Hondoh, T., H. Iwamatsu and S. Mae (1990). Dislocation mobility for nonbasal glide in ice measured by in situ X-ray topography. *Phil. Mag. A*, **62**, 89–102.

Hull, D. and D. J. Bacon (1984). *Introduction to Dislocations*. International Series on Materials Science and Technology, 3rd edn. New York: Pergamon Press.

Humphreys, F. J. and Hatherly, M. (1996). *Recrystallization and Related Annealing Phenomena*. Oxford: Pergamon, Elsevier Science Ltd.

Iliescu, D., I. Baker and X. Li (2003). The effects of sulfuric acid on the creep, recrystallization, and electrical properties of ice. *Can. J. Phys.*, **81**, 395–400.

Jones, S. J. and N. K. Gilra (1973). X-ray topographical study of dislocations in pure and HF-doped ice. *Phil. Mag.*, **27**, 457–472.

Kamb, W. B. (1961). The glide direction in ice. *J. Glaciol.*, **3**, 1097–1106.

Ketcham, W. M. and P. V. Hobbs (1969). An experimental determination of the surface energies of ice. *Phil. Mag.*, **19**, 1161–1173.

Kobayashi, T. and Y. Furukawa (1975). On twelve-branched snow crystals. *J. Cryst. Growth*, **28**, 21–28.

Krausz, A. S. and L. W. Gold (1967). Surface features observed during thermal etching of ice. *J. Colloid Interface Sci.*, **25**, 255–262.

Liu, F. and I. Baker (1995). Thermally induced dislocation loops in polycrystalline ice. *Phil. Mag. A*, **71**, 1–14.

Liu, F., I. Baker and M. Dudley (1993). Dynamic observations of dislocation generation at grain boundaries in ice. *Phil. Mag. A*, **67**, 1261–1276.

Liu, F., I. Baker and M. Dudley (1995). Dislocation-grain boundary interactions in ice crystals. *Phil. Mag. A*, **71**, 15–42.

Louchet, F. (2004). Dislocations and plasticity in ice. *C. R. Physique*, **5**, 687–698.

Makkonen, L. (1997). Surface melting of ice. *J. Phys. Chem. B*, **101**, 6197–6200.

Mishima, O. and H. E. Stanley (1998). The relationship between liquid, supercooled and glassy water. *Nature*, **396**, 329–335.

Montagnat, M., Duval, P., Bastie, P. and Hamelin, B. (2003). Strain gradients and geometrically necessary dislocations in deformed ice single crystals. *Scripta Mater.*, **49**, 411–415.

Muguruma, J. (1961). Electron microscope study of etched ice surface. *J. Electron. Microsc.*, **10**, 246–250.

Nabarro, F. R. N. (2003). *Dislocations in Solids*, vol. 11. Amsterdam: Elsevier Science Publishing Co.

Nickolayev, O. Y. and E. M. Schulson (1995). Grain-boundary sliding and across-column cracking in columnar ice. *Phil. Mag. Lett.*, **72**, 93–97.

Nye, J. F. (1992). Water veins and lenses in polycrystalline ice. In *Physics and Chemistry of Ice*, eds. N. Maeno and T. Hondoh. Sapporo: Hokkaido University Press, pp. 200–205.

Oguro, M. (1988). Dislocations in artifically grown single crystals. In *Lattice Defects in Ice Crystals*, ed. A. Higashi. Sapporo: Hokkaido University Press, pp. 27–47.

Oguro, M., T. Hondoh and K. Azuma (1988). Interactions between dislocations and point defects in ice crystals. In *Lattice Defects in Ice Crystals*, ed. A. Higashi. Sapporo: Hokkaido University Press, pp. 97–128.

Pauling, L. (1935). The structure and entropy of ice and of other crystals with some randomness of atomic arrangement. *J. Amer. Chem. Soc.*, **57**, 2680–2684.

Perez, J., C. Mai, J. Tatibouet and L. F. Vassoille (1979). Study on grain-boundaries in 1H ice using internal-friction values. *Phys. Stat. Sol. A*, **52**, 321–330.

Petrenko, V. F. and R. W. Whitworth (1999). *Physics of Ice*. New York: Oxford University Press.

Picu, R. C. and V. Gupta (1995a). Crack nucleation in columnar ice due to elastic anistropy and grain boundary sliding. *Acta. Metall. Mater.*, **43**, 3783–3789.

Picu, R. C. and V. Gupta (1995b). Observations of crack nucleation in columnar ice due to grain boundary sliding. *Acta. Metall. Mater.*, **43**, 3791–3797.

Read, W. T. (1953). *Dislocations in Crystals*. New York: McGraw-Hill.

Röttger, K., A. Endriss, J. Ihringer, S. Doyle and W. F. Kuhs (1994). Lattice constants and thermal expansion of H_2O and D_2O ice Ih between 10 and 265 K. *Acta Crystallogr. B*, **50**, 644–648.

Schmitt, B., C. deBergh and M. Festou, Eds. (1998). *Solar System Ices: Based on Reviews Presented at the International Symposium "Solar System Ices" held in Toulouse, France on March 27–30, 1995*. Dordrecht: Kluwer Academic Publishers.

Shearwood, C. and R. W. Whitworth (1989). X-ray topographic observations of edge dislocation glide on non-basal planes in ice. *J. Glaciol.*, **35**, 281–283.

Shearwood, C. and R. W. Whitworth (1991). The velocity of dislocations in ice. *Phil. Mag. A*, **64**, 289–302.

Sinha, N. K. (1977). Dislocations in ice revealed by etching. *Phil. Mag.*, **36**, 1385–1404.

Sinha, N. K. (1978). Observation of basal dislocations in ice by etching and replicating. *J. Glaciol.*, **21**, 385–395.

Swalin, R. A. (1962). *Thermodynamics of Solids*. New York: John Wiley & Sons.

Tammann, G. (1900). Ueber die Grenzen des festen Zustandes IV. *Ann. Phys.*, **2**, 1–31.

Tatibouet, J., J. Perez and R. Vassoille (1987). 7th Symposium on the Physics and Chemistry of Ice, 1–5 September 1987 Grenoble (France). *J. Physique*, **48**, R3–R4.

Taupin, V., *et al.* (2007). Effects of size on the dynamics of dislocations in ice single crystals. *Phys. Rev.. Lett.*, **99**, 155507-1–155507-4.

van der Veen, J. F., B. Pluis and A. W. Denier van der Gon (1988). Surface melting. In *Chemistry and Physics of Solid Surface VII*, eds., R. Vanselow and R. F. Howe. Berlin: Springer.

van Oss, C. J., R. F. Giese, R. Wentzek, J. Norris and E. M. Chuvilin (1992). Surface tension parameters of ice obtained from contact angle data and from positive and negative particle adhesion to advancing freezing fronts. *J. Adhesion Sci. Technol.*, **6**, 503–516.

Weertman, J. and J. R. Weertman (1992). *Elementary Dislocation Theory*. New York: Oxford University Press.

Weiss, J. and E. M. Schulson (2000). Grain boundary sliding and crack nucleation in ice. *Phil. Mag.*, **80**, 279–300.

Whitworth, R. W. (1980). The influence of the choice of glide plane on the theory of the velocity of dislocations in ice. *Phil. Mag. A*, **41**, 521–528.

3

Microstructure of natural ice features

3.1 Introduction

The microstructure of a natural ice feature is a direct result of its thermal-mechanical history. Glaciers, for instance, form from snow through the processes of sintering and densification. At the surface, sintering of dry snow is driven by a reduction in surface and grain boundary energy. Below a depth of 2–3 m, density increases under the combined effects of particle rearrangement and sintering. Snow gradually changes into firn, a form of porous ice of relative density of about 0.6. The transition marks the end of particle packing as the dominant densification process. Firn densifies mainly through creep. Closed pores progressively form, leading to bubbly ice. In polar ice sheets, depending upon the snow accumulation rate and temperature, the transition from firn to bubbly ice occurs at a relative density between 0.82 and 0.84 (Arnaud, 1997). At the close-off density, ice contains cylindrical and spherical pores. The density of bubbly ice increases with depth, and the pressure within bubbles progressively increases. In polar sheets, bubbles transform into hydrate crystals below a depth of 500 m (Miller, 1973). The evolution of the microstructure with depth depends on temperature, strain rate and impurities, and depends as well on grain growth and recrystallization. The grain size of the ice that eventually forms is typically between a millimeter and several centimeters (Chapter 6). During the evolution, the random *c*-axis orientation in snow progressively disappears and a preferential orientation of grains develops under the combined effects of the rotation of the lattice by dislocation glide and recrystallization. In other words, the microstructure evolves as the body ages and deforms plastically over a period of hundreds to thousands of years.

A floating ice sheet, in comparison, forms through the process of solidification, generally over a period weeks to months. The process initiates on or near the surface, and then continues unidirectionally through the downward movement of the ice–water interface as heat is transferred from the relatively warm water below to the colder atmosphere above. The product is a polycrystalline cover that, in the case of sea ice, is composed of columnar-shaped grains of 5–10 mm diameter and of relatively large aspect ratio (10–20). The cover acts as a thermal barrier: the thicker the ice, the slower it grows. A sheet on the Arctic Ocean, for instance, can reach a thickness of 15–20 cm in a few days (Perovich and Richter-Menge, 2000). In the absence of deformation, it can grow to an average thickness of about two meters during an arctic winter (Rothrock *et al.*, 1999). Wind-driven movement and fracturing can lead to large-scale pressure ridging and to over-riding or rafting and thus to much greater thicknesses in certain locations. Thermal thickening, whether on a lake or the ocean – or occasionally on a river – is accompanied by the development of a crystallographic growth texture. The *c*-axes of crystals that make up sea ice, for instance, are more or less horizontally oriented (i.e., perpendicular to the growth direction), and are occasionally aligned within this plane (Weeks and Ackley, 1986). Lake ice, in comparison, generally possesses a texture in which the *c*-axes are vertically oriented (Gow, 1986). Growth texture is important because, like the deformation texture of glaciers, it imparts pronounced mechanical anisotropy.

The other terrestrial features of consequence are hailstones and atmospheric icing. Hailstones originate from ice nuclei in clouds, and can grow to a size >50 mm in diameter through the accretion of super-cooled water droplets. The physics of the process is described by Ludlam (1951) and by Knight and Knight (1968a,b; 1971). Dense, hard stones result when water is absorbed before freezing, while less dense and softer ones result from freezing upon contact. The microstructure consists of finely grained polycrystalline layers within which air bubbles are entrapped. Similarly, the atmospheric icing of airplanes, ships, electrical transmission lines and other structures generally occurs through the accretion of water droplets (Poots, 2000). The product is again a finely grained (<1 mm) polycrystal (Druez *et al.*, 1987; Ryerson and Gow, 2000; Kermani *et al.*, 2008) which is classified as either *glaze* (small super-cooling of ~1 K) or *rime* (larger super-cooling of ~5 K). Glaze is hard, highly adhesive and dense; rime is softer, less adhesive and less dense (Lock, 1990). Icing can also occur via direct deposition from the vapor state when the temperature and pressure are below the H_2O triple point (0.16 °C and 611 Pa). Frost forms in that manner.

Our objective in this chapter is to offer a short account of the microstructure of ice within snow, firn, glaciers and floating sheets. Each body is essentially a polycrystalline aggregate. The character, however, varies greatly, as will become apparent.

3.2 Experimental techniques

We begin with a few comments on experimental techniques. Blackford (2007) offers further comments. X-ray topography was discussed in Section 2.4.5. Notable by its absence is transmission electron microscopy. Although commonly employed to examine other materials, TEM has proven difficult to apply to ice, for the reasons mentioned earlier (Chapter 2).

3.2.1 Optical microscopy (OM)

Historically, OM coupled with a universal tilting stage has been used to characterize the microstructure and texture of ice of any provenance (Chapters 6 and 7). Exploited in such work is the optical anisotropy of ice Ih. A beam of light generally dissociates into two waves that travel through the crystal at different velocities. When emerging, the waves are out of phase. As a result of this effect and of the fact that the two waves are polarized at 90° to each other, individual grains can be distinguished by viewing between crossed polarizing filters. Grains within columnar ice can be easily detected within sections as thick as 25 mm or more, when viewed along the columns. Sections of granular ice should be thinner than the grain size. When sufficiently thin ($\sim$1 mm), grains of different orientations exhibit interference colors.

The microstructure of ice can be characterized also using reflected light. Figure 3.1 shows two-dimensional images that were obtained in coaxial reflected light. This technique reveals the pore network, grains and grain boundaries (Arnaud *et al.*, 1998).

3.2.2 X-ray microtomography (XMT)

X-ray microtomography is well suited to snow and firn. It provides information on the three-dimensional pore network and on the specific surface area (Flin *et al.*, 2004). Resolution better than 10 μm can be easily obtained on cm^3 samples. Low-density material is impregnated with chloronaphthalene to preserve the microstructure and to enhance contrast (Pieritz *et al.*, 2004). Figure 3.2 shows an XMT image of snow of density $\sim$260 kg m^{-3}.

3.2.3 Scanning electron microscopy (SEM)

Scanning electron microscopy reveals surface topography through imaging using secondary electrons. This technique can resolve structure with a spatial resolution two orders of magnitude or more better than light microscopy.

When coupled with X-ray microanalysis or energy dispersive spectroscopy (EDS), SEM-EDS allows impurities to be identified. The first attempts were made by Mulvaney *et al.* (1988) and by Wolff *et al.* (1988) who examined ice from Dolleman Island, Antarctica. To prevent charging, the ice was coated with

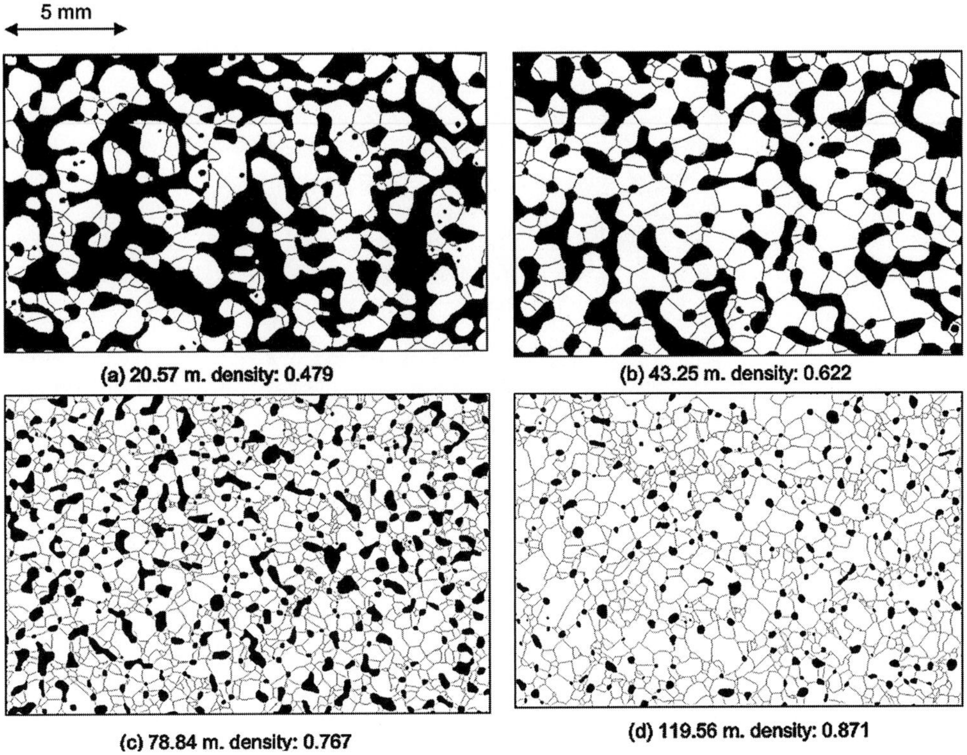

Figure 3.1. Binary images that illustrate the 2D structure of snow, firn and ice, obtained by image processing of photographs taken in reflected light at Dome Concordia (East Antarctica). (With the permission of Hélène Brunjail; LGGE/CNRS.)

aluminum. The analysis revealed sulphur concentrated at grain boundary triple junctions, presumably in the form of a sulphate or sulphuric acid.

The problems with the early work are that the coating absorbed X-rays emitted from the specimen and Al Kα X-rays emitted from the coating obscured rays from the specimen. Cullen and Baker (2000) and Baker and Cullen (2003) resolved these issues. Using a low-vacuum SEM they examined specimens of ice from Greenland Ice Sheet Project 2 (GISP2) and from Byrd Station, Antarctica, without using a coating, at temperatures between −115 °C and −60 °C. Sublimation produced enough ionized molecules to neutralize charging. Sublimation actually aids such analysis, because it concentrates impurities both within the grains and at grain boundaries. For instance, after sublimation of GISP2 ice, filaments rich in Na and Cl decorated most of the grain boundaries (Baker and Cullen, 2003). The filaments are actually artifacts in that they formed *in situ* through the coalescence of impurities along the boundary.

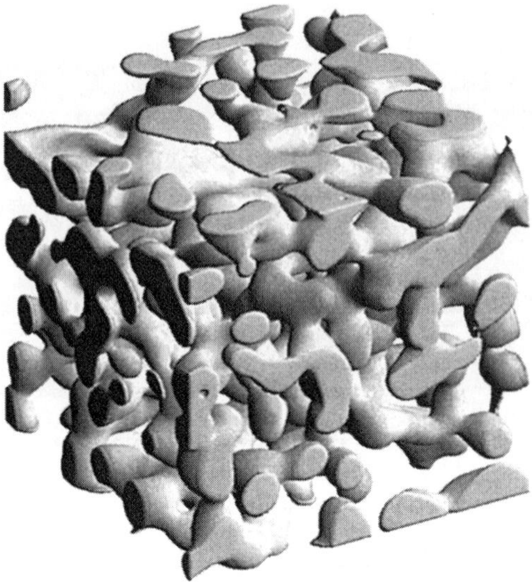

Figure 3.2. X-ray tomography image of a dry natural snow sample after isothermal metamorphism at −2 °C for 2000 hours. Sample width = 2.5 mm. (From Pieritz *et al.*, 2004; with permission of the International Glaciological Society.)

3.2.4 Raman spectroscopy

To identify chemical compounds and not just chemical elements, Fukazawa *et al.* (1998) applied Raman spectroscopy. They examined ice from two sites in Antarctica by passing a laser beam down an optical microscope. The beam excited atomic bonds and this excitation allowed nitrates and hydrogen sulphates to be identified. Subsequently, Baker *et al.* (2005) coupled Raman spectroscopy with scanning con-focal optical microscopy to reveal $NaNO_3$ at grain boundaries in specimens taken from GISP2 ice cores. There is a shortcoming, however. Raman spectroscopy cannot identify ions of elements such as Na, K, Ca, Al and Cl. Those elements can be identified by SEM-EDS. Raman spectroscopy may then be viewed as a complementary method.

3.2.5 Electron back-scatter diffraction (EBSD)

EBSD in SEM is an established technique in materials science. The method reveals the orientation of individual grains and was first applied to ice by Iliescu *et al.* (2004). It was later used to determine both the overall texture of polycrystalline aggregates from GISP2 and to identify features that result from recrystallization (Obbard *et al.*, 2006). As well as allowing rapid determination of texture, the method is superior to optical microscopy in another way: it allows both the *c*-axes and *a*-axes of individual grains to be determined (Obbard *et al.*, 2006).

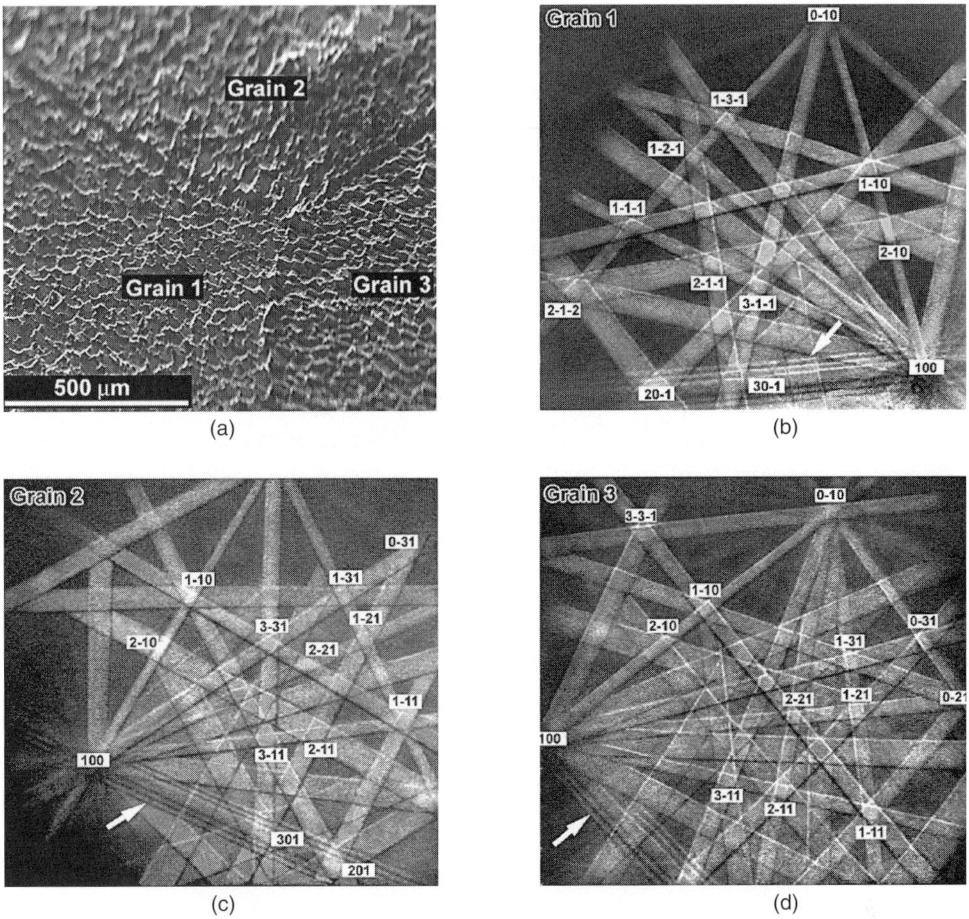

Figure 3.3. SEM micrographs showing (a) a triple junction in fresh-water ice and (b–d) EBSD patterns from each of the three grains. (From Iliescu *et al.*, 2004.)

SEM-EBSD also allows misorientation across grain boundaries to be determined, as well as the frequency of a given misorientation (Obbard and Baker, 2007). This technique can be used also to determine the lattice distortion with spatial and angular resolution of 0.25 μm and 0.3°, respectively (Piazolo *et al.*, 2008).

Figure 3.3 shows three grains meeting at a triple junction within a polycrystal of laboratory-grown fresh-water ice, accompanied by the EBSD pattern from each grain (Iliescu *et al.*, 2004). The clarity of the pattern is such that it permits the orientation of the grains (relative to the plane of viewing) to be determined to within 1°. The crystallographic information originates within the upper part (~50 nm) of the surface. The lateral resolution depends mainly on the type of electron gun used in the SEM, but in general is <0.5 μm.

3.3 The microstructure of ice within glaciers and polar ice sheets

3.3.1 Sintering and densification of snow

In the absence of external pressure, solid-state snow sintering is driven by a reduction in surface energy and grain boundary energy. At the surface of glaciers and ice sheets, snow grains are subjected to water vapor gradients generated by differences in curvature and/or temperature. The vapor pressure in terms of the mean radius of curvature is given by the Kelvin equation, whereas the temperature dependence of the vapor pressure is given by the Clausius–Clapeyron equation (Colbeck, 1983). Temperature gradients in the first meters of ice sheets, which can be higher than 10 °C/m, cause vapor motion at a much greater rate than could occur due to gradients in curvature. Thus, sintering by differences in curvature is limited to the first centimeters (Colbeck, 1998). The geometry of the neck between two ice particles is shaped as a groove with the presence of a grain boundary, Figure 3.4, and not by concavity as usually described (see Colbeck, 1997). From Maeno and Ebinuma (1983), mass transport by surface diffusion could be important only at small neck sizes ($<2\ \mu m$) and high temperatures (>-6 °C). Non-isothermal metamorphism of snow can create weak layers that can result in avalanches. Settling or the displacement under gravity of individual particles relative to their neighbors with some mechanical destruction of snow grains becomes important below the first few meters (Alley, 1987; Salamatin *et al.*, in press).

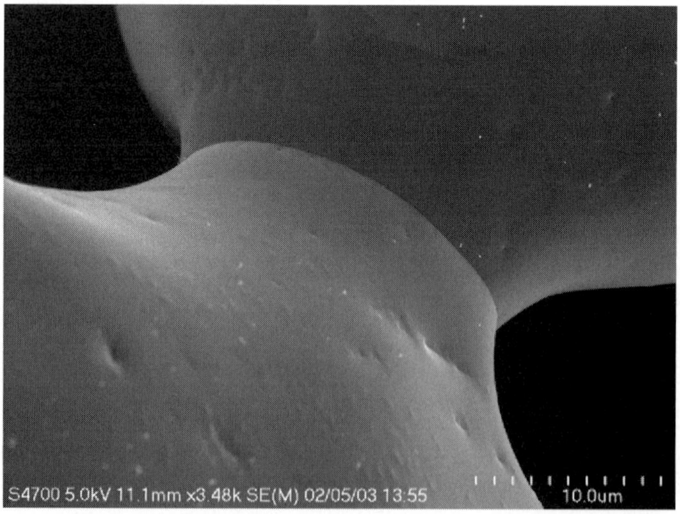

Figure 3.4. Low-temperature SEM image of the microstructure of two ice single crystals. Grain boundary can be seen at the neck between particles. (From Blackford, 2007; with permission of Institute of Physics Publishing Limited.)

In liquid-phase sintering, mass transport occurs predominantly via water by diffusion and convection (Blackford, 2007). In temperate glaciers, where water is present, the process occurs much more rapidly than solid-state sintering, for several reasons. Melt-water lubricates the grains and so accelerates packing. Also, through the effect of surface tension, water tends to draw the individual crystals together. In addition, molecular diffusion within the liquid state occurs much more rapidly than within the solid state (Chapter 4).

Further discussion can be found in reviews by Colbeck (1983, 1997), Maeno and Ebinuma (1983), Gubler (1985), Paterson (2000) and Blackford (2007).

3.3.2 Densification of dry firn

The densification of firn starts when dislocation creep becomes the dominant process. Particle rearrangement can locally occur down to a relative density (with respect to fully dense ice) of about 0.7 (Salamatin *et al.*, in press). As seasonal variations of surface temperature disappear below 5–10 m, the densification occurs at constant temperature. This stage corresponds to relative densities between about 0.6 and 0.92 (Arnaud, 1997). The number of contacts (bonds) per grain and the fraction of the surface area of grains involved are essential parameters (Arnaud *et al.*, 2000; Salamatin *et al.*, in press). A significant fraction of the surface area lies in grain boundaries at the snow–firn transition (Arnaud *et al.*, 2000). Hence, densification by grain sliding involves groups of crystals. Two-dimensional images clearly show the significant surface fraction of grains involved in bonds at the snow–firn transition (Figure 3.1). The densification of snow and firn is associated with grain growth and/or rotation recrystallization (Gow, 1969; Arnaud, 1997).

The snow and firn layer is typically 50–100 m thick and the age of ice at the firn–ice transition (close-off) can actually reach 3000 years in the coldest parts of the Antarctic ice sheet (Arnaud *et al.*, 2000). Sintering and densification processes in a temperate glacier occur more rapidly by melting/refreezing and by packing. The transformation into ice can occur within 10 years (Paterson, 2000).

Reviews of the densification and structure of firn can be found in Maeno and Ebinuma (1983), Wilkinson (1988), Arnaud (1997) and in Blackford (2007).

3.3.3 Bubbly ice and air-hydrate crystals

Densification of ice containing closed bubbles mainly proceeds by dislocation creep and is driven by the pressure difference between bubbles and ice. The bubble pressure increases more rapidly than the ice pressure and the bubble-pressure lag is very small from a depth of about 200 m at Vostok, East Antarctica (Lipenkov *et al.*, 1997). The number and mean radius of bubbles at Vostok are, respectively,

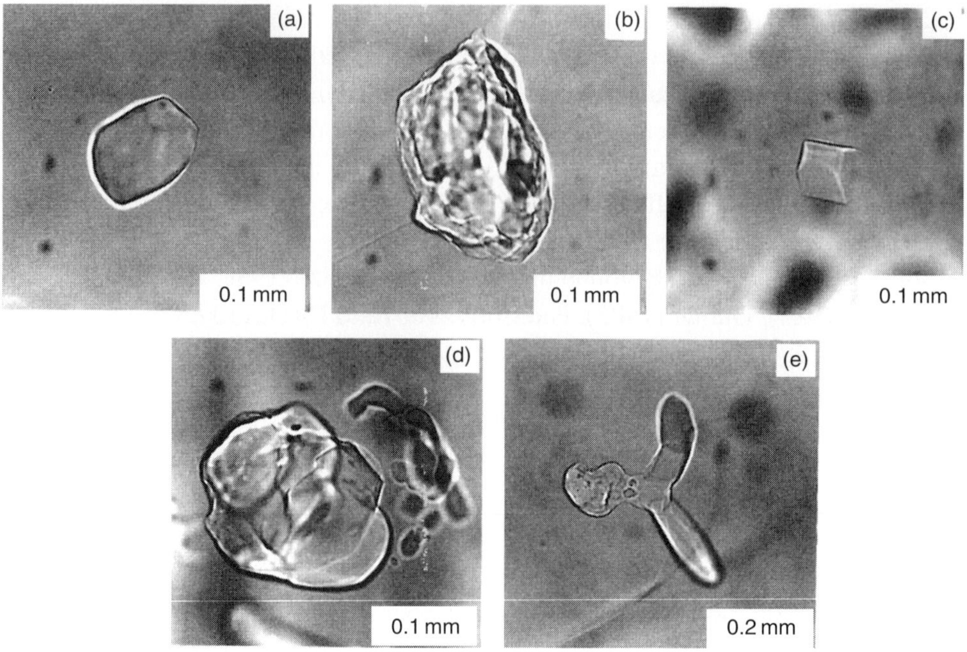

Figure 3.5. Morphological types of air-hydrate crystals observed in the Vostok ice core. (a) Rounded crystal, (b) cloud-like crystal, (c) faceted crystal, (d) collapsed hydrate and (e) cluster of several hydrate crystals. (From Lipenkov, 2000; with permission of Hokkaido University Press.)

380/g and 0.15 mm at a depth of about 200 m (Lipenkov, 2000). The higher number of bubbles found under glacial climate is related to the smaller size of ice grains.

Air bubbles trapped in polar ice transform into hydrate crystals of various shapes as soon as the pressure exceeds the hydrate dissociation pressure of air molecules at the *in-situ* temperature (Miller, 1973). The transition zone where air bubbles and hydrates crystals co-exist covers several hundred meters. At Vostok, the bubble–hydrate transition is between the depths of 500 and 1250 m (Lipenkov, 2000). Rounded and partly faceted crystals are common, Figure 3.5.

The mean hydrate radius at the end of the bubble–hydrate transition at Vostok is about 0.07 mm for interglacial ice and 0.048 mm for glacial ice (Salamatin and Lipenkov, 2003). A clear increase in the hydrate crystal size is observed along deep ice cores (Uchida *et al.*, 1994; Pauer *et al.*, 1999). The difference in size between hydrate crystals induces gas-concentration gradients in the ice matrix, thus creating a driving force for the diffusion of air molecules from the smaller to the larger crystals (Salamatin and Lipenkov, 2003). The process is known as *Ostwald ripening*.

Air bubbles and atmospheric particulates are important microstructural constituents of ice that has formed through the consolidation of snow, particularly of polar ice sheets (Chapter 2). Their importance lies not so much in relation to creep and fracture, although their role there cannot be ignored (Chapter 6), but more in the chemical composition of the air.

3.4 Floating ice sheets

3.4.1 Nucleation and early stage of development

Imagine a large body of quiet water whose surface is cooled, say, by radiation to the sky at night. Classical nucleation theory (Turnbull and Fisher, 1949) holds that solidification begins when super-cooling reaches a threshold. At that point, the decrease in volume free energy accompanying the water-to-ice transformation is just sufficient to overcome the increase in surface free energy that accompanies the creation of an interface. Embryonic ice crystals are now stabilized against re-solution and thus constitute stable nuclei. Upon further cooling the nuclei begin to grow: molecules, driven by the reduction in volume free energy, cross the interface from the liquid to the solid and become attached to the crystal. Growth can occur four to five orders of magnitude more rapidly along the basal plane than perpendicular to it (Hillig, 1959), owing to the greater availability of attachment sites on molecularly rougher edge facets. This kind of anisotropy leads to the formation of thin crystals of a variety of shapes and crystallographic orientations, from nearly circular discs with c-axes vertical to needles with c-axes horizontal (Fletcher, 1970).

Nucleation generally occurs heterogeneously, either at the air–water interface, on snow crystals, or on foreign particles or films. This greatly lowers the nucleation free energy barrier. It also reduces the size of the critical nucleus, to a thickness of only one monolayer layer in some instances (Seeley and Seidler, 2001). As a result, super-cooling of only a few hundredths to a few tenths of a degree is usually required to initiate the process (Hillig, 1958; Weeks and Ackley, 1986).

The nucleation of ice on the sea occurs in essentially the same manner as on cold lakes. There are, however, differences in the actual temperature at which nucleation begins and in the distribution of nuclei. Both differences are related to solute concentration and to its effects on the freezing point and on the temperature at which the density of water reaches a maximum. The ocean contains sodium chloride ($NaCl$) (mainly) as well as sodium sulphate (Na_2SO_4), magnesium chloride ($MgCl_2$), calcium carbonate ($CaCO_3$), potassium chloride (KCl) and other salts, to a total concentration of around 35 g/kg (or 35 parts per thousand or ppt). The sea salts lower the temperature for onset of freezing, from near zero to −1.8 °C. They also lower the temperature of maximum density, from +3.8 °C for zero salinity to −3.7 °C for a salinity of 35 ppt. This means that before salt-water begins to freeze by cooling from above, the density of near-surface water increases, creating an unstable vertical density distribution. This instability results in free convection within the salt-water column and in cooling of the column to the freezing point to depths as great as tens of meters (Weeks and Ackley, 1986). In comparison, cooling fresh-water to temperatures below that of maximum density leads to the formation of a stable surface layer which can freeze more rapidly. That is why lakes and rivers begin to freeze before the sea.

Once begun, nucleation and crystal growth continue until a skim forms. The skim consists of crystals of a variety of shapes (Weeks and Ackley, 1986; Koo *et al.*, 1991). The most common, particularly on lakes (Hallett, 1960; Knight, 1966), are needle-shaped crystals with *c*-axes inclined in many directions (Gow, 1986). The needles interlock. Between them platelets form with their *c*-axes nearly vertical. From this composite the sheet begins to thicken.

A similar process occurs under windy conditions. However, turbulence mixes the water and so nuclei form throughout the upper layer. The nuclei-cum-crystals then aggregate into slush, which solidifies into a finely grained polycrystalline layer upon further cooling.

The details of nucleation and crystal growth need not concern us here. The reader will find clear accounts in books by Fletcher (1970) and Hobbs (1974) and in a review by Oxtoby (1999). For our purposes, the important point is the formation of a thin layer of ice that contains crystals of many different orientations. This means that the subsequent development of growth texture is not simply a reflection of the absence of crystals of certain orientations. Instead, it means that texture develops through the preferential selection of a specific orientation.

3.4.2 Growth

As the ice–water interface advances, favorably oriented crystals expand at the expense of less favorably oriented ones (Weeks and Wettlaufer, 1996) which are wedged out, Figure 3.6. The process is termed *geometric selection* and accounts for the development of *growth texture* (Tiller, 1957; Perey and Pounder, 1958; Walton and Chalmers, 1959; Hellawell and Herbert, 1962; Flemings, 1974; Weeks and Ackley 1986). The general rule is that the larger is the angle between the *c*-axis and the growth direction, the greater is the probability that that grain will be preferred (Ketcham and Hobbs, 1967; Ramseier, 1968; Kawamura, 1987). This selection is a result of the easy growth direction being parallel to the basal planes (Hillig, 1958; Macklin and Ryan, 1965). Thus, when sheets grow from thin layers in which crystals of all orientations are present, they are expected to possess predominantly the *c*-axis horizontal texture. First-year sea ice and fresh-water ice formed through

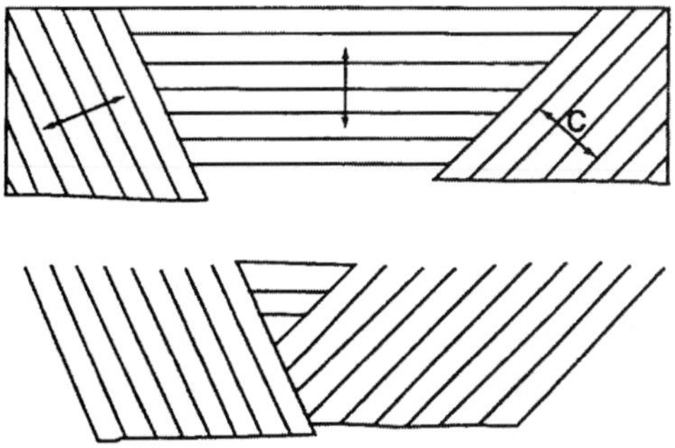

Figure 3.6. A schematic sketch showing the process of geometrical selection during the unidirectional solidification of water. The plate-like structure of the crystals denotes the orientation of the basal (0001) planes. (From Weeks and Wettlaufer, 1996.)

unidirectional solidification possess this characteristic (Perey and Pounder, 1958; Weeks and Ackley, 1986), Figure 3.7.

Lake ice, in comparison, generally possesses a texture in which the *c*-axes are preferentially oriented in the vertical direction; i.e., parallel to the direction of growth. (Knight, 1966; Gow, 1986). One interpretation is that crystals within the skim are not randomly oriented. The process may be more complicated than that, and could be related to the depth to which the water column is super-cooled and to thermally constrained geometric selection in the absence of mechanical disturbance (Weeks and Wettlaufer, 1996).

Ice sheets formed through unidirectional solidification also possess a characteristic *fabric* (Figure 3.7). As already noted, sheets of first-year sea ice, for instance, are generally composed of columnar-shaped grains 5–10 mm in diameter, elongated in the growth direction. River ice, too, can possess this fabric (e.g., Gold and Krausz, 1971), although examples are rare owing to more turbulent growth conditions. Lake ice, on the other hand, is generally composed of irregular and sometimes quite massive grains that can reach a meter or more in diameter near the bottom (Gow, 1986). Within floating ice covers, the average grain diameter increases with depth, at a rate in sea ice of approximately 0.033 m/m (Weeks and Ackley, 1986). The effect is analogous to the widening of grains within the columnar zone of metallic ingots (Walton and Chalmers, 1959) and is caused presumably by the migration of grain boundaries.

3.4.3 Sea ice growth

In a general sense, sheets of salt-water ice and fresh-water ice grow in a similar manner. Consequently, their microstructure is similar, as evident from Figure 3.7.

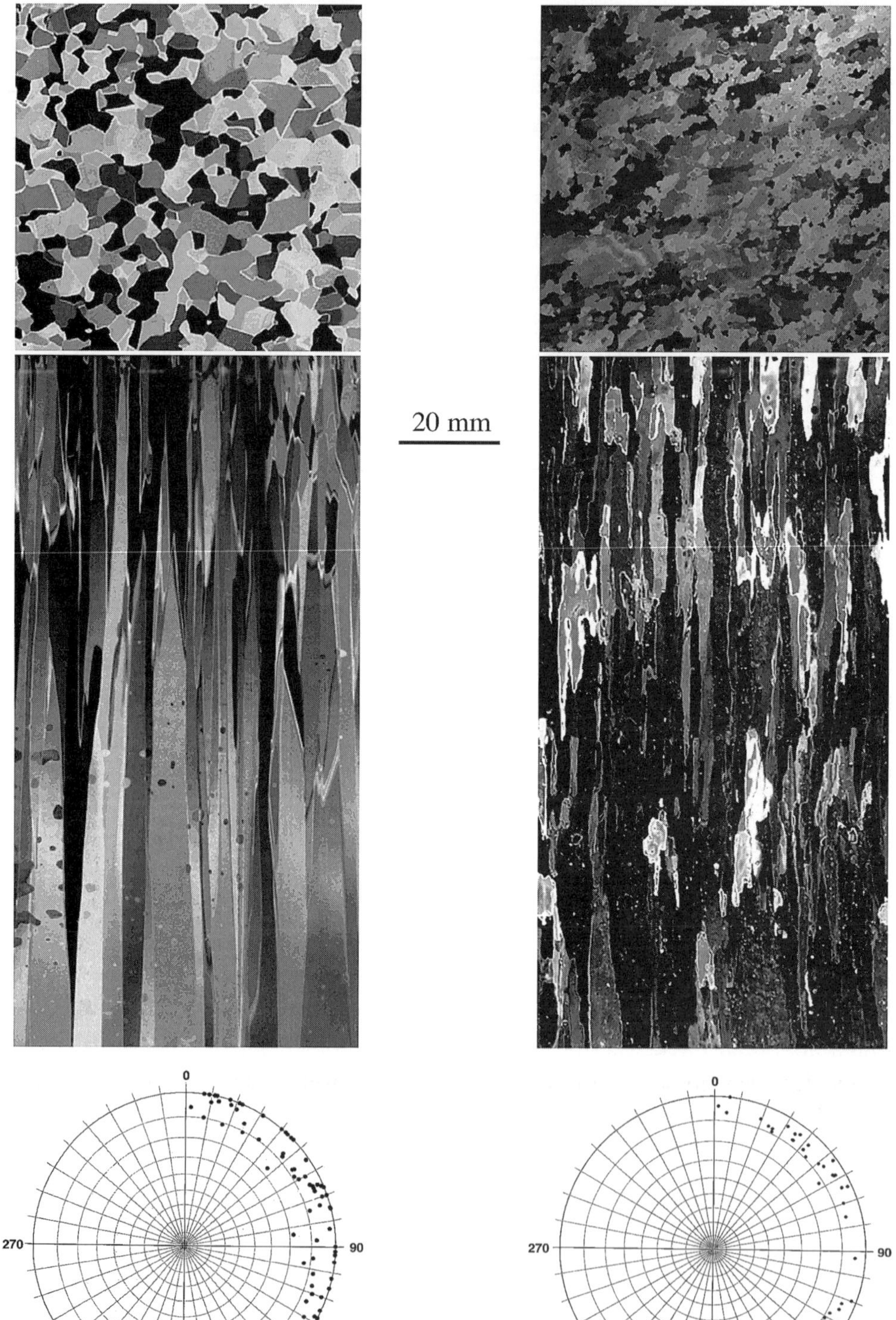
20 mm
0
270
90
180
0
270
90
180

There is, however, one significant difference. Within sea ice, brine becomes entrapped. This leads to an intragranular substructure and to inter-digitated as opposed to faceted grain boundaries. The entrapment is related to the intolerance of the ice Ih crystal lattice for solutes (Gross and Svec, 1997) and to their rejection into the liquid as the ice–water interface advances. It arises as follows.

As salt-water freezes, very little NaCl or any other salt is dissolved in the ice. Instead, most salts are rejected. The rejection raises the salt concentration at the ice–water interface, and within the adjacent boundary zone, to a level above the bulk concentration. As a result of the varying solute concentration ahead of the advancing interface, there is a corresponding variation in the equilibrium freezing temperature, which is given by the water-rich liquidus of the water–salt phase diagram. The temperature at the interface is fixed by the local equilibrium conditions, but the actual temperature of the water ahead of the interface is set by the thermal gradient near the surface of the ocean. As a result, the water within the boundary zone ahead of the interface becomes cooled to below its equilibrium freezing temperature. The process is termed *constitutional super-cooling* (as opposed to *thermal super-cooling*, which precedes nucleation) and was first described by Rutter and Chalmers (1953) to explain interfacial instability during the solidification of metallic alloys. Constitutional super-cooling leads to a dendritic instead of a planar interface and, during the freezing of seawater, to the mechanical entrapment of brine between the dendrites. The shape of dendrites in sea ice is plate-like with the large faces parallel to the {0001} basal plane. Brine entrapment creates a *substructure* in which sub-millimeter-sized brine pockets are distributed in planar-like arrays, Figure 3.8. A typical spacing is 0.5–1 mm (Lofgren and Weeks, 1969; Nakawo and Sinha, 1984) and a typical amount of entrapped brine at −10 °C is 2–3% by volume. This leads to an average melt-water salinity of about 4–6 ppt (Cox and Weeks, 1974). The ice also contains about the same volume of entrapped air as brine, the two phases often occurring together within the substructure. The brine-related porosity of sea ice increases with increasing temperature and seems to reach a percolation threshold around −5 °C (Weeks and Ackley, 1986; Golden *et al.*, 1998). The substructure is important because it defines an easy path for the propagation of cracks.

Sea ice also possesses macrostructure. It consists of an array of irregularly shaped, tree-like features termed *brine-drainage channels*, Figure 3.9. They are

---

Figure 3.7. Photographs showing the microstructure of unidirectionally solidified S2 fresh-water ice (left-hand side) and first-year arctic sea ice (right-hand side), as viewed in thin sections through crossed-polarizing filters in an optical microscope. The top row shows horizontal sections; the middle shows vertical sections. The lowest row shows the preferred orientation of the crystallographic *c*-axes in Schmidt nets, where the center corresponds to the direction of growth. (From Fortt and Schulson, unpublished.)

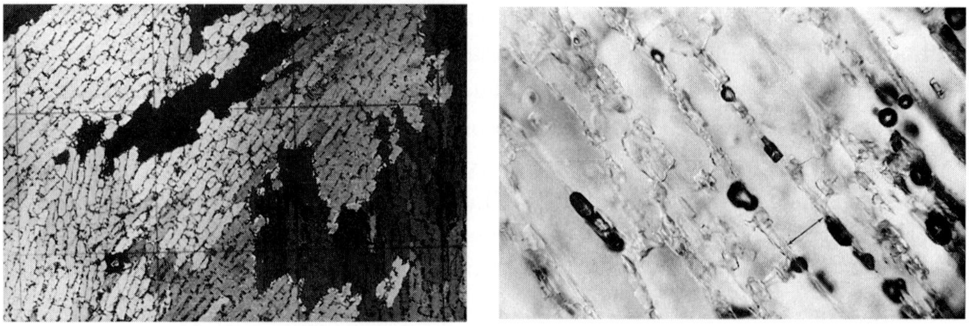

Figure 3.8. Photographs showing substructure in the form of arrays of brine-pockets, as viewed within a horizontal section of first-year sea ice, showing: (a) arrays within several grains (grid side is 10 mm) and (b) a planar array within a single grain (plane spacing 0.6 mm). (From Weeks and Ackley, 1986.)

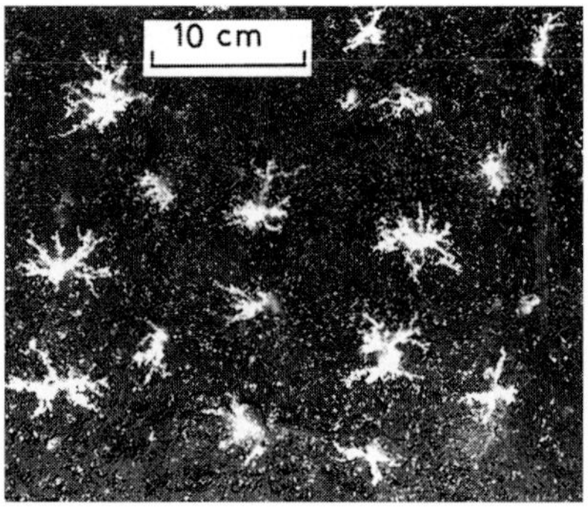

Figure 3.9. Photograph of brine drainage channels in first-year sea ice as viewed within a horizontal section. (From Wakatsuchi and Kawamura, 1987).

composed of vertically oriented tubes accompanied by a number of smaller tributaries. These features are typically centimeters in diameter and tens of centimeters in average spacing (Wakatsuchi and Kawamura, 1987; Cole and Shapiro, 1998). Like the substructure, the channels serve as short-circuit fracture paths (Shapiro and Weeks, 1993).

Crystalline salts constitute the other microstructural constituent of sea (Sinha, 1977). The major one is $NaCl.2H_2O$. It precipitates out of the brine upon cooling to −21.2 °C, the eutectic temperature in the H_2O–NaCl system. Other hydrated species precipitate out at higher and lower temperatures (e.g., $Na_2SO_4.10H_2O$ at −8.7 °C) (Killawee *et al.*, 1998; Tison *et al.*, 2002), but in lesser quantities. The volume fraction of precipitated sea salt is much lower than the brine volume and is probably too small to affect mechanical behavior.

We mentioned the preference for *c*-axis confinement to the horizontal plane of a first-year sea ice cover. In addition, within certain regions of the ice cover on the Arctic Ocean the *c*-axes are often aligned within this plane (Cherepanov, 1971; Weeks and Gow, 1980). Alignment appears to develop slowly as the cover thickens – in one instance near Barrow, Alaska, not until the sheet had thickened to ~30 cm (Cole and Shapiro, 1998) – but can extend over large regions, tens to hundreds of kilometers across. The direction of alignment correlates with the direction of average oceanic current at the ice/water interface (Weeks and Wettlaufer, 1996). The origin of *c-axis alignment* is not fully understood. The prevalent view (Weeks and Wettlaufer, 1996) is that it results from a lessening of solute buildup at the tips of dendritic plates whose normals are parallel to the current. This is in keeping with the suggestion (Langhorne and Robinson, 1986) that the current must exceed a threshold velocity of ~0.01 m/s before the feature can develop.

The microstructure of antarctic sea ice appears to be quite similar to that of arctic ice (Gow *et al.*, 1982). Although regionally diverse, the grains are largely columnar in shape, at least in some locations; they possess the substructure described above; and they exhibit *c*-axis horizontal texture, both unaligned and aligned (Jeffries *et al.*, 1993). However, there is one significant difference: platelets of ice are created in the Antarctic, some as thick as 5 mm and as wide as 150 mm (Jeffries *et al.*, 1993). The platelets are transported via advection to the underside of land-fast ice, forming in the process an open lattice through which unaligned columnar ice subsequently grows. The platelets are single crystals and originate through either the flow of melt-water off the subcontinent or the flow of low-density seawater from beneath ice shelves and the subsequent super-cooling of this water as it rises.

3.5 Terminology and classification

Terminology In this book we use the term "texture" to describe the preferred orientation of grains within polycrystals, in keeping with its traditional usage in materials science and its growing usage in structural geology. We use the term "fabric" to denote the shape of the grains. Some investigators reverse these terms.

Classification In describing the microstructure of floating ice sheets, we follow the classification suggested by Michel and Ramseier (1971). Accordingly, the bulk of the sheet is termed secondary (S) ice. The preferred orientation of the grains is then conveyed by the digits 1, 2 or 3. The S1 class denotes secondary ice whose *c*-axes are predominantly vertical; S2 denotes ice whose *c*-axes are oriented in a predominantly horizontal direction and randomly oriented within the horizontal plane (Figure 3.7); and S3 denotes ice whose *c*-axes are confined more or less to the horizontal plane of the sheet and aligned in a particular direction within that plane.

3.6 Summary

Ice exhibits a rich and varied microstructure that depends sensitively upon its thermal-mechanical history and, in the case of sea ice, upon the presence of oceanic salts. With the exception of snow-flakes and some icicles, all other natural bodies are polycrystalline aggregates. The firn and bubbly ice of glaciers and polar ice sheets are generally composed of equiaxed and randomly oriented grains that have sintered together from snow under the action of heat and pressure, assisted in some instances by liquid-phase diffusion. Such material can be considered to be mechanically isotropic. Upon slow deformation over time, crystallographic texture develops. This change in microstructure imparts plastic anisotropy, as we discuss in Chapters 6 and 7. In comparison, the ice cover on the Arctic Ocean is composed of columnar-shaped grains, like those within a metallic casting. This fabric is a direct result of unidirectional solidification. Brine and air are entrapped within sub-millimeter-sized pores and larger channels. The additional phases create a composite material characterized by both substructure and macrostructure. The ice cover possesses a growth texture in which the crystallographic *c*-axes are more or less confined to the horizontal plane, owing to geometric selection and to the preferential growth along the crystallographic *a*-axes. In some cases the *c*-axes are randomly oriented within that plane (S2 ice), while in others they are aligned (S3 ice). In either case, the ice is mechanically anisotropic, as we discuss in Chapters 4, 6, 7, 11 and 12. More complex microstructures are expected within the icy material of the outer Solar System where environmental conditions and the presence of impurities may favor high-pressure ice and clathrate hydrates (Chapter 8).

References

Alley, R. B. (1987). Firn densification by grain boundary sliding: a first model. *J. Physique*, **48**, C1-249–C1-256.

Arnaud, L. (1997). Modélisation de la transformation de la neige en glace à la surface des calottes polaires; Etude du transport des gaz dans ces milieux poreux. Thesis, Université Joseph Fourier, Grenoble.

Arnaud, L., V. Lipenkov, J. M. Barnola, M. Gay and P. Duval (1998). Modelling of the densification of polar firn: characterization of the snow-firn transition. *Ann. Glaciol.*, **26**, 39–44.

Arnaud, L., J. M. Barnola and P. Duval (2000). Physical modeling of the densification of snow-firn and ice in the upper part of polar ice sheets. In *Physics of Ice Core Records*, ed. T. Hondoh. Sapporo: Hokkaido University Press, pp. 285–305.

Baker, I. and D. Cullen (2003). SEM/EDS observations of impurities in polar ice: Artifacts or not? *J. Glaciol.*, **49**, 184–190.

Baker, I., D. Iliescu, R. Obbard *et al.* (2005). Microstructural characterization of ice cores. *Ann. Glaciol.*, **42**, 441–444.

Blackford, J. R. (2007). Sintering and microstructure of ice: a review. *J. Phys. D: Appl. Phys.*, **40**, R355–R385.

Cherepanov, N. V. (1971). Spatial arrangement of sea ice crystal structure. *Prob. Arkt. i Antarkt.*, **38**, 176–181.

Colbeck, S. C. (1983). Theory of metamorphism of dry snow. *J. Geophys. Res.*, **88**, 5475–5482.

Colbeck, S. C. (1997). A review of sintering in seasonal snow. *CRREL Report*, **97**-10, 1–11.

Colbeck, S. C. (1998). Sintering in a dry snow cover. *J. Appl. Phys.*, **84**, 4585–4589.

Cole, D. M. and L. H. Shapiro (1998). Observations of brine drainage networks and microstructure of first-year sea ice. *J. Geophys. Res.*, **103**, 21739–21750.

Cox, G. F. N. and W. F. Weeks (1974). Salinity variations in sea ice. *J. Glaciol.*, **13**, 109–120.

Cullen, D. and I. Baker (2000). The chemistry of grain boundaries in Greenland ice. *J. Glaciol.*, **46**, 703–706.

Druez, J., J. Cloutier and L. Claveau (1987). Etude comparative de la resistance a la traction et a la compression de la glace atmospherique. *J. Physique Coll. C1*, **49**, C1-337–C1-343.

Flemings, M. C., Ed. (1974). *Solidification Processing.* Materials Science and Engineering Series. New York: McGraw-Hill, Inc.

Fletcher, N. H. (1970). *The Chemical Physics of Ice.* Cambridge: Cambridge University Press.

Flin, F., J. B. Brzoska, B. Lesaffre, C. Coleou and R. Pieritz (2004). Three-dimensional geometric measurements of snow microstructural evolution under isothermal conditions. *Ann. Glaciol.*, **38**, 39–44.

Fukazawa, H., K. Sugiyama, S. Mae, H. Narita and T. Hondoh (1998). Acid ions at triple junction of Antarctic ice observed by Raman scattering. *Geophys. Res. Lett.*, **25**, 2845–2848.

Gold, L. W. and A. S. Krausz (1971). Investigation of the mechanical properties of St. Lawrence river ice. *Can. Geotech. J.*, **8**, 163–169.

Golden, K. M., S. F. Ackley and V. I. Lytle (1998). The percolation phase transition in sea ice. *Science*, **282**, 2238–2241.

Gow, A. J. (1969). On the rates of growth of grains and crystals in south polar firn. *J. Glaciol.*, **8**, 241–252.

Gow, A. J. (1986). Orientation textures in ice sheets of quietly frozen lakes. *J. Cryst. Growth*, **74**, 247–258.

Gow, A. J., S. F. Weeks and J. W. Govoni (1982). Physical and structural characteristics of Antarctic sea ice. *Ann. Glaciol.*, **3**, 113–117.

Gross, G. W. and R. K. Svec (1997). Effect of ammonium on anion uptake and dielectric relaxation in laboratory-grown ice columns. *J. Phys. Chem. B*, **101**, 6282–6284.

Gubler, H. (1985). Model for dry snow metamorphism by interparticle vapor flux. *J. Geophys. Res.*, **90**, 8081–8092.

Hallett, J. (1960). Crystal growth and the formation of spikes in the surface of supercooled water. *J. Glaciol.*, **3**, 698–704.

Hellawell, A. and P. M. Herbert (1962). The development of preferred orientations during the freezing of metals and alloys. *Proc. R. Soc. A*, **269**, 560–573.

Hillig, W. B. (1958). The kinetics of freezing of ice in the direction perpendicular to the basal plane. In *Growth and Perfection of Crystals*, eds. R. H. Doremus, B. W. Roberts and D. Turnbull. New York: Wiley & Sons, pp. 350–359.

Hillig, W. B. (1959). Kinetics of solidification from nonmetallic liquids. In *Kinetics of High Temperature Processes*, ed. W. D. Kingery. New York: Wiley & Sons, pp. 127–135.

Hobbs, P. V. (1974). *Ice Physics*. Oxford: Clarendon Press.

Iliescu, D., I. Baker and H. Chang (2004). Determining the orientations of ice crystals using electron backscatter patterns. *Microsc. Res. Tech.*, **63**, 184–187.

Jeffries, M. O., W. F. Weeks, R. Shaw and K. Morris (1993). Structural characteristics of congelation and platelet ice and their role in the development of Antarctic land-fast sea ice. *J. Glaciol.*, **39**, 223–238.

Kawamura, T. (1987). Studies on preferred growth of sea ice grain. In *Contributions of the Institute of Low Temperature Science*. Sapporo: Hokkaido University Press, pp. 1–29.

Kermani, M., M. Farzaneh and R. Gagnon (2008). Bend strength and effective modulus of atmospheric ice, *Cold Reg. Sci. Technol.*, **53**, 162–169.

Ketcham, W. M. and P. V. Hobbs (1967). The preferred orientation in the growth of ice from the melt. *J. Cryst. Growth*, **1**, 263–270.

Killawee, J. A., I. J. Fairchild, J. -L. Tison, L. Janssens and R. Lorrain (1998). Segregation of solutes and gases in experimental freezing of dilute solutions: implications for natural glacial systems. *Geochem. Cosmochim. Acta*, **62**, 3637–3655.

Knight, C. A. (1966). Grain boundary migration and other processes in the formation of ice sheets on water. *J. Appl. Phys.*, **37**, 568–574.

Knight, C. A. and N. C. Knight (1968a). Spongy hailstone growth criteria I. Orientation fabrics. *J. Atmos. Sci.*, **25**, 445–452.

Knight, C. A. and N. C. Knight (1968b). Spongy hailstone growth criteria II. Microstructures. *J. Atmos. Sci.*, **25**, 453–459.

Knight, C. A. and N. C. Knight (1971). Hailstones. *Sci. Am.*, **224**, 96–103.

Koo, K. -k., R. Ananth and W. N. Gill (1991). Tip splitting in dendritic growth of ice crystals. *Phys. Rev. A*, **44**, 3782–3790.

Langhorne, P. J. and W. H. Robinson (1986). Alignment of crystals in sea ice due to fluid motion. *Cold Reg. Sci. Technol.*, **12**, 197–214.

Lipenkov, V. (2000). Air-bubbles and air-hydrate crystals in the Vostok ice core. In *Physics of Ice Core Records*, ed. T. Hondoh. Sapporo: Hokkaido University Press, pp. 327–385.

Lipenkov, V., A. N. Salamatin and P. Duval (1997). Bubble-densification in ice sheets. *J. Glaciol.*, **43**, 397–407.

Lock, G. S. H. (1990). *The Growth and Decay of Ice*. Cambridge: Cambridge University Press.

Lofgren, G. and W. F. Weeks (1969). Effect of growth parameters on the substructure spacing in NaCl ice crystals. *J. Glaciol.*, **8**, 153–164.

Ludlam, F. H. (1951). The heat economy of a rimed cylinder. *Q. J. R. Meteorol. Soc.*, **77**, 663–666.

Macklin, W. C. and B. F. Ryan (1965). The structure of ice grown in bulk supercooled water. *J. Atmos. Sci.*, **22**, 452–459.

Maeno, N. and T. Ebinuma (1983). Pressure sintering of ice and its implication to the densification of snow at polar glaciers and ice sheets. *J. Phys. Chem.*, **87**, 4103–4110.

Michel, B. and R. O. Ramseier (1971). Classification of river and lake ice. *Can. Geotech. J.*, **8**, 36–45.

Miller, S. L. (1973). The clathrate hydrates – their nature and occurrence. In *Physics and Chemistry of Ice*, eds. E. Whalley, S. J. Jones and L. W. Gold. Ottawa, Canada: Royal Society of Canada, pp. 42–50.

Mulvaney, R., E. W. Wolff and K. Oates (1988). Sulfuric-acid at grain-boundaries in Antarctic ice. *Nature*, **331**, 247–249.

Nakawo, M. and N. K. Sinha (1984). A note on brine layer spacing in first-year sea ice. *Atmosphere-Ocean*, **22**, 193–206.

Obbard, R. and I. Baker (2007). The microstructure of meteroric ice from Vostok, Antarctica. *J. Glaciol.*, **53**, 41–62.

Obbard, R., I. Baker and K. Sieg (2006). Using electron backscatter diffraction patterns to examine recrystallization in polar ice sheets. *J. Glaciol.*, **52**, 546–557.

Oxtoby, D. W., Ed. (1999). *Nucleation and Surface Melting of Ice*. NATO ASI Series, Ice Physics and the Natural Environment. Heidelberg: Springer-Verlag.

Paterson, W. S. (2000). *Physics of Glaciers*, 3rd edn. Oxford: Butterworth Heineman.

Pauer, F., J. Kipfstuhl, W. F. Kuhs and H. Shoji (1999). Air clathrate crystals from the GRIP deep ice core: a number-, size- and shape-distribution study. *J. Glaciol.*, **45**, 22–30.

Perey, F. G. J., and E. R. Pounder (1958). Crystal orientation in ice sheets. *Can. J. Phys.*, **36**, 494–502.

Perovich, D. K. and J. A. Richter-Menge (2000). Ice growth and solar heating in springtime leads. *J. Geophys. Res.*, **105**, 6541–6548.

Piazolo, S., M. Montagnat and J. R. Blackford (2008). Sub-structure characterization of experimentally and naturally deformed ice using Cryo-EBSD. *J. Microscopy*, **230**, 509–519.

Pieritz, R., J. B. Brzoska, F. Flin, B. Lesaffre and C. Coleou (2004). From snow X-ray tomography raw volume data to micromechanics modeling: first results. *Ann. Glaciol.*, **38**, 52–58.

Poots, G. (2000). Ice and snow accretion on structures: Introductory remarks. *Phil. Trans. R. Soc. Lond. A*, **358**, 2803–2810.

Ramseier, R. O. (1968). Origin of preferred orientation in columnar ice. *J. Cryst. Growth*, **3**, 621–624.

Rothrock, D. A., U. Yu and G. A. Maykut (1999). Thinning of the Arctic sea-ice cover. *Geophys. Res. Lett.*, **26**, 3469–3472.

Rutter, B. F. and B. Chalmers (1953). A prismatic substructure formed during solidification of metals. *Can. J. Phys.*, **1**, 15–39.

Ryerson, C. C. and A. J. Gow (2000). Crystalline structure and physical properties of ship superstructure spray ice. *Phil. Trans. R. Soc. Lond. A*, **358**, 2847–2871.

Salamatin, A. N. and V. Lipenkov (2003). Air-hydrate crystal growth in polar ice. *J. Cryst. Growth*, **257**, 412–426.

Salamatin, A. N., V. Lipenkov, J. M. Barnola, A. Hori and P. Duval (in press). Snow-firn densification in polar ice sheets. In *Physics of Ice Core Records II*, ed. T. Hondoh. Sapporo: Hokkaido University Press.

Seeley, L. H. and G. T. Seidler (2001). Two-dimensional nucleation of ice from supercooled water. *Phys. Rev. Lett.*, **87**, 057702.

Shapiro, L. H. and W. F. Weeks (1993). The influence of crystallographic and structural properties on the flexural strength of small sea ice beam. In *Ice Mechanics-1993; 1993 Joint ASME Applied Mechanics and Materials Summer Meeting AMD*. New York: American Society of Mechanical Engineers.

Sinha, N. K. (1977). Technique for studying structure of sea ice. *J. Glaciol.*, **18**, 315–323.

Tiller, W. A. (1957). Preferred growth direction of metals. *J. Metals*, **9**, 847–855.

Tison, J. -L., C. Haas, M. M. Gowing, S. Sleewaegen and A. Bernard (2002). Tank study of physico-chemical controls on gas content composition during growth of young sea ice. *J. Glaciol.*, **48**, 177–190.

Turnbull, D. and J. C. Fisher (1949). Rate of nucleation in condensed systems. *J. Chem. Phys.*, **17**, 71–73.

Uchida, T., T. Hondoh, S. Mae, V. Lipenkov and P. Duval (1994). Air-hydrate crystals in deep-ice core samples from Vostok station, Antarctica. *J. Glaciol.*, **40**, 79–86.

Wakatsuchi, M. and T. Kawamura (1987). Formation processes of brine drainage channels in sea ice. *J. Geophys. Res.*, **92**, 7195–7197.
Walton, D. and B. Chalmers (1959). The origin of preferred orientation in the columnar zone of metal ingots. *Trans. AIME*, **215**, 447–457.
Weeks, W. F. and S. F. Ackley (1986). The growth, structure, and properties of sea ice. In *The Geophysics of Sea Ice*, ed. N. Untersteiner. New York: Plenum, pp. 9–164.
Weeks, W. F. and A. J. Gow (1980). Crystal alignments in the fast ice of arctic Alaska. *J. Geophys. Res.*, **84**, 1137–1146.
Weeks, W. F. and J. S. Wettlaufer, Eds. (1996). Crystal orientations in floating ice sheets. *The Johannes Weertman Symposium*, The Minerals, Metals & Materials Society.
Wilkinson, D. S. (1988). A pressure-sintering model for the densification of polar firn and glacier ice. *J. Glaciol.*, **34**, 40–45.
Wolff, E. W., R. Mulvaney and K. Oates (1988). The location in impurities in Antarctic ice. *Ann. Glaciol.*, **11**, 194–197.

4

Physical properties: elasticity, friction and diffusivity

4.1 Introduction

In this chapter we review the elastic behavior of ice, friction of ice on ice and mass diffusion. In terms of creep, elastic properties allow the applied stress to be normalized and thus the behavior to be analyzed within the context of physical mechanisms (Chapters 5–8). The mass diffusion coefficient plays a similar role in creep under low stresses. It is important, as well, to the transformation from snow to ice (Chapter 3). In terms of fracture, elastic constants affect fracture toughness (Chapter 9) and, through that property, both the tensile (Chapter 10) and the compressive strength (Chapters 11, 12). Elasticity is also relevant to the ductile-to-brittle transition (Chapters 13) and to ice loads on structures (Chapter 14). Friction is a factor in the DB transition under compression and is a major consideration in brittle compressive failure, on scales small (Chapters 11, 12) and large (Chapter 15). Friction is also fundamental to tidally driven, strike-slip-like tectonic activity on a number of icy satellites within the outer Solar System, including Jupiter's moon Europa (Greenberg *et al.*, 1998; Hoppa *et al.*, 1999; Schulson, 2002; Kattenhorn, 2004), Neptune's Triton (Prockter *et al.*, 2005) and Saturn's Enceladus (Nimmo *et al.*, 2007; Smith-Konter and Pappalardo, 2008). Thermal properties play a less direct role, but we list them for completeness, Table 4.1.

4.2 Elastic properties of ice Ih single crystals

Elastic properties have been relatively well studied. Our objectives are to outline the basic theory to make clear the meaning of the tensorial components that make up stiffness and compliance, and then to present numerical values for both single crystals and polycrystals. We consider both intact as well as damaged material, and include the role of porosity and its effect on sea ice. For comparison, we include a short discussion of snow and firn.

Table 4.1. *Thermal properties if ice Ih at −20 °C*

Property	Units	Value
Thermal conductivity	$W\,m^{-1}\,K^{-1}$	2.4
Specific heat	$kJ\,kg^{-1}\,K^{-1}$	1.96
Latent heat of fusion	$kJ\,kg^{-1}$	333.5

4.2.1 Fundamental elastic constants

Under sufficiently small stresses the amount of strain imparted is recoverable and is directly proportional to the magnitude of the stress. Following Nye (1985), the relationship is expressed by Hooke's law, which states that $\boldsymbol{\varepsilon} = \boldsymbol{S\sigma}$ where $\boldsymbol{\varepsilon}$ denotes strain, $\boldsymbol{\sigma}$ stress and $\boldsymbol{S}$, compliance. Alternatively, $\boldsymbol{\sigma} = \boldsymbol{C\varepsilon}$ where $\boldsymbol{C}$ is stiffness. Both stress and strain are specified by second-order tensors, and so $\boldsymbol{S}$ and $\boldsymbol{C}$ are specified by fourth-order tensors. Hooke's law may then be written:

$$\varepsilon_{ij} = S_{ijkl}\sigma_{kl} \tag{4.1}$$

or

$$\sigma_{ij} = C_{ijkl}\varepsilon_{kl}. \tag{4.2}$$

ε_{ij} and σ_{ij} are the components of strain and stress, respectively, with respect to a given rectangular coordinate system X_i, and S_{ijkl} and C_{ijkl} are the elastic compliance constants and the elastic stiffness constants, respectively, with respect to that same coordinate system. The subscripts $i, j, k, l = 1, 2, 3$ and repeated subscripts in any term of the equations are summed over. Owing to the symmetry of the stress and strain tensors $S_{ijkl} = S_{jilk}$ and $C_{ijkl} = C_{jilk}$ and so the number of constants reduces from 81 to 36.

The constants are generally given as components of 6 × 6 matrices, after introducing the following change in notation:

tensor subscript	matrix subscript
11	1
22	2
33	3
23, 32	4
13, 31	5
12, 21	6

Stress and strain are then written:

$$\begin{bmatrix} \sigma_{11} & \sigma_{12} & \sigma_{13} \\ \sigma_{12} & \sigma_{22} & \sigma_{23} \\ \sigma_{13} & \sigma_{23} & \sigma_{33} \end{bmatrix} \rightarrow \begin{bmatrix} \sigma_1 & \sigma_6 & \sigma_5 \\ \sigma_6 & \sigma_2 & \sigma_4 \\ \sigma_5 & \sigma_4 & \sigma_3 \end{bmatrix} \text{and} \begin{bmatrix} \varepsilon_{11} & \varepsilon_{12} & \varepsilon_{13} \\ \varepsilon_{12} & \varepsilon_{22} & \varepsilon_{23} \\ \varepsilon_{13} & \varepsilon_{23} & \varepsilon_{33} \end{bmatrix} \rightarrow \begin{bmatrix} \varepsilon_1 & \varepsilon_{6/2} & \varepsilon_{5/2} \\ \varepsilon_{6/2} & \varepsilon_2 & \varepsilon_{4/2} \\ \varepsilon_{5/2} & \varepsilon_{4/2} & \varepsilon_3 \end{bmatrix}$$

and Hooke's law becomes:

$$\varepsilon_i = S_{ij}\,\sigma_j(i,j = 1,2\ldots6) \qquad (4.3)$$

or

$$\sigma_i = C_{ij}\,\varepsilon_j(i,j = 1,2\ldots6), \qquad (4.4)$$

where S_{ij} and C_{ij} denote the components of the matrices. (It is important not to confuse the components of stress and strain in the matrix notation with the principal stresses and principal strains of the tensor notation, which are also expressed in terms of single subscripts 1, 2 and 3.) The S's and C's are related through the matrix product $SC = I$ where I is the unit matrix. Upon introducing the matrix notation, factors of 2 and 4 are also introduced, but only to the compliance constants via the scheme:

$$\begin{aligned} S_{ijkl} &= S_{mn} \text{ when both } m \text{ and } n \text{ are } 1,2 \text{ or } 3,\\ 2S_{ijkl} &= S_{mn} \text{ when either } m \text{ or } n \text{ is } 4,5 \text{ or } 6,\\ 4S_{ijkl} &= S_{mn} \text{ when both } m \text{ and } n \text{ are } 4,5 \text{ or } 6. \end{aligned}$$

Of the 36 components of S and C, only five are independent for crystals based upon the hexagonal system. Hence, ice Ih possesses only five independent components. This may be seen following Nye (1985): From consideration of the change in internal energy density during elastic deformation and from the fact that energy is a state variable, $S_{ij} = S_{ji}$ and $C_{ij} = C_{ji}$. This matrix symmetry thus reduces the number to 21. Crystal symmetry arguments for the hexagonal system reduce the number of independent constants still further, to five: S_{11}, S_{12}, S_{13}, S_{33}, S_{44} or C_{11}, C_{12}, C_{13}, C_{33}, C_{44}. Thus, the compliance and stiffness matrices for crystals of ice Ih become:

$$S_{ij} = \begin{pmatrix} S_{11} & S_{12} & S_{13} & 0 & 0 & 0\\ S_{12} & S_{11} & S_{13} & 0 & 0 & 0\\ S_{13} & S_{13} & S_{33} & 0 & 0 & 0\\ 0 & 0 & 0 & S_{44} & 0 & 0\\ 0 & 0 & 0 & 0 & S_{44} & 0\\ 0 & 0 & 0 & 0 & 0 & 2(S_{11} - S_{12}) \end{pmatrix},$$

$$C_{ij} = \begin{pmatrix} C_{11} & C_{12} & C_{13} & 0 & 0 & 0\\ C_{12} & C_{11} & C_{13} & 0 & 0 & 0\\ C_{13} & C_{13} & C_{33} & 0 & 0 & 0\\ 0 & 0 & 0 & C_{44} & 0 & 0\\ 0 & 0 & 0 & 0 & C_{44} & 0\\ 0 & 0 & 0 & 0 & 0 & 1/2(C_{11} - C_{12}) \end{pmatrix}.$$

Because the value of S_{14}, S_{15}, etc. and of C_{14}, C_{15} etc. is zero, the axes of principal stress and principal strain are coincident.

The fundamental constants are expressed with respect to a rectangular coordinate system whose axes are specified in terms of the unit cell of the crystal. For ice Ih, axis X_3 is parallel to the *c*-axis (Chapter 2) and X_1 and X_2 may be taken as any pair of axes within the basal plane (because the elastic properties are isotropic in that plane, more below). On this basis the elastic compliance constants may be interpreted as follows:

S_{11} gives the normal strain perpendicular to the *c*-axis owing to a normal stress acting along X_1;

S_{33} gives the normal strain parallel to the *c*-axis owing to a normal stress acting along the *c*-axis;

S_{12} gives the normal strain perpendicular to the *c*-axis owing to a normal stress also perpendicular to the *c*-axis and perpendicular to the direction of interest;

S_{13} gives the normal strain perpendicular to the *c*-axis owing to a normal stress acting along the *c*-axis, as well as the normal strain along the *c*-axis owing to a normal stress along a direction perpendicular to the *c*-axis;

S_{44} gives the shear strain in a plane parallel to the *c*-axis owing to a shear stress in the same plane.

4.2.2 Numerical values of fundamental constants

Two different approaches have been taken to measure the elastic properties of ice. The simpler, but more problematic approach, is the measurement of displacement under an applied stress, termed the "static" method. The problem is that unless the stress is low and is applied and then released very rapidly – e.g., less than 1 MPa at a rate of 0.5 MPa s^{-1} for less than 10 s at −40 to −3 °C (Gold, 1958) – time-dependent deformation (i.e. creep) contributes to the deformation and this leads to an underestimate of stiffness or to an overestimate of compliance (Traetteberg *et al.*, 1975). For instance, we have found that the apparent Young's modulus, when measured from the slopes of compressive stress–strain curves obtained from single crystals loaded to about 10 MPa by deforming at a rate of 10^{-2} s^{-1} at −10 °C, is about a factor of two lower than the true elastic modulus. Similarly, Arakawa and Maeno (1997) found that the maximum tangent modulus obtained from stress–strain curves for finely grained (d = 0.9 mm), fresh-water ice deformed at strain rates of the order of 10^{-4} s^{-1} at −10 °C is ~1.4 GPa, a factor of about seven lower than Young's modulus (see below). The better approach to obtaining elastic moduli exploits the propagation of sound waves, and leads to the so-called dynamic constants (also termed the adiabatic constants at constant entropy). We quote below only dynamic constants, unless otherwise noted.

Table 4.2. *Fundamental elastic constants for ice Ih at −16 °C*

Property and units	Symbol	Value
Elastic stiffness (10^9 N m^{-2})	C_{11}	13.93 ± 0.04
	C_{12}	7.08 ± 0.04
	C_{13}	5.76 ± 0.02
	C_{33}	15.0 ± 0.05
	C_{44}	3.01 ± 0.01
Elastic compliance (10^{-12} m^2 N^{-1})	S_{11}	103 ± 0.05
	S_{12}	-42.9 ± 0.4
	S_{13}	-23.2 ± 0.2
	S_{33}	84.4 ± 0.4
	S_{44}	331.8 ± 0.2
Compressibility (10^{-12} m^2 N^{-1}) $2S_{11} + S_{33} + 2(S_{12} + 2S_{13})$	K	112.4 ± 0.2
Bulk modulus (10^9 N m^{-2})	$B = 1/K$	8.90 ± 0.02
Poisson's ratio	υ	$\upsilon_{12} = -S_{12}/S_{11} = 0.415$
		$\upsilon_{13} = -S_{13}/S_{11} = 0.224$
		$\upsilon_{31} = -S_{13}/S_{33} = 0.274$

Gammon *et al.* (1983)

The most accurate values to date have been obtained by Gammon *et al.* (1983). They applied the method of Brillouin spectroscopy in which incident laser light was scattered from thermally induced acoustic waves and Doppler shifted (Gammon *et al.*, 1980, 1983; Gagnon *et al.*, 1988, 1990). The velocity, v, so measured, from regions as small as 0.2 mm in diameter, was then used to obtain stiffness constants via the relationship $C = \rho v^2$ where ρ denotes mass density. Table 4.2 lists numerical values at −16 °C. The constants are termed the characteristic constants because they are given with respect to the crystallographic axes of the Ih unit cell. A different set arises when the orientation of the coordinate axes changes with respect to the axes of crystal symmetry. Included in Table 4.2 are compressibility K and Poisson's ratio υ where K is defined from the relationship $\Delta V/V = Kp$, where V is volume, $\Delta V/V = \varepsilon_1 + \varepsilon_2 + \varepsilon_3$ and pressure $p = (\sigma_1 + \sigma_2 + \sigma_3)/3$ where the subscripts denote principal strains and stresses; υ is defined as the negative of the ratio of transverse strain to longitudinal strain measured under a uniaxial, longitudinal load. The constants are not affected by impurities in water at the time of freezing (Gammon *et al.*, 1983). The elastic constants of several high-pressure ices have also been obtained using Brillouin spectroscopy and are given elsewhere (Gagnon *et al.*, 1990; Tulk *et al.*, 1994, 1996, 1997; Shimizu *et al.*, 1996; Baer *et al.*, 1998).

As temperature decreases from near the melting point and approaches 0 K, the stiffness of ice increases (or compliance decreases) by about 25% (Proctor, 1966; Dantl, 1968, 1969; Gagnon *et al.*, 1988). Over the range normally encountered in the natural environment on Earth, however, say from 0 °C to −50 °C, the relative increase in stiffness is only about 5%. Over this more limited range of temperature, the value $V(T)$ of any of the fundamental constants at temperature T may be given by the relationship (Gammon *et al.*, 1983; Petrenko and Whitworth, 1999):

$$V(T) = V(T_r)[1 \pm a(T - T_r)] \tag{4.5}$$

where T_r is the reference temperature at which the constant was measured, $a = 1.42 \times 10^{-3}\,K^{-1}$, "+" is for compliance and "−" is for stiffness.

Again, values obtained from the static method greatly misrepresent the true effect of temperature on elastic behavior. For instance, such measures show that the tangent modulus increases by a factor of about five upon decreasing temperature from −10 °C to −85 °C (Arakawa and Maeno, 1997), whereas over this temperature range Young's modulus is expected to increase only by about 10%.

4.2.3 Young's modulus and torsional modulus

A quantity of interest is Young's modulus in an arbitrary direction X_3'. Imagine a cylindrical crystal whose longitudinal axis is parallel to new direction X_3' that differs from direction X_3 by the acute angle θ. The crystal is loaded by applying a tensile stress along its axis. (In practice the modulus should not be obtained in this way, for the reason mentioned above.) Young's modulus is defined as the ratio of the longitudinal stress to the longitudinal strain, and so from Equation (4.3) its value along X_3' is given by:

$$E_3' = \frac{1}{S_{33}'} \tag{4.6}$$

where S_{33}' is given with respect to the new coordinate system. To find its value, one must first re-express the component in tensor form and then transform the tensor element using the transformation law for fourth-order tensors; i.e., $S_{33}' = S_{3333}' = a_{3i}a_{3j}a_{3k}a_{3l}S_{ijkl}$ where (a_{ij}) denotes the transformation matrix and elements a_{31}, a_{32} and a_{33} define the cosine of the angle between new direction X_3' and reference directions X_1, X_2 and X_3, respectively. It follows that:

$$E_3'(\theta) = \frac{1}{S_{33}'} = \frac{1}{S_{3333}'} = \frac{1}{(S_{11}\sin^4\theta + S_{33}\cos^4\theta + (S_{44} + 2S_{13}\sin^2\theta\cos^2\theta)}. \tag{4.7}$$

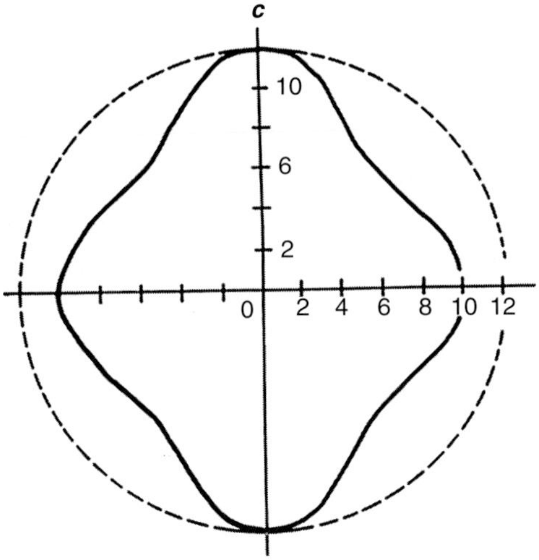

Figure 4.1. Young's modulus (in GPa) of ice Ih single crystal at −16 °C vs. orientation of the loading direction with respect to the *c*-axis of the unit cell. (From Michel, 1978 after Fletcher, 1970.)

Figure 4.1 shows this relationship graphically. Young's modulus reaches a minimum at $\theta \sim 50°$ of $E = 8.42$ GPa at −16 °C. This compares with $E(0) = 11.8$ GPa along the *c*-axis (the highest value) and with $E(90) = 9.71$ GPa along any direction within the basal plane. These differences are relatively small and indicate that the elastic behavior of ice Ih is only moderately anisotropic.

Another quantity of interest is the torsional or shear modulus. Again imagine a cylindrical crystal, in this case twisted about its axis. Through a similar analysis, one can show that the torsional modulus is given by the expression (Nowick and Berry, 1972):

$$G'(\theta) = \frac{1}{S_{44} + (S_{11} - S_{12} - S_{44}/2)\sin^2\theta + 2(S_{11} + S_{33} - 2S_{13} - S_{44})\cos^2\theta\sin^2\theta}. \tag{4.8}$$

The torsional modulus reaches a maximum at $\theta \sim 48°$ of $G = 3.67$ GPa at −16 °C. This compares with $G(0) = 3.01$ GPa and $G(90) = 3.21$ GPa.

Both E and G depend only on the angle between the crystal axis and the *c*-axis of the unit cell. This means that the elastic properties of ice Ih are invariant with respect to rotation about that axis.

4.3 Elastic properties of polycrystals

In principle the elastic properties of polycrystals free from porosity, inclusions and other defects can be calculated from the fundamental elastic constants and from the orientations, sizes and shapes of the grains, using a number of different methods. For instance, Voigt's (1910) method assumes that all grains are strained equally and gives an upper bound on the stiffness moduli. Reuss' (1929) method assumes that all grains are equally stressed and gives a lower bound on stiffness. Hill's method (1952) averages the V–R approaches. Self-consistent methods (for references see Nemat-Nasser and Horii (1993)) take into account the actual distribution of stress and strain within a grain. Each method as it relates to ice has been discussed in the literature (Michel, 1978; Sinha, 1989a; Nanthikesan and Sunder, 1994). It turns out that, owing to the relatively low elastic anisotropy of the constituent crystals, the maximum difference between the methods, defined as the difference between the V–R limits, is only about 4.2% (Nanthikesan and Sunder, 1994). It is practically immaterial, therefore, which approach one takes. As long as the number of grains exceeds about 230 (Elvin, 1996), polycrystalline behavior can be approximated using any of them. Further discussion on the calculation of polycrystalline behavior from the fundamental elastic constants is beyond the scope of this book.

4.3.1 Randomly oriented polycrystals

Although the constituent crystals are anisotropic, homogeneous aggregates composed of randomly oriented grains are elastically isotropic. As a result, their elastic properties are completely described by only two independent constants (Nye, 1957; Gammon *et al.*, 1983), chosen from Young's modulus E, the shear modulus G, Poisson's ratio υ and the bulk modulus B. For instance:

$$G = \frac{E}{2(1+\upsilon)} \tag{4.9}$$

and

$$B = \frac{E}{3(1-2\upsilon)}. \tag{4.10}$$

Values measured at −16 °C are listed in Table 4.3. Again, the effect of temperature may be obtained from Equation (4.5). Figure 4.2 compares Young's modulus of ice with that of a number of other materials, on the basis of density.

4.3.2 Polycrystals with growth textures

As described in Chapter 3, a crystallographic growth texture develops within a floating ice sheet as it thickens. As a result, the structure of the sheet has three mutually perpendicular mirror planes (two in the vertical plane) and so possesses orthotropic symmetry. This means that when the three orthogonal loading axes

Table 4.3. *Elastic properties of homogeneous polycrystals of isotropic ice Ih at −16 °C*

Property	Units	Value
Young's modulus, *E*	$N\,m^{-2}$	9.33×10^{9}
Compressibility, *K*	$N^{-1}\,m^{2}$	112.4×10^{-12}
Bulk modulus, *B*	$N\,m^{-2}$	8.90×10^{9}
Shear modulus, *G*	$N\,m^{-2}$	3.52×10^{9}
Poisson's ratio, *υ*	n/a	0.325

From Gammon *et al.* (1983)

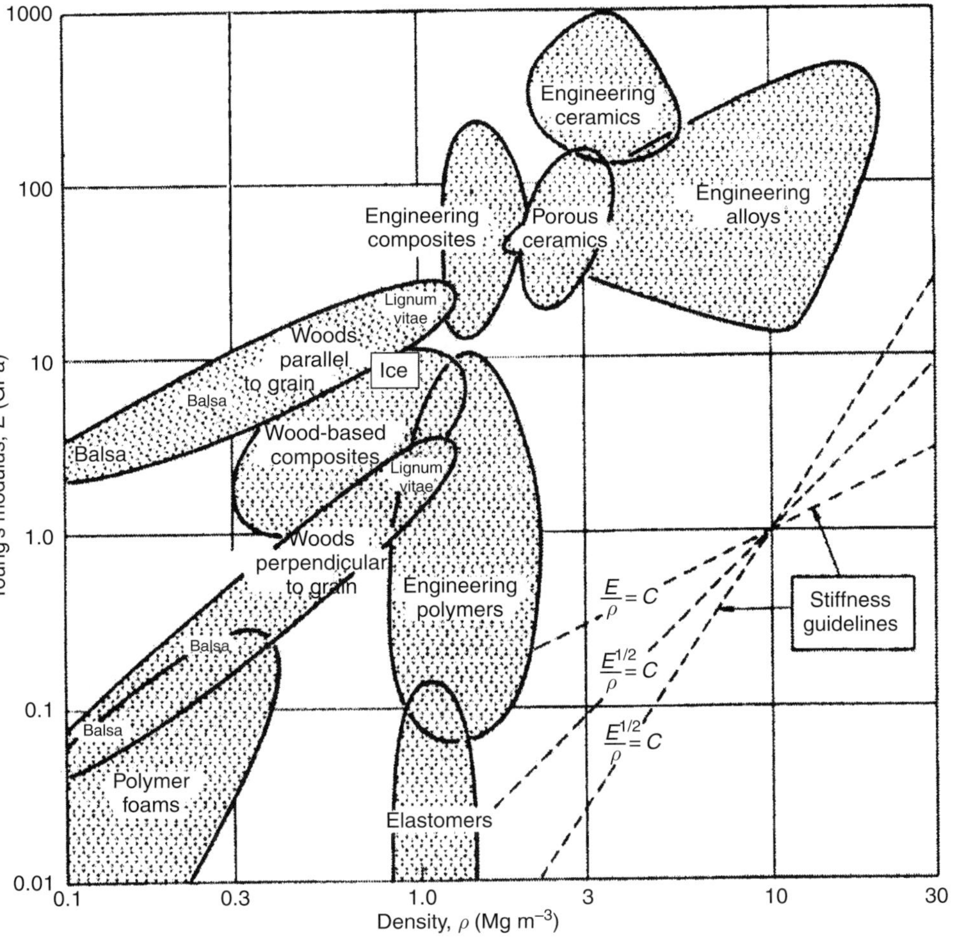

Figure 4.2. Young's modulus of a variety of engineering materials. (From Wegst and Ashby, 2004; reprinted by permission of Taylor & Francis Ltd, www.infoworld.com.)

X_i are aligned with the normals to these planes, the sheet has in the most general case nine independent compliances (Hearmon, 1961), given by the matrix:

$$S_{ij} = \begin{pmatrix} S_{11} & S_{12} & S_{13} & 0 & 0 & 0 \\ S_{12} & S_{22} & S_{23} & 0 & 0 & 0 \\ S_{13} & S_{23} & S_{33} & 0 & 0 & 0 \\ 0 & 0 & 0 & S_{44} & 0 & 0 \\ 0 & 0 & 0 & 0 & S_{55} & 0 \\ 0 & 0 & 0 & 0 & 0 & S_{66} \end{pmatrix}.$$

The polycrystalline compliances are generally different from the fundamental or single crystalline compliances discussed above. The polycrystalline compliances are related to three Young's moduli E_1, E_2, E_3, to three shear moduli G_{12}, G_{13}, G_{23} and to six Poisson's ratios $\upsilon_{12}, \upsilon_{21}, \upsilon_{13}, \upsilon_{31}, \upsilon_{23}, \upsilon_{32}$ (where $\upsilon_{ij} \stackrel{\Delta}{=} -\left(\varepsilon_j/\varepsilon_i\right)_{\sigma_i}$; i.e., v_{ij} is defined as the negative of the strain in direction j divided by the strain in direction i for uniaxial loading in direction i.) The relationships are:

$$E_1 = \frac{1}{S_{11}}, \qquad E_2 = \frac{1}{S_{22}}, \qquad E_3 = \frac{1}{S_{33}}; \tag{4.11}$$

$$G_{12} = \frac{1}{S_{66}}, \qquad G_{13} = \frac{1}{S_{55}}, \qquad G_{23} = \frac{1}{S_{44}}; \tag{4.12}$$

$$\frac{\upsilon_{12}}{E_1} = \frac{\upsilon_{21}}{E_2} = -S_{12}; \tag{4.13}$$

$$\frac{\upsilon_{13}}{E_1} = \frac{\upsilon_{31}}{E_3} = -S_{13}; \tag{4.14}$$

$$\frac{\upsilon_{23}}{E_2} = \frac{\upsilon_{32}}{E_3} = -S_{23}. \tag{4.15}$$

The compliance matrix is then written as:

$$S_{ij} = \begin{pmatrix} \frac{1}{E_1} & -\frac{\nu_{21}}{E_2} & -\frac{\nu_{31}}{E_3} & 0 & 0 & 0 \\ -\frac{\nu_{12}}{E_1} & \frac{1}{E_2} & -\frac{\nu_{32}}{E_3} & 0 & 0 & 0 \\ -\frac{\nu_{13}}{E_1} & -\frac{\nu_{23}}{E_2} & \frac{1}{E_3} & 0 & 0 & 0 \\ 0 & 0 & 0 & \frac{1}{G_{23}} & 0 & 0 \\ 0 & 0 & 0 & 0 & \frac{1}{G_{13}} & 0 \\ 0 & 0 & 0 & 0 & 0 & \frac{1}{G_{12}} \end{pmatrix}.$$

Table 4.4 lists the calculated values at −16 °C of each element in the compliance matrix for the three variants of textured sheet ice described in Chapter 3, denoted S1, S2 and S3 ice (Michel and Ramseier, 1971). The values are specified with respect

Table 4.4. *Elastic compliance constants (calculated) for homogeneous orthotropic sheets of ice Ih at −16 °C*

Texture[a] $S_{ij} \times 10^{-12}$ (m^2/N)	S1 *c*-axes ‖ to X_3	S2 *c*-axes random in X_1–X_2 plane	S3 *c*-axes ‖ to X_1
S_{11}	103	104[b]	84.4
S_{12}	−42.9	−34.0[c]	−23.2
S_{13}	−23.2	−35.8[c]	−23.2
S_{22}	103	104[b]	103
S_{23}	−23.2	−35.8[c]	−42.9
S_{33}	84.4	103	103
S_{44}	332	312[d]	292
S_{55}	332	312[d]	332
S_{66}	292	277[d]	332

The compliances are given with respect to the orthogonal axes X_i where X_1 and X_2 lie within the horizontal plane of the sheet and X_3 is parallel to the vertical direction.
[a] After Michel and Ramseier (1971).
[b] Obtained from the Voigt–Reuss average (Nanthikesan and Sunder, 1994).
[c] Estimated from the range of Poisson's ratio values for monocrystals.
[d] Obtained from the average shear modulus in the X_1–X_2 plane (Nanthikesan and Sunder, 1994).

to a rectangular coordinate system where X_1 and X_2 lie within the horizontal plane of the sheet and X_3 is parallel to the vertical direction. X_1 may be chosen as any horizontal direction within the S1 and S2 variants, owing to the circular symmetry of the microstructure. Within the S3 variant, however, X_1 is specified as the direction in which the *c*-axes are aligned. The compliances for S3 ice are taken to be the same as those for the ice Ih single crystal (see Table 4.2). The corresponding calculated values of the elastic moduli and of Poisson's ratios, with respect to the same coordinate system, are given in Table 4.5. The values of elastic moduli along other directions may be obtained by transforming the compliances, as described above for single crystals. The S2 texture imparts the least degree of elastic anisotropy.

Poisson's ratios listed in Table 4.5 differ significantly from values obtained using "static" methods, where values as low as 0.02 and as high as 1.2 have been reported from measurements of transverse and longitudinal strain in warm ice under longitudinal loading (Saeki *et al.*, 1981; Wang, 1981). The difference is attributed to creep deformation (Sinha, 1989b). In some instances, creep is accompanied by cracking along planes perpendicular to the transverse direction, which accounts for ratios greater than unity. In other words, the low and the high values should not be taken to be Poisson's ratios, but to be measures of highly anisotropic plastic flow. Caution is thus appropriate when reading the literature.

Table 4.5. *Elastic properties (calculated) of homogeneous orthotropic sheets of ice Ih at −16 °C*

	Young's modulus, E ($\times 10^9$ N/m^2)			Shear modulus, G ($\times 10^9$ N/m^2)			Poisson's ratio					
Texture	E_1	E_2	E_3	G_{12}	G_{13}	G_{23}	ν_{12}	ν_{21}	ν_{13}	ν_{31}	ν_{23}	ν_{32}
S1	9.71	9.71	11.8	3.42	3.01	3.01	0.415	0.415	0.224	0.274	0.224	0.274
S2	9.58[a]	9.58[a]	9.71	3.61[b]	3.21[b]	3.21[b]	0.327[c]	0.327[c]	0.344[c]	0.320[c]	0.344[c]	0.320[c]
S3	11.8	9.71	9.71	3.01	3.01	3.42	0.274	0.224	0.274	0.224	0.415	0.415

The properties are given with respect to orthogonal axes X_i where X_1 and X_2 lie within the horizontal plane of the sheet; X_3 is parallel to the vertical direction.
[a] Voigt–Reuss average (Nanthikesan and Sunder, 1994).
[b] Average shear modulus in X_1–X_2 plane (Nanthikesan and Sunder, 1994).
[c] Estimated from range of Poisson's ratio for monocrystal.

In engineering practice an effective Poisson's ratio is often of interest when loads are applied at pseudo-static rates. Then, temperature, loading rate and even grain size are important factors. Gold (1988, 1994) discusses this point further.

4.3.3 Polycrystals with deformation and recrystallization textures

Three kinds of texture can develop within glaciers as they creep (Chapter 6). The multi-maxima texture that forms as the result of discontinuous dynamic recrystallization is considered not to impart a significant degree of elastic anisotropy, and so the elastic properties of glacier ice with this texture are taken to be those for homogeneous polycrystals, listed in Table 4.3. The strong single maximum can be compared to S1 ice and the homogeneous distribution of *c*-axes within a plane can be compared to S2 ice. Consequently, the elastic properties of glacier ice with these textures are expected to be similar to those of the corresponding orthotropic sheet ice.

4.3.4 Sea ice

Owing to the presence of brine pockets (Chapter 3), sea ice is more compliant than fresh-water ice. Its compliance is also more sensitive to temperature. The region between the pores is composed of fresh-water ice whose fundamental elastic constants agree within experimental error with those obtained from non-saline ice (Gammon *et al.*, 1983). The greater thermal sensitivity of sea ice, particularly within about 10 degrees of the melting point, arises because the porosity/brine volume is a unique function of temperature and salinity (Cox and Weeks, 1983, 1986), provided the volume of entrapped air is negligible. This means that a floating sea ice sheet is

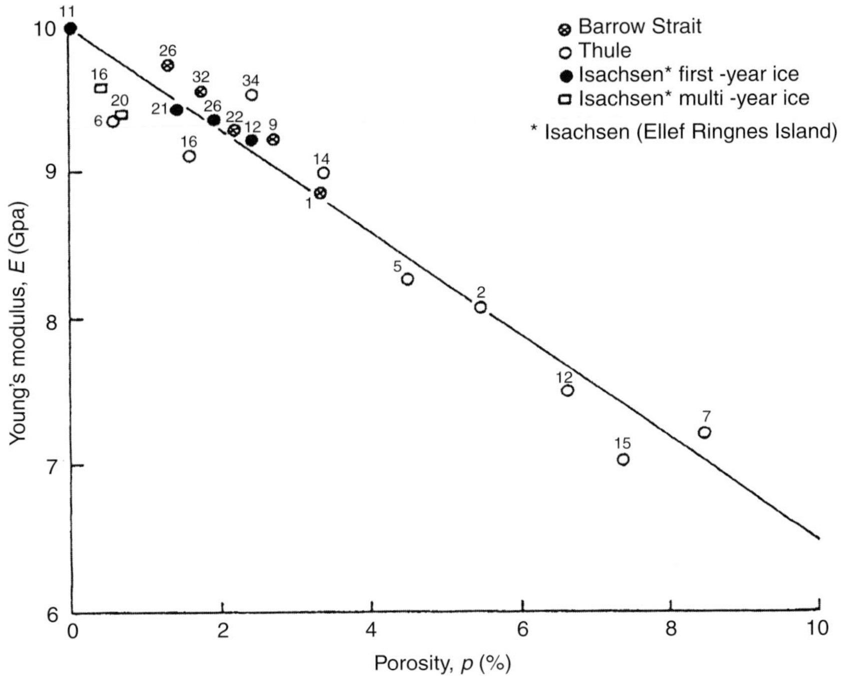

Figure 4.3. Young's modulus (dynamic) vs. porosity of cold, arctic sea ice. (From Langleben and Pounder, 1963.)

significantly stiffer near the top than the bottom, at least during the growth season. This difference can affect the mode of failure, particularly during indentation (Chapter 14). The thermal/brine sensitivity also means that winter ice (colder, lower brine volume) is stiffer than summer ice.

Figure 4.3 shows measurements by Langleben and Pounder (1963) of dynamic Young's modulus versus porosity (defined as the volume fraction of both brine and air) of cold, first-year arctic sea ice. No account is given of the intrinsic anisotropy described above, probably because the effect of porosity dominates. For low-porosity ice ($\phi < 0.1$) the modulus may be described by:

$$E(\mathrm{GPa}) = 10.00 - 35.1\phi. \tag{4.16}$$

The effect appears to be essentially independent of water content, as evident from Mellor's (1983) compilation, which shows that both saline and non-saline ice follow a similar relationship.

Concerning Poisson's ratio, Langleben and Pounder (1963) obtained a dynamic value of $\upsilon = 0.295$ for arctic ice, with no evidence of any variation with temperature (from -3.6 to -15 °C), salinity (from 0.113 to 6.11‰) or type of ice (first-year ice or multi-year ice).

4.3.5 Snow and firn

The transition from snow to firn is taken to occur when the rate of densification by packing and particle rearrangement becomes lower than that induced by plastic flow around the particle contacts. The transition density is about 600 kg m^{-3} at −20 °C and about 530 kg m^{-3} at −55 °C (Anderson and Benson, 1963). In polar ice sheets, the depth at which the snow-to-firn transition occurs is between 10 and 30 m. Depending upon temperature and accumulation rate, the age of ice within the transition zone is between about 10 years (−20 °C) and about 200 years at −55 °C.

A comprehensive review of the mechanical properties of snow was written by Mellor (1975, 1977). Figure 4.4 summarizes the dynamic Young's modulus and Poisson's ratio of dry snow and firn versus density. The modulus exhibits a large dispersion, owing to variations in microstructure. Like ice per se, measurement under elastic conditions requires rapid loading to prevent creep deformation. Despite the dispersion, the data clearly show that E increases sharply as ρ increases. The effect was tentatively analyzed by considering snow as an open foam (Kirchner *et al.*, 2001). From Gibson and Ashby (1988), the relationship between the Young's modulus and density of foam is given by:

$$\frac{E_{\text{snow}}}{E_{\text{ice}}} = \left(\frac{\rho_{\text{snow}}}{\rho_{\text{ice}}}\right)^2. \tag{4.17}$$

Considering data given in Figure 4.4, Kirchner *et al.* (2001) show that the elastic modulus of snow depends upon density to a power significantly higher than 2. This exponent is found to be of about 6.5 in the (relative) density range between 0.15 and 0.6. Variations with density in the average fraction of the surface area of ice grains involved in bonds probably account for this discrepancy.

4.3.6 Cracked/damaged polycrystals

Cracks reduce stiffness. They form in ice, as they do in ceramics and rocks, as a product of differential thermal straining and/or during deformation. When induced by deformation, they tend to be oriented, parallel to the direction of greatest applied stress under compression and perpendicular to the direction of applied stress under tension. The preferential orientation is expected to impart additional elastic anisotropy. However, studies on ice have not been published, and so for now we must turn to damage theory (Budiansky and O'Connell, 1976; Hoenig, 1979; Reyes-Morel and Chen, 1990) for insight. We outline below the essential elements of the theory for the case of randomly oriented polycrystals loaded under compression and subjected to axis-symmetric confinement.

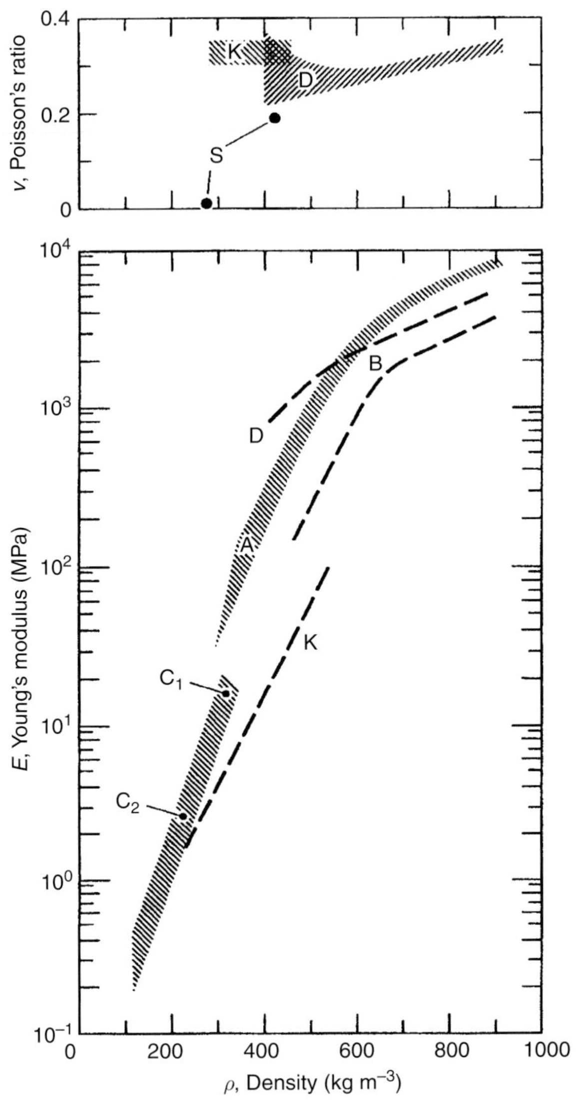

Figure 4.4. Young's modulus and Poisson's ratio vs. density for dry and coherent snow. Data were obtained between −10 and −25 °C, from pulse propagation at high frequencies (A, D), from uniaxial compression at strain rates between 10^{-5} and 10^{-2} s^{-1} (B and C_1) and from static creep tests (C_2, K, S). (From Shapiro *et al.*, 1997.)

The crack-free body is elastically isotropic, as described above. When compressed to the extent that an axis-symmetric array of cracks develops, oriented along the direction of maximum compressive stress, the initial three-dimensional isotropy is degraded. Transverse isotropy, however, remains. Elastic behavior of the damaged polycrystal may then be described in a manner similar to that used to describe the behavior of the Ih single crystal. There, orthogonal directions X_1 and X_2 were taken to lie in the basal (0001) plane and direction X_3, to be parallel to the c-axis. Here, directions X_1 and X_2 are taken to be parallel to the confining stress and X_3 to be parallel to the direction of greatest shortening or most compressive stress. Once again Hooke's law may be employed. Now, however, instead of the fundamental elastic constants we use damage-dependent elastic stiffness constants, C^{d}_{ij}, and elastic compliance constants, S^{d}_{ij}. Thus:

$$\varepsilon_i = S^{\mathrm{d}}_{ij}\sigma_j \qquad (i, j = 1, 2 \ldots 6) \tag{4.18}$$

or

$$\sigma_i = C^{\mathrm{d}}_{ij}\varepsilon_j \qquad (i, j = 1, 2 \ldots 6) \tag{4.19}$$

where again only five components are independent, 11 = 22, 33, 12, 13 = 23, 44 = 55, and where the S^{d}_{ij} may be obtained from the C^{d}_{ij} through the relationship $S^{\mathrm{d}}C^{\mathrm{d}} = I$. Damage may be expressed in terms of the dimensionless parameter δ, which can be expressed in terms of the number of cracks per unit volume N and the crack radius, r; i.e., $\delta = Nr^3$. In the limit of small crack density ($N < 0.1$) we assume a linear relationship between the elastic constants and damage. Then, the stiffness constants may be written (Reyes-Morel and Chen, 1990):

$$C^{\mathrm{d}}_{ij}(\delta) = C^{\mathrm{d}}_{ij}(0)(1 + \Delta_{ij}\delta) \tag{4.20}$$

where $C^{\mathrm{d}}_{ij}(0)$ is the value for the crack-free body and Δ_{ij} are components of a deformation matrix that has the values $\Delta_{11} = \Delta_{22} = -3.9$, $\Delta_{33} = -1.2$, $\Delta_{12} = -6.4$, $\Delta_{13} = -5.7$, $\Delta_{44} = -1.1$ and $\Delta_{66} = -2.0$ for Poisson's ratio of $\upsilon = 0.3$ (Reyes-Morel and Chen, 1990).

Table 4.6 lists the expected values of the elastic constants at −16 °C for lightly damaged ice of $\delta = 0.1$. Also listed for comparison are values for isotropic, undamaged material that were obtained from the two independent constants $S_{11} = 1/E$ and $S_{12} = -\upsilon/E$ (where E and υ are given in Table 4.3); the corresponding C values were obtained from $CS = I$. Damage is not expected to affect Young's modulus along the direction of the most compressive stress (i.e., parallel to the crack plane) where $E^{\mathrm{d}}_3 = 1/S^{\mathrm{d}}_{33} = 9.39 \times 10^9\ \mathrm{N\,m^{-2}}$, but is expected to lower the modulus along the orthogonal directions to $E^{\mathrm{d}}_1 = E^{\mathrm{d}}_2 = 1/S^{\mathrm{d}}_{11} = 6.94 \times 10^9\ \mathrm{N\,m^{-2}}$. The damage is also expected to lower the resistance to shear across the plane of the cracks from $C^{\mathrm{d}}_{66} = (C^{\mathrm{d}}_{11} - C^{\mathrm{d}}_{12})/2 = 3.53 \times 10^9\ \mathrm{N\,m^{-2}}$ for crack-free ice to $2.70 \times 10^9\ \mathrm{N\,m^{-2}}$ for the cracked material. The damage is not expected to affect one of the Poisson's ratios, $\upsilon^{\mathrm{d}}_{31} = -\varepsilon^{\mathrm{d}}_1/\varepsilon^{\mathrm{d}}_3 = -S^{\mathrm{d}}_{13}/S^{\mathrm{d}}_{33}$, but is expected to lower the other from $\upsilon^{\mathrm{d}}_{12} = -\varepsilon^{\mathrm{d}}_2/\varepsilon^{\mathrm{d}}_1 = -S^{\mathrm{d}}_{12}/S^{\mathrm{d}}_{11} = 0.325$ to 0.283. Finally, the low level of $\delta = 0.1$ is expected to raise the compressibility (for open cracks and thus for ice under

Table 4.6. *Expected elastic constants for undamaged ($\delta = 0$) and lightly damaged ($\delta = 0.1$) equiaxed and randomly oriented polycrystalline ice at −16 °C*

Elastic property	$\delta = 0$	$\delta = 0.1$
S^d_{11}	107	144
S^d_{33}	107	107
S^d_{12}	−34.8	−40.8
S^d_{13}	−34.8	−34.8
S^d_{44}	284	315
C^d_{11}	13.6	8.43
C^d_{33}	13.6	12.2
C^d_{12}	6.56	3.02
C^d_{13}	6.56	3.61
C^d_{44}	3.52	3.17

Compliance constants S^d_{ij} are given in units of $10^{-12}\,m^2\,N^{-1}$; stiffness constants C^d_{ij} in units of $10^9\,N\,m^{-2}$.

low compressive loads) by ~50%, from $K^d = 112 \times 10^{-12}\,m^2\,N^{-1}$ to $K^d = 174 \times 10^{-12}\,m^2\,N^{-1}$, where $K^d = 2S^d_{11} + 2S^d_{12} + 4S^d_{13} + S^d_{33}$.

These expectations, we repeat, are rough predictions. They are based solely on a model and need to be tested through experiment.

4.4 Friction coefficient of ice on ice at low sliding speeds

We consider next the kinetic coefficient of friction of ice sliding upon itself. Attention is limited to relatively low sliding speeds and to smooth interfaces (~2 μm) prepared using a microtome. Later (Chapter 12) we describe frictional sliding across rougher interfaces that were produced through the creation of shear faults.

4.4.1 Characteristics

Sliding is characterized by slip-stick behavior. This characteristic is evident from experiments on both fresh-water and salt-water (4.3 ± 0.2 ppt melt-water salinity) aggregates of granular and columnar ice that were slid at speeds from $10^{-6}\,m\,s^{-1}$ to $10^{-2}\,m\,s^{-1}$ at temperatures from −3 °C to −40 °C under normal pressures from 0.007 MPa to 1 MPa using a double-shear apparatus (Kennedy *et al.*, 2000; Montagnat and Schulson, 2003). The variation in the frictional force about the mean increases with increasing speed, from ~5% at the lower speed of $5 \times 10^{-5}\,m\,s^{-1}$ to ~20% at the higher speed of $10^{-3}\,m\,s^{-1}$. In double-shear tests, the kinetic

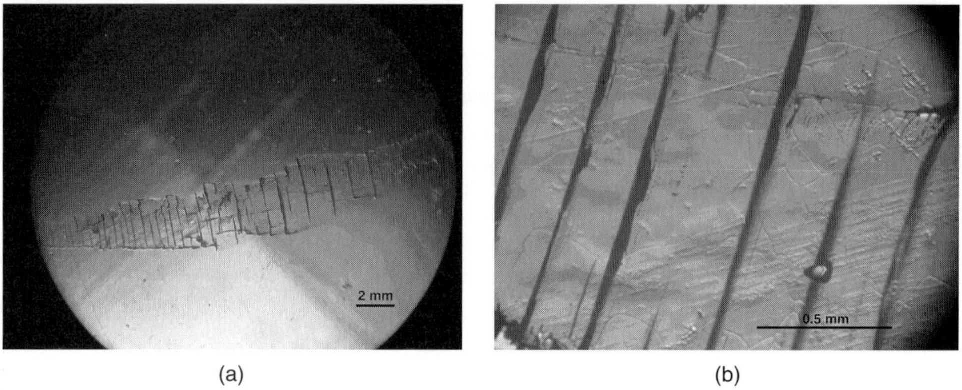

Figure 4.5. Photographs showing surface cracks in S2 ice after sliding across the grains against itself at −10 °C at a speed of 10^{-5} m s^{-1}: (a) lower magnification and (b) higher magnification. Note in (b) the recrystallized grains and the cracks that cut through them. (From Montagnat and Schulson, 2003.)

coefficient of friction is obtained by dividing the average sliding force by twice the normal force applied across the interface. Slip-stick behavior is attributed to dynamic adhesion and to rupture across the sliding interface.

Frictional sliding is generally accompanied by interfacial damage. This is particularly evident within the more optically transparent, fresh-water ice. Damage creates a nearly opaque zone that extends to ~1 mm on either side of the interface, and often includes sets of cracks that are spaced at ~0.5 mm and are oriented on planes perpendicular to both the interface and the direction of sliding, Figure 4.5. In addition to structural change, sliding can also impart recrystallization within localized regions. For instance, within the scene shown in Figure 4.5b (please consult the original paper for details) the average size of the new grains is ~ 0.3 mm (Montagnat and Schulson, 2003). This is reminiscent of the sliding-induced recrystallization that occurs when single crystals of ice slide slowly over granite (Barnes *et al.*, 1971). The process appears to occur dynamically, which is evident from the fact that the new grain boundaries are intersected by the friction-induced cracks.

4.4.2 Kinetic coefficient of friction

Figure 4.6 shows the kinetic coefficient of friction vs. sliding speed and temperature, obtained using the double-shear method. The values are independent of both grain size and grain shape (granular vs. columnar). They are also independent of pressure across the sliding interface, at least over the range 0.007 to 1 MPa (Kennedy *et al.*, 2000; Montagnat and Schulson, 2003). Interestingly, the friction coefficients for fresh-water and salt-water ice (of 4.3 ppt salinity) are almost indistinguishable at higher temperatures (−3 °C and −10 °C); at lower temperatures, the

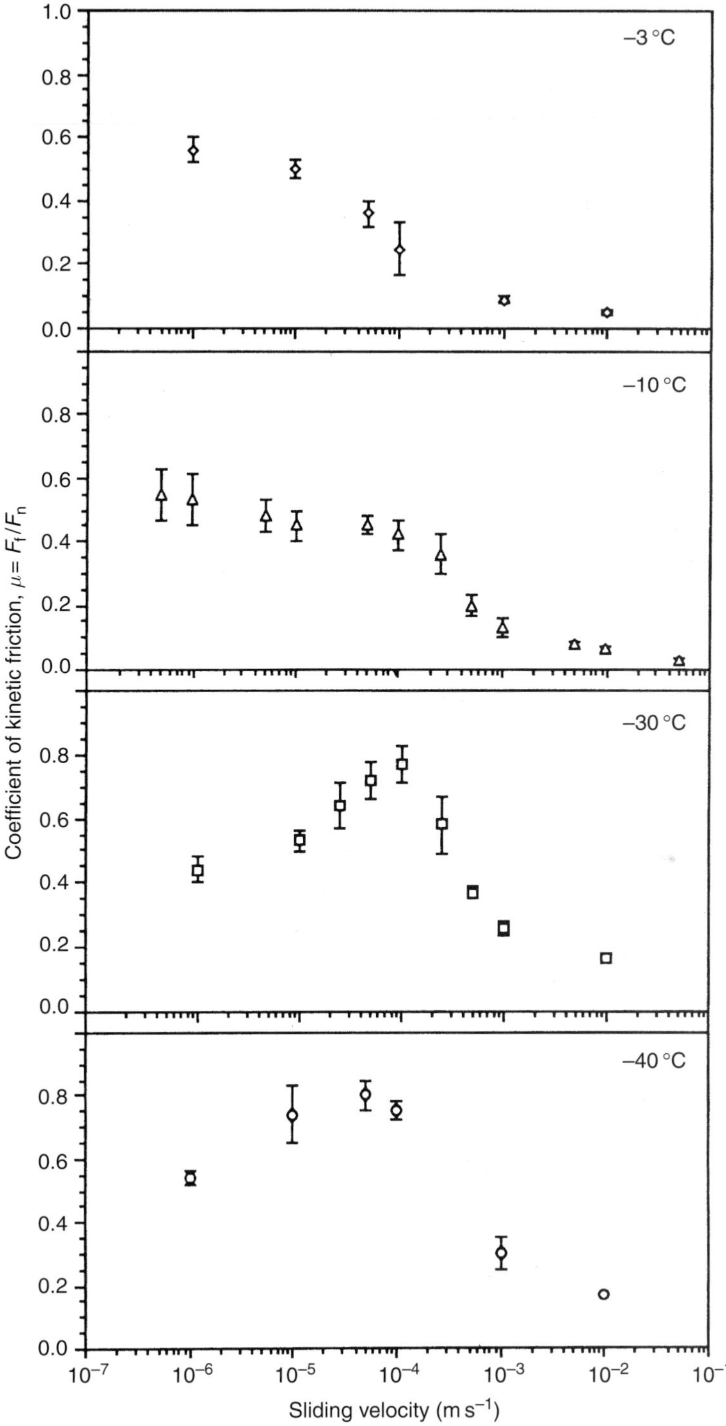

Figure 4.6. The kinetic friction coefficient of fresh-water granular ice sliding across itself across a smooth interface. (From Kennedy *et al.*, 2000.)

coefficients are somewhat greater for fresh-water ice (Kennedy *et al.*, 2000). The double-shear values are close to those obtained from sliding along an inclined plane under similar conditions (Fortt, 2006), implying that the measured coefficient is independent of the method of measurement. The figure shows that the coefficient varies from $\mu = 0.8$ to $\mu = 0.05$, depending upon temperature and sliding speed. Generally, it decreases with increasing temperature and with increasing speed, although at the lower temperatures (−30 °C and −40 °C) there is evidence that the coefficient reaches a maximum at an intermediate velocity.

We interpret this behavior in Chapter 12 where we show that the friction coefficient for rougher surfaces, although greater, exhibits similar dependencies upon sliding speed and temperature. At this juncture, suffice it to say that the behavior can be explained in terms of a combination of creep, fracture and frictional melting, where creep dominates under lower speeds and fracture under higher speeds (Kennedy *et al.*, 2000). The fact that the initial microstructure plays little role implies that sliding induces its own material state within the region of the interfaces.

4.5 Molecular diffusion in ice

Water molecules experience frequent displacements from their temporary positions of equilibrium and can diffuse through the ice crystal under a gradient of concentration. In one dimension, mass diffusion is based upon Fick's law:

$$\frac{\partial c}{\partial t} = D\left(\frac{\partial^2 c}{\partial x^2}\right) \tag{4.21}$$

where c and t, respectively, denote the concentration of water molecules (in units of molecules per unit volume) and time, D is the diffusion coefficient and x the direction of diffusion. In this equation, D is assumed to be independent of c. The diffusion of molecules is termed *self-diffusion* and occurs by either a vacancy or an interstitial mechanism. Diffusion of molecules is directly dependent upon the crystallographic structure and the number of defects. Diffusion along dislocations, along grain boundaries and across the surface is expected to be easier than through a perfect lattice (see Shewmon (1989) for a comprehensive introduction). In addition to the diffusion of H_2O itself, the *diffusion of gases* in ice is important in relation to the interpretation of paleoclimatic records in deep ice cores, and so we include a short discussion of that process.

4.5.1 Self-diffusion

Self-diffusion can impact many processes involved in the deformation of ice: the climb of dislocations, diffusional creep and grain boundary sliding accommodated by diffusion. Bulk diffusion measurements have focused primarily on H_2O self-diffusion kinetics. Early diffusion studies used scintillation tracer techniques to determine self-diffusion coefficients of $H_2{}^{18}O$, D_2O and T_2O (see Hobbs, 1974). These isotopic tracer diffusion studies were performed at relatively high temperature between −40 °C and the melting point. The measured H_2O diffusion coefficients were very similar for all the isotope probe molecules with a mean value of $1.49 \pm 0.06 \times 10^{-15}\,m^2\,s^{-1}$ at −10 °C (Ramseier, 1967). For comparison, the self-diffusion coefficient of water at 0 °C is about $10^{-9}\,m^2\,s^{-1}$ (Eisenberg and Kauzmann, 1969).

The temperature dependence may be described by the relationship:

$$D = D_0 \exp(-Q/RT) \tag{4.22}$$

where D_0 and Q (the activation energy for diffusion) are independent of temperature in the temperature range −5 to −30 °C. The activation energy deduced by Ramseier is 0.62 ± 0.04 eV or 60 ± 3.8 kJ/mol. A slight anisotropy has been observed: the diffusion coefficient perpendicular to the *c*-axis is about 10% higher than parallel to this axis (Ramseier, 1967). The close similarity of the diffusion coefficient for the isotopes of hydrogen (^3H) and oxygen ($^{18}0$) confirms that self-diffusion is due to the migration of water molecules.

The diffusion of molecules through the ice crystal involves the migration of vacancies and/or interstitials. X-ray topography was used by Goto *et al.* (1986) for determining diffusion coefficients of point defects. The self-diffusion coefficient was determined by Goto *et al.* (1986) by using the *in-situ* X-ray topography technique, which permits the determination of the variation with time of the size of dislocation loops. This method is based on the relationship between the climb velocity of a dislocation (Chapter 5) and on the excess concentration of point defects around interstitial dislocation loops. From Goto *et al.* (1986), the predominant point defects above −50 °C are not vacancies but self-interstitials. The self-diffusion coefficient can thus be deduced from the diffusion and the concentration of interstitials. The concentration of interstitials at a given temperature was determined from the number and size of newly formed dislocation loops after cooling. Values of the self-diffusion coefficient obtained in this way are in close agreement with values deduced by tracer techniques. This supports the assumption of an interstitial process for the self-diffusion in ice above −50 °C.

Laser-induced thermal desorption techniques have been used to study HDO and $H_2{}^{18}O$ diffusion in pure ice over the temperature range −120 °C to −103 °C (Brown and George, 1996; Livingston *et al.*, 1997, 1998). Similar diffusion kinetics were found for HDO and $H_2{}^{18}O$ confirming that diffusion occurs via a molecular mechanism. The water molecule, in other words, diffuses as a unit through the ice lattice. Values of the laser-based self-diffusion coefficients obtained by extrapolation to high temperature appear to be more than two orders of magnitude higher than those given by Ramseier (1967) and others (see Hobbs, 1974). The experiments were performed on ultrathin ice films and that feature of technique could account for this discrepancy (Livingston *et al.*, 1997).

There is no experimental value of the coefficient of *grain boundary diffusivity* of water molecules. Diffusion of water molecules along grain boundaries in polycrystalline ice may control diffusion creep and grain boundary migration.

4.5.2 Volume diffusion of gases and other molecules

The fractionation of N_2 and O_2 in deep polar ice sheets was observed in air bubbles (Bender, 2002) and in clathrate hydrates from deep polar ice cores (Ikeda *et al.*, 1999). This was attributed to differences in the diffusion coefficient of these gases. This result has important implications in the evolution of the air composition in bubbles and clathrate hydrates in polar ice sheets (Chapter 3) and for the reconstructions of the paleoatmosphere from the analysis of polar ice cores.

The first measurements of the diffusion of helium in ice single crystals were made by Haas *et al.* (1971). The diffusion coefficient is about $1 \times 10^{-9}\,m^2\,s^{-1}$ at $-10\,°C$ with an activation energy of 0.12 eV. The diffusion constants of O_2, N_2 and CH_4 were estimated by using molecular dynamics simulations (Hori and Hondoh, 2003). Calculations were made by allowing the distortion of the ice lattice and the breaking of hydrogen bonds. Upper limits of the diffusion constants for O_2, N_2 and CH_4 were estimated to be 1.8×10^{-11}, 2.5×10^{-12} and $2.0 \times 10^{-14}\,m^2\,s^{-1}$ at $-3\,°C$. The diffusion coefficients of these gases were found to depend upon their molecular size.

4.5.3 Diffusion of acids

Diffusion coefficients larger than the self-diffusion coefficients are found for HF (Hobbs, 1974) and HNO_3 (Thibert and Domine, 1998). Molecules of HF can be introduced in the ice lattice in place of H_2O molecules and move easily into and between interstitial sites (Petrenko and Whitworth, 1999). Values of the diffusion coefficients of HF and HNO_3, respectively, are about 8×10^{-11} and $2 \times 10^{-14}\,m^2\,s^{-1}$ at $-10\,°C$. These values should be considered as upper limits because of the possibility that diffusion occurred along sub-boundaries in the samples that were studied (Thibert and Domine, 1998). At the opposite extreme, the diffusion rate of HCl appears to be lower than the diffusion rate of water molecules (Thibert and Domine, 1997). An upper limit of $3 \times 10^{-16}\,m^2\,s^{-1}$ at $-10\,°C$ is expected. From Dominé and Xueref (2001), the large values of the diffusion coefficients of HCl found by Livingston and George (2001) could be related to the presence of amorphous or crystalline hydrates of HCl in their samples or to diffusion along arrays of dislocations.

4.5.4 Surface diffusion

Diffusion of water molecules on ice surfaces is important for ice sintering, crystal growth and atmospheric gas–ice reactions (Hobbs, 1974; Livingston *et al.*, 2002). Measurements of the surface diffusion coefficient D_s represent a serious experimental challenge. Evaporation–condensation processes or the sensitivity of the ice surface to contamination can introduce artifacts. The liquid-like layer on ice (Dash *et al.*, 1995) is associated with rapid surface diffusion. Values of D_s obtained by molecular dynamics simulation are in the range of 10^{-9} to 10^{-11} $m^2 s^{-1}$ near the melting temperature (Furukawa and Nada, 1997). A value of about 10^{-9} $m^2 s^{-1}$ was obtained at −2 °C by Nasello *et al.* (2007) by studying the formation of grain boundary grooves on the surface of polycrystalline ice. The upper limit roughly corresponds to the self-diffusion in water. At lower temperature, studies of mass transport in the near surface of ice are very limited. Laser-induced thermal desorption experiments at −133 °C yielded an upper limit on surface diffusivity of $D_s = 10^{-13}$ $m^2 s^{-1}$ (Livingston *et al.*, 1998). This value seems rather high for this temperature, but may reflect an enhancement owing to surface defects.

References

Anderson, D. L. and C. S. Benson (1963). The densification and diagenesis of snow. In *Ice and Snow*, ed. W. D. Kingery. Cambridge, Mass.: MIT Press, pp. 391–411.

Arakawa, M. and N. Maeno (1997). Mechanical strength of polycrystalline ice under uniaxial compression. *Cold Reg. Sci. Technol.*, **26**, 215–229

Baer, B. J., J. M. Brown, J. M. Zaug, D. Schiferl and E. L. Chronister (1998). Impulsive stimulated scattering on ice VI and ice VII. *J. Chem. Phys.*, **108**, 4540–4544.

Barnes, P., D. Tabor, F. R. S. Walker and J. F. C. Walker (1971). The friction and creep of polycrystalline ice. *Proc. R. Soc. Lond. A*, **324**, 127–155.

Bender, M. L. (2002). Orbital tuning chronology for the Vostok climate record supported by trapped gas composition. *Earth Planet. Sci. Lett.*, **204**, 275–289.

Brown, D. E. and S. M. George (1996). Surface and bulk diffusion of $H_2{}^{18}O$ on single-crystal $H_2{}^{16}O$ ice multilayers. *J. Phys. Chem.*, **100**, 15460–15469.

Budiansky, B. and R. J. O'Connell (1976). Elastic-moduli of a cracked solid. *Int. J. Solids Struct.*, **12**, 81–97.

Cox, G. F. N. and W. F. Weeks (1983). Equations for determining the gas and brine volumes in sea-ice samples. *J. Glaciol.*, **29**, 306–316.

Cox, G. F. N. and W. F. Weeks (1986). Changes in the salinity and porosity of sea-ice samples during shipping and storage. *J. Glaciol.*, **32**, 371–375.

Dantl, G. (1968). Die elastischen Moduln von Eis-Einkristallen. *Phys. Kondens. Mater.*, **7**, 390–397.

Dantl, G. (1969). Elastic moduli of ice. In *Physics of Ice: Proceedings of the International Symposium on Physics of Ice, Munich, Germany, September 9–14, 1968*, eds. N. Riechl, B. Bullemer and H. Engelhardt. New York: Plenum Press, 223–230.

Dash, J. G., H. Fu and J. S. Wettlaufer (1995). The pre-melting of ice and its environmental consequences. *Rep. Prog. Phys.*, **58**, 115–167.

Domine, F. and I. Xueref (2001). Evaluation of depth profiling using laser resonant desorption as a method to measure diffusion coefficients in ice. *Anal. Chem.*, **73**, 4348–4353.

Eisenberg, D. and W. Kauzmann (1969). *The Structure and Properties of Water*. New York: Oxford University Press.

Elvin, A. A. (1996). Number of grains required to homogenize elastic properties of polycrystalline ice. *Mech. Mater.*, **22**, 51–64.

Fletcher, N. H. (1970). *The Chemical Physics of Ice*. Cambridge: Cambridge University Press.

Fortt, A. (2006). The resistance to sliding along coulombic shear faults in columnar S2 ice. Ph.D. thesis, Thayer School of Engineering, Dartmouth College.

Furukawa, Y. and H. Nada (1997). Anisotropic surface melting of an ice crystal and its relationship to growth forms. *J. Phys. Chem. B.*, **101**, 6167–6170.

Gagnon, R. E., H. Kiefte, M. J. Clouter and E. Whalley (1988). Pressure dependence of the elastic constants of ice Ih to 2.8 kbar by Brillouin spectroscopy. *J. Chem. Phys.*, **89**, 4522–4528.

Gagnon, R. E., H. Kiefte, M. J. Clouter and E. Whalley (1990). Acoustic velocities and densities of polycrystalline ice-Ih, Ice-II, Ice-III, Ice-V, and Ice-VI by Brillouin spectroscopy. *J. Chem. Phys.*, **92**, 1909–1914.

Gammon, P. H., H. Kiefte and M. J. Clouter (1980). Elastic-constants of ice by Brillouin spectroscopy. *J. Glaciol.*, **25**, 159–167.

Gammon, P. H., H. Kiefte, M. J. Clouter and W. W. Denner (1983). Elastic constants of artificial ice and natural ice samples by Brillouin spectroscopy. *J. Glaciol.*, **29**, 433–460.

Gibson, L. G. and M. F. Ashby (1988). *Cellular Solids: Structure and Properties*, 1st edn. Oxford: Pergamon Press.

Gold, L. W. (1958). Some observations on the dependence of strain on stress for ice. *Can. J. Phys.*, **36**, 1265–1275.

Gold, L. W. (1988). On the elasticity of ice plates. *Can. J. Civil Eng.*, **15**, 1080–1084.

Gold, L. W. (1994). The elastic modulus of columnar-grain fresh-water ice. *Ann. Glaciol.*, **19**, 13–18.

Goto, K., T. Hondoh and A. Higashi (1986). Determination of diffusion coefficients of self-interstitials in ice with a new method of observing climb of dislocations by X-ray topography. *Jpn. J. Appl. Phys.*, **25**, 351–357.

Greenberg, R., P. Geissler, G. Hoppa *et al.* (1998). Tectonic processes on Europa: tidal stresses, mechnaical response and visible features. *Icarus*, **135**, 64–78.

Haas, J., Bullemer, B. and Kahane, A. (1971). Diffusion of helium in monocrystalline ice. *Solid State Commun.*, **9**, 2033–2035.

Hearmon, R. F. S. (1961). *Introduction to Applied Anisotropic Elasticity*. Oxford: Oxford University Press.

Hill, R. (1952). The elastic behavior of a polycrystalline aggregate. *Proc. Phys. Soc. Lond. A*, **65**, 349–354.

Hobbs, P. V. (1974). *Ice Physics*. Oxford: Clarendon Press.

Hoenig, A. (1979). Elastic-moduli of a non-randomly cracked body. *Int. J. Solids Struct.*, **15**, 137–154.

Hoppa, G., B. R. Tufts, R. Greenberg and P. Geissler (1999). Strike-slip faults on Europa: global shear patterns driven by tidal stress. *Icarus*, **141**, 287–298.
Hori, A. and T. Hondoh (2003). Theoretical study on the diffusion of gases in hexagonal ice by the molecular orbital method. *Can. J. Phys.*, **81**, 251–259.
Ikeda, T., H. Fukazawa, S. Mae *et al.* (1999). Extreme fractionation of gases caused by formation of clathrate hydrates in Vostok Antarctic ice. *Geophys. Res. Lett.*, **26**, 91–94.
Kattenhorn, S. A. (2004). Strike-slip fault evolution on Europa: evidence from tailcrack geometries. *Icarus*, **172**, 582–602.
Kennedy, F. E., E. M. Schulson and D. Jones (2000). Friction of ice on ice at low sliding velocities. *Phil. Mag. A*, **80**, 1093–1110.
Kirchner, H. O. K., G. Michot, N. Narita and T. Suzuki (2001). Snow as a foam of ice: plasticity, fracture and the brittle-to-ductile transition. *Phil. Mag. A*, **81**, 2161–2181.
Langleben, M. P. and E. R. Pounder (1963). Elastic parameters of sea ice. In *Ice and Snow Processes, Properties and Application*, ed. W. E. Kingery. Cambridge, Mass.: MIT Press, pp. 69–78.
Livingston, F. E. and S. M. George (2001). Diffusion kinetics of HCl hydrates in ice measured using infrared laser resonant desorption depth-profiling. *J. Phys. Chem. A*, **105**, 5155–5164.
Livingston, F. E., G. C. Whipple and S. M. George (1997). Diffusion of HDO into single-crystal $H_2{}^{16}O$ ice multilayers: Comparison with $H_2{}^{18}O$. *J. Phys. Chem. B.*, **101**, 6127–6131.
Livingston, F. E., G. C. Whipple and S. M. George (1998). Surface and bulk diffusion of HDO on ultrathin single-crystal ice multilayers on Ru (001). *J. Chem. Phys.*, **108**, 2197–2207.
Livingston, F. E., J. A. Smith and S. M. George (2002). General trends for bulk diffusion in ice and surface diffusion on ice. *J. Phys. Chem. A*, **106**, 6309–6318.
Mellor, M. (1975). A review of basic snow mechanics. In *International Association of Hydrological Sciences Publication* 114, pp. 251–291.
Mellor, M. (1977). Engineering properties of snow. *J. Glaciol.*, **19**, 15–66.
Mellor, M. (1983). Mechanical behavior of sea ice. *US Army Cold Regions Research and Engineering Laboratory (CRREL) Report*, M 83-1.
Michel, B. (1978). *Ice Mechanics*. Laval, Quebec: Laval University Press.
Michel, B. and R. O. Ramseier (1971). Classification of river and lake ice. *Can. Geotech. J.*, **8**(36), 36–45.
Montagnat, M. and E. M. Schulson (2003). On friction and surface cracking during sliding. *J. Glaciol.*, **49**, 391–396.
Nanthikesan, S. and S. S. Sunder (1994). Anisotropic elasticity of polycrystalline ice IH. *Cold Reg. Sci. Technol.*, **22**, 149–169.
Nasello, O. B., S. Navarro de Juarez and C.L. Di Prinzio (2007). Measurement of self-diffusion on ice surface. *Scr. Mater.*, **56**, 1071–1073.
Nemat-Nasser, S. and M. Horii (1993). *Micromechanics: Overall Properties of Heterogeneous Materials*. Amsterdam, The Netherlands: Elsevier Science Publishers.
Nimmo, F., J. R. Spencer, R. T. Pappalardo and M. E. Mullen (2007). Shear heating as the origin of the plumes and heat flux on Enceladus. *Nature*, **447**, 289–291.
Nowick A. S. and B. S. Berry (1972). *Anelastic Relaxation in Crystalline Solids*. New York: Academic Press.
Nye, J. F. (1957). The distribution of stress and velocity in glaciers and ice-sheets. *Proc. R. Soc. Lond. Ser. A*, **239**, 113–133.
Nye, J. F. (1985). *Physical Properties of Crystals: Their Representation by Tensors and Matrices*. Oxford: Oxford University Press.

Petrenko, V. F. and R. W. Whitworth (1999). *Physics of Ice*. New York: Oxford University Press.
Prockter, L. M., F. Nimmo and R. T. Pappalardo (2005). A shear heating origin for ridges on Triton. *Geophys. Res. Lett.*, **32**, L14202, doi: 10.1029/2005GL022832.
Proctor, T. M. (1966). Low-temperature speed of sound in single-crystal ice. *J. Acoust. Soc. Amer.*, **39**, 972–977.
Ramseier, R. O. (1967). Self-diffusion of tritium in natural and synthetic ice monocrystals. *J. Appl. Phys.*, **38**, 2553–2556.
Reuss, A. Z. (1929). Berechnung der Fliessgrenze von Mischkristallen auf Grund der Pastizitatsbedingung fur Einkristalle. *Z. Angew. Math. Mech.*, **9**, 49.
Reyes-Morel, P. E. and I.-W. Chen (1990). Stress-biased anisotropic microcracking in zirconia polycrystals. *J. Am. Ceram. Soc.*, **73**, 1026–1033.
Saeki, H., A. Ozaki and Y. Kubo (1981). Experimental study on flexural strength and elastic modulus of sea ice. In *Proceedings of the 6th International Conference on Port and Ocean Engineering under Arctic Conditions*. Quebec, Canada, Laval University, pp. 536–547.
Schulson, E. M. (2002). On the origin of a wedge crack within the icy crust of Europa. *J. Geophys. Res.*, **107**, doi:10.1029/2001JE001586.
Shapiro, L. H., J. B. Johnson, M. Sturm and G. L. Blaisdell (1997). Snow mechanics: review of the state of knowledge and applications. *CRREL Report*, 97-3, 1–43.
Shewmon, P. G. (1989). *Diffusion in Solids*. Warrendale, PA: Minerals, Metals and Materials Society.
Shimizu, H., T. Niabetani, T. Nishiba and S. Sasaki (1996). High-pressure elastic properties of the VI and VII phase of ice in dense H_2O and D_2O. *Am. Phys. Soc.*, **53**, 6107–6110.
Sinha, N. K. (1989a). Elasticity of natural types of polycrystalline ice. *Cold Reg. Sci. Technol.*, **17**, 127–135.
Sinha, N. K. (1989b). Experiments on anisotropic and rate-sensitive strain ratio and modulus of columnar-grained ice. *Trans. ASME*, **111**, 354–560.
Smith-Konter, B. and R. T. Pappalardo (2008). Tidally driven stress accumulation and shear failure of Enceladus's tiger stripes. *Icarus* (in press).
Thibert, E. and F. Domine (1997). Thermodynamics and kinetics of the solid solution of HCl in ice. *J. Phys. Chem. B.*, **101**, 3554–3565.
Thibert, E. and F. Domine (1998). Thermodynamics and kinetics of the solid solution of HNO_3 in ice. *J. Phys. Chem. B.*, **102**, 4432–4439.
Traetteberg, A., L. Gold and R. Frederking (1975). The strain rate and temperature dependence of Young's modulus of ice. In *3rd International Symposium on Ice*, Hanover, New Hampshire, International Association for Hydraulic Research.
Tulk, C. A., R. E. Gagnon, H. H. Kieffer and M. J. Clouter (1994). Elastic constants of ice III by Brillouin spectroscopy. *J. Chem. Phys.*, **101**, 2350–2354.
Tulk, C. A., R. E. Gagnon, H. H. Kieffer and M. J. Clouter (1996). Elastic constants of ice VI by Brillouin spectroscopy. *J. Chem. Phys.*, **104**, 7854–7859.
Tulk, C. A., H. H. Kieffer, M. J. Clouter and R. E. Gagnon (1997). Elastic constants of ice III, V, and VI by Brillouin spectroscopy. *J. Phys. Chem. B.*, **101**, 6154–6157.
Voigt, W. (1910). *Lehrbuch der Kristallphysik*, Berlin: B. G. Teubner.
Wang, Y. S. (1981). Uniaxial compression testing of Arctic sea ice. In *6th International Conference on Port and Ocean Engineering under Arctic Conditions*. Quebec, Canada, Laval University.
Wegst, U. G. K., and M. F. Ashby (2004). The mechanical efficiency of natural materials. *Phil. Mag.*, **84**, 2167–2181.

5

Plastic deformation of the ice single crystal

5.1 Introduction

Single crystals undergo plastic deformation as soon as there is a component of shear stress on the basal plane. Basal slip is observed for shear stresses in the basal plane lower than 0.02 MPa (Chevy, 2005). Evidence of easy basal slip was first shown by McConnel (1891) and confirmed by many authors (Glen and Perutz, 1954; Griggs and Coles, 1954; Steinemann, 1954; Readey and Kingery, 1964; Higashi, 1967; Montagnat and Duval, 2004; see Weertman, 1973, for a review). A clear illustration of basal slip was obtained by Nakaya (1958) by performing bending experiments. Traces of the basal slip lines were made clearly visible by shadow photography (Fig. 5.1). Though some prismatic glide is observed for orientations close to those that inhibit basal slip, no clear observation of any deformation is reported in crystals loaded along the [0001] direction, which inhibits both basal and prismatic slip.

Basal slip takes place from the motion of basal dislocations with the $a/3\langle 11\bar{2}0\rangle$ Burgers' vector (Hobbs, 1974). The macroscopic slip direction corresponds to the maximum shear direction in the basal plane (Glen and Perutz, 1954). The slip direction is therefore always close to the direction of the maximum shear stress. From Kamb (1961), the failure to detect a slip direction in ice is explained by the fact that slip can occur in the three possible glide directions on the basal plane with a value of the stress exponent, which relates strain rate to stress, between 1 and 3.

The plastic deformation of the ice crystal is generally analyzed by considering the individual motion of dislocations. Accordingly, strain rate is related to the dislocation velocity v and the mobile dislocation density ρ_m by the famous Orowan equation. We will give results on the mobility of individual dislocations in the ice crystal obtained from *in-situ* X-ray topography measurements. The difficulty in deforming single crystals when basal slip is inhibited is related to the low density of non-basal edge dislocations. We will show that the cross-slip of basal dislocations on non-basal planes can be initiated by the internal stress field induced by basal

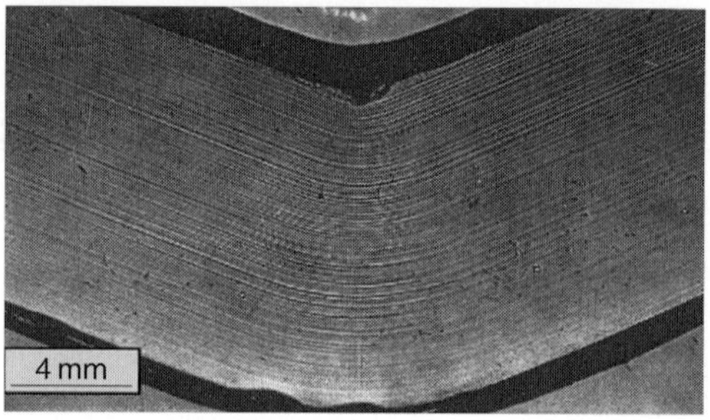

Figure 5.1. Basal slip lines observed by shadow photography on a single crystal bent with the c-axis parallel to the loading direction; height of the sample: 10 mm. (From Nakaya, 1958.)

screw dislocations and that this is invoked for dislocation multiplication (Chevy *et al.*, 2007).

In crystalline materials with high dislocation mobility, plastic deformation is a consequence of the collective motion of a large number of dislocations that interact at long distance. Plastic deformation takes place through isolated bursts or dislocation avalanches, characterized by very high instantaneous strain rates, whereas dislocations are almost immobile everywhere. This means that most dislocations at a given time are practically at rest and the instantaneous strain rate may exceed the average strain rate by several orders of magnitude (Zaiser, 2006). These temporal dynamics of plastic flow were clearly identified in ice single crystals by Weiss and Grasso (1997) by using the technique of acoustic emission. We show in this chapter that the plasticity of the ice crystal is characterized by large spatio-temporal fluctuations with scale-invariant characteristics. An overview of experimental and theoretical investigations of these phenomena in crystalline solids can be found in Zaiser (2006).

5.2 Plastic deformation of the ice crystal

5.2.1 Basal slip

Typical stress–strain curves obtained in uniaxial tension at −15 °C are shown in Figure 5.2. Each curve shows a marked yield drop and no work hardening is observed up to 4% of strain. Similar behavior is observed when ice is deformed in compression (Jones and Glen, 1969a). Such large drops in yield stresses were also observed in lithium fluoride by Johnston and Gilman (1959) and in several semiconductors (Alexander and Haasen, 1968). From Johnston and Gilman, the multiplication of mobile dislocations is at the origin of this softening, and hardening by dislocation interactions is not significant at relatively small strain. In such tests, the shear strain rate $\dot{\varepsilon}$ is given by:

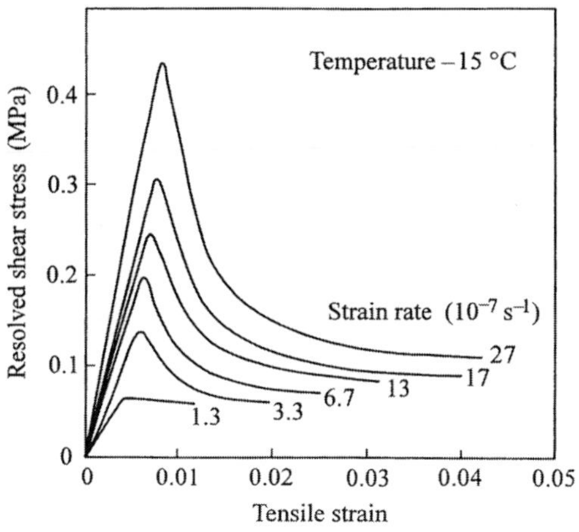

Figure 5.2. Stress–strain curves of ice single crystals with the *c*-axis at 45° to the tensile stress at −15 °C. (From Higashi *et al.*, 1964.)

$$\dot{\varepsilon} = \dot{\tau}/G + \rho_{\mathrm{m}} b v \tag{5.1}$$

where τ is the shear stress, G the elastic shear modulus, ρ_{m} the density of mobile dislocations, b the Burgers' vector and v the average velocity of dislocations on the basal plane.

The yield point is observed when the initial dislocation density is low. It is worth noting that the dislocation density in freshly grown ice crystals can be significantly less than $10^8\,\mathrm{m}^{-2}$ (Oguro, 1988) whereas a density of mobile dislocations slightly higher than $10^{10}\,\mathrm{m}^{-2}$ is expected after a deformation of 4% (Mansuy, 2001).

A significant reduction of the yield stress was observed when ice crystals were mechanically polished (Muguruma, 1969). This observation indicates that the operation of dislocation sources near the surface or the number of mobile dislocations depends on surface conditions.

The apparent steady state observed after the peak stress could correspond to the balance between dislocation multiplication and recovery processes, the emergence of dislocations at the surface of samples or the immobilization of dislocations. The hardening observed at −70 °C from a strain higher than 18% by Jones and Glen (1969a) could be due to the inhomogeneous deformation of the ice crystal rather than to the establishment of an internal stress field producing hardening.

The creep of single crystals deformed so that basal slip can take place is characterized by a continuously increasing strain rate. Typical creep curves obtained in tension by Jones and Glen (1969a) at −50 °C are shown in Figure 5.3. A steady state with a constant strain rate seems to have been observed by Higashi *et al.* (1965) for a strain of about 20% from bending creep tests. But deformation gradients associated with bending can have induced hardening. As for constant strain rate

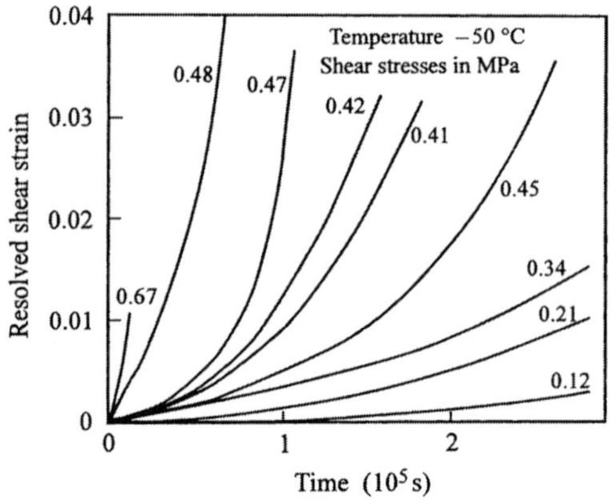

Figure 5.3. Tensile creep curves for single crystals oriented with the *c*-axis at 45° to the loading direction under various shear stresses. (From Jones and Glen, 1969a; with permission of the International Glaciological Society.)

tests, the increase of creep rate is explained by the multiplication of mobile dislocations.

The orientation dependence of the strength

Most mechanical tests have been performed in tension or compression with the *c*-axis oriented around 45° to the loading direction. Considering that slip occurs in the direction of the maximum resolved shear stress, the shear stress on the basal plane is:

$$\tau = \sigma \cos\theta \sin\theta \tag{5.2}$$

where σ is the applied stress in compression or tension and θ the angle between the loading direction and the normal to the basal plane. Figure 5.4 shows variation of the peak stress with strain for several values of θ. A constant peak shear stress was found when results were normalized to a constant shear strain rate on the basal plane (Trickett *et al.*, 2000a). This is in agreement with the Schmid law for single slip.

The effect of impurities

Creep and stress–strain curves for ice crystals doped with HF, HCl and H_2SO_4 show a softening effect of impurities (Jones and Glen, 1969b; Nakamura and Jones, 1973; Trickett *et al.*, 2000b). Figure 5.5 shows results obtained at −70 °C for pure and HF-doped crystals by Jones and Glen (1969b). A small amount of HF produces a significant softening. A similar effect was found on HCl-doped ice (Nakamura and Jones, 1970). These results were discussed in terms of the mechanisms of dislocation glide. From Glen (1968), the mobility of dislocations is dependent on the

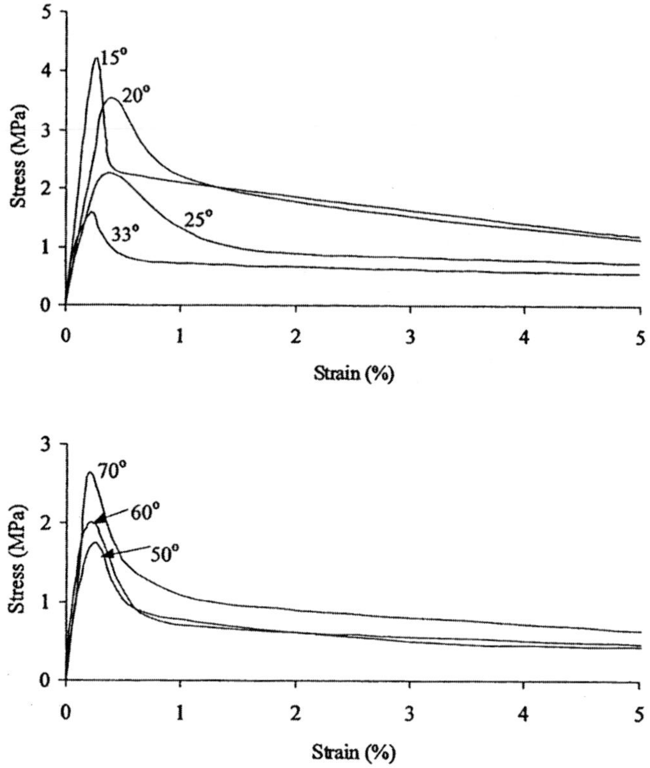

Figure 5.4. Stress–strain curves of ice single crystals at the strain rate of $1.0 \times 10^{-5}\ s^{-1}$ and −20 °C for various values of the angle between the compression direction and the *c*-axis. (From Trickett *et al.*, 2000a; with permission of the International Glaciological Society.)

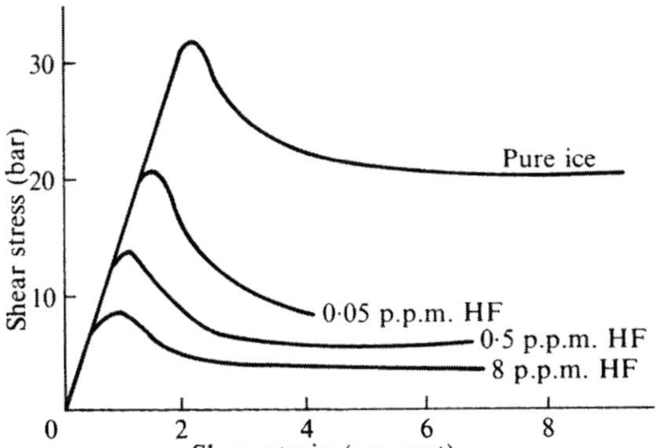

Figure 5.5. Stress–strain curves for single crystals at the strain rate of $2.7 \times 10^{-7}\ s^{-1}$ and −70 °C for various concentrations of HF dissolved in the ice lattice. (From Jones and Glen, 1969b; reprinted by permission of Taylor & Francis Ltd, www.infoworld.com.)

reorientation of water molecules, which occurs by the migration of proton point defects. HF and HCl would dissolve in the ice lattice, inducing an increase of Bjerrum L-defects and H_3O^+ ions (described in Chapter 2). The dislocation mobility would therefore increase with the impurity content. This assumption was also put forward by Baker *et al.* (2002) to explain the softening effect of H_2SO_4. The softening effect of dissolved impurities has also been discussed in terms of the increase of the density of dislocations (Jones and Gilra, 1973, Trickett *et al.*, 2000b). This last assumption is acceptable since basal slip appears to be not controlled by the mobility of dislocations along the basal plane (see Section 5.3).

Internal friction measurements

When subjected to small stresses and for a short time, an ice crystal may not suffer permanent deformation. The reversible motion of dislocations can induce anelasticity and energy dissipation. The associated internal friction δ is given by:

$$\delta = \frac{1}{2\pi}\frac{\Delta W}{W} \tag{5.3}$$

where ΔW and W, respectively, are the dissipated energy during one cycle of the loading stress and the maximum elastic energy during this cycle.

Internal friction experiments performed over a wide frequency range and at high temperature have been used to study the dynamics of dislocations in a single crystal associated with basal slip (Tatibouet *et al.*, 1983, 1986). Variation of the internal friction at the frequency of 0.8 Hz with temperature is shown in Figure 5.6. The peak in the internal friction or relaxation peak is considered in terms of the reorientation of water molecules as for dielectric relaxation (Bass, 1958; Gosar, 1974). The relaxation time decreases when the number of Bjerrum defects increases and it does not depend on the dislocation density (Kuroiwa, 1964; Tatibouet *et al.*, 1983).

The large increase of the internal friction above 200 K is related to the motion of basal dislocations, which is temperature dependent. Internal friction increases with the density of mobile dislocations, as illustrated by the effect of the deformation imposed on the ice sample before internal friction measurements (Fig. 5.6). The deformation associated with cyclic loading is generally too low to produce dislocations. From Tatibouet *et al.* (1986), at relatively high frequencies ($\geq 10^{-2}$ Hz), dislocations move in a reversible way under the applied stress, whereas at lowest frequencies, the non-reversible motion of dislocations must be considered. The climb of 60° basal dislocations on prismatic planes was assumed to explain this behavior (Tatibouet *et al.*, 1986).

Internal friction at high temperature was modeled by Cole (1995) by considering dislocation relaxation processes characterized by a distribution of relaxation times. The distribution of restoring forces acting in opposition to the motion of dislocations

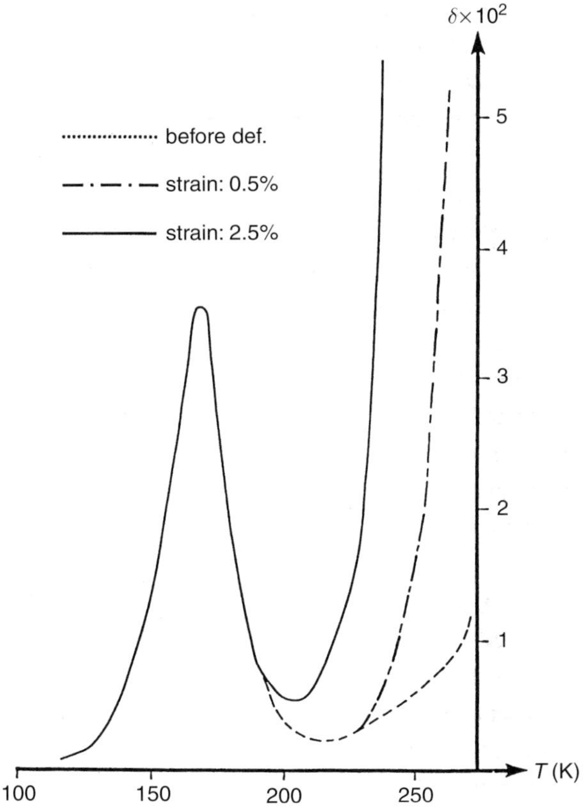

Figure 5.6. Internal friction of single crystals vs. temperature at 0.8 Hz as a function of strain. The high-temperature internal friction is related to the motion of dislocations on the basal plane. (From Vassoille, 1978.)

could be at the origin of the distribution of relaxation times. Mobile dislocations would be expected to be pinned by other quasi-immobile dislocations (Tatibouet *et al.*, 1986).

In conclusion, internal friction measurements on ice crystals give information on the mean density of mobile dislocations and the dynamics of basal dislocations under cyclic loading. However, this technique cannot provide information on the dynamics of interacting dislocations because that interaction is associated with the collective motion of dislocations, as discussed in Section 5.3.2.

5.2.2 Non-basal slip

Non-basal slip can be studied when ice crystals are loaded such that there is no resolved shear stress on the basal plane. But tests are difficult to perform, since a slight misorientation ($\leq 1°$) can give basal slip. Stress–strain curves obtained

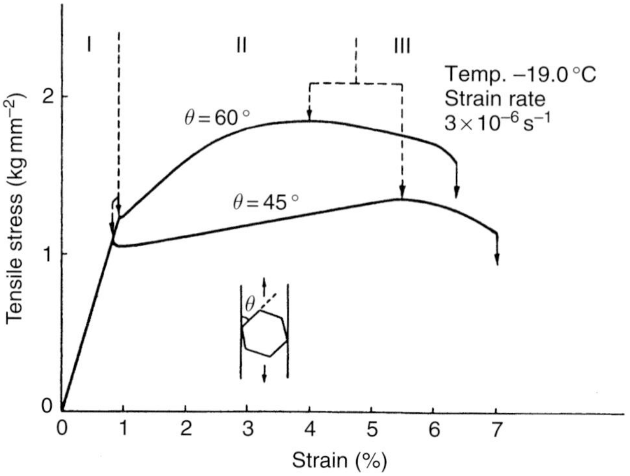

Figure 5.7. Stress–strain curves of ice single crystals with the *c*-axis at 90° to the tensile stress. Stage II is associated with work-hardening; fracture is occurring at the end of stage III. (From Higashi, 1967.)

at −19 °C on single crystals oriented with the *c*-axis perpendicular to the tensile axis are shown in Figure 5.7. The tensile stress was more than 10 MPa, which is about two orders of magnitude larger than that found at the same strain rate for basal slip (Duval *et al.*, 1983).

Figure 5.8 shows the relationship between strain rate and stress for basal and non-basal slip. The non-basal deformation of ice crystals requires a stress at least 60 times larger than that for basal slip at the same strain rate. It is worth noting that strain rate for basal slip was determined before reaching steady state. For example, data from Wakahama (1967) were obtained for a strain less than 1%.

5.2.3 The flow law for basal slip

The strain rate determined after a strain of about 5% in creep tests can be expressed by:

$$\dot{\varepsilon} = A\sigma^n \exp(-Q/RT) \tag{5.4}$$

where R is the gas constant and T the temperature (K). Almost all authors report a stress exponent n of 2 ± 0.3 and an activation energy close to 65 kJ/mole (Higashi *et al.*, 1965; Jones and Glen, 1969a; Mellor and Testa, 1969; Ramseier, 1972). It is important to point out that the value of n was determined for finite strain between 3 and 20%. There is no clear indication that steady state was reached in these experiments. A value of Q of about 40 kJ/mole was found by Jones and Glen (1969a) between −90 and −50 °C. From Jones and Brunet (1978), there is no indication of any increase of the activation energy near the melting point.

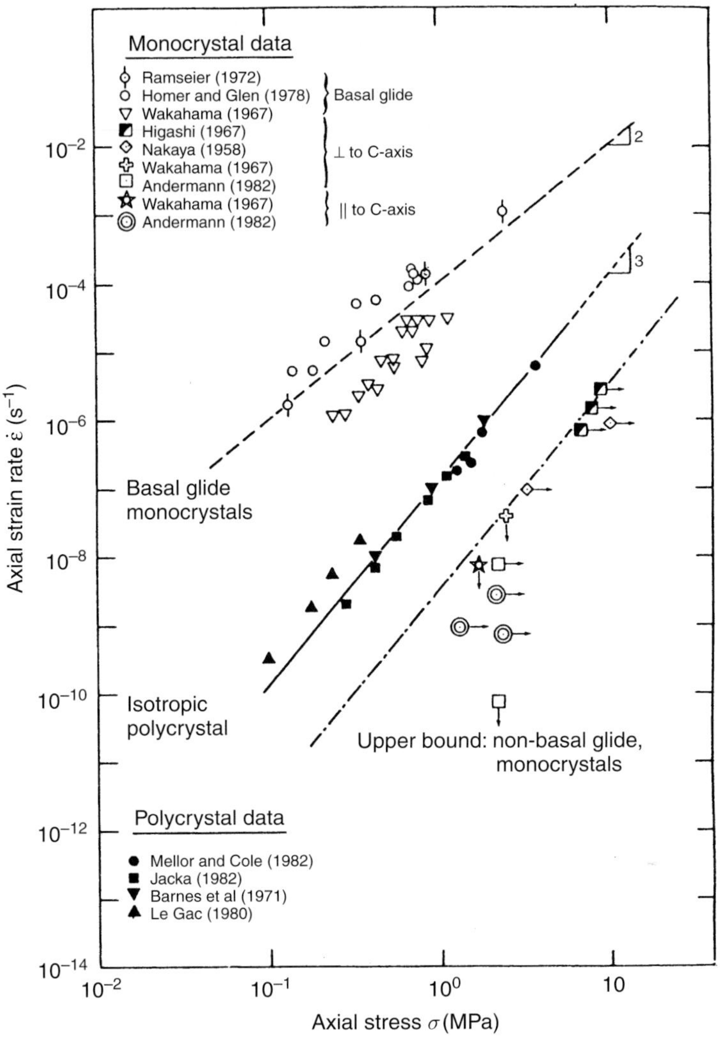

Figure 5.8. Stress–strain rate relationship for steady-state basal and non-basal slip in single crystals at −10 °C. The behavior of isotropic polycrystalline ice during secondary creep is also shown. All data for basal slip were obtained before reaching a steady state; data from Homer and Glen were obtained at a strain of about 5%, whereas data from Wakahama (1967) were obtained at a strain lower than 1%. (From Duval *et al.*, 1983 and references therein.)

At imposed constant strain rate, the dependence of the peak stress on strain rate also gives a value of n between 1.6 and 2 (Higashi *et al.*, 1964; Jones and Brunet, 1978). It is worth noting that the peak stress depends on the initial dislocation density and, as mentioned above, on the number of dislocation sources (Muguruma, 1969). It is reached for a strain lower than 2% (Higashi *et al.*, 1964; Jones and Glen,

1969a), as shown in Figure 5.2. Parameters of the flow law determined at the peak stress therefore cannot be directly used to determine rate-controlling processes for basal slip.

5.3 Dynamics of dislocations

5.3.1 Dynamics of individual dislocations

Experimental measurements of dislocation velocity

High-quality measurements have been made of the velocity at which dislocations glide on both basal and prismatic planes in ice (Higashi *et al.*, 1985; Ahmad and Whitworth, 1988; Shearwood and Whitworth, 1991; Okada *et al.*, 1999). Results obtained for stresses lower than 1 MPa show that the average velocity on the basal plane scales linearly with the shear stress resolved on that plane (Petrenko and Whitworth, 1999). A linear dependence was also found for the slip of edge dislocations on the $(10\bar{1}0)$ prismatic plane (Hondoh *et al.*, 1990). Dislocation mobility, defined as velocity divided by stress, is more than five times greater for non-basal edge dislocations than for basal dislocations, at least at temperatures around 263 K (Fig. 5.9). The difference in mobility between the two slip systems

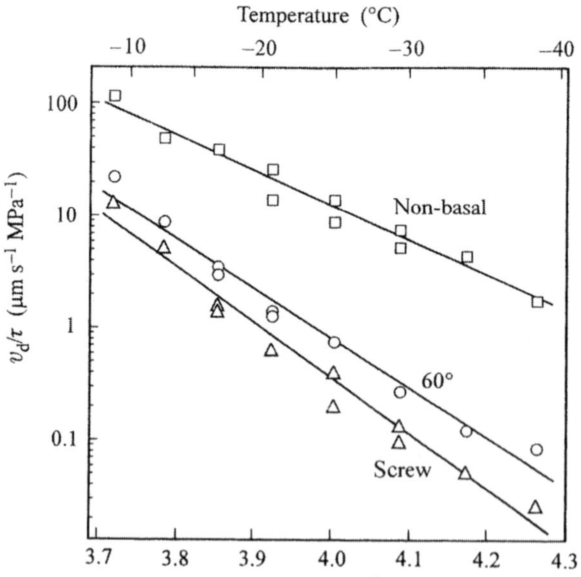

Figure 5.9. Dislocation velocities per unit stress in pure ice as a function of temperature for straight screw and 60° dislocation segments on the basal plane and for edge dislocation segments on non-basal planes. (From Shearwood and Whitworth, 1991; reprinted by permission of Taylor & Francis Ltd, www.infoworld.com.)

Table 5.1. *Mobility of dislocations with the Burgers' vector $1/3\langle 11\bar{2}0\rangle$*

	Q (eV)	Mobility at −20 °C ($\mu m\, s^{-1}\, MPa^{-1}$)	References
Basal screw	0.75	1.3 ± 0.1	Okada *et al.* (1999)
	0.95	1.0 ± 0.1	Shearwood and Whitworth (1991)
Basal 60°	0.69	2.6 ± 0.1	Okada *et al.* (1999)
	0.87	2.0 ± 0.1	Shearwood and Whitworth (1991)
Non-basal edge	0.61	14.5 ± 2	Hondoh *et al.* (1990)
	0.63	22.0 ± 1	Shearwood and Whitworth (1991)

From Hondoh (2000)

increases with decreasing temperature. Dislocation mobility is a thermally activated process and obeys a Boltzmann-type expression with an activation energy, Q, that characterizes the rate-controlling mechanism. Values are given in Table 5.1. The Q-values vary significantly, for reasons not understood (Hondoh, 2000), thus impeding the search for the rate-limiting mechanism.

The dislocation mobility in ice is very low compared with other materials. For instance, the velocity of basal screw dislocations in ice loaded under a normalized stress (shear stress/shear modulus) of $\sigma/G = 10^{-4}$ at a homologous temperature (absolute temperature/absolute melting temperature) of $T/T_f = 0.85$ is about $0.01\ \mu m\, s^{-1}$ (Shearwood and Whitworth, 1991). Under the same normalized conditions, the velocity in germanium and silicon, two other tetrahedrally bonded solids, is more than $100\ \mu m\, s^{-1}$ (Caillard and Martin, 2003). The reason for the sluggish kinetics in ice is probably related to the requirement for protonic rearrangement.

Rate-limiting processes for dislocation motion

The intrinsic resistance of the lattice to dislocation glide is referred to as the Peierls barrier or lattice friction stress. The energy of dislocations depends on their position in the lattice. From Hondoh *et al.* (1990), the Peierls barrier for the non-basal dislocations is much less important than that for basal dislocations. Ahmad and Whitworth (1988) arrived at the same conclusion by noting straight basal dislocations and curved non-basal dislocations. The large variation in the mobility between basal and non-basal dislocations could be related to the level of the Peierls barrier (Louchet, 2004a,b).

Dislocation glide is assumed to be limited by the propagation of kinks or steps along a dislocation line between adjacent Peierls valleys. Accordingly, the dislocation velocity is given by:

$$V_d = hC_k V_k \tag{5.5}$$

where C_k and V_k are the concentration (in units of number per unit length) and velocity of kinks, respectively, and h the distance between Peierls troughs (Hirth and Lothe, 1982). In the low stress limit (<1 MPa) the dislocation velocity is given by (Shearwood and Whitworth, 1991; Petrenko and Whitworth, 1999):

$$V_d = \nu \frac{2\tau b a h^2}{kT} \exp\left(-\frac{F_k + F_m}{kT}\right) \tag{5.6}$$

where ν is the attempt frequency that cannot exceed the Debye cutoff frequency, a is the lattice parameter, b is the magnitude of the Burgers' vector and k is Boltzmann's constant. F_k and F_m are, respectively, the free energy for the formation and migration of a kink. From Hondoh (2000), the activation energy for kink migration is $F_m = 0.62$ eV. This value corresponds to that found for the motion of non-basal edge dislocations (Table 5.1), which is in accordance with the low Peierls barrier for the motion of non-basal dislocations (Shearwood and Whitworth, 1991). The difference in the activation energy for the motion of basal screw and 60° dislocations (Table 5.1) could be due to differences in the kink formation energy (Okada *et al.*, 1999). A higher concentration of thermal kinks on the dissociated 60° dislocations is therefore expected, which would explain their greater mobility (Table 5.1).

The rate-limiting process for the migration of kinks could be related to the proton disorder. Glen (1968) argued that the motion of dislocations in ice is hindered by the disorder of hydrogen atoms on the O–H–O bonds. Accordingly, the rate at which a dislocation can move would be dependent on the rate of the reorientation of water molecules, which is directly related to the concentration and mobility of Bjerrum defects (D and L defects). The time for the reorientation of bonds is itself related to the dielectric relaxation time (Hobbs, 1974). Based on the role of proton disorder, several theories have been developed to simulate the motion of dislocations (Whitworth *et al.*, 1976; Frost *et al.*, 1976; Shearwood and Whitworth, 1991). The activation energy for Debye relaxation is close to that for the motion of non-basal edge dislocations, which would be controlled by kink migration. The calculated dislocation velocity deduced from such models, however, is much lower than those measured. A larger concentration of Bjerrum defects near the dislocation core would give a lower value of the relaxation time and therefore would explain the relatively high dislocation mobility (Shearwood and Whitworth, 1991). A non-crystalline core with highly mobile molecules, as suggested by Whitworth (1978), could explain the relatively high mobility of kinks. But, from Petrenko and Whitworth (1999), this extended non-crystalline core is not consistent with the observed straight dislocation segments.

The relationship between the concentration of protonic defects and the relaxation time for dislocation motion was studied by Shearwoood and Whitworth (1992) by measuring the dislocation velocity in HCl-doped ice, which was supposed to introduce Bjerrum defects and therefore to increase the rate of reorientation of bonds. No significant change in the dislocation velocity was detected. These authors arrived at the conclusion presented above that the time for the reorientation of hydrogen bonds for dislocation glide is not the time that is typical of the bulk material.

In conclusion, the rate-controlling mechanism for the glide of dislocations on both basal and non-basal planes in ice is not yet well defined. Measurements indicate that the glide dislocation velocity is higher than that given by theoretical models.

Cross-slip of screw dislocations

Cross-slip is a process through which screw dislocations move from one slip plane to another in the zone of the Burgers' vector. It is an important process in the plasticity of materials. For example, it is involved in dislocation multiplication by the double cross-slip mechanism (Friedel, 1964) and can allow screw dislocations to bypass obstacles on the primary slip plane. In ice, the dissociation of basal screw dislocations into two widely extended partial dislocations should make difficult the cross-slip process, as already noted (Chapter 2). These dislocations must first constrict over a short segment to permit the process to begin. From Duesbery (1998), the classical model in which the dislocation cannot leave its slip plane unless it is fully constricted is incorrect. Duesbery asserts that cross-slip can occur, regardless of the state of constriction of the dislocation, provided that the glide stress in the cross-slip plane is sufficiently large. It is worth noting that the stress in the cross-slip plane (prismatic or pyramidal) induced by an array of like-signed basal screw dislocations can induce cross-slip even if the applied stress in the cross-slip plane is very low (Chevy *et al.*, 2007). Cross-slip of basal screw dislocations to a prismatic plane was observed by Fukuda *et al.* (1987) using X-ray topography. These authors assumed that constriction of extended dislocations occurs at the surface and leads to the formation of edge dislocation segments mobile in the prismatic plane. Such constriction can also occur at grain boundaries.

Cross-slip of basal screw dislocations is probably at the origin of the formation of basal slip bands from individual slip lines, as observed by Nakaya (1958) and by Montagnat *et al.* (2006). Multiplication of basal screw dislocations by the occurrence of the double cross-slip mechanism was suggested by Chevy *et al.* (2007) and by Taupin *et al.* (2007) to explain the deformation of ice single crystals deformed in torsion. The role of the long-range internal stress field in the initiation of cross-slip is discussed below in relation to the collective motion of dislocations and the formation of slip bands.

Dislocation climb

Dislocation climb involves the motion of dislocations in planes that do not contain their Burgers' vector. It is a dominant deformation mechanism at high temperature in most materials. The climb motion requires the diffusion of point defects under a force oriented perpendicular to the dislocation line. Under high temperature and low stresses, the climb velocity of an isolated dislocation with a high jog density is given by:

$$V_{\mathrm{cl}} = \frac{D\Omega\sigma}{bkT} \tag{5.7}$$

where D is the self-diffusion coefficient, Ω is the molecular volume, k the Boltzman constant and T the temperature. σ is the normal stress perpendicular to the climb plane (Friedel, 1964). From Hondoh (2000), the climb mobility $D\Omega/bkT$ at $-10\,°\mathrm{C}$ is more than two orders of magnitude smaller than the mobility of dislocations gliding on the basal plane. Although slow, dislocation climb allows the ice crystal to deform along the c-axis and will be invoked in Chapter 6 when discussing the deformation of the polycrystal.

The widely extended basal dislocations with stacking faults lying on the (0001) plane restrict the climb of dislocations out of the basal plane. However, the formation of tilt sub-boundaries, consisting of walls of parallel edge basal dislocations aligned perpendicular to the basal slip plane, probably involves some dislocation climb on prismatic planes (Fukuda *et al.*, 1987).

In conclusion, dislocation climb is a very slow deformation mechanism compared with basal slip. But, as we show in Chapter 6, it can be invoked in the deformation of polycrystalline ice as an additional process to satisfy stress equilibrium and deformation compatibility between ice grains.

5.3.2 Collective motion of dislocations and dislocation patterns

Investigations of the plastic deformation of ice crystals in terms of the dynamics of dislocations generally involve computing the velocity of one dislocation segment under stress in the presence of various obstacles, and the macroscopic plasticity is directly related to the uncorrelated motion of a large number of dislocations (Weertman, 1973; Tatibouet *et al.*, 1986; Louchet, 2004a).

Recent work questions the concept of uncorrelated motion and the modeling that stems from it. Experiments on ice single crystals reveal that the strain response to a constant load consists of a series of dislocation avalanches (Weiss and Grasso, 1997; Miguel *et al.*, 2001; Weiss and Miguel, 2004). The size distribution of energy bursts associated with dislocation avalanches and monitored by acoustic

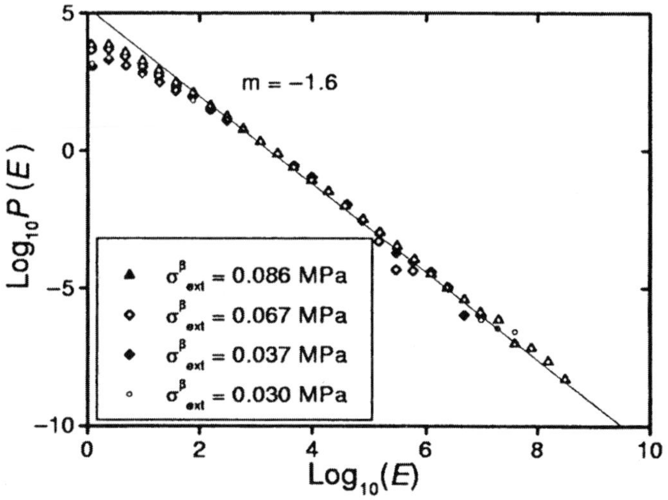

Figure 5.10. Distribution of acoustic energy bursts recorded in ice single crystals under creep tests at −10 °C. The resolved shear stress on the basal plane is indicated in the inset. The data follow a power law $P(E) \approx E^{-1.6}$ over several decades. This result indicates that a large number of dislocations move cooperatively in an intermittent fashion. (From Miguel *et al.*, 2001; with permission of Nature Publishing group.)

emission[1] follows a power law over several decades (Fig. 5.10). As Louchet (2006) noted, this scale invariance means that there is no characteristic size of avalanche or energy burst. Instead, a relatively large number of dislocations appear to move cooperatively in an intermittent fashion where the movement is correlated over large distances. This suggests a close-to-critical state for the dislocation ensemble during plastic deformation; "critical" in the sense that a small perturbation may lead to a large event (Louchet *et al.*, 2006). This also means that, at fixed location, the plastic strain rate is close to zero most of the time and that deformation takes place as isolated bursts where the instantaneous strain rate may significantly exceed the average strain rate. In such a situation, the transition between the microscopic and the macroscopic scales cannot be carried out by applying simple averaging methods over all dislocations. The motion of individual dislocations, as presented above, could not represent the real situation for the deformation of the ice crystal. Long-range dislocation interaction is associated with this collective behavior involving a large number of correlated events.

The collective motion of dislocations, which appears to characterize the deformation of ice single crystals, is not compatible with large friction stresses. Such stresses

[1] Ice is a good material model to study dislocation glide by acoustic emission because of its transparency, the facility to fix transducers by fusion/freezing and the facility to detect acoustic signals. The frequency bandwidth of piezoelectric transducers used for ice is 0.1–1 MHz (Weiss and Grasso, 1997).

would not allow one dislocation "to be seen" by other dislocations located far from it. Lattice friction, as an obstacle to dislocation motion, would be of little importance in comparison with long-range dislocation interactions (Zaiser and Seeger, 2002).

Slip line observations

The collective behavior manifests itself in the form of dislocation patterns. There is clear evidence that the motion of dislocations in ice crystals is associated with the formation of slip lines and slip bands, as already noted. Slip bands parallel to the basal plane were revealed by shadow photography (Nakaya, 1958) and by observations of replicas under an optical microscope (Readings and Bartlett, 1968). Slip bands have also been revealed by the X-ray topography technique (Liu *et al.*, 1995). They are seen to form successively with a decreasing spacing as strain increases (Hamelin *et al.*, 2004). Depending on strain, slip band spacing is typically between 10 and 500 mm (Faria and Kipfstuhl, 2004; Hamelin *et al.*, 2004). But the possibility to relate the slip line spacing to strain and stress is disputed, with results showing the fractal character of the slip band spacing (Montagnat *et al.*, 2006). This is in accordance with long-range interactions between dislocations located in different slip bands and with their collective motion (Neuhauser, 1983; Weiss and Miguel, 2004).

The relation between the formation of glide bands and dislocation avalanches was found by Mendelson in 1963 in several ionic single crystals, by using a photoelastic technique which allows one to follow the birefringence induced by dislocation arrays. Each glide band is seen to form in an avalanche in about 0.1 s. The estimated dislocation velocity is several orders of magnitude greater than that deduced from the Orowan relation, which expresses strain rate as a function of the density of mobile dislocations and an average dislocation velocity. Spatial and temporal fluctuations of the dislocation velocity v and of the density of mobile dislocations ρ_m would be too high for applicability of the averaging procedure implicitly assumed in the Orowan equation (Weiss and Montagnat, 2007). This equation can only give rough information on average values of ρ_m and v at large spatial and temporal scales.

Scale invariance of slip patterns in ice crystals deformed in torsion

Synchrotron X-ray topography experiments were performed at the European Synchrotron Radiation Facility (ESRF) on ice single crystals deformed under pure torsion (Montagnat *et al.*, 2006). The torsion axis was parallel to the *c*-axis, so that basal slip was the main deformation mode. A constant torque was applied to samples in the form of cylinders. The applied shear stress on the outer surface of cylinders was 0.05 MPa. As expected, creep curves showed the increase of strain rate with strain noted above.

Figure 5.11. Synchrotron X-ray topograph obtained on a radial slice of a sample deformed in pure torsion. The torsion strain at the outer surface was 0.71%. The torsion axis was along the *c*-axis. (From Montagnat *et al.*, 2006.)

X-ray topography is an imaging technique based on Bragg diffraction. It provides a two-dimensional intensity mapping of the diffracted beam. White-beam radiation was used to detect the non-homogeneous lattice distortion field in the whole sample. In this configuration, with the torsion axis along the *c*-axis, the orientation contrast is mainly related to the arrangement of basal screw dislocations. Diffraction intensity variations provide a map of dislocation density fluctuations within the bulk. Figure 5.11 shows an example of a diffraction pattern on the $[10\bar{1}0]$ prismatic plane with the long dimension corresponding to the *c*-axis (and the torsion axis). Darker zones correspond to high dislocation density regions. The pattern reveals a strong heterogeneity of the dislocation density along the *c*-axis, associated with basal slip lines and slip bands.

The intensity profile along the *c*-axis was analyzed by means of a spectral analysis (Montagnat *et al.*, 2006). A power-law regime for the power spectrum of the intensity record over the scale range of the studied sample (Fig. 5.11) was found. The power spectrum E scales with the frequency f as $E(f) \approx f^{-\mu}$ with $\mu = 1/3 \pm 0.1$. This power law reveals the scale invariance of the intensity records, i.e., of the dislocation density arrangement along the *c*-axis (Zaiser and Seeger, 2002). This invariance indicates that the dislocation density is spatially correlated over large distances with a correlation length of the order of the sample size. These large-scale correlations are the fingerprint of strong interactions between dislocations. As already noted, low lattice friction is a condition for long-range interactions between dislocations (Louchet, 2006). The stress barrier associated with lattice friction, when too high, can screen the elastic interaction between dislocations and consequently hinder the long-range interaction.

Dislocations are not distributed randomly and are not concentrated on well-defined "slip bands" more or less regularly spaced. Instead, their scale-free pattern is in qualitative agreement with the scale-free power law distribution of the size of dislocation avalanches recorded by acoustic emission during the creep of ice single crystals (Miguel *et al.*, 2001). It is important to note that the hard X-ray intensity record from non-deformed samples does not show such behavior. The scale invariance is therefore induced by plastic deformation. We discuss again such long-range spatial correlation in dislocation patterns when we consider the deformation of ice polycrystals (Chapter 6).

These observations on deformed ice crystals question the pertinence of classical and widely used concepts such as slip bands and dislocation densities in other materials. The collective dislocation motion is not restricted to ice. Simulations demonstrate that intermittent dislocation avalanches with scale-invariant characteristics are a general feature of crystal plasticity. The paradigm of plastic flow as a smooth and steady-state process has been challenged from both experimental and theoretical points of view (Zaiser and Seeger, 2002; Zaiser, 2006). Dislocation avalanches were seen to lead to jumps in the stress–strain curves in deformed nickel microcrystals, whereas plasticity appears as a smooth process in macroscopic samples since the strain produced by an avalanche is inversely proportional to the specimen size (Dimiduk *et al.*, 2006). In polycrystals, dislocation avalanches are limited by grain boundaries and plasticity appears as a smooth process (Richeton *et al.*, 2005; Csikor *et al.*, 2007). According to Dimiduk *et al.* (2006), dislocation motion and crystal deformation have more in common with plate tectonics than with viscous fluid flow. A discussion on the scale invariance in plastic flow of crystalline solids in general can be found in Zaiser (2006).

5.4 Rate-controlling processes for basal slip

The general behavior of the deformation of the ice crystal can be understood by following the analysis made on several semiconductors. The yield point corresponds with the onset of extensive dislocation multiplication. According to Johnston and Gilman (1959), the dislocation density is found to be proportional to strain and the hardening is assumed negligible for the peak stress. The application of this analysis to ice was questioned by Higashi *et al.* (1964, 1965) on account of a lack of knowledge of the dislocation multiplication law and the difficulty in estimating the role of hardening.

According to Equation (5.4) and considering that the dislocation velocity is proportional to the resolved shear stress, the density of mobile dislocations should depend linearly on stress. Such a relationship was deduced from acoustic emission measurements during the deformation of ice single crystals (Weiss and Grasso, 1997).

Values of the activation energy for basal slip are significantly lower than those for the motion of basal dislocations given in Table 5.1. One reason for this difference is the possible increase of the dislocation density when temperature decreases (Petrenko and Whitworth, 1999). The rate-controlling processes for basal slip could be directly related to the mobility of basal dislocations. Considering the possible role of dislocation climb (Louchet, 2004b) and cross-slip (Taupin *et al.*, 2007) in dislocation multiplication and the collective motion of basal dislocations, the activation energy for basal slip should not be related to the mobility of basal dislocations, but rather to mechanisms that induce dislocation multiplication.

From Louchet (2004b), strain rate should be proportional to the density of non-basal edge dislocations and the velocity of such dislocations. A stress exponent equal to 2 is obtained for steady state and the activation energy predicted by this model is that for dislocation climb, in agreement with experimental values.

The validity of this model, however, can be questioned since the same flow law was deduced from torsion tests for which the slip of screw dislocations on the basal plane is the main deformation process (Montagnat *et al.*, 2003). As discussed in Section 5.3.1, cross-slip processes, through which screw dislocations move from the basal plane to another plane, can be assumed for the successive operation of dislocation sources. This process cannot be ignored when looking for dislocation multiplication. The role of long-range internal stresses, induced by basal dislocations, in the initiation of cross-slip in ice single crystals is a sound assumption (Montagnat *et al.*, 2006). On that basis, the cross-slip process would control strain rate. A steady state could be reached by assuming isotropic and kinematic hardening giving a constant density of mobile dislocations. The isotropic hardening could be induced by the non-basal edge dislocations located in prismatic planes and produced by dislocation climb or cross-slip, whereas kinematic hardening could be associated with the accumulation of basal dislocations in each basal plane. These physical processes have been incorporated in a dislocation field theory (Varadhan *et al.*, 2006) in terms of the coupled dynamics of excess dislocations (or geometrically necessary dislocations) gliding on basal planes and statistically distributed dislocations developed through cross slip on prismatic planes (Taupin *et al.*, 2007). The concept of excess dislocations was introduced by Nye (1953) and Ashby (1970) to account for the accommodation of gradients of plastic strain by these dislocations. As a consequence, statistical dislocations result in compatible deformation.

These two concepts are scale dependent and, at a sufficiently fine scale of spatial resolution, all dislocations are excess dislocations (Taupin *et al.*, 2007). The creep behavior of the ice crystal is well reproduced by this field theory and a steady state is expected after an equivalent strain higher than 20% (Taupin, 2007). The softening associated with the production of mobile excess dislocations is of greater importance than the increase of the internal stress during the primary creep.

5.5 Conclusions

In conclusion, ice deforms mainly by the glide of dislocations on the basal plane and dislocation multiplication appears to occur through either the cross-slip of basal screw dislocations on prismatic planes and/or the climb of basal dislocations. It is suggested that the edge jogs formed in prismatic planes by the cross-slip process can act as obstacles to slip of screw dislocations on the basal plane. These non-basal dislocations could be responsible for isotropic hardening at high finite strain. The accumulation of basal screw dislocations on the basal plane participates in kinematic hardening and to the likely occurrence of a steady state at relatively high strain (Taupin *et al.*, 2007). The activation energy for basal slip could be that of either cross-slip or dislocation climb. The simulation of the dynamics of dislocations in a single crystal based on the glide of dislocations on the basal plane and the cross-slip and/or the climb of dislocations on prismatic planes appears to be a promising method to advance our understanding of deformation and rate-controlling processes (Louchet, 2004a; Chevy *et al.*, 2007).

It is important to point out that experimental data have not established that a steady state was unambiguously reached even for a strain higher than 20%. One reason could be that the symmetry associated with tension or compression tests is not respected with single crystals oriented for basal slip and a non-uniform state of stress within the sample should develop. This is not the case in torsion tests with the *c*-axis along the torsion axis. Deformation gradients associated with this test can obscure the steady state if it exists.

Macroscopic slip on $(1\bar{1}00)$ planes can occur via the motion of short non-basal edge dislocation segments along the prismatic plane. This slip can be produced either by the climb of 60° basal dislocations or by the cross-slip of basal screw dislocations induced by the long-range internal stress field produced by basal dislocations. This non-basal slip is therefore associated with basal slip. Another situation is found when the ice crystal is loaded in such a way that the applied shear stress on the basal plane is very low. With this situation, the density of basal dislocations will be too low to produce mobile prismatic dislocations. Non-basal slip is therefore of small importance when basal slip is inhibited. This analysis is supported by results shown in Figure 5.8, which show the difficulty in initiating non-basal slip when an isolated ice crystal is loaded such that the shear stress on the basal plane is very low. At this stage, it is important to point out that the behavior of ice crystals embedded within a polycrystal matrix can not be exactly that of an isolated single crystal. The long-range internal stress that develops during primary creep should induce non-basal slip.

There is no convincing information on the deformation of ice crystals loaded parallel to the *c*-axis. In this orientation, deformation is expected to result from the glide of dislocations on pyramidal planes with a component of the Burgers' vector in the [0001] direction. This pyramidal slip was not suggested by X-ray topography and its contribution to non-basal slip, if it exists, must be very low.

References

Ahmad, S. and R. W. Whitworth (1988). Dislocation motion in ice: a study by synchrotron X-ray topography. *Phil. Mag.*, **A57**, 749–766.

Alexander, H. and P. Haasen (1968). Dislocations and plastic flow in diamond structure. In *Solids State Physics*, eds. F. Seitz, D. Turnbull and H. Ehrenreich. New York: Academic Press.

Ashby, M. F. (1970). The deformation of plastically non-homogeneous materials. *Phil. Mag.*, **21**, 399–424.

Baker, I., Y. L. Trickett, D. Iliescu and P. M. S. Pradham (2002). The effects of H_2SO_4 on the mechanical behavior of ice single crystals. In *Creep Deformation Fundamentals and Applications*, eds. R. S. Mishra, J. C. Earthman and S. V. Raj. Warrendale, Pa.: TMS, 85–94.

Bass, R. (1958). Zur theorie der mechanischen relaxation des Eises. *Z. Phys.*, **153**, 16–37.

Caillard, D. and J. L. Martin (2003). *Thermally Activated Mechanisms in Crystal Plasticity.* Pergamon Materials Series, Vol. 8. Oxford: Pergamon Press.

Chevy, J. (2005). Mécanismes de déformation de la glace monocristalline en torsion; approche expérimentale et modélisation. Unpublished Master en Matériaux et Génie des Procédés, INPG, Grenoble, France.

Chevy, J., M. Montagnat, P. Duval, M. Fivel and J. Weiss (2007). Dislocation patterning and deformation processes in ice single crystals deformed by torsion. In *Physics and Chemistry of Ice*, ed. W. F. Kuhs. Cambridge: Royal Society of Chemistry, pp. 141–146.

Cole, D. M. (1995). A model for the anelastic straining of saline ice subjected to cyclic loading. *Phil. Mag.*, **A72** (1), 231–248.

Csikor, F. F., C. Motz, D. Weygand, M. Zaiser and S. Zapperi (2007). Dislocation avalanches, strain bursts, and the problem of plastic forming at the micrometer scale. *Science*, **318**, 251–254.

Dimiduk, D. M., C. Woodward, R. LeSar and M. D. Uchic (2006). Scale-free intermittent flow in crystal plasticity. *Science*, **312**, 1188–1190.

Duesbery, M. S. (1998). Dislocation motion, constriction and cross-slip in fcc metals. *Modeling Simul. Mater. Sci. Eng.*, **6**, 35–49.

Duval, P., M. F. Ashby and I. Anderman (1983). Rate-controlling processes in the creep of polycrystalline ice. *J. Phys. Chem.*, **87**, 4066–4074.

Faria, S. H. and S. Kipfstuhl (2004). Preferred slip-band orientations and bending observed in the Dome Concordia (East Antarctica) ice core. *Ann. Glaciol.*, **39**, 386–390.

Friedel, J. (1964). *Dislocations*. Oxford: Pergamon Press.
Frost, H. J., D. J. Goodman and M. F. Ashby (1976). Kink velocities on dislocations in ice. A comment on the Whitworth, Paren and Glen model. *Phil. Mag.*, **33**, 951–961.
Fukuda, A., T. Hondoh and A. Higashi (1987). Dislocation mechanisms of plastic deformation of ice. *J. Physique*, **48** (Colloque C1), 163–173.
Glen, J. W. (1968). The effect of hydrogen disorder on dislocation movement and plastic deformation of ice. *Phys. Kondens. Mater.*, **7**, 43–51.
Glen, J. W. and M. F. Perutz (1954). The growth and deformation of ice crystals. *J. Glaciol.*, **2**, 397–403.
Gosar, P. (1974). Theory of the anelastic relaxation of cubic and hexagonal ice. *Phil. Mag.*, **29**, 221–240.
Griggs, D. T. and N. E. Coles (1954). Creep of single crystals of ice. *U.S. Snow, Ice and Permafrost Research Establishment*, *Report*, **11**.
Hamelin, B., P. Bastie, P. Duval, J. Chevy and M. Montagnat (2004). Lattice distortion and basal slip bands in deformed ice crystals revealed by hard X-ray diffraction. *J. Phys. IV*, **118**, 27–33.
Higashi, A. (1967). Mechanisms of plastic deformation in ice single crystals. In *Physics and Snow and Ice*, ed. H. Oura. Sapporo: Hokkaido University Press, pp. 277–289.
Higashi, A., S. Koinuma and S. Mae (1964). Plastic yielding in ice crystals. *Jpn J. Appl. Phys.*, **3**, 610–616.
Higashi, A., S. Koinuma and S. Mae (1965). Bending creep of ice single crystals. *Jpn. J. Appl. Phys.*, **4**, 575–582.
Higashi, A., A. Fukuda, T. Hondoh, K. Goto and S. Amakai (1985). Dynamical dislocation processes in ice crystal. In *Dislocations in Solids*, eds. H. Susuki, T. Ninomiya, K. Sumino and S. Takeuchi. Tokyo: University of Tokyo Press, pp. 511–515.
Hirth, J. P. and J. Lothe (1982). *Theory of Dislocations*. New York: Wiley.
Hobbs, P. V. (1974). *Ice Physics*. Oxford: Clarendon Press.
Hondoh, T. (2000). Nature and behavior of dislocations in ice. In *Physics of Ice Core Records*, ed. T. Hondoh. Sapporo: Hokkaido University Press, 3–24.
Hondoh, T., H. Iwamatsu and S. Mae (1990). Dislocation mobility for non-basal glide in ice measured by in situ X-ray topography. *Phil. Mag.*, **A62**, 89–102.
Johnston, W. G. and J. J. Gilman (1959). Dislocation velocities, dislocation densities, and plastic flow in lithium fluoride crystals. *J. Appl. Phys.*, **30**, 129–144.
Jones, S. J. and J. G. Brunet (1978). Deformation of ice single crystals close to the melting point. *J. Glaciol.*, **21**, 445–455.
Jones, S. J. and N. K. Gilra, (1973). X-ray topographical study of dislocations in pure and HF-doped ice. *Phil. Mag.*, **27**, 457–472.
Jones, S. J. and J. W. Glen (1969a). The mechanical properties of single crystals of pure ice. *J. Glaciol.*, **8**, 463–473.
Jones, S. J. and J. W. Glen (1969b). The effect of dissolved impurities on the mechanical properties of ice crystals. *Phil. Mag.*, **19**, 13–24.
Kamb, W. B. (1961). The glide direction in ice. *J. Glaciol.*, **3**, 1097–1106.
Kuroiwa, D. (1964). Internal friction of ice. *Contrib. Inst. Low Temp. Sci.*, **A18**, 1–62.
Liu, F., I. Baker and M. Dudley (1995). Dislocation-grain boundary interactions in ice crystals. *Phil. Mag.*, **A71**, 15–42.
Louchet, F. (2004a). Dislocations and plasticity of ice. *C. R. Physique*, **5**, 687–698.
Louchet, F. (2004b). A model for steady state plasticity of ice single crystals. *Phil. Mag. Lett.*, **84**, 797–802.
Louchet, F. (2006). From individual dislocation motion to collective behavior. *J. Mater. Sci.*, **41**, 2641–2646.

Louchet, F., J. Weiss and Th. Richeton (2006). Hall-Petch law revisited in terms of collective dislocation dynamics. *Phys. Rev. Lett.*, **97**, 075504-1–075504-4.

Mansuy, Ph. (2001). Contribution à l'étude du comportement viscoplastique d'un multicristal de glace: hétérogénéité de la déformation et localisation, expériences et modèles. Thèse de l'Université Joseph Fourier, Grenoble, France.

McConnel, J. C. (1891). On the plasticity of ice. *Proc. R. Soc. London*, **49**, 323–343.

Mellor, M. and R. Testa (1969). Creep of ice under low stress. *J. Glaciol.*, **8**, 147–152.

Mendelson, S. (1963). Glide band formation and broadening in ionic single crystals. *Phil. Mag.*, **8**, 1633–1648.

Miguel, C. M., A. Vespignani, S. Zapperi, J. Weiss and J. R. Grasso (2001). Intermittent dislocation flow in viscoplastic deformation. *Nature*, **410**, 667–671.

Montagnat, M. and P. Duval (2004). The viscoplastic behavior of ice in polar ice sheets: experimental results and modelling. *C. R. Physique*, **5**, 699–708.

Montagnat, M., P. Duval, P. Bastie and B. Hamelin (2003). Strain gradients and geometrically necessary dislocations in deformed ice single crystals. *Scr. Mater.*, **49**, 411–415.

Montagnat, M., J. Weiss, P. Duval *et al.* (2006). The heterogeneous nature of slip in ice single crystals deformed under torsion. *Phil. Mag.*, **86**, 4259–4270.

Muguruma, J. (1969). Effects of surface condition on the mechanical properties of ice crystals. *Brit. J. Appl. Phys.*, **2**, 1517–1525.

Nakamura, T. and S. Jones (1970). Softening effect of dissolved hydrogen chloride in ice crystals. *Scr. Metall.*, **4**, 123–126.

Nakamura, T. and S. Jones (1973). Mechanical properties of impure ice crystals. In *Physics and Chemistry of Ice*, eds. E. Whalley, S. J. Jones and L. W. Gold. Ottawa: Royal Society of Canada, pp. 365–369.

Nakaya, U. (1958). Mechanical properties of single crystals of ice. Part 1. Geometry of deformation. *U.S. Army Snow Ice and Permafrost Establishment Research Report*, **28**.

Neuhauser, H. (1983). Slip line formation and collective dislocation motion. In *Dislocations in Solids*, Vol. 6, ed. F. R. N. Nabarro. North Holland, pp. 319–440.

Nye, J. F. (1953). Some geometrical relations in dislocated crystals. *Acta Metall.*, **1**, 153–162.

Oguro, M. (1988). Dislocations in artificially grown single crystals. In *Lattice Defects in Ice Crystals*, ed. A. Higashi. Sapporo: Hokkaido University Press, pp. 27–47.

Okada, Y., T. Hondoh and S. Mae (1999). Basal glide of dislocations in ice observed by synchrotron radiation topography. *Phil. Mag.*, **79**, 2853–2868.

Petrenko, V. F. and R. W. Whitworth (1999). *Physics of Ice*. New York: Oxford University Press.

Ramseier, R. O. (1972). Growth and mechanical properties of river and lake ice. Ph.D. thesis, Laval University, Canada.

Readey, D. W. and W. D. Kingery (1964). Plastic deformation of single crystal ice. *Acta Metall.*, **12**, 171–178.

Readings, C. J. and J. T. Bartlett (1968). Slip in single crystals of ice. *J. Glaciol.*, **7**, 479–491.

Richeton, Th., J. Weiss and F. Louchet (2005). Dislocation avalanches: role of temperature, grain size and strain hardening. *Acta Mater.*, **53**, 4463–4471.

Shearwood, C. and R. W. Whitworth (1991). The velocity of dislocations in ice. *Phil. Mag.*, **A64**, 289–302.

Shearwood, C. and R. W. Whitworth (1992). The velocity of dislocations in crystals of HCl-doped ice. *Phil. Mag.*, **A65**, 85–89.

Steinemann, S. (1954). Results of preliminary experiments on the plasticity of ice crystals. *J. Glaciol.*, **2**, 404–413.

Tatibouet, J., J. Perez and R. Vassoille (1983). Study of lattice defects in ice by very low frequency internal friction measurements. *J. Phys. Chem.*, **87**, 4050–4054.

Tatibouet, J., J. Perez and R. Vassoille (1986). High-temperature internal friction and dislocations in ice Ih. *J. Physique*, **47**, 51–60.

Taupin, V. (2007). Incompatibilité de réseau et organization collective des dislocations. Thèse de l'Université Paul Verlaine-Metz, France.

Taupin, V., S. Varadhan, J. Chevy *et al.* (2007). Effects of size on the dynamics of dislocations in ice single crystals. *Phys. Rev. Lett.*, **99**, 155507-1–155507-4.

Trickett, Y. L., I. Baker and P. M. S. Pradham (2000a). The orientation dependence of the strength of ice single crystals. *J. Glaciol.*, **46**, 41–44.

Trickett, Y. L., I. Baker and P. M. S. Pradham (2000b). The effects of sulfuric acid on the mechanical properties of ice single crystals. *J. Glaciol.*, **46**, 239–243.

Varadhan, S., A. J. Beaudoin and C. Fressengeas (2006). Coupling the dynamics of statistically distributed and excess dislocations. *Proc. Sci.*, 1–11.

Vassoille, R. (1978). Comportement anélastique et microplastique de la glace Ih à basse temperature. Thèse de l'INSA, Lyon, France.

Wakahama, G. (1967). On the plastic deformation of single crystal of ice. In *Physics of Snow and Ice*, ed. H. Oura. Sapporo: Hokkaido University Press, pp. 291–311.

Weertman, J. (1973). Creep of ice. In *Physics and Chemistry of Ice*, ed. E. Whalley, S. J. Jones and L. W. Gold. Ottawa: Royal Society of Canada, pp. 320–337.

Weiss, J. and J. R. Grasso (1997). Acoustic emission in single crystals of ice. *J. Phys. Chem.*, **101**, 6113–6117.

Weiss, J. and C. M. Miguel (2004). Dislocation avalanches correlations. *Mat. Sci. Eng.*, **A387–389**, 292–296.

Weiss, J. and M. Montagnat (2007). Long-range spatial correlations and scaling in dislocation and slip patterns. *Phil. Mag.*, **87**, 1161–1174.

Whitworth, R. W. (1978). The core structure and the mobility of dislocations in ice. *J. Glaciol.*, **21**, 341–359.

Whitworth, R. W., J. G. Paren and J. W. Glen (1976). The velocity of dislocations in ice – a theory based on proton disorder. *Phil. Mag.*, **33**, 409–426.

Zaiser, M. (2006). Scale invariance in plastic flow of crystalline solids. *Adv. Phys.*, **55**, 185–245.

Zaiser, M. and A. Seeger (2002). Long-range internal stresses, dislocation patterning and work-hardening in crystal plasticity. In *Dislocations in Solids*, eds. F. R. N. Nabarro and M. S. Duesbery. Amsterdam: Elsevier Science B.V., pp. 1–99.

6

Ductile behavior of polycrystalline ice: experimental data and physical processes

6.1 Introduction

Understanding how polar ice sheets interact with the climatic system is of the highest importance to predict sea-level changes. Ice sheets contain information on the climate and the atmospheric composition over the last 800 000 years (EPICA Community Members, 2004). Interpretation of ice core data is directly dependent on the accuracy of ice sheet flow models used for ice core dating. Knowledge of the rheological properties of ice in the low stress conditions of glaciers and polar ice sheets is therefore needed to improve the constitutive laws that are incorporated in flow models. Due to very high viscoplastic anisotropy of the crystal (Chapter 5), ice is considered as a model material to validate micro-macro polycrystal models used to simulate the behavior of anisotropic viscoplastic materials (Gilormini *et al.*, 2001; Lebensohn *et al.*, 2007).

Ice displays a wide range of mechanical properties, including elasticity, visco-elasticity, viscoplasticity, creep rupture and brittle failure (Schulson, 2001). In glaciers and ice sheets, ice is generally treated as a heat-conducting *non-linear viscous fluid*.

Ice is assumed here to be incompressible. It will be shown that the main effect of hydrostatic pressure on the ductile behavior of ice is to modify the melting temperature of pure ice T_f with $dT_f/dP \approx 0.074$ °C/MPa (Lliboutry, 1971).

In this chapter, we focus the analysis on the mechanical behavior of *granular* glacier ice. We assume that the behavior is ductile without the formation of cracks. This deformation regime typically corresponds to applied shear stresses lower than 1 MPa and to higher stresses when ice is subjected to confining pressure. The mechanical behavior of glacier ice is essentially studied by performing creep tests. Depending on the finite strain, the applied load or the applied stress is maintained at a constant value; however, a short discussion of tests performed at constant strain rates is given.

Knowledge of the physical processes that occur during the viscous (i.e., time-dependent) deformation of ice in glaciers and ice sheets is needed to establish constitutive laws: relations between the macroscopic variables of strain rate, strain, stress and temperature. Despite a large amount of theoretical and experimental effort, many questions remain unanswered. There is no doubt that a deformation regime with a stress exponent lower than 2 exists at low shear stresses. However, mechanisms that produce the deformation and are rate-controlling are still under discussion, even though we have strong arguments to consider basal slip as the main deformation process in the large range of deviatoric stresses involved in glaciers and polar ice sheets. The deformation of glacier ice can also be associated with diffusional processes and grain boundary sliding. The importance of these deformation processes will be discussed. Emphasis will be placed on the role of impurities in the deformation of glacier ice. Dynamic recrystallization is very active in glaciers and ice sheets. Transitions amongst grain growth, rotation (or continuous) and migration (or discontinuous) recrystallization are not well defined. One objective in this chapter is to describe these processes and to list the critical points for the understanding of the mechanical behavior of glacier ice.

Dislocations mainly carry out the viscous deformation of ice in glaciers and ice sheets. They are stored during the plastic deformation and cause hardening. At relatively high temperature, dislocations rearrange and can be removed by recovery and dynamic recrystallization. For the reasons presented in Chapter 5, the possibility to express strain rate as a function of a mean density of mobile dislocations that move independently with an average velocity is questioned. Indeed, dislocations appear to move cooperatively in groups with long-range correlations both in space and in time (see Chapter 5).

This chapter is mainly devoted to the creep behavior. Earlier reviews on the creep of granular ice can be found in: Weertman (1973, 1983); Hobbs (1974); Michel (1978); Mellor (1980); Sanderson (1988); Duval *et al.* (1983); Budd and Jacka (1989); Paterson (1994); Petrenko and Whitworth (1999).

6.2 Deformation behavior of isotropic granular ice

Polycrystalline ice can be deformed to large strains. In the flow of glaciers and ice sheets, plastic strains of more than 1 are usual. There is a clear evidence that intracrystalline dislocation glide is the main deformation process in the deformation of polycrystalline glacier ice (Glen and Perutz, 1954; Glen, 1955; Steinemann, 1958; Barnes *et al.*, 1971). The glide of dislocations on the basal plane and the cross-slip of basal dislocations were invoked to explain the behavior of single crystals (see Chapter 5). The climb of basal dislocations will be also considered as a complementary deformation mode of polycrystalline ice. Strain can also be achieved by transport of matter by diffusion or by shear along grain boundaries (grain boundary sliding or GBS). GBS creates voids or overlaps that can be

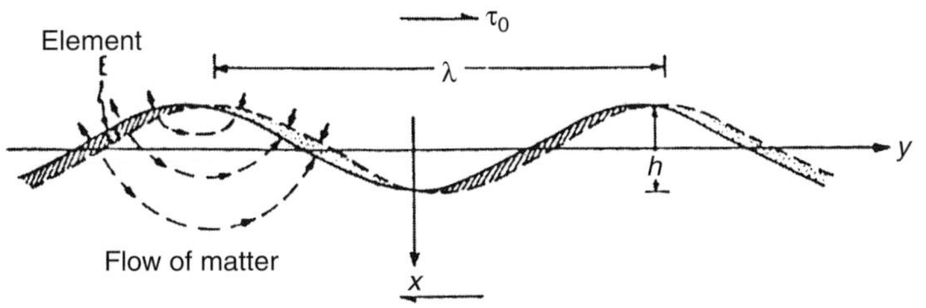

Figure 6.1. Grain boundary sliding with diffusional accommodation. (From Raj and Ashby, 1971.)

accommodated by the diffusion of water molecules from compressed parts of the boundary to those in tension. Inversely, diffusion creep creates the driving force for GBS (Fig. 6.1). These two deformation processes, GBS and diffusion creep, are strongly coupled and mutually accommodating (Raj and Ashby, 1971). GBS can also be associated with intracrystalline dislocation slip as suggested by several authors for superplastic materials in the regime corresponding to a stress exponent equal to 2. Some of these deformation modes include a grain size effect. They are discussed here in the context of the deformation of ice.

6.2.1 Creep tests

Consider what happens as a constant load is applied to a granular ice sample that is homogeneous and initially isotropic. A typical creep curve, giving strain as a function of time, is shown in Figure 6.2. There is first an initial instantaneous *elastic* strain ε_e, which follows Hooke's law. This is followed by a period of primary or transient creep where the creep rate decreases continuously. This primary creep occurs up to about 1% strain; it includes a *time-dependent recoverable strain* or *delayed elastic strain* when the load is removed. In stage II, the creep rate appears to remain constant, indicating a nearly steady-state condition. A minimum creep rate can be defined in the range of strain of about 1 to 2%. This secondary creep corresponds to the transition between the decelerating primary creep and the accelerating tertiary creep. Beyond about 10% strain, a steady state can be reached when cracking does not occur. In this case, the increase of creep rate after secondary creep is related to the development of preferred crystal orientation (*textures* or *fabrics*) and dynamic recrystallization (Duval *et al.*, 1983; Budd and Jacka, 1989). Steady state during tertiary creep implies stable textures and the equilibrium between hardening associated with the development of internal stresses and softening associated with recovery and recrystallization processes. It is worth noting that such behavior implies relatively high temperature and the development of recrystallization textures (Jacka and Maccagnan, 1984; Duval and Castelnau, 1995). Indeed, a plastic

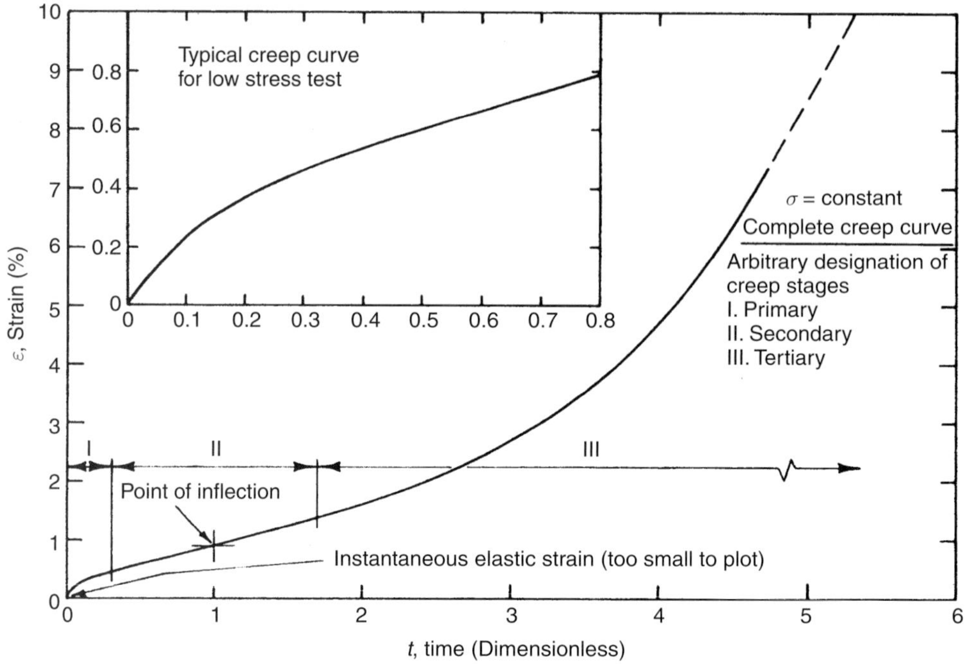

Figure 6.2. Typical creep curve for a constant applied stress test on isotropic granular ice. (From Mellor, 1980.)

strain of 10% is far too small to produce an important rotation of crystals by slip. As discussed in Chapter 7, most textures in polar ice sheets (where plastic strain can exceed unity) are induced by slip since recrystallization is not very active at relatively low temperature (Alley, 1992; de La Chapelle *et al.*, 1998).

A comprehensive description of the creep of isotropic ice can be found in Jacka (1984a). Figure 6.3 shows smoothed plots of creep rate as a function of strain for tests conducted at different levels of stress and temperature. Samples with a grain size of about 1.7 mm were tested between −30 °C and −5 °C under axial compressive stresses ranging from 0.1 to 1.3 MPa. The duration of experiments was too short to reach the steady-state tertiary creep. The minimum creep rates are well defined and they are reached approximately at the same strain whatever the stress and temperature.

Primary creep

The primary creep strain ε_p is generally well described by the Andrade law (Glen, 1955; Duval, 1976):

$$\varepsilon_p = \varepsilon_0 + \beta t^{1/3} + \dot{\varepsilon}_{min} t \tag{6.1}$$

where ε_0 is "instantaneous" strain. The transient strain $\varepsilon_t = \beta t^{1/3}$ is made up of a recoverable "delayed elastic" strain ε_d plus an irreversible viscous strain. The strain

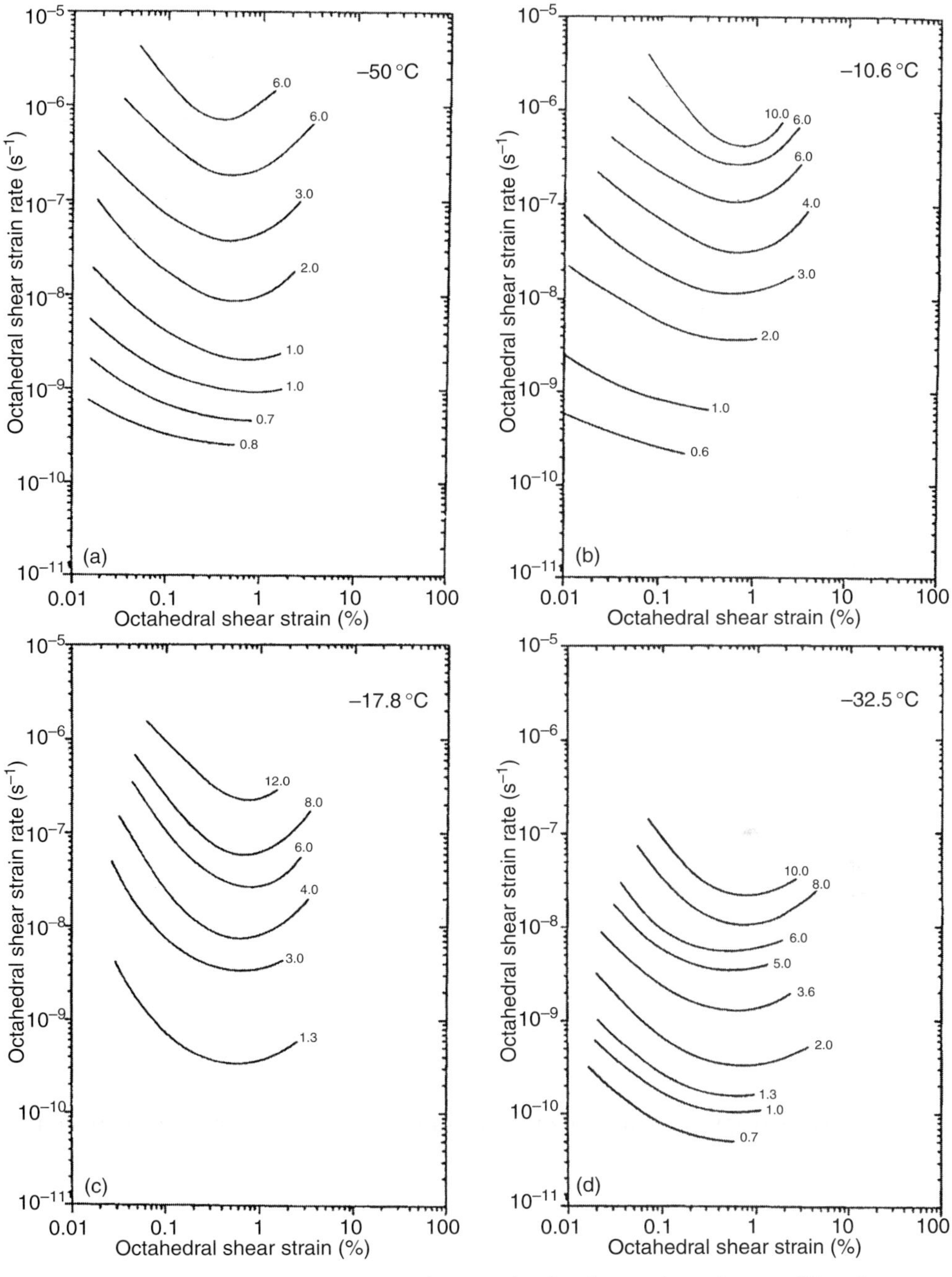

Figure 6.3. Strain rate plotted against strain for isotropic polycrystalline ice deformed in uniaxial compression. Values of the applied stress given in the figure correspond to the octahedral stress with $\tau_{\text{oct}} = \left(\sqrt{2}/3\right)\sigma_{zz}$; the octahedral shear strain rate is $\dot{\varepsilon}_{\text{oct}} = \dot{\varepsilon}_{zz}/\sqrt{2}$. (From Jacka, 1984a.)

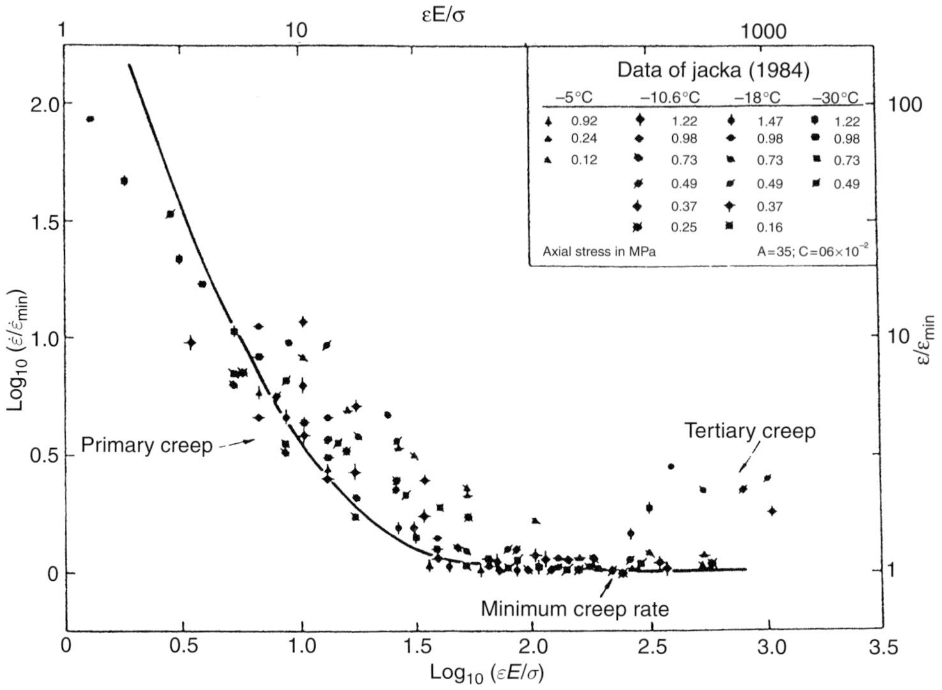

Figure 6.4. Strain rate plotted against strain using the reduced variables $\dot{\varepsilon}/\dot{\varepsilon}_{\min}$ and $\varepsilon E/\sigma$. (From Ashby and Duval, 1985.)

ε_d can be one order of magnitude higher than the theoretical elastic strain (Duval, 1978) whereas the transient creep strain can be more than 50 times the elastic strain (Ashby and Duval, 1985). It is worth noting that the transient strain is sometimes considered as fully recoverable by some authors (Sinha, 1978; Sunder and Wu, 1990), which is the case as long as the transient strain is lower than 10^{-4} (Tatibouet *et al.*, 1986; Cole and Durell, 1995).

Based on high-quality creep data of Jacka (1984a) obtained over a range of temperature and stress, a single master curve was obtained (Fig. 6.4). From Ashby and Duval (1985), the transient strain is given by:

$$\varepsilon_t = \frac{\sigma}{E}\left[1 + A\left(1 - \exp\left(-\left(\frac{C\dot{\varepsilon}_{\min}t}{\sigma/E}\right)^{1/n}\right)\right)\right]. \qquad (6.2)$$

Values of $A = 70$, $C = 1.6 \times 10^{-2}$ and $n = 3$ give a good description of all Jacka data up to the tertiary creep.

The relation (6.1) has been shown to be obeyed by many crystalline and non-crystalline materials (Andrade, 1914; Dysthe *et al.*, 2002). In polycrystalline ice, a rapid "instantaneous" plastic strain follows the loading. As creep relaxes the resolved shear stress on the basal plane on each grain, load is transferred to the harder non-basal slip systems. A long-range internal stress develops, which promotes the cooperative glide of dislocations and dislocation avalanches, as described in Chapter 5. Even if the Andrade law is not completely explained, it is clear that the decrease in creep rate with increasing time is related to the long-range interaction between dislocations. The hardening associated with primary creep is essentially kinematic or directional: it is represented by a tensor and it is responsible for the large deformation recovered after unloading (Duval, 1978; Meyssonnier and Goubert, 1994). The short-range dislocation interactions producing isotropic hardening appear to be of less importance.

Effect of grain size Transient creep is sensitive to variations in grain size (Duval, 1973; Duval and Le Gac, 1980). Figure 6.5 shows the variation of ε_p with

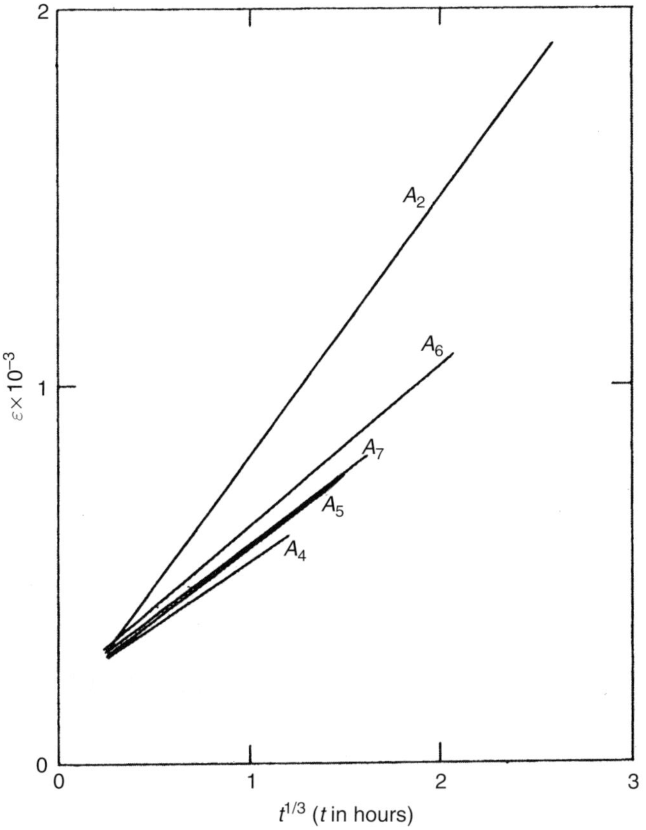

Figure 6.5. Transient creep plotted against $t^{1/3}$ for isotropic granular ice deformed in uniaxial compression at −7.2 °C with different mean grain size. A_6: 7.0 mm; A_2: 5.0 mm; A_7: 1.9 mm; A_5: 1.4 mm; A_4: 1.1 mm; the applied compression stress was 0.5 MPa. (From Duval and Le Gac, 1980; with permission of the International Glaciological Society.)

$t^{1/3}$ for grain size varying between 1.1 and 7.0 mm. The Andrade creep appears to increase with grain size. This grain size dependence could be associated with the Hall–Petch law and was analyzed in terms of the correlated motion of interacting dislocations by Louchet *et al.* (2006). The incompatibility stress or the back stress, which increases during transient creep and which is associated with the development of long-range internal stress field, would increase with decreasing grain size. This assertion is supported by results giving the variation of the distribution of dislocation avalanche size with grain size.

From Richeton *et al.* (2005), a cutoff of the distribution of avalanche size is observed at the largest amplitudes of the acoustic emission and this cutoff is linked to the average grain size: the smaller the grain size, the smaller the cutoff amplitude (Fig. 6.6). The duration of dislocation avalanches decreases with hardening and decreasing grain size. The damping of avalanches is therefore dependent on grain size. Grain boundaries appear to act as barriers to the propagation of avalanches, in keeping with X-ray topographical observations (Liu *et al.*, 1995), but transmit internal stresses.

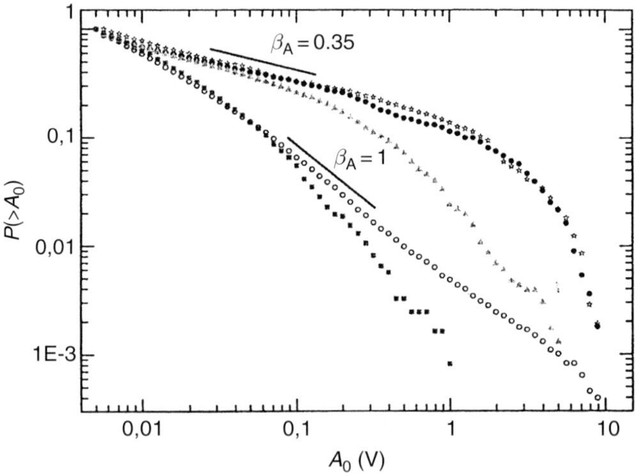

Figure 6.6. Distribution of dislocation avalanche size in polycrystals with different mean grain size. *Squares*: $d = 0.26$ mm, applied compression stress $\sigma_{zz} = 0.67$ MPa, $T = -10$ °C; *triangles*: $d = 1.10$ mm, $\sigma_{zz} = 0.57$ MPa, $T = -10$ °C; *closed circles*: $d = 1.92$ mm, $\sigma_{zz} = 0.54$ MPa, $T = -10$ °C; *stars*: $d = 1.81$ mm, $\sigma_{zz} = 1.41$ MPa, $T = -3$ °C. These distributions are compared with a typical cumulative probability distribution showing no cutoff and an exponent $\beta = 1$ (open circles). The cumulative probability distributions for acoustic emission amplitudes can be approximated by the relation $P(>A_0) \approx A_0^{-\beta} \exp(-A_0/A_c)$ where A_c is the cutoff amplitude. (From Richeton *et al.*, 2005.)

Secondary creep

There is general agreement on the form of the constitutive law for the minimum creep rate. It is well described by Glen's law (Glen, 1955) and is known in the engineering literature as Norton's law. For multi-axial stress states, the strain rate $\dot{\varepsilon}_{ij}$ is related to the deviatoric stresses σ'_{ij} by:

$$\dot{\varepsilon}_{ij} = \frac{B}{2}\tau^{n-1}\sigma'_{ij} \tag{6.3}$$

where B and n are creep constants and τ the effective shear stress defined by $\tau^2 = 1/2\sum_{i,j}(\sigma'_{ij})^2$. Data for axial stresses in the range of 0.2 to 2 MPa suggest a stress exponent $n=3$ (Barnes *et al.*, 1971; Ramseier, 1972; Budd and Jacka, 1989). Equation (6.3) was verified with tests performed in shear and compression (Duval, 1976; Budd and Jacka, 1989; Li Jun *et al.*, 1996). It is worth noting that the value of $n=4$ given by Kirby *et al.* (1987) and Durham *et al.* (1997) was deduced not for secondary creep but from steady-state creep that was generally reached for a strain higher than 10%, from experiments on granular ice with initial grain size 0.6–1.0 mm and for temperatures between 195 and 240 K. Thus, this higher-n regime does not correspond to secondary creep, but probably to tertiary creep.

Behavior at low stresses The behavior of ice at low stresses (<0.2 MPa) is the subject of extensive studies over several decades. This interest is mainly related to the flow of ice sheets for which deviatoric stresses are generally lower than 0.1 MPa and to the deformation of ice in planetary conditions (Durham *et al.*, 2001). The difficulty of obtaining reliable data at strain rates lower than $10^{-10}\,\mathrm{s}^{-1}$ is at the origin of contradictory results. Laboratory experiments at low stresses and low temperatures can require several years to reach secondary creep, making questionable several sets of data obtained from short-time experiments. A clear indication of the decrease of the stress exponent for deviatoric stress below 0.1 MPa is found from both the analysis of field data and laboratory tests (Mellor and Testa, 1969; Barnes *et al.*, 1971; Dahl-Jensen and Gundestrup, 1987; Pimienta and Duval, 1987; Lipenkov *et al.*, 1997). Figure 6.7 shows this transition between a behavior with a stress exponent of about 3 at *high stresses* and a behavior with $n \leq 2$ at *low stresses*. This figure also shows the large difference between the behavior of single crystals oriented for basal slip and the polycrystal.

A low-stress regime with $n=1.8$ was obtained on finely grained ice by Goldsby and Kohlstedt (1997, 2001). Figure 6.8 shows results obtained from compression creep tests at 236 K and grain size smaller than 100 μm. These finely grained ices, assumed to be isotropic, were prepared by pressure-sintering ice powders. The most finely grained ices were obtained by pressurizing samples into the ice II stability

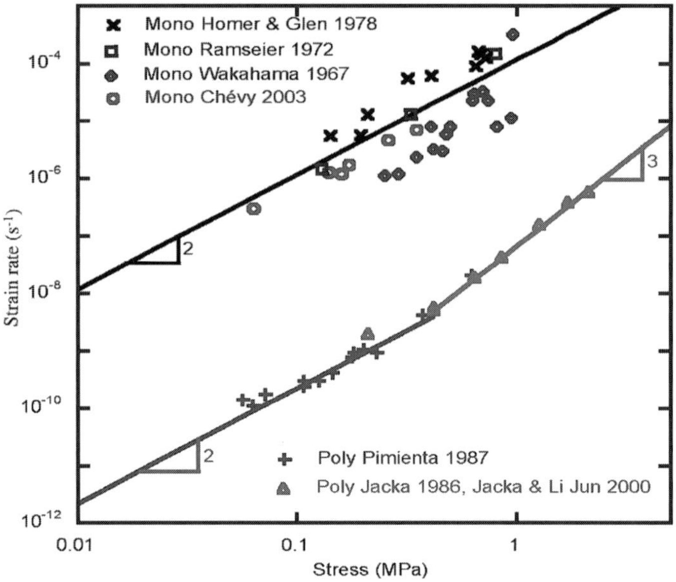

Figure 6.7. Equivalent secondary strain rate $\dot{\varepsilon}_{eq}$ plotted against equivalent stress σ_{eq} for single crystals deformed by basal slip and for isotropic polycrystalline ice. Strain rates were derived at a strain between 0.04 and 0.05. Data from Wakahama (1967) were obtained at a strain of about 2%. $\dot{\varepsilon}^2_{eq} = (2/3)\dot{\varepsilon}_{ij}\dot{\varepsilon}_{ij}$ and $\sigma^2_{eq} = (3/2)\sigma'_{ij}\sigma'_{ij}$. (From Montagnat and Duval, 2004 and references therein.)

field ($\geq$200 MPa) and then quenching under pressure back into the ice I stability field, causing homogeneous nucleation to produce grains with grain size 2–5 μm.

The strain rate in the regime with $n = 1.8$ is clearly dependent upon grain size with a grain size exponent of about 1.4. The activation energy is about 49 kJ mol^{-1}. Good agreement with the data cited above (Fig. 6.7) was obtained, but at the expense of extrapolations for grain size (from about 50 μm to about 1 mm) and temperature (from 236 K to about 260 K) (Goldsby and Kohlstedt, 2001).

A Newtonian viscosity $\eta = \sigma'_{ij}/2\dot{\varepsilon}_{ij}$ was assumed for effective shear stresses less than 0.1 MPa (Langdon, 1973; Doake and Wolff, 1985; Alley, 1992). But there is no evidence of this behavior even if polynomial laws with three exponents from $n = 1$ to $n = 5$ are suggested for glacier flow modeling (Lliboutry, 1969; Colbeck and Evans, 1973; Hutter, 1983; Pettit and Waddington, 2003). A grain-size-sensitive creep regime with $n = 1$ was suggested for shear stresses lower than 75 kPa from measurements of the tilting of several boreholes in an isothermal temperate glacier (Marshall *et al.*, 2002).

Effect of temperature The variation of the material constant B (Eq. 6.3) with temperature has been studied by many authors (see review of Weertman, 1973, and Homer and Glen, 1978) in the regime with $n = 3$. The activation energy, determined between −10 °C and −48 °C and for an axial stress of 1 MPa by Barnes *et al.* (1971), is about

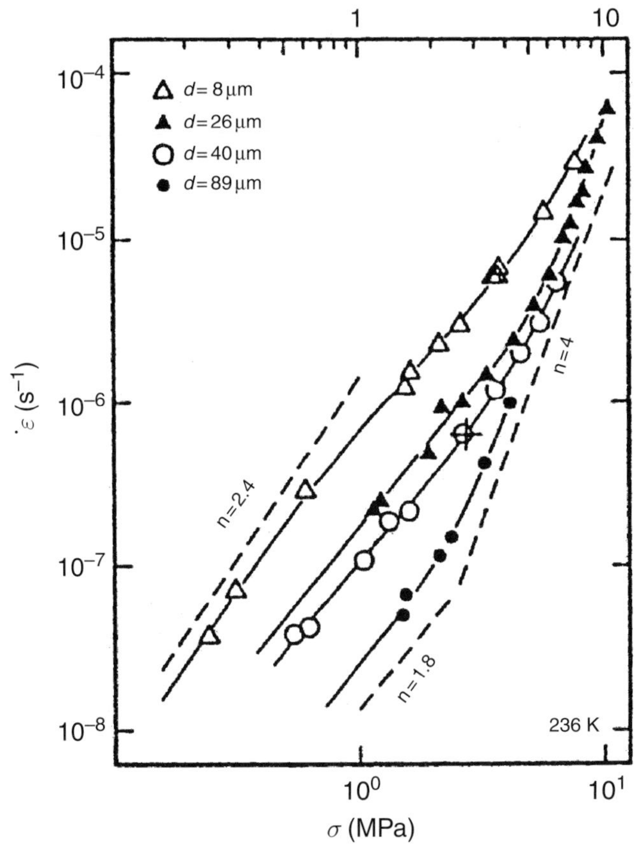

Figure 6.8. Steady-state strain rate as a function of the applied compression stress. The symbol + is the point for the sample with $d = 40$ μm deformed to $\varepsilon \approx 0.80$. (From Goldsby and Kohlstedt, 1997.)

78 kJ mol^{-1}. This value is in agreement with several studies (Hooke, 1981, 1998). At −10 °C, depending on several parameters (grain size, density, impurity content, etc.), B is typically between 5×10^{-7} and 2×10^{-6} (MPa)$^{-3}$ s^{-1} for pure isotropic granular ice. Values outside this margin for pure glacier ice are questionable.

Above about −10 °C, the creep rate becomes more temperature dependent and only an apparent activation energy can be determined (Glen, 1955; Mellor and Testa, 1969; Barnes *et al.*, 1971). For example, a value of 120 kJ mol^{-1} was found by Barnes *et al.* between −8 °C and −2 °C. Higher values are found when approaching the ice melting point (Morgan, 1991). A value of B of about 4×10^{-5} (MPa)$^{-3}$ s^{-1} was obtained at the temperature of 0.01 °C by Morgan (1991). Measurements made at the melting point can be used to determine the variation of the creep rate with the amount of liquid present within the ice sample. From Duval (1977), the creep rate increases by a factor of four when the water content rises from less than 0.1% to 1%.

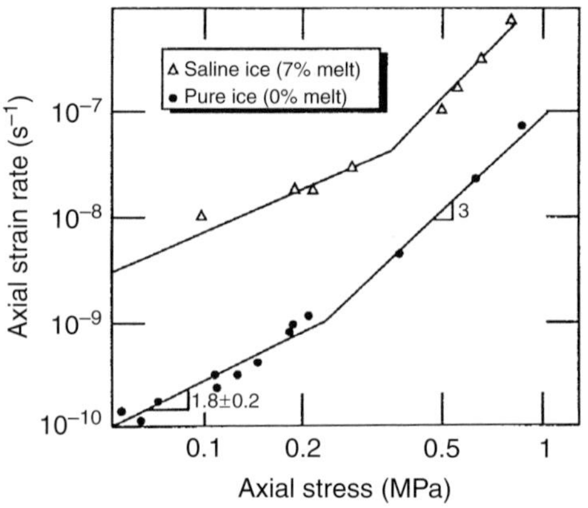

Figure 6.9. Plots of creep rate versus equivalent stress for pure and saline polycrystalline ice at −13 °C. The creep rate was determined for a strain of about 1%. (From de La Chapelle *et al.*, 1999.)

Creep data on melt-free pure ice and on saline ice with 7% melt are shown in Figure 6.9. Results obtained over a range of axial stresses between 0.1 and 0.7 MPa show that secondary creep rates are enhanced by more than one order of magnitude by the presence of the liquid phase (de La Chapelle *et al.*, 1999), in keeping with the observation by Cole (1995) that sea ice of ~3% brine content has a creep rate an order-of-magnitude greater than fresh-water ice of the same grain size and shape. The increase of the activation energy above −10 °C is not found in ice single crystals (Jones and Brunet, 1978). This shows that processes at grain boundaries must be invoked to explain these results.

Effect of grain size Grain size in glaciers and ice sheets is typically in the range of 1 to 10 mm (Chapter 3). Considering that several deformation mechanisms imply a grain-size effect, flow laws incorporating a grain-size dependence were suggested (Cuffey *et al.*, 2000a, 2000b; Peltier *et al.*, 2000; Goldsby and Kohlstedt, 2001; Durham and Stern, 2001). In the stress range with $n=3$ and providing that the sample size is more than about 12 times greater than grain size, secondary creep does not appear to depend on grain size (Duval and Le Gac, 1980; Jones and Chew, 1981, 1983a; Jacka, 1984b, 1994). At *low stresses*, the minimum creep rate seems to increase with decreasing grain size (Duval, 1973). This result is in agreement with *in-situ* deformation measurements obtained by Cuffey *et al.* (2000a,b) and with physical models of deformation. A significant grain size effect was found by Goldsby and Kohlstedt (1997, 2001) for the regime with $n=1.8$ (Fig. 6.8). For grain size between 10 and 100 μm and a compressive stress equal to 1.4 MPa, the grain size exponent appears to be $p=1.4$ (with $\dot{\varepsilon} \propto d_g^{-p}$). Superplastic flow for which grain boundary sliding is an important deformation process was assumed to explain these results.

Effect of particles and impurities The issue of the mechanical behavior of ice containing particles and soluble impurities is highly significant for glacier dynamics and for the study of ice of the outer Solar System (Durham *et al.*, 1992; Mangold *et al.*, 2002). The effect of particles on the creep behavior is dependent on both the concentration and the size. Insoluble atmospheric impurities in polar ice sheets typically consist of particles in the size range 0.1 to 5 μm with a concentration between 10 and 5000 ng per gram of ice. For these "low" impurity concentrations, the effect on creep is difficult to assess. Indeed, glacial ice, which contains many more particles than interglacial ice (Paterson, 1991), also has a higher concentration of soluble impurities (Legrand and Mayewski, 1997). On the other hand, crystal size and the orientation of crystals are also well correlated with climate (Gow and Williamson, 1976). From Paterson (1991), soluble impurities at the concentrations normally found in glacial ice, less than 200 ppb for chloride, the major impurity with sulfate (Wolff *et al.*, 2006), have no direct effect on the deformation rate. From Fischer and Koerner (1986), there is a clear indication that particles increase creep rate. For these low impurity contents, it is difficult to separate the effects of particles, soluble impurities, crystal size and grain orientation. Thus, from Durand *et al.* (2006), particles exert a retarding force on moving grain boundaries and have a profound effect on grain growth in polar ice sheets. Variations in grain size between glacial ice and interglacial ice, caused by the pinning of grain boundaries by particles, could be at the origin of the difference in viscosity between these ices.

At the base of glaciers and ice sheets, debris-laden ice can contain more than 70% particles by weight with particle sizes between 1 μm and 1 mm (Echelmeyer and Zhongxiang, 1987). The effect of a high concentration of fine particles on the creep behavior is well documented. The creep rate generally decreases with increasing volume fraction of particles (Hooke *et al.*, 1972; Durham *et al.*, 1992; Mangold *et al.*, 2002). For example, from Hooke *et al.* (1972), the secondary creep rate of ice samples with 0.35 volume fraction fine sand tested in uniaxial compression at −9.1 °C was about a factor of 10 lower than that determined with clean ice. Strengthening by dispersed *fine* particles can be explained by the interaction between dislocations and particles (Ashby, 1966). For a high volume fraction of relatively *large* particles, results can be interpreted in the context of the effect of hard phases in composite materials (Durham *et al.*, 1992).

Several studies indicate that particles with a size between 1 and 100 μm and a concentration higher than 0.5 by weight increase creep rate as soon as temperature is above −20 °C (Baker and Gerberich, 1979; Cuffey *et al.*, 1999; Song *et al.*, 2005a, b). From deformation measurements at the base of a sub-polar glacier, the effective viscosity of debris-laden ice at −4 °C, containing 21–39% ice by weight, is lower, by a factor of about 100, than that of adjacent clean ice (Echelmeyer and Zhongxiang, 1987). The softening of ice above −10 °C by such a high concentration of sediment particles was found in several glaciers (Lawson, 1996; Fitzsimons *et al.*, 1999; Cohen, 2000). From these authors, the presence of a liquid-like layer surrounding solid particles is probably at the origin of the reduction of the viscosity. It is worth noting that the apparent softening of "amber" ice deduced from velocity profiles at the base of the Suess and Rhone glaciers (Dry Valleys, Antarctica) is caused not only by particles, but also by the strong concentration of *c*-axes along the vertical direction (Samyn *et al.*, 2005). The enhanced deformation of ice containing particles could be also related to the small size of ice grains when ice is deformed at low stresses (Cuffey *et al.*, 2000a).

In conclusion, the effect of particles is dependent on their size, concentration and temperature. Ice softening should be related to the presence of a liquid film surrounding particles. Strengthening of ice containing fine dispersed particles is related to the interaction between dislocations and particles. The hardening observed with a high particle concentration should be analyzed by considering the effect of hard phases in composite materials (Durham *et al.*, 1992).

Effect of confining pressure The effect of hydrostatic pressure P on the mechanical behavior of ice deserves some interest when considering the deformation in ice sheets where the ice pressure can be higher than 30 MPa in East Antarctica. The creep of ice Ih at planetary conditions involves hydrostatic pressures that can reach 200 MPa, which is near the transition between ice I, ice II and ice III (Chapter 8). High confining pressures are also found during ice–structure interaction (Chapter 14). Pressures can be as high as 50 MPa in the central zone of the damaged ice (Jordaan, 2001).

In spite of the paucity of data, the behavior of ice under confining pressure can be described in a way that can be easily understood. The secondary creep rate is found to slightly decrease as the hydrostatic pressure increases at lower levels and then to increase at higher levels of pressure (Jones and Chew, 1983b; Barrette and Jordaan, 2003). Taking into account the hydrostatic pressure, the minimum creep rate can be given by:

$$\dot{\varepsilon}_{\min} = A\sigma^n \exp[-(E + PV)/RT] \tag{6.4}$$

where E is the activation energy and V the activation volume. In uniaxial compression, σ corresponds to the equivalent stress σ_{eq} defined by $\sigma_{\text{eq}}^2 = (3/2)\sigma'_{ij}\sigma'_{ij}$ and P the confining pressure. To conform to this behavior, V should be positive at low pressures and negative at high pressures. This means that several processes are controlling the deformation. At high pressure, the variation of the melting temperature with pressure and some melting at grain boundaries explain the increase of creep rate with pressure (Jones and Chew, 1983b). From Kirby *et al.* (1987), this pressure effect is consistent with a negative activation volume for creep of $-13 \times 10^{-6}\,\text{m}^3\,\text{mol}^{-1}$. A value higher than $50 \times 10^{-6}\,\text{m}^3\,\text{mol}^{-1}$ is given by

Jones and Chew (1983b) at −10 °C. The positive value of V at low pressure should be related to the activation volume of the rate-controlling process. But data at low pressure are too few to permit an estimation of this volume.

Recovery creep

When unloaded, ice exhibits significant creep recovery or delayed elasticity, probably in connection with kinematic hardening. The buildup of directional internal stresses during creep is at the origin of the reverse motion of dislocations and explains the effect (Duval, 1978; Cole, 1995, 2004; Richeton *et al.*, 2005). The large deformation recovered after unloading cannot be explained by grain boundary sliding (Duval, 1978; Cole, 1991). Indeed, the recoverable deformation associated with grain boundary sliding is only of the order of the elastic strain deduced from the Young's modulus (Raj and Ashby (1971). From Cole (2004), the creep recovery strain in granular fresh-water ice is proportional to σ^2, where σ is the applied stress before unloading.

Tertiary creep

After the minimum creep rate corresponding to secondary creep, strain rate increases into tertiary creep. At high temperature (above about −15 °C) and at relatively low stresses to avoid cracking, a quasi-steady state is reached at a strain slightly above 0.1 (Fig. 6.10). The increase of strain rate during tertiary creep is explained by softening processes directly associated with recrystallization processes and by the formation of "recrystallization textures" (Duval, 1981; Alley, 1992). The finite plastic strain is far too small to produce any important rotation of the crystal lattice and thus the observed textures form by recrystallization (de La Chapelle *et al.*, 1998).

Tertiary creep is associated with the rapid migration of grain boundaries between dislocation-free nuclei and deformed grains, inducing a new orientation of grains. This recrystallization regime is termed *migration recrystallization* or *discontinuous*

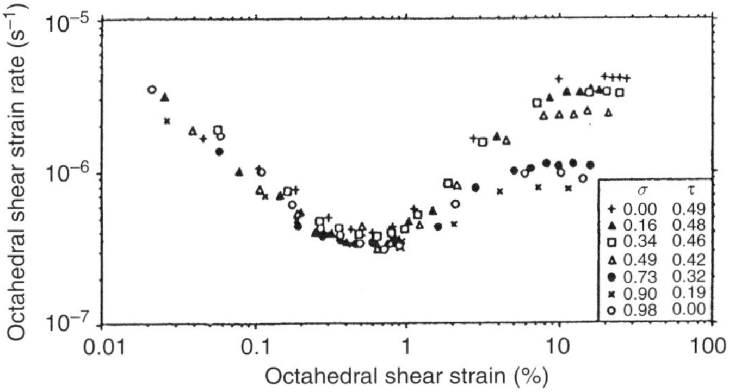

Figure 6.10. Strain rate plotted against strain for various combinations of compression and shear at −2.0 °C. (From Li Jun *et al.*, 1996; with permission of the International Glaciological Society.)

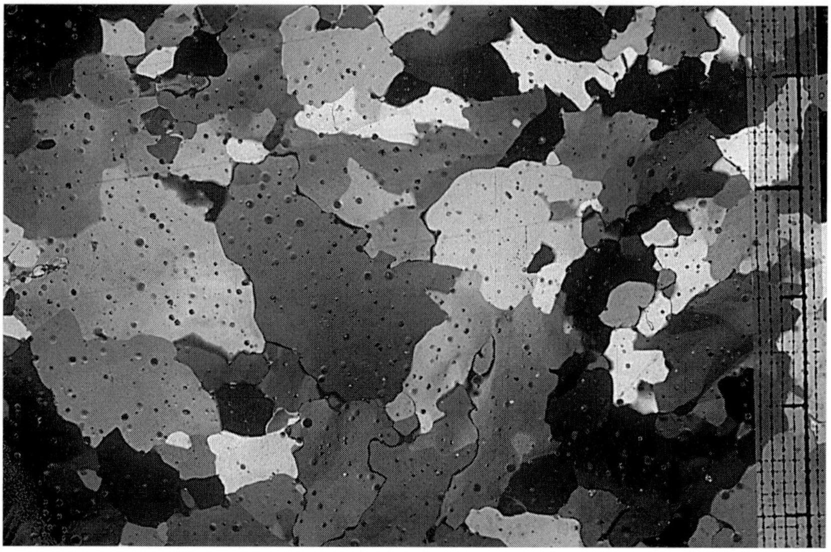

Figure 6.11. Thin section of ice photographed between crossed polaroids from Terre Adélie (East Antarctica) at the depth of 70 m. Bubbles can be seen within grains. Finest division on scale is 1 mm. (From Duval and Castelnau, 1995.)

recrystallization. Grain boundary migration rates are typically between 10^{-12} m^2 s^{-1} and 10^{-10} m^2 s^{-1} (Duval and Castelnau, 1995). This relatively high velocity of grain boundaries tends to produce an interlocking grain structure (Fig. 6.11). Migration recrystallization occurs extensively in temperate glaciers and in polar ice sheets near the bottom where a temperature above −15 °C is found. But the stored energy associated with dislocations must be high enough to induce this recrystallization regime. Thus, there is no evidence of migration recrystallization along the Vostok ice core, even though the temperature reaches the melting point in the deepest layers of the ice sheet (de La Chapelle *et al.*, 1998). Tertiary creep rate is about three times the secondary creep rate in compression and about eight times in simple shear (Budd and Jacka, 1989). The number of data is too small to give, for a given state of stress, a value of the stress exponent. But a value of the stress exponent of about 4 was deduced from tests performed in compression at −11.5 °C by Steinemann (1958). This value of $n = 4.0 \pm 0.6$ was also obtained by Kirby *et al.* (1987) from tests performed in triaxial compression at constant displacement rates in the temperature range 240–258 K and under a confining pressure of 50 MPa. Steady-state strain rates, between 3.5×10^{-7} and 3.5×10^{-4} s^{-1}, were determined after a finite strain generally higher than 10%. These results unambiguously correspond to tertiary creep.

In spite of the lack of reliable data on tertiary creep, there is, however, a clear indication that steady state is associated with an equilibrium grain size (Jacka and Li Jun, 1994). The grain size d appears to be related to the equivalent stress via

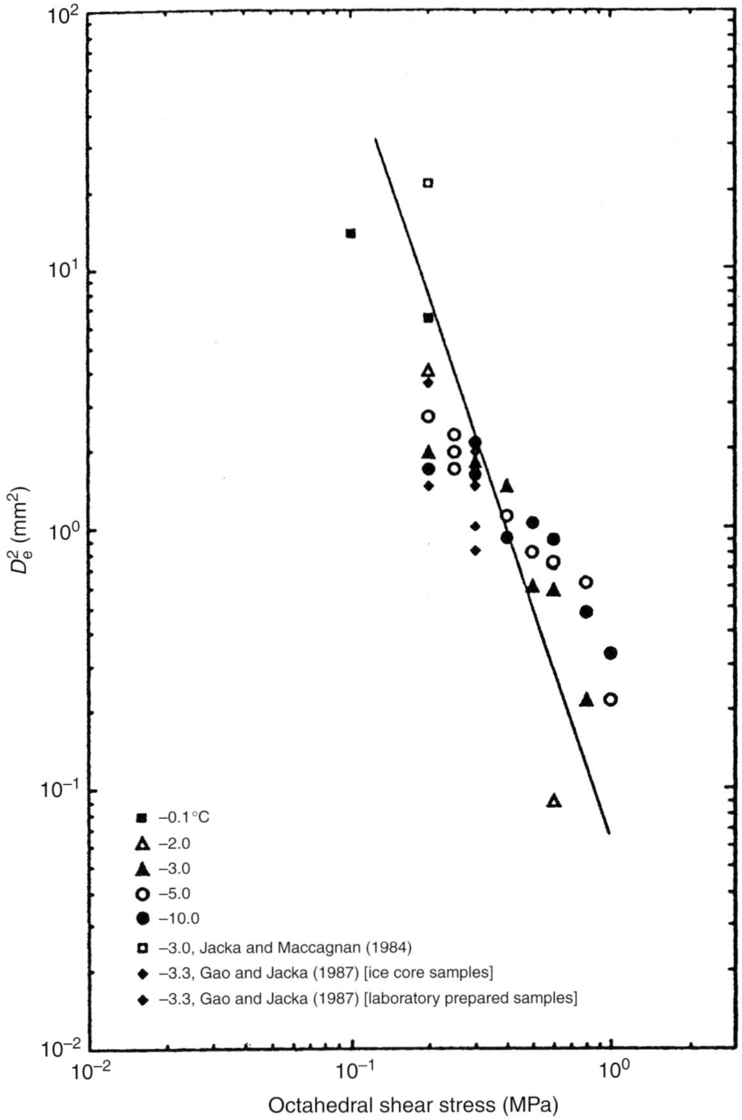

Figure 6.12. Relation between recrystallized grain size and stress from creep tests performed at various temperatures. (From Jacka and Li Jun, 1994, and references therein; with permission of the International Glaciological Society.)

the relationship $d^2 \propto \tau^{-3}$, without any trend with temperature between −10 °C and 0.1 °C (Fig. 6.12). This grain size–stress relation is also found in several other materials (Derby, 1991). A dynamical balance between grain nucleation and grain boundary migration rate can explain this universal relationship. But it must still be considered as empirical in spite of a good physical basis (Poirier, 1985).

6.2.2 Deformation at constant strain rate

Tests at constant strain rate are used in ice engineering in connection with short-term transient loading to failure. Strain rates are generally higher than $10^{-7}\,s^{-1}$ and the deformation can be associated with the formation of cracks. Figure 6.13 shows a typical stress–strain curve obtained below the ductile-to-brittle transition (see Chapter 13) at a strain rate of $10^{-7}\,s^{-1}$. The secant modulus determined from the slope of the curve $\sigma(\varepsilon)$ at the origin is lower than the true Young's modulus at low strain rates since the material is plastic from the beginning (as discussed in Chapter 4). A slight decrease of stress is observed after the peak, which is attained at a strain of about 0.4%. This softening is associated with dynamic recrystallization.

From Mellor and Cole (1982), the peak stress is observed in a strain range between 0.2 and 1.5%. Comparison of results for tests carried out at applied load and constant strain rate suggests that the constant stress test (creep) and the constant strain-rate test (strength) give much the same information when the analysis is restricted to the ductile behavior (Mellor and Cole, 1982). But cracking, which cannot be completely excluded in such tests, limits the comparison to a limited range of strain rates. From Drouin and Michel (1971), the value of the activation energy determined from the peak stress is about 80 kJ/mole for strain rates between 10^{-8} and $10^{-7}\,s^{-1}$ and temperatures between −30 and −5 °C. This value is near that determined for secondary creep (see Section 6.2.1).

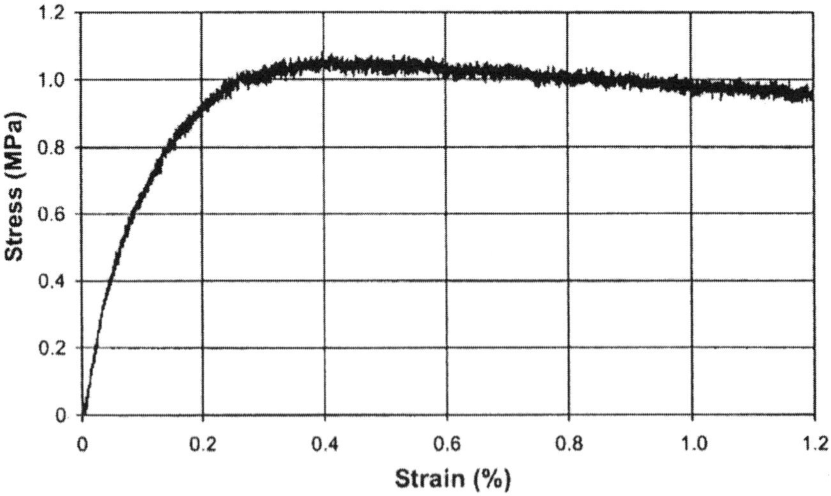

Figure 6.13. Typical stress–strain curve for a sample tested at −10 °C and $10^{-7}\,s^{-1}$. (From Jones *et al.*, 2003; with permission of NRC Research Press.)

6.3 Deformation of columnar ice

The analysis is limited to S2 ice with c-axes perpendicular to the direction of the columnar-shaped grains (see Chapter 3 for the formation and the description of columnar ices). These ices can be considered as transversely isotropic. To reproduce conditions in ice covers, laboratory mechanical tests are generally performed in uniaxial compression with the load applied in the direction normal to the columns. From Gold (1972), crack formation occurs in type S2 ice during compressive creep when the applied stress exceeds 0.6 MPa. This value is significantly lower than the value found for granular ice in uniaxial compression (Kalifa, 1988). Results on the ductile behavior are therefore limited to a narrow range of stresses and few correct data are available.

The creep curve resembles that for granular ice with primary and secondary creep and with tertiary creep associated with dynamic recrystallization. Primary creep was assumed by Sinha (1978) to be associated with grain boundary sliding with $\varepsilon_p \propto 1/d$. However, from Plé and Meyssonnier (1997), there is no clear indication of a grain size effect on either primary or secondary creep in the regime where the stress exponent is equal to 3. The value of the creep constant B (Eq. 6.3) at $-10\,°\mathrm{C}$ is about $2 \times 10^{-6}\,(\mathrm{MPa})^{-3}\,\mathrm{s}^{-1}$, i.e., near the higher bound for isotropic granular ice. Results obtained by Drouin and Michel (1971) on granular and S2 columnar ice with tests performed at a strain rate of about $5 \times 10^{-8}\,\mathrm{s}^{-1}$ show that the viscosity of granular ice is slightly higher than that of columnar ice.

Ductile failure of columnar S2 ice We consider the ductile behavior of columnar saline ice with the objective of giving information on the failure envelope and the failure surface when ice is loaded under biaxial and triaxial compression. This analysis is specially centered on saline ice to understand the interaction between sea ice and a wide engineered structure. Depending on confinement, the ductile-to-brittle transition strain rate of salt-water ice at $-10\,°\mathrm{C}$ is typically between 5×10^{-4} and $5 \times 10^{-3}\,\mathrm{s}^{-1}$ (Chapter 13).

Given that S2 ice is an orthotropic material, experimental data on the strength of saline ice are analyzed by reference to the criterion of Hill (1950) for the plastic yielding of orthotropic materials. The criterion is given by:

$$F(\sigma_{22} - \sigma_{33})^2 + G(\sigma_{33} - \sigma_{11})^2 + H(\sigma_{11} - \sigma_{22})^2 = 1 \tag{6.5}$$

with

$$F = G = \frac{1}{2(\sigma^{\mathrm{d}}_{\mathrm{u},3})^2}, \qquad H = \frac{1}{(\sigma^{\mathrm{d}}_{\mathrm{u},1})^2} - \frac{1}{2(\sigma^{\mathrm{d}}_{\mathrm{u},3})^2}$$

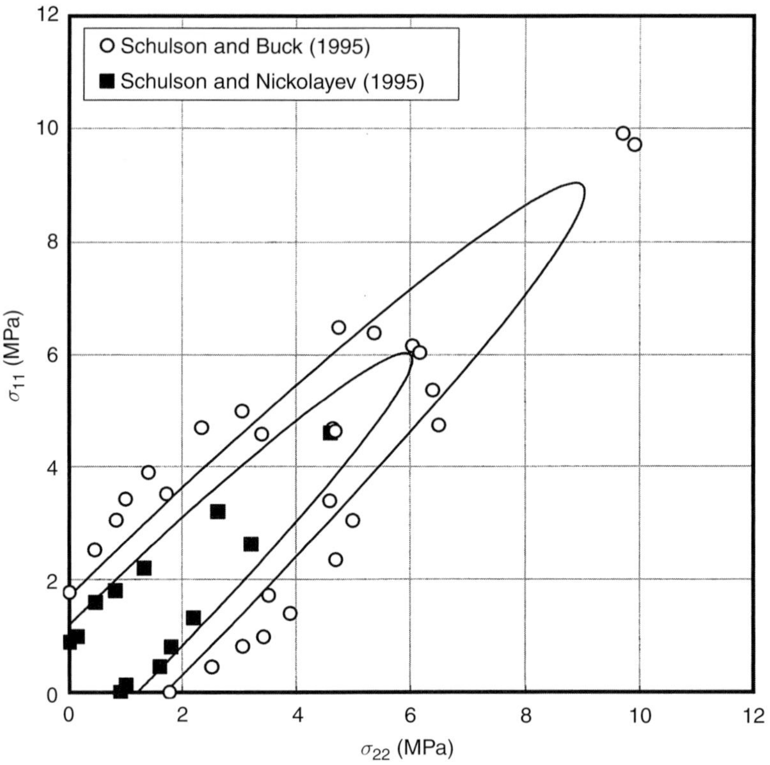

Figure 6.14. Ductile compressive failure envelopes for both fresh-water (open points) and salt-water (4.3 ± 0.7 ppt melt-water salinity, closed points) columnar-grained S2 ice proportionally loaded biaxially across the columns at $-10\,°C$ and $10^{-5}\,s^{-1}$. The points denote experimental measurements. The yield loci were calculated using Hill's criterion for plastically orthotropic materials. (Re-plotted using data from Schulson and Nickolayev, 1995, and Schulson and Buck, 1995.)

where $\sigma^{d}_{u,1}$ and $\sigma^{d}_{u,3}$ are the unconfined across-column and along-column compressive yield stresses, respectively (Melton and Schulson, 1998).

Figure 6.14 shows the failure envelopes for both fresh-water and salt-water S2 ice proportionally loaded across the columns at $-10\,°C$ at $\dot{\varepsilon}_{11} = 10^{-5}\,s^{-1}$. The data were obtained from experimental measurements and the envelopes were computed from Hill's criterion (Schulson and Nickolayev, 1995; Schulson and Buck, 1995). The agreement between measurements and calculation is good. Experimental results obtained over a range of temperature ($-40\,°C$ to $-5\,°C$) and a range of strain rate ($10^{-6}\,s^{-1}$ to $10^{-4}\,s^{-1}$) are also in keeping with Hill's criterion. The agreement is compatible with dislocation-induced deformation. Hill's criterion appears to describe the behavior of S2 ice under low confinement even with the presence of stable micro-cracks at these relatively high strain rates within the ductile regime.

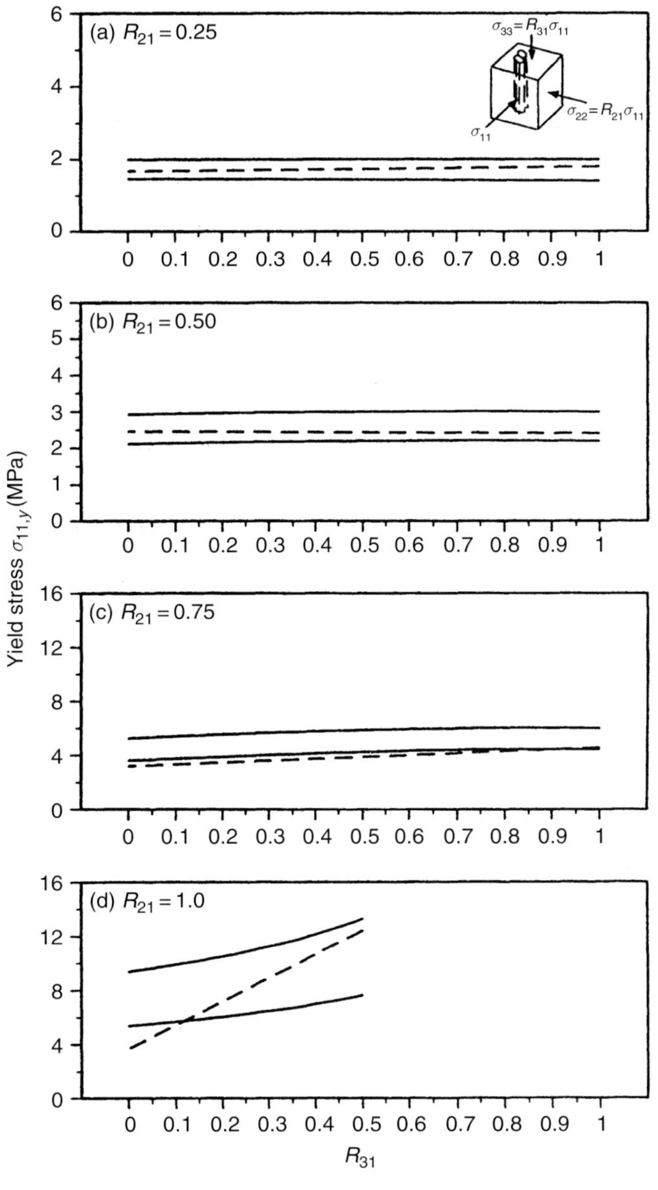

Figure 6.15. Variation of the measured yield stress (dashed curves) with the upper and lower bounds estimates (solid curves) from Hill criterion versus $R_{31} = \sigma_{33}/\sigma_{11}$ for different values of the ratio $R_{21} = \sigma_{22}/\sigma_{11}$. (From Melton and Schulson, 1998.)

Figure 6.15 compares the measured triaxial yield stress at −10 °C at $4 \times 10^{-5}\ \mathrm{s}^{-1}$ with the yield stresses calculated from Hill's criterion. The criterion appears to bracket the data at $R_{21} = \sigma_{22}/\sigma_{11} = 0.25$ but to underestimate slightly the yield stress at higher values of across-column confinement. The triaxial yield stress is not affected by the along-column stress σ_{33} because this stress component has almost no shear component on the basal plane. From Melton and Schulson (1998), the triaxial yield stress is expected to increase with the along-column stress when the across-column confinement approaches unity. The data are too few, however, to say that S2 ice, loaded under triaxial compression, obeys Hill's criterion over wide ranges of strain rate, temperature and salinity, although we expect it does.

6.4 Rate-controlling processes in the creep of polycrystalline ice

6.4.1 Secondary creep with n = *3*

For deviatoric stresses roughly between 0.1 and 1 MPa, the viscous deformation of glacier ice is essentially produced by intracrystalline dislocation creep. In analogy with metals deformed at high temperature, the steady-state creep rate $\dot{\varepsilon}$ of polycrystalline ice deformed in either uniaxial tension or compression can be expressed through a relationship of the form (Frost and Ashby, 1982):

$$\dot{\varepsilon} = A\frac{D_{\mathrm{v}}Gb}{kT}\left(\frac{\sigma}{G}\right)^{n} \tag{6.6}$$

where D_{v} is the volume diffusion coefficient, G the shear modulus, b the Burgers' vector magnitude, k the Boltzmann constant, T the temperature, n the stress exponent of about 3 for secondary creep (Budd and Jacka, 1989) and σ is the applied stress. A is a dimensionless constant. As discussed above, no grain size effect is expected for power-law secondary creep.

Power-law creep or Glen's law with a stress exponent of about 3 characterizes secondary creep (Barnes *et al.*, 1971; Weertman, 1973; Budd and Jacka, 1989). This *natural* creep law with $n = 3$ can be derived by assuming that a density $(\sigma/Gb)^2$ of dislocations moves at the velocity of climb, $D\Omega\sigma/bkT$ (Poirier, 1985). This velocity calculated at $-10\,°\mathrm{C}$ with $D = 1.5 \times 10^{-15}\,\mathrm{m}^2\,\mathrm{s}^{-1}$ (Petrenko and Whitworth, 1999) and 0.1 MPa is about $10^{-9}\,\mathrm{m\,s}^{-1}$. As noted in Chapter 5, this climb velocity is more than two orders of magnitude lower than that for the glide of basal dislocations (Fig. 5.9). However, this is in qualitative agreement with the large difference between basal slip and the behavior of the isotropic polycrystal (Fig. 6.7). Then, basal slip could be accommodated by the climb of basal dislocations on prismatic planes. Shear on the basal plane generates dislocations, which then climb on planes normal to the basal plane.

To satisfy stress equilibrium and strain compatibility between grains, deformation must occur on a number of independent slip or climb systems. In the framework of the uniform-strain approximation, the minimum of independent systems is five and the polycrystal strength is at its upper bound (Hutchinson, 1977). Calculations based on the self-consistent approach, which does not presuppose uniform strain, show that extensive plasticity is possible with only four

Table 6.1. *Slip systems for the ice crystal*

	Slip systems	Number of slip systems	Resulting deformation	RRSS
Basal	$\{0001\}\langle 11\bar{2}0\rangle$	3	$\varepsilon_{13}, \varepsilon_{23}$	τ_a
Prismatic	$\{01\bar{1}0\}\langle 11\bar{2}0\rangle$	3	ε_{11} or $\varepsilon_{22}, \varepsilon_{12}$	τ_b
Pyramidal	$\{11\bar{2}2\}\langle 11\bar{2}\bar{3}\rangle$	6	$\varepsilon_{11}, \varepsilon_{22}, \varepsilon_{33}, \varepsilon_{23}, \varepsilon_{13}, \varepsilon_{12}$	τ_c

τ_a, τ_b, τ_c are the reference resolved shear stresses (RRSS) for the basal, prismatic and pyramidal slip systems, respectively.
Strains are given in a Cartesian coordinate set X_1, X_2, X_3 with X_1 and X_2 in the basal plane and X_3 normal to it.

independent systems (Hutchinson, 1977). Basal slip provides three slip systems, but only two are independent. The slip of basal dislocations on prismatic planes $\{01\bar{1}0\}\langle 11\bar{2}0\rangle$ gives two more independent systems. Shear on the pyramidal system $\{11\bar{2}2\}\langle 11\bar{2}\bar{3}\rangle$ gives a fifth independent system. These slip systems are summarized in Table 6.1. It is worth noting that the climb of basal dislocations also gives the deformation ε_{11} or ε_{22} and ε_{12}.

Cross-slip of dislocations on the prismatic or pyramidal planes induced by the internal stress field of basal dislocations cannot be excluded as an accommodation process of basal slip (Taupin *et al.*, 2007). But it is worth noting that cross-slip of basal dislocations on prismatic planes does not provide deformation along the *c*-axis. The cross-slip of basal screw dislocations on pyramidal planes could be invoked for this deformation. The glide of non-basal dislocations on pyramidal planes was suggested by several authors (Muguruma *et al.*, 1966; Duval *et al.*, 1983; Hondoh, 2000). However, this hypothetical deformation mode requires stresses several orders of magnitude larger than for basal slip (Fig. 5.8) and there is no experimental evidence of the presence of such non-basal dislocations in deformed ice (Hondoh, 2000). Another possibility is the accommodation of basal slip by grain boundary sliding. This deformation mode gives a stress exponent close to 1 (Langdon, 1993).

There is strong evidence, as mentioned in Section 6.2.1, that the activation energy for secondary creep is about 78 kJ mol^{-1} (Barnes *et al.*, 1971), whereas the activation energy for the self-diffusion process is about 60 kJ mol^{-1} (Petrenko and Whitworth, 1999). From Homer and Glen (1978), this difference could be explained by variation with temperature of the elastic modulus. But variation of the elastic modulus with temperature is far too small to explain this high value of the activation energy (Chapter 4). Another possibility is that the density of jogs, where the emission or the absorption of point defects takes place, is in thermal equilibrium. In this case, the activation energy for the climb velocity is higher than that corresponding to the self-diffusion since it includes the formation energy of jogs,

Q_j (Friedel, 1964). From Hondoh (2000), Q_j should be smaller than 20 kJ mol^{-1} below −10 °C.

In conclusion, a combination of deformation systems operates during the deformation of polycrystalline ice in the power-law creep regime. On first loading, the stress state is uniform. As creep relaxes the resolved shear stress on the basal plane on each grain, load is transferred to the harder non-basal systems. An increasingly non-uniform state of internal stress develops. These internal stresses oppose further deformation; that is, they give directional or kinematic hardening. They can be relieved by reversing the stress, so they are responsible for the large recoverable strain that appears when ice is unloaded (Duval, 1978). The development of this internal stress field during primary creep is associated with a creep rate decrease by a factor of 100 or more (Fig. 6.4). The initial creep rate is largely determined by that of the softest system, i.e., basal slip. The steady-state creep rate is determined by an appropriate average of the resistances of all systems. For this reason, the creep rate of polycrystals lies between the extremes of the single crystal (Fig. 5.8). The internal stress on the basal plane almost cancels the applied stress (Ashby and Duval, 1985). As a consequence, recovery processes, which reduce the level of internal stresses, probably control the near-steady creep rate. The climb and/or cross-slip of basal dislocations on non-basal planes are likely candidates for the rate-controlling processes (Ashby and Duval, 1985). A dislocation glide-controlled mechanism was suggested by Cole and Durell (2001) and Cole (2003). Following the Orowan equation, strain rate should be governed by the product of a mobile dislocation density and an average dislocation velocity of basal dislocations. Considering the importance of the long-range internal stress field induced by the large plastic anisotropy of the ice crystal, the creep of polycrystalline ice cannot be controlled by basal slip. Deformation processes in series with basal slip must be rate controlling since they are slower.

6.4.2 Secondary creep with $n \leq 2$

The study of the rheology of ice at deviatoric stresses lower than 0.1 MPa is important since these stress conditions are found in glaciers, ice sheets and in several icy planetary bodies. There is a clear indication that a creep regime with a stress exponent of about $n = 1.8$ characterizes the behavior of ice at such low stresses (see Section 6.2.1). This creep regime is in agreement with the behavior of superplastic materials subjected to a tensile stress corresponding to region II in the stress–strain rate relationship (Langdon, 1994; Mukherjee, 2002). The creep regime in region I observed at the lowest stresses in superplastic materials and giving a stress exponent higher than 2.5 should be related to the presence of a threshold stress whose significance would increase with decreasing stress (Mohamed, 2005). Grain boundary sliding appears to be an important deformation mode in superplasticity when finely grained samples are deformed under low stresses.

Other creep mechanisms, such as diffusion creep and Harper–Dorn creep, can operate in ice deformed at low stresses. Finally, the role of grain boundary migration, as a softening process, was also introduced by Montagnat and Duval (2000) to relate strain rate in the first hundred meters of polar ice sheets to a mean dislocation density and grain size. Accordingly, the following section provides a description of these creep mechanisms.

Grain boundary sliding One of the dominant creep mechanisms in superplasticity is grain boundary sliding (GBS). The grain compatibility during GBS is maintained by some accommodation processes, which may involve some slip within grains, grain rotation, and/or diffusion. An important characteristic of GBS is that grains retain essentially their original shape so that, at high strains, there is an increase in the total number of grains lying along the tensile axis (Langdon, 1993).

In the superplastic region II, associated with $n \approx 2$, GBS appears to be accommodated by slip in the grains and finally by the climb of dislocations into grain boundaries (Langdon, 1993). Strain rate is given by:

$$\dot{\varepsilon} = A' \frac{DGb}{kT} \left(\frac{b}{d_g} \right)^p \left(\frac{\sigma}{G} \right)^n \tag{6.7}$$

where d_g is the mean grain size and p is the grain size dependence of strain rate. The activation energy corresponds to that for the grain boundary diffusion process.

Results obtained by Goldsby and Kohlstedt (1997, 2001) on finely grained ice with $d_g \leq 89$ μm are in agreement with these deformation modes even if their tests were performed in compression. Indeed, superplasticity is defined by the materials science community as the ability of a material to be deformed in tension to large strains. This term was adopted by Goldsby and Kohlstedt with the objective to compare deformation mechanisms. Values of $n = 1.8$ and $p = 1.4$ were used to suggest the preponderance of GBS in this deformation regime. Extrapolation of these results to glacier ice samples with $d > 1$ mm gives results more or less in agreement with data on glacier ice.

It is worth noting that, under optimum superplastic conditions, GBS accounts for essentially all the deformation (Langdon, 1994). This result makes questionable such a contribution of GBS in the deformation of ice in glaciers and polar ice sheets since there intracrystalline glide plays a fundamental role in the formation of textures (Azuma and Higashi, 1985; Alley, 1988, 1992; Duval and Montagnat, 2002). On the other hand, as shown in Figure 6.7, basal slip is by far the easier deformation mode of the ice crystal.

Diffusion creep When a stress is applied to a polycrystalline sample, there is an excess of vacancies along those grain boundaries experiencing a tensile stress and a corresponding depletion of vacancies along grain boundaries experiencing a compressive stress. Diffusion creep refers to the stress-directed flow of vacancies that takes place in order to restore an equilibrium condition. The flow of defects may occur either through the crystalline lattice via the process known as Nabarro–Herring diffusion creep, or along the grain boundaries via the process known as Coble diffusion creep. These two deformation mechanisms occur independently, and give a Newtonian viscosity. Strain rate is given by:

$$\dot{\varepsilon} = 14\frac{\sigma\Omega}{kTd_g^2}\left\{D_v + \frac{\pi\delta D_b}{d_g}\right\} \tag{6.8}$$

where Ω is the molecular volume, D_v is the lattice diffusion coefficient, D_b that for boundary diffusion and δ is the grain boundary thickness (Frost and Ashby, 1982).

By considering only volume diffusion, the strain rate deduced from Equation (6.8) with $\sigma = 0.1$ MPa, $T = 263$ K, $d_g = 2$ mm and $D_v = 1.5 \times 10^{-15}$ m^2 s^{-1} at -10 °C is about 1.3×10^{-11} s^{-1}. For the same conditions, the strain rate deduced from Figure 6.7 is about 1.5×10^{-10} s^{-1}. Unfortunately, there is no information on the grain boundary diffusivity and the grain boundary thickness to deduce the creep rate induced by this process. But, from Goodman *et al.* (1981), the contribution to diffusion along grain boundaries is expected to be about two orders of magnitude lower than that associated with volume diffusion.

In conclusion, considering the behavior of single crystals deformed by basal slip (Fig. 6.7), diffusion creep must not be considered as an efficient deformation mode in glaciers and polar ice sheets. However, diffusion processes could be invoked to accommodate basal slip and GBS at very low stresses.

Harper–Dorn creep Harper–Dorn creep refers to the anomalous behavior in the creep of coarsely grained polycrystals. The creep rate is linearly related to the stress with no grain size dependence. Strain rates associated with Harper–Dorn creep are more than two orders of magnitude greater than those predicted by diffusion creep (Ruano *et al.*, 1991). Although a number of deformation mechanisms have been proposed for Harper–Dorn creep (Langdon and Yavari, 1982), several studies question its occurrence (Ginter *et al.*, 2001; Blum and Maier, 1999). From Nabarro (2002), there is some indication that this mechanism is not a genuine steady-state process. Strain rates are determined at very low strain, i.e., in the primary stage.

The occurrence of this deformation mode in polycrystalline ice deformed under low stresses has been discussed by several authors (Lliboutry and Duval, 1985; Alley, 1992; Pettit and Waddington, 2003; Song *et al.*, 2005a). Harper–Dorn creep with $n=1$ is expected when the density of mobile dislocations is independent of stress. Such behavior was observed by Song *et al.* (2006) with samples containing a high initial dislocation density. This can be considered as non-steady-state behavior and not an indication that the dislocation density is independent of stress. A flow law with $n=1$ was inferred for the shallow ice of polar ice sheets where normal grain growth occurs (Pimienta and Duval, 1987; Alley, 1992). The production of dislocations by deformation would be balanced by the absorption of dislocations by the moving grain boundaries. From Pettit (2006), the flow law for ice sheet flow modeling is expected to include a linear term and a non-linear term with $n=3$. The crossover stress at which the linear and non-linear terms contribute equally to strain rate would be about 0.02 MPa. It is worth emphasizing that several parameters change with depth or time (strain rate and/or stress, temperature, grain size, impurities) in ice sheets making it difficult to deduce a flow law from *in-situ* deformation measurements.

Effect of a liquid phase The mechanical response of partially molten polycrystalline ice is discussed by considering the dislocation creep regime. Results given in Figure 6.9 clearly indicate that basal slip is the main deformation process. The effect of a liquid phase must therefore be viewed in considering this constituent as an element in the accommodation process of basal slip. Following the discussion given above, the occurrence of grain boundary sliding, which would be promoted by the liquid phase (Kohlstedt and Zimmerman, 1996), cannot be invoked with a stress exponent of about 1.8 (de La Chapelle *et al.*, 1999). To explain the large enhancement in strain rate by melt, we assume that the liquid phase promotes basal glide by reducing the internal stress field. This assumption is based on the large variation of the transient strain rate during a creep test, which is induced by the development of the long-range internal stress field (Fig. 6.4).

Intracrystalline slip accommodated by grain boundary migration (GBM)

In the low stress conditions, $\sigma_{eq} < 0.1$ MPa, strain rates in polycrystalline ice are, as at high stresses, significantly lower than those for single crystals deforming by basal slip (Fig. 6.7). Hence, basal slip can be considered as the dominant deformation mode. The activity of basal slip systems, given by self-consistent approaches (Chapter 7), which do not presuppose uniform stress and strain rate within the polycrystal, is about 90% of the activity of all slip systems (Castelnau *et al.*, 1996a). Other slip systems or deformation modes are required to satisfy stress equilibrium and the compatibility of deformation (Hutchinson, 1976). The importance of other deformation modes can be reduced by recovery processes that accommodate the mismatch of slip at grain boundaries and hence reduce the internal stress field induced by the strong plastic anisotropy of the ice crystal. An efficient recovery or softening mechanism in glaciers and ice sheets is grain boundary migration (GBM) associated with grain growth and dynamic recrystallization (Montagnat and Duval, 2000). During *grain growth*, dislocations located in the volume swept by the mobile grain boundaries disappear. Using the density of dislocations ρ as an internal variable, the evolution of the dislocation density with time can be deduced from the equation:

$$\frac{\mathrm{d}\rho}{\mathrm{d}t} = \frac{\dot{\varepsilon}}{bd_{\mathrm{g}}} - \alpha\frac{\rho K}{d_{\mathrm{g}}^2} \tag{6.9}$$

where K is the grain boundary migration rate associated with grain growth and α is a coefficient exceeding 1 to take into account a higher dislocation density near grain boundaries. The efficiency of grain boundary migration as a recovery process in polar ice sheets was clearly shown by de La Chapelle *et al.* (1998).

During *continuous recrystallization*, the reduction of the dislocation density is due to GBM and the formation of boundaries (Montagnat and Duval, 2000). If grain size does not change with time, the evolution of the dislocation density is given by:

$$\frac{\mathrm{d}\rho}{\mathrm{d}t} = \frac{\dot{\varepsilon}}{bd_{\mathrm{g}}} - \alpha\frac{\rho K}{d_{\mathrm{g}}^2} - \frac{K\theta_{\mathrm{c}}}{bd_{\mathrm{g}}^3} \tag{6.10}$$

where θ_{c} is the angle for which sub-boundaries transform into grain boundaries (Montagnat and Duval, 2000). Application of this physical model to the behavior of shallow ice at Summit (Greenland) has permitted Montagnat and Duval (2000) to obtain the evolution of the dislocation density in the first 1500 m of the GRIP ice core with $\alpha = 1$ (Fig. 6.16). A steady state with a constant dislocation density and grain size can be reached with:

$$\dot{\varepsilon} = \frac{bK}{d_{\mathrm{g}}}\left(\alpha\rho + \frac{\theta_{\mathrm{c}}}{bd_{\mathrm{g}}}\right). \tag{6.11}$$

Knowing strain rate ($\approx 2.5 \times 10^{-12}\,\mathrm{s}^{-1}$) and the grain boundary migration rate, K ($1.2 \times 10^{-16}\,\mathrm{m}^2\,\mathrm{s}^{-1}$), it is possible to deduce the total dislocation density at equilibrium. A dislocation density of $7 \times 10^{10}\,\mathrm{m}^{-2}$ was found at equilibrium with $\alpha = 2$, $d_{\mathrm{g}} = 4\,\mathrm{mm}$ and $\theta_{\mathrm{c}} = 5°$.

By assuming a stress exponent of 2, as found at low stresses (Fig. 6.7), grain size could vary inversely with stress and the dislocation density would be proportional to stress.

This deformation mode with intracrystalline slip accommodated by dynamic recrystallization is in accordance with the development of textures and the occurrence of continuous recrystallization in polar ice sheets. By considering that GBM

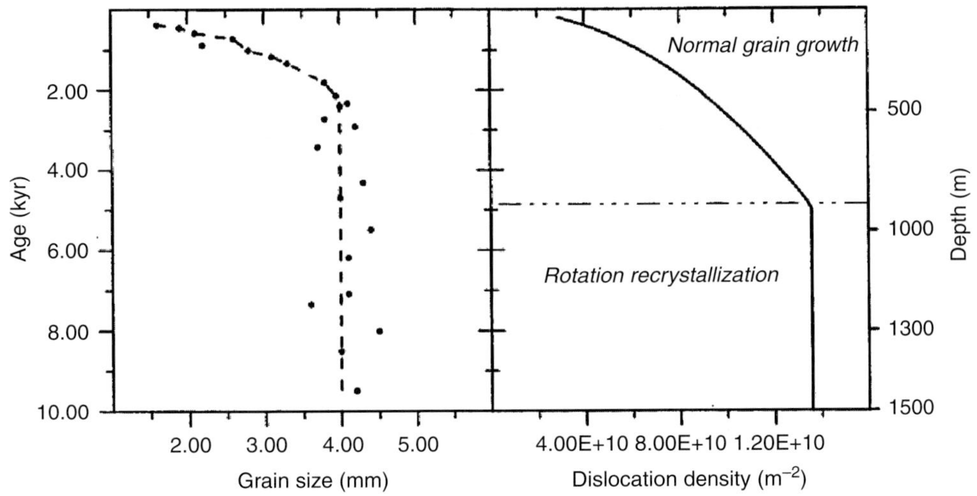

Figure 6.16. Grain size and dislocation density as a function of age and depth in the GRIP ice core. (From Montagnat and Duval, 2000.)

is not a deformation process, basal slip must be associated with other deformation modes. But the softening of the material induced by GBM is expected to minimize the importance of accommodation processes such as non-basal slip, GBS and/or diffusion. This view on the role of GBM in the deformation of polar ice is essential since GBM significantly reduces the dislocation density and, consequently, hardening. The question for which we do not yet have a response is the importance of deformation modes in addition to basal slip.

6.4.3 Rate-controlling processes for tertiary creep

Beyond the minimum corresponding to secondary creep, the creep rate accelerates into tertiary creep. An almost steady state is reached for a strain higher than 10% (Fig. 6.10). The acceleration of creep rate after secondary creep is explained by both the softening processes associated with dynamic recrystallization and the formation of *recrystallization* textures, since migration recrystallization leads to textures that promote basal slip (Duval, 1981). Tertiary creep rates are stress-configuration dependent. The enhancement factor defined by the ratio between strain rate during tertiary creep and strain rate during secondary creep is about 3 in compression or pure shear and between 8 and 10 in simple shear (Duval, 1981; Li Jun *et al.*, 1996). This result can be analyzed by considering the instantaneous mechanical behavior of such anisotropic ices under these stress configurations. Calculations with the VPSC model (Chapter 7) on textured ices similar to those produced by recrystallization give an enhancement factor of about 5 in simple shear and slightly higher than 1 in uniaxial compression (Castelnau *et al.*, 1996a). Hence, the direct effect of texture on the mechanical behavior cannot explain the measured enhancement factors. The softening processes associated with dynamic recrystallization should contribute to an enhancement factor between 2 and 3.

6.5 Grain growth and recrystallization

Recrystallization is a process that involves the formation and the growth of new grains within the deformed structure. *Dynamic* recrystallization is the phenomenon by which the stored energy present in a deforming microstructure is used to generate new dislocation-free grains. It takes place during deformation at elevated temperature. Grain nucleation and grain boundary migration are two processes that contribute to the reduction of the dislocation density. The effect of recrystallization is therefore to soften the material. The increase of creep rate after secondary creep is partially caused by this softening process (Fig. 6.10). New grains are in turn deformed until they undergo recrystallization. This continuous sequence of deformation and recrystallization gradually results in a steady state and is associated with tertiary creep (Poirier, 1985). Recrystallization readily occurs in ice because dynamic recovery is slow and it is difficult to activate non-basal slip systems. When the stored energy within grains is too low to induce grain nucleation, grain boundary migration driven by differences in dislocation density between adjacent grains or by the reduction in the energy of grain boundaries can occur. This process can induce grain growth.

Recrystallization and grain growth significantly influence the microstructure as well as the mechanical properties. They affect the whole behavior of the material by reducing the level of the long-range internal stress field. Static recrystallization, which occurs during high-temperature annealing following deformation at relatively low temperature, is not a significant process in glaciers and ice sheets since ice is not annealed in the *in-situ* conditions.

6.5.1 Grain growth in polar ice sheets

Grain growth driven by the free energy of grain boundaries is mainly observed at the surface of polar ice sheets where fine-grained ice is present. Thus, the occurrence of grain growth during deformation is possible as long as the energy of dislocations does not reach the critical value for grain nucleation. But some grain nucleation can locally occur and, therefore, affect the grain growth law. An important review of the grain growth process with emphasis put on the influence of impurities, bubbles and microparticles can be found in Alley *et al.* (1986a,b) and Durand *et al.* (2006).

In the upper part of polar ice sheets, the average grain size increases with depth (Gow, 1969, 1974). At the South Pole, crystal size increases from 0.18 mm^2 at 0.1 m depth to 0.63 mm^2 at 50 m. It is worth noting that the rapid crystal growth in the top 4 m is attributed to the effect of temperature gradients in the snowpack, which promote the diffusion of water vapor. The continuous increase of crystal size is observed down to several hundred meters in Central Greenland (Gow *et al.*, 1997; Thorsteinsson *et al.*, 1997) and Antarctica (Gow and Williamson, 1976; Duval and Lorius, 1980; Li Jun *et al.*, 1998; Azuma *et al.*, 2000).

This grain growth process is driven by grain boundary curvature. The driving pressure P is given by:

$$P = \gamma_{gb}\left(\frac{1}{R_1} - \frac{1}{R_2}\right) \tag{6.12}$$

where R_1 and R_2 are the principal radii of curvature of a boundary of energy γ_{gb}. If the boundary is part of a sphere of radius R, then $R = R_1 = R_2$ and

$$P = \frac{2\gamma_{gb}}{R}. \tag{6.13}$$

By assuming that γ_{gb} is the same for all boundaries and the radius R is proportional to the mean radius $\bar{R}$, the boundary velocity is given by:

$$\frac{d\bar{R}}{dt} = M\frac{\alpha\gamma_{gb}}{\bar{R}} \tag{6.14}$$

where M is the boundary mobility and α a geometric constant (Humphreys and Hatherly, 1996).

With grain sizes in polar ice in the range from 1 to 10 mm and $\gamma_{gb} = 0.065$ J m^{-2}, P is between 20 and 200 J m^{-3}. By comparison, the driving force in metals with higher γ_{gb} and lower $\bar{R}$ is typically 10^4 J m^{-3} (Humphreys and Hatherly, 1996).

From Equation (6.14), the evolution with time of the mean grain size is given by:

$$\bar{R}^n - \bar{R}_0^n = Kt \tag{6.15}$$

where K again can be considered as the grain boundary migration rate.

The range of K between −50 and −10 °C is roughly between 10^{-17} and 5×10^{-15} m^2 s^{-1} (Duval *et al.*, 1983). From data of grain growth in firn and shallow ice in Greenland and Antarctica, the activation energy is about 50 kJ mol^{-1} (Gow, 1969). A much higher value of the activation energy is found above −10 °C (Jacka and Li Jun, 1994). The mobility of grain boundaries should increase rapidly when the temperature approaches the melting point.

The parabolic law with $n = 2$ was derived from mean field approximations which consider the behavior of a single grain embedded in an environment that is some average representation of the whole assembly (Hillert, 1965). Measurements of grain growth kinetics in several materials give a value of the exponent n higher than 2 (Humphreys and Hatherly, 1996). However, Equation (6.15) with $n=2$ was largely adopted to describe grain growth in polar ice sheets (Gow, 1969; Duval and Lorius, 1980; Alley *et al.*, 1986a,b; Alley, 1992). Values of $n>2$ were obtained from several ice cores. The profile of grain growth in the first 400 m of the GRIP core is well reproduced with $n=2.5$ (Thorsteinsson *et al.*, 1997). A value of $n=3.2$ was found between 100 and 400 m depth in the Dome Concordia ice core (Weiss *et al.*, 2002). From Durand *et al.* (2006), the departure from the theoretical parabolic law is likely the result of bubble pinning. These authors assume that the grain boundary mobility is significantly higher than the mobility of bubbles. At Dome Concordia, the pinning pressure by bubbles can, indeed, be significant since most of the 400 bubbles per gram of ice in the Holocene ice would be located on grain boundaries (Durand *et al.*, 2006). This is not the general case since a uniform distribution of bubbles within grains was found in the Byrd ice core by Gow (1968). Another explanation for the value of $n>2$ in deep ice cores is the progressive occurrence of rotation recrystallization (see Section 6.5.2).

Grain growth in glacial ice Grain growth rate in ice from glacial climatic periods is significantly lower than that for interglacial periods (Gow and Williamson, 1976; Duval and Lorius, 1980; Lipenkov *et al.*, 1989; Paterson, 1991; Gow *et al.*, 1997; Thorsteinsson *et al.*, 1997). Several explanations have been given for this correlation between grain size and climate, including impurity drag (Alley, 1986b) and the pinning of grain boundaries by microparticles (Gow *et al.*, 1997; Durand *et al.*, 2006).

The pinning effect of microparticles has been modeled by Zener (cited by Smith (1948)). The maximum restraining force F_Z by a particle of radius r is:

$$F_Z = \pi r \gamma_{gb}. \tag{6.16}$$

The pinning pressure P_Z exerted on unit area of the grain boundary depends on the location of particles. Assuming that all particles have the same radius with a random distribution of particles, the pinning pressure is given by:

$$P_Z = 2\pi\gamma_{gb} N_v r^2 \tag{6.17}$$

where N_v is the number of particles per unit volume. If all particles are located on grain boundaries, the pinning pressure is:

$$P_Z = \pi r \gamma_{gb} R N_v / 3. \tag{6.18}$$

Table 6.2. *Mean concentration of solid impurities in the Antarctic and Greenland ice sheets*

Station	Climatic period	Typical concentration (ng/g)	Reference
Byrd	Holocene	10	Paterson (1991)
	LGM	100	
EPICA/Dome C	Holocene	15	Delmonte *et al.* (2002)
	LGM	800	
Vostok	Holocene	35	Delmonte *et al.* (2002)
	LGM	850	
Camp century	Holocene	80	Paterson (1991)
	LGM	1200	
GRIP	Holocene	50	Steffensen (1997)
	LGM	5000	

LGM means the Last Glacial Maximum about 20 000 years before present.

The rate of grain growth in the presence of the particles is:

$$\frac{\mathrm{d}\bar{R}}{\mathrm{d}t} = M(P - P_{\mathrm{Z}}). \tag{6.19}$$

Taking into account the particle content in the Holocene and the Last Glacial Maximum (LGM) along several deep ice cores (Table 6.2), the grain size decrease associated with LGM can be explained by the pinning effect by assuming most of the particles are located on grain boundaries (Durand *et al.*, 2006). The mean value of the diameter of atmospheric particles in polar ice cores is about 2 μm with a lognormal distribution (Delmonte *et al.*, 2002).

When the pressure on boundaries due to particle pinning P_{Z} equals the driving force for grain growth, the growth of grains ceases and a limiting grain size is reached. A Zener limiting average grain size (mean radius) of about 1.5 mm was found in glacial ice in the Dome Concordia ice core (Durand *et al.*, 2006). Considering the reduction of this driving force by particles, the formation of boundaries during deformation could contribute to maintain grain size to this relatively small value.

With regard to the effect of soluble impurities, the drag of impurities segregated to grain boundaries could reduce the grain boundary mobility (Alley *et al.*, 1986a,b). Based on correlations between soluble impurity content and grain size along the GISP2 ice core in Greenland, Alley and Woods (1996) suggested that the impurity drag could affect the grain growth rate. This effect was excluded for the Dome Concordia ice core on the basis of observations including grain size and impurity analyses (Durand *et al.*, 2006).

In conclusion, the relatively low grain growth rate in glacial ice is probably the result of grain boundary pinning by particles located there. It is worth noting that significant variations in the particle and impurity content are found in the Antarctic and Greenland ice sheets, most of them well correlated with climate (Legrand and Mayewski, 1997; De Angelis *et al.*, 1997; Delmonte *et al.*, 2002). Bubbles could account for a small part of the grain size change at the last climatic transition (Durand *et al.*, 2006).

6.5.2 Rotation recrystallization in glaciers and ice sheets

A commonly used classification scheme in minerals is to divide dynamic recrystallization into rotation and migration recrystallization (Poirier, 1985). With the production of dislocations by deformation, sub-boundaries progressively form with depth. The misorientation of sub-boundaries increases as deformation proceeds and high angle boundaries develop, leading to the nucleation of new grains. This mechanism is termed *rotation recrystallization* by the geological community and *continuous recrystallization* by the materials science community. In this recrystallization regime, grain boundaries migrate in the same low-velocity regime as the one associated with grain growth (Duval *et al.*, 1983). The grain boundary migration rate corresponding to this low-velocity regime is lower than $10^{-14}\,m^2\,s^{-1}$ below $-10\,°C$ (Duval *et al.*, 1983). The sub-boundaries observed in ice by optical microscopy (Fig. 5.1) are probably tilt boundaries with basal edge dislocations. According to Read (1953), the energy of a tilt boundary increases with increasing misorientation, but the energy per dislocation decreases with increasing misorientation. Therefore, there is a driving force to form more highly misoriented boundaries as recovery proceeds. Twist boundaries composed of basal screw dislocations can be seen by X-ray diffraction (Montagnat *et al.*, 2003).

The rotation recrystallization mechanism appears to occur extensively in polar ice sheets (Alley, 1992; de La Chapelle *et al.*, 1998, Hamann *et al.*, in press). Depending on deformation conditions, rotation recrystallization can be associated with a reduction of grain size by the formation of boundaries and an increase of grain size by grain boundary migration. This can make difficult the discussion on the occurrence of grain growth driven by the energy of grain boundaries and recrystallization. According to Alley *et al.* (1995), polygonization processes associated with rotation recrystallization counteract grain growth below 400 m in the Byrd ice core. The same explanation was given by Castelnau *et al.* (1996b) for the constant grain size in the GRIP core between 650 and 1625 m. This recrystallization process is very active in most parts of polar ice sheets (Gow *et al.*, 1997; Thorsteinsson *et al.*, 1997; Durand *et al.*, 2006). This process gives a microstructure similar to that resulting from grain growth with a lognormal grain size distribution and an almost equiaxed grain structure (Thorsteinsson *et al.*, 1997; Durand *et al.*, 2006). A typical

Figure 6.17. Thin section of ice photographed between crossed polaroids from the Vostok ice core (East Antarctica) at the depth of 2351 m; scale in mm. (Courtesy of V. Lipenkov.)

structure of polar ice (East Antarctica) undergoing rotation recrystallization can be seen in Figure 6.17.

6.5.3 Migration recrystallization in glaciers and ice sheets

Migration recrystallization gives coarse and interlocking grains (Fig. 6.11). Figure 6.18 shows the large difference in the grain boundary migration rate between rotation and migration recrystallization. This recrystallization regime is known to occur in the laboratory after a critical strain of about 1% when ice is deformed at a temperature higher than about −15 °C (Steinemann, 1958; Kamb, 1972; Duval, 1981; Jacka and Maccagnan, 1984). In the central areas of polar ice sheets, where strain rates are generally lower than $10^{-10}\,s^{-1}$, strain energy can be too low to initiate migration recrystallization (Alley, 1992; de La Chapelle *et al.*, 1998).

Migration recrystallization can be a discontinuous process. As soon as recrystallization starts, a wave of accelerating and decelerating creep passes through the sample (Steinemann, 1958). The main difference with metals and ceramics lies in the small critical strain needed to initiate recrystallization. This difference is attributed to the large anisotropy of the ice crystal, which is expected to rapidly induce the required variation of the stored energy between *soft* and *hard* grains to initiate recrystallization.

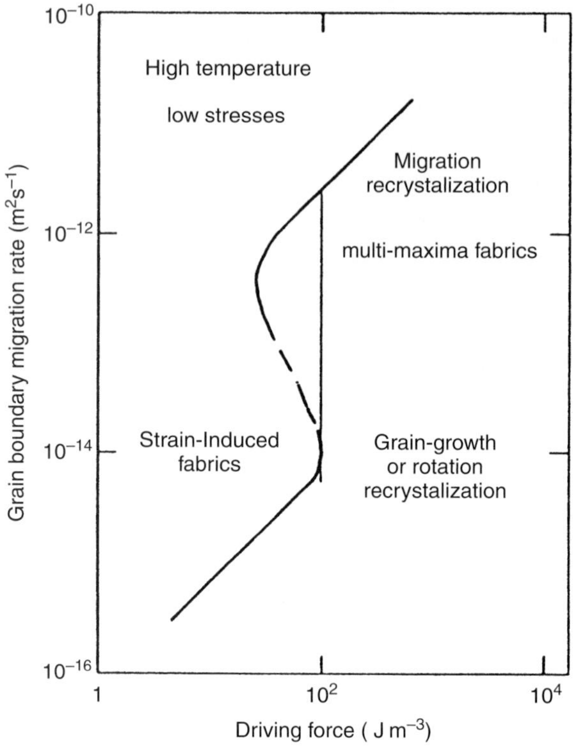

Figure 6.18. Grain boundary migration rate as a function of driving force in the temperature range −10 to −15 °C. (From Duval and Castelnau, 1995.)

6.5.4 Energetics of dynamic recrystallization

There exists a driving force on a grain boundary if the free energy G is reduced during the motion of the boundary. The driving force P can be given by:

$$P = \mathrm{d}G/\mathrm{d}V \tag{6.20}$$

where $\mathrm{d}V$ is the volume swept by the moving grain boundary. P has the dimension of a gained free energy per unit volume (J m^{-3}) or a force acting per unit area of the grain boundary (N m^{-2}).

The driving force for *grain growth* results from the reduction in the energy stored in the material in the form of grain boundaries. The driving force P is typically between 20 and 200 J m^{-3} in polar ice sheets (see Section 6.5.1).

The driving force for recrystallization P_d in glacier ice is the stored energy of the dislocations. As an approximation to the relationship given in Chapter 2, the energy of a dislocation per unit length is given by $\frac{1}{2}Gb^2$, and so the driving force associated with a dislocation density ρ is given by:

$$P_{\mathrm{d}} = \frac{1}{2}\rho G b^2 \qquad (6.21)$$

where G is the shear modulus, b the Burgers' vector.

For a mean dislocation density of $10^{11}\,\mathrm{m}^{-2}$, typical of polar ice (Montagnat and Duval, 2000), P_{d} is of the order of $30\,\mathrm{J\,m}^{-3}$. A higher value of P_{d} is obviously expected near grain boundaries. It is worth noting that the driving force for grain growth and for dynamic recrystallization is of the same order of magnitude in polar ice sheets, which explains the importance of grain growth at the surface of ice sheets even if some grain nucleation can occur locally.

Differences in the elastic energy between two grains can be also considered as the driving force for grain boundary migration. But variations of the elastic modulus with orientation and the level of stress are far too small in glaciers and ice sheets to induce grain boundary migration. In compressive loading of an ice mass with an indenter or structure, deviatoric stresses can be higher than 10 MPa in high pressure zones (Johnston *et al.*, 1998, Jordaan, 2001; see Chapter 14). Differences in elastic energy between grains can be high enough to induce grain boundary migration. In these high stress situations, a dislocation density probably higher than $10^{12}\,\mathrm{m}^{-2}$ is expected, giving a value for P_{d} much higher than $300\,\mathrm{J\,m}^{-3}$. Thus, both the energy associated with the elastic deformation and that associated with dislocations must be taken into account to analyze dynamic recrystallization when ice is deformed under high triaxial stresses.

6.5.5 The nucleation of recrystallization

Nucleation is a critical factor determining both the size and orientation of the recrystallized grain. Nucleation resulting from the progressive misorientation of subgrain boundaries is a well-accepted mechanism for *rotation recrystallization* (Poirier, 1985; Alley, 1992; Humphreys and Hatherly, 1996). The orientation of new grains is therefore near the orientation of parent grains.

The nucleation and growth of new grains during *migration recrystallization* can be analyzed by considering mechanisms occurring during the primary recrystallization of metals and alloys. Because of the limited slip systems in the ice crystal, the distribution of deformation is inhomogeneous on both the intragranular and intergranular scales (Wilson and Zhang, 1996; Castelnau *et al.*, 1996a). It is therefore likely that recrystallization originates in the regions of large orientation gradients, which must be present near grain boundaries. The bulging of a part of a pre-existing grain boundary into a deformed neighboring grain can be the second step (Humphreys and Hatherly, 1996). A viable recrystallization nucleus requires a decrease of the free energy even if the total surface energy increases. For growth, the size of critical nucleus radius R_r must respect the condition:

$$R_r \geq \frac{2\gamma_{gb}}{P_d} \geq 4\frac{\gamma_{gb}}{\rho G b^2}. \tag{6.22}$$

With a dislocation density of $10^{12}\,m^{-2}$, R_r must be greater than 0.4 mm. A smaller value is expected with a larger dislocation density near grain boundaries. These values are completely compatible with observations and grain size in glaciers.

Nucleation of new grains at intragranular sites within kink bands (a part of a crystal limited by two almost parallel polygonization walls giving a double orientation change of opposite sign) was observed by Wilson (1986) from optical microscopy observations during the deformation of thick sections under polarized light. These results support nucleation in regions of large orientation gradients by limited subgrain growth and grain boundary bulging.

At this stage, it is important to point out that nucleation conditions during dynamic recrystallization can be different from nucleation during static recrystallization. Indeed, grain nucleation during deformation leads to a redistribution of stresses within the polycrystal and, as a consequence, must be associated with a reduction of the internal stress field. The orientation of new grains must respect this condition.

Nucleation of grains in ice containing dispersed particles

Particles can have important effects on recrystallization:

- the strain incompatibility between ice and particles is accommodated by dislocations inducing an increase of the stored energy;
- particles can act as nucleation sites;
- particles may exert a pinning pressure on moving grain boundaries, as discussed in Section 6.5.1.

The first two effects tend to promote recrystallization whereas the last tends to hinder grain growth.

Nucleation in the presence of a large orientation gradient at the ice–particle interface in two-phase alloys is well described by Humphreys and Hatherly (1996). However, this mechanism is invoked in metals during static recrystallization. During dynamic recrystallization, dislocations that accommodate strain incompatibility are "necessary" and can be used for nucleation only if strain incompatibility is reduced by recrystallization.

A significant effect of particles on dynamic recrystallization was found for ice samples containing uniformly distributed 1 wt. % silt-sized particles (Song *et al.*, 2005b). From these authors, particle-stimulated nucleation (PSN) was suggested.

Recrystallized grain size was found to be small in comparison with that in particle-free ice. This single result does not allow conclusion on the role of PSN in grain nucleation. The first effect of particles is probably to slow down the rate of grain boundary migration.

6.6 Textures in glaciers and polar ice sheets

The viscous behavior of glacier ice depends on the mechanical properties of individual grains in the aggregate, the orientation of such anisotropic grains and the mechanical interaction between grains. Ice in glaciers and ice sheets is seen to be built of ice crystals with *c*-axes mostly oriented along specific directions. The non-random distribution of the orientations of ice grains or the textures makes the polycrystal anisotropic. The macroscopically anisotropic ices greatly influence the flow of ice sheets. This was shown by bore-hole inclinometry surveys (Russell-Head and Budd, 1979; Gundestrup and Hansen, 1984) and by ice-sheet flow modeling (Van der Veen and Whillans, 1990; Mangeney *et al.*, 1997; Gagliardini and Meyssonnier, 2000; Gillet-Chaulet *et al.*, 2006, Pettit *et al.*, 2007).

Textures in glaciers and ice sheets basically develop as the result of lattice rotation by intracrystalline slip (Azuma and Higashi, 1985; Alley, 1988, 1992; de La Chapelle *et al.*, 1998). But recrystallization textures develop in temperate glaciers (ice is at its melting point except for a surface layer) and in the deepest hundred meters of polar ice sheets where temperature is above about −15 °C (Alley, 1992; Duval and Castelnau, 1995) when migration recrystallization is occurring. From Alley (1992), this recrystallization process reflects the instantaneous stress configuration, whereas deformation textures reflect the whole deformation history. Due to the heterogeneous deformation between and within grains, the stored energy can locally be sufficiently high to initiate some migration recrystallization, whereas textures are still induced by deformation.

6.6.1 Deformation textures

The orientation changes that take place during deformation when migration recrystallization is not active are related to the lattice rotation by intracrystalline slip. It is well known that, for a single crystal under uniaxial compression, the normal to the slip plane rotates toward the compression axis, whereas this axis rotates away from the tension axis under uniaxial tension (Schmid and Boas, 1936). Figure 6.19 shows typical tension and compression textures observed along the Vostok and GRIP deep ice cores. When looking for the evolution of textures along these ice cores, there is a conclusive relation between the strength of textures and the finite strain (Azuma and Higashi, 1985; Castelnau *et al.*, 1996a; Gagliardini and Meyssonnier, 1999;

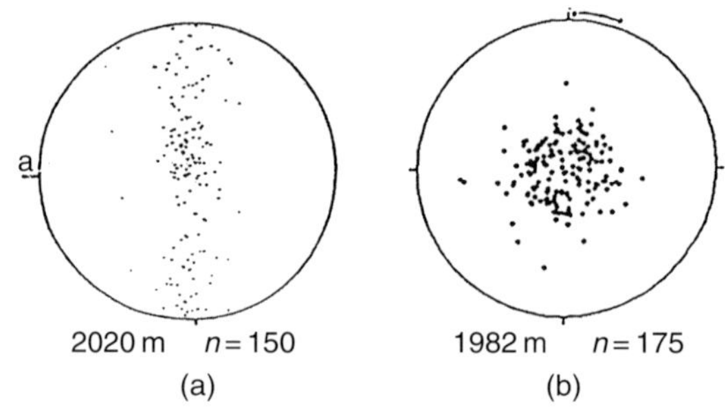

Figure 6.19. (a) c-axes texture in the Vostok ice core at the depth of 2020 m; ice is assumed to be deformed in tension. (From Lipenkov *et al.*, 1989.) (b) c-axes texture in the GRIP ice core at the depth of 1982 m; ice is assumed to be deformed in compression. The center of the diagram represents the vertical direction. (From Thorsteinsson *et al.*, 1997.)

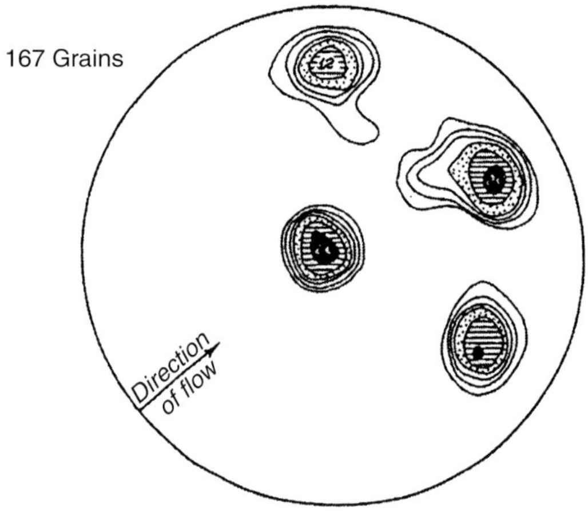

Figure 6.20. Texture diagram of ice from Emmons Glacier (USA) with four maxima. (From Rigsby, 1951.)

Thorsteinsson, 2002). Modeling of deformation texture development can be found in Chapter 7.

6.6.2 Recrystallization textures

Recrystallization textures are the result of the nucleation of grains and their *rapid* growth. In glaciers and ice sheets, they are mostly associated with dynamic recrystallization. Multi-pole textures are generally found in temperate glaciers (Fig. 6.20) and in warm ice near the bed of polar ice sheets (Gow and Williamson, 1976; Alley, 1992). From Budd and Jacka (1989), these textures could also be the result of reduced stress at high temperature or some annealing processes.

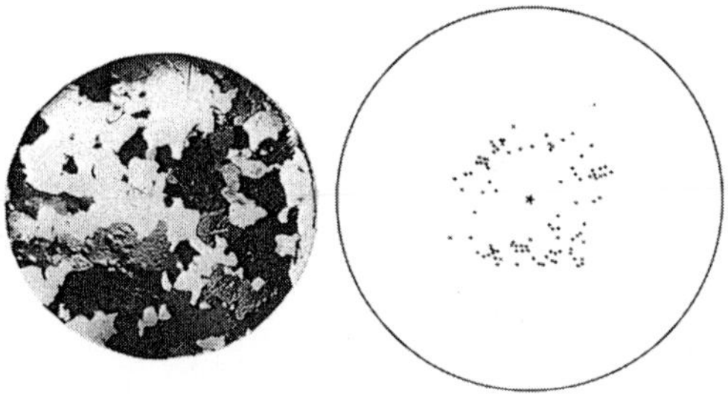

Figure 6.21. Photographs of a thin section under polarized light and c-axes texture after deformation in uniaxial compression at −3 °C and for an octahedral strain of 0.32. (From Jacka and Maccagnan, 1984.)

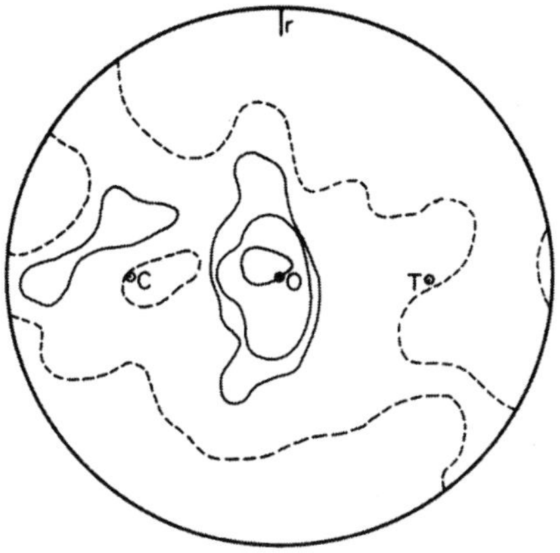

Figure 6.22. c-axes texture of an ice sample deformed in torsion at −1 °C for a strain of 0.45. C (compression), T (tension) and r are the axes of principal stresses; O is the torsion axis. Contours are at densities of 0.5, 3, 5 and 10% for a total of 144 grains. (From Duval, 1981; with permission of the International Glaciological Society.)

The relation between the stress state and textures has been clarified by laboratory tests carried out above −10 °C. In uniaxial compression, a girdle texture develops around the compression axis with c-axes at about 30° from the compression axis (Fig. 6.21), whereas slip induces a maximum centered on the compression direction (Fig. 6.19).

In simple shear, a texture with two maxima forms. At first, the first maximum is located near the normal to the permanent shear plane, whereas the second one is located near the normal to the second plane of maximum shear stress (Kamb, 1972). Thus, these textures appear to reflect the stress state. Then, the second maximum rotates with deformation toward the principal direction of compression (Fig. 6.22) and progressively disappears (Kamb, 1972; Bouchez and Duval, 1982).

These laboratory results show that most crystals during dynamic migration recrystallization are preferably oriented for basal slip and that textures are mainly stress-controlled (Duval, 1981). Indeed, the mean shear stress on basal planes is between 0.8 and 0.9 times the applied shear stress (Duval, 1981). On the other hand, the critical strain for the development of recrystallization textures being less than 10% (Jacka and Maccagnan, 1984), textures cannot be explained by the rotation of the lattice by slip.

Thus, textures observed during *dynamic* migration recrystallization are very different from those resulting from slip. As an example, Figure 6.23 shows deformation and recrystallization textures observed at about the same depth in the GRIP ice core. The pinning of grain boundaries by particles at the depth of 2806 m impedes migration recrystallization (de La Chapelle *et al.*, 1998). Recrystallized grains are roughly well oriented for basal slip and, owing to the redistribution of stresses within the polycrystal during deformation, the *soft* grains appear to be the less stressed (Castelnau *et al.*, 1996a).

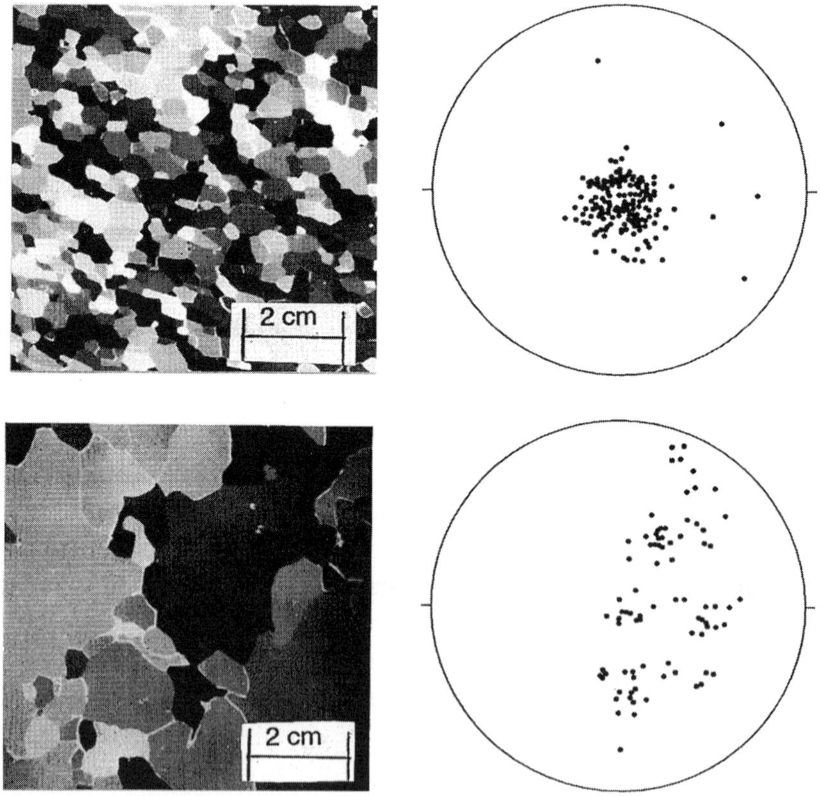

Figure 6.23. Photographs of thin sections under polarized light and *c*-axes textures in the GRIP ice core. (a) Depth: 2806 m; (b) depth 2860 m (from de La Chapelle *et al.*, 1998.)

Due to the long-range internal stress field induced by the mismatch of slip at grain boundaries, the nucleation and growth of grains modify the stress distribution within the whole polycrystal and not only in the neighborhood of the new grain (Richeton *et al.*, 2005). The hardness of the polycrystal during dynamic recrystallization depends on recovery processes associated with recrystallization and obviously on the orientation of new grains. The growth of grains during dynamic migration recrystallization, not well oriented for basal slip (*hard grains*), cannot extensively occur since the material would become harder, whereas the long-range internal stress field would develop with an increasing density of geometrically necessary dislocations to accommodate strain heterogeneities. On the contrary, the formation and the growth of *soft* grains should reduce internal stresses and give a microstructure that minimizes stress and strain heterogeneities within and between grains. Since the behavior of each grain is related to the orientation of each grain in the polycrystal, the multi-maxima textures typical of temperate glaciers could be seen to be those which minimize the long-range internal stress field. Thus, the strength of the polycrystal would be mainly controlled by the density of geometrically necessary dislocations or "excess dislocations", which are a continuous manifestation of lattice incompatibility. These dislocations are non-uniformly distributed. In contrast, "statistical dislocations" result in compatible deformation. Such distinction between dislocations is needed as soon as long-range internal stresses partially control the deformation of materials (Ashby, 1970). The strong dislocation interactions lead to the self-organization of dislocation ensembles (Fig. 5.11) and collective behavior involving large numbers of correlated events (Richeton *et al.*, 2005). The dynamic coupling of both types of dislocations through a continuum description of dislocations was shown to describe well the deformation of the ice crystal by basal slip (Taupin *et al.*, 2007). This field dislocations mechanics theory (Acharya, 2001) is promising to describe the deformation of the polycrystal. This approach has the potential to give a dynamic description of dislocation patterning. The evolution with strain of the mobile and excess dislocations is obtained and the stored energy associated with excess and statistical dislocations can be estimated.

In conclusion, recrystallization textures can be understood by considering the nucleation and growth of grains in the long-range internal stress field. Nucleation in regions of large orientation gradients can be seen as a prerequisite. In the low stress conditions of ice sheets, the orientation of new grains should be not very different from that of deformed grains. The preferred growth of grains of specific orientations in the presence of the long-range internal stress field should control recrystallization textures. The classic view on the direct relationship between the orientation of recrystallized grains, stress, subgrain size and local dislocation density should be revised. Recrystallization textures observed in temperate glaciers (Fig. 6.20) and in the laboratory (Figs. 6.21 and 6.22) should be interpreted in this framework.

6.6.3 Recrystallization textures after annealing

Recrystallization that occurs during annealing treatment following deformation is referred to as static recrystallization. This process has been studied largely in metals and alloys. Static recrystallization almost never occurs in glaciers and ice sheets since the time scale for any temperature change is very long compared with the time for recrystallization. Annealing by stress release can result from flow over an irregular bed-rock. From Budd and Jacka (1989), multi-maxima textures observed in temperate glaciers could form in such conditions. These processes are very far from those occurring during static recrystallization since textures before such annealing are textures produced during dynamic recrystallization and not deformation textures. From Wilson and Russell-Head (1982), little texture change is expected when dynamic recrystallization occurs before annealing. Analysis of recrystallization textures formed in the laboratory after annealing of ice exhibiting deformation textures should be done to determine processes controlling the nucleation and growth of grains during recrystallization. The use of high-resolution electron back-scatter diffraction (EBSD) to follow the nucleation and the growth of grains will give useful results. This technique provides full *c*- and *a*-axis orientation of individual grains (Iliescu *et al.*, 2004) and has already been used to examine recrystallization in the GRIP ice core (Obbard *et al.*, 2006).

References

Acharya, A. (2001). A model of crystal plasticity based on the theory of continuously distributed dislocations. *J. Mech. Phys. Sol.*, **49**, 761–784.

Alley, R. B. (1988). Fabrics in polar ice sheets: development and prediction. *Science*, **240**, 493–495.

Alley, R. B. (1992). Flow-law hypotheses for ice-sheet modeling. *J. Glaciol.*, **38**, 245–256.

Alley, R. B. and G. A. Woods (1996). Impurity influence on normal grain growth in the GISP2 ice core, Greenland. *J. Glaciol.*, **42**, 255–260.

Alley, R. B., J. H. Perepezko and C. R. Bentley (1986a). Grain growth in polar ice: I. Theory. *J. Glaciol.*, **32**, 415–424.

Alley, R. B., J. H. Perepezko and C. R. Bentley (1986b). Grain growth in polar ice: II. Application. *J. Glaciol.*, **32**, 425–433.

Alley, R. B., A. J. Gow and D. A. Meese (1995). Mapping *c*-axis fabrics to study physical processes in ice. *J. Glaciol.*, **41**, 197–203.

Andrade, E. N. da C. (1914). The flow of metals under large constant stresses. *Proc. R. Soc. London A*, **90**, 329–340.

Ashby, M. F. (1966). Work hardening of dispersion-hardened crystals. *Phil. Mag.*, **14**, 1157–1178.

Ashby, M. F. (1970). The deformation of plastically non-homogeneous materials. *Phil. Mag.*, **21**, 399–424.

Ashby, M. F. and P. Duval (1985). The creep of polycrystalline ice. *Cold Reg. Sci. Technol.*, **11**, 285–300.

Azuma, N. and A. Higashi (1985). Formation processes of ice fabric pattern in ice sheets. *Ann. Glaciol.*, **6**, 130–134.

Azuma, N., Y. Wang, K. Mori *et al.* (2000). Textures and fabrics in the Dome F (Antarctica) ice core. *Ann. Glaciol.*, **29**, 163–168.

Baker, R. W. and W. W. Gerberich (1979). The effect of crystal size and dispersed-solid inclusions on the activation energy for creep of ice. *J. Glaciol.*, **24**, 179–194.

Barnes, P., D. Tabor and J. C. F. Walker (1971). The friction and creep of polycrystalline ice. *Proc. R. Soc. London A*, **324**, 127–155.

Barrette, P. D. and I. J. Jordaan (2003). Pressure-temperature effects on the compressive behavior of laboratory-grown and iceberg ice. *Cold Reg. Sci. Technol.*, **36**, 25–36.

Blum, W. and W. Maier (1999). Harper-Dorn creep – a myth? *Phys. Stat. Sol. A*, **171**, 467–474.

Bouchez, J. L. and P. Duval (1982). The fabric of polycrystalline ice deformed in simple shear: experiments in torsion, natural deformation and geometrical interpretation. *Textures Microstruct.*, **5**, 171–190.

Budd, W. F. and T. H. Jacka (1989). A review of ice rheology for ice sheet modelling. *Cold Reg. Sci. Technol.*, **16**, 107–144.

Castelnau, O., P. Duval, R. A. Lebensohn and G. R. Canova (1996a). Viscoplastic modelling of texture development in polycrystalline ice with a self-consistent approach: comparison with bound estimates. *J. Geophys. Res.*, **101** (B6), 13,851–13,868.

Castelnau, O., Th. Thorsteinsson, J. Kipfstuhl, P. Duval and G. R. Canova (1996b). Modelling fabric development along the GRIP ice core, central Greenland. *Ann. Glaciol.*, **23**, 194–201.

Cohen, D. (2000). Rheology of ice at the bed of Engabreen, Norway. *J. Glaciol.*, **46**, 611–621.

Colbeck, S. C. and R. J. Evans (1973). A flow law for temperate glacier ice. *J. Glaciol.*, **12**, 71–86.

Cole, D. M. (1991). Anelastic straining in polycrystalline ice. In *Proceedings of the 6th International Specialty Conference on Cold Regions Engineering*, ed. D. Sodhi. Lebanon, N.H.: American Society of Civil Engineers, pp. 504–518.

Cole, D. M. (1995). A model for the anelastic straining of saline ice subjected to cyclic loading. *Phil. Mag. A*, **72** (1), 231–248.

Cole, D. M. (2003). A dislocation-based analysis of the creep of granular ice: preliminary experiments and modeling. *Ann. Glaciol.*, **37**, 18–22.

Cole, D. M. (2004). A dislocation-based model for creep recovery in ice. *Phil. Mag.*, **84**, 3217–3234.

Cole, D. M. and G. D. Durell (1995). The cyclic loading of saline ice. *Phil. Mag.*, **72**, 209–229.

Cole, D. M. and G. D. Durell (2001). A dislocation-based analysis of strain history effects in ice. *Phil. Mag. A*, **81**, 1849–1872.

Cuffey, K. M., H. Conway, B. Hallet, A. M. Gades and C. F. Raymond (1999). Interfacial water in polar glaciers and glacier sliding at −17 °C. *Geophys. Res. Lett.*, **26**, 751–754.

Cuffey, K. M., T. Thorsteinsson and E. D. Waddington (2000a). A renewed argument for crystal size control of ice sheet strain rates. *J. Geophys. Res.*, **105** (B12), 27,889–27,894.

Cuffey, K. M., H. Conway, A. Gades *et al.* (2000b). Deformation properties of subfreezing glacier ice: role of crystal size, chemical impurities, and rock particles inferred from *in situ* measurements. *J. Geophys. Res.*, **105** (B12), 27,895–27,915.

Dahl-Jensen, D. and N. S. Gundestrup (1987). Constitutive properties of ice at Dye 3, Greenland. In *The Physical Basis of Ice Sheet Modeling*, eds. E. D. Waddington and J. S. Walder. Vancouver: IAHS Publication 170, pp. 31–43.

De Angelis, M., J. P. Steffensen, M. Legrand, H. Clausen and C. Hammer (1997). Primary aerosol (sea salt and soil dust) deposited in Greenland ice during the last climatic cycle: comparison with east Antarctic records, *J. Geophys. Res.*, **102** (C12), 26,681–26,698.

de La Chapelle, S., O. Castelnau, V. Lipenkov and P. Duval (1998). Dynamic recrystallization and texture development in ice as revealed by the study of deep ice cores in Antarctica and Greenland. *J. Geophys. Res.*, **103** (B3), 5091–5105.

de La Chapelle, S., H. Milsch, O. Castelnau and P. Duval (1999). Compressive creep of ice containing a liquid intergranular phase: ratecontrolling processes in the dislocation creep regime. *Geophys. Res. Lett.*, **26**, 251–254.

Delmonte, B., J. R. Petit and V. Maggi (2002). Glacial to Holocene implications of the new 27 000-year dust record from the EPICA Dome C (East Antarctica) ice core. *Climate Dynamics*, **18**, 647–660.

Derby, B. (1991). The dependence of grain size on stress during dynamic recrystallization. *Acta Metall. Mater.*, **39**, 955–962.

Doake, C. S. M. and E. W. Wolff (1985). Flow law for ice in polar ice sheets. *Nature*, **314**, 255–257.

Drouin, M. and B. Michel (1971). Les poussées d'origine thermique exercées par les couverts de glace sur les structures hydrauliques. Rapport S-23, Laboratoire de Mécanique des Glaces, Université Laval, Canada.

Durand, G. and 10 others (2006). Effect of impurities on grain growth in cold ice sheets. *J. Geophys. Res.*, **111**, FO1015, 1–18.

Durham, W. B. and L. A. Stern (2001). Rheological properties of water ice – Applications to satellites of the outer planets. *Ann. Rev. Earth Planet. Sci.*, **29**, 295–330.

Durham, W. B., S. H. Kirby and L. A. Stern (1992). Effect of dispersed particulates on the rheology of water ice at planetary conditions. *J. Geophys. Res.*, **97** (E12), 20,883–20,897.

Durham, W. B., S. H. Kirby and L. A. Stern (1997). Creep of water ices at planetary conditions: a compilation. *J. Geophys. Res.*, **102** (E7), 16,293–16,302.

Durham, W. B., L. A. Stern and S. H. Kirby (2001). Rheology of ice I at low stresses and elevated confining pressure. *J. Geophys. Res.*, **106**, 11,031–11,042.

Duval, P. (1973). Fluage de la glace polycrystalline pour les faibles contraintes. *C.R. Acad. Sci. Paris*, **277**, 703–706.

Duval, P. (1976). Lois de fluage transitoire ou permanent de la glace polycristalline pour divers états de contrainte. *Ann. Géophys.*, **IV**, **32**, 335–350.

Duval, P. (1977). The role of the water content on the creep rate of polycrystalline ice. In *Isotopes and Impurities in Snow and Ice*. IAHS Publication, **118**, 29–33.

Duval, P. (1978). Anelastic behavior of polycrystalline ice. *J. Glaciol.*, **21**, 621–628.

Duval, P. (1981). Creep and fabrics of polycrystalline ice under shear and compression. *J. Glaciol.*, **27**, 129–140.

Duval, P. and O. Castelnau (1995). Dynamic recrystallization of ice in polar ice sheets. *J. Phys. IV*, **5** (Colloque N°3), C3-197–C3-205.

Duval, P. and H. Le Gac (1980). Does the permanent creep-rate of polycrystalline ice increase with crystal size? *J. Glaciol.*, **25**, 151–157.

Duval, P. and C. Lorius (1980). Crystal size and climatic record down to the last ice age from Antarctic ice. *Earth Planet. Sci. Lett.*, **48**, 59–64.

Duval, P. and M. Montagnat (2002). Comments on "Superplastic deformation of ice: experimental observations" by D. L. Goldsby and D. L. Kohlstedt. *J. Geophys. Res.*, **107B**, ECV4, 1–2.

Duval, P., M. F. Ashby and I. Anderman (1983). Rate-controlling processes in the creep of polycrystalline ice. *J. Phys. Chem.*, **87**, 4066–4074.

Dysthe, D. K., Y. Podladchikov, F. Renard, J. Feder and G. Jamtveit (2002). Universal scaling in transient creep. *Phys. Rev. Lett.*, **89**, 246102-1–246102-4.

Echelmeyer, K. and W. Zhongxiang (1987). Direct observation of basal sliding and deformation of basal drift at sub-freezing temperatures. *J. Glaciol.*, **33**, 83–98.

EPICA Community Members (2004). Eight glacial cycles from an Antarctic ice core. *Nature*, **429**, 623–628.

Fischer, D. A. and R. M. Koerner (1986). On the special rheological properties of ancient microparticles-laden Northern Hemisphere ice as derived from bore-hole and core measurements. *J. Glaciol.*, **32**, 501–510.

Fitzsimons, S. J., R. Lorrain and K. J. McManus (1999). Structure and strength of basal ice and substrate of a dry-based glacier: evidence for substrate deformation at sub-freezing temperatures. *Ann. Glaciol.*, **28**, 236–240.

Friedel, J. (1964). *Dislocations*. Oxford: Pergamon Press.

Frost, H. J. and M. F. Ashby (1982). *Deformation-Mechanism Maps for Metals and Alloys*. Oxford: Pergamon Press.

Gagliardini, O. and J. Meyssonnier (1999). Analytical derivations for the behavior and fabric evolution of a linear orthotropic ice polycrystal. *J. Geophys. Res.*, **104** (B8), 17,797–17,809.

Gagliardini, O. and J. Meyssonnier (2000). Simulation of anisotropic ice flow and fabric evolution along the GRIP-GISP2 flowline, Central Greenland. *Ann. Glaciol.*, **30**, 503–509.

Gao, X. Q. and T. H. Jacka, (1987). The approach to similar tertiary creep rates for Antarctic core ice and laboratory prepared ice. *J. Physique Coll.*, **48**, C1-289–C1-296.

Gillet-Chaulet, F., O. Gagliardini, J. Meyssonnier, T. Zwinger and J. Ruokolainen (2006). Flow-induced anisotropy in polar ice and related ice-sheet flow modeling. *J. Non-Newtonian Fluid Mech.*, **134**, 33–43.

Gilormini, P., M. V. Nebozhyn and P. Ponte Castaneda (2001). Accurate estimates for the creep behavior of hexagonal polycrystals. *Acta Mater.*, **49**, 329–337.

Ginter, T. J., P. K. Chaudhury and F. A. Mohamed (2001). An investigation of Harper-Dorn creep at large strains. *Acta Mater.*, **49**, 263–272.

Glen J. W. (1955). The creep of polycrystalline ice. *Proc. R. Soc. London A*, **228**, 519–538.

Glen, J. W. and M. F. Perutz (1954). The growth and deformation of ice crystals. *J. Glaciol.*, **2**, 397–403.

Gold, L. W. (1972). *The Failure Process in Columnar-Grained Ice*. Tech. Paper No. 369 of the Division of Building Research. Ottawa: National Research Council of Canada.

Goldsby, D. L. and D. L. Kohlstedt (1997). Grain boundary sliding in fine-grained ice I. *Scr. Mater.*, **37**, 1399–1406.

Goldsby, D. L. and D. L. Kohlstedt (2001). Superplastic deformation of ice: experimental observations. *J. Geophys. Res.* **106** (B6), 11,017–11,030.

Goodman, D. J., H. J. Frost and M. F. Ashby (1981). The plasticity of polycrystalline ice. *Phil. Mag.*, **43**, 665–695.

Gow, A. J. (1968). Bubbles and bubble pressure in Antarctic glacier ice. *J. Glaciol.*, **7**, 167–182.

Gow, A. J. (1969). On the rates of growth of grains and crystals in south polar firn. *J. Glaciol.*, **8**, 241–252.

Gow, A. J. (1974). Time-temperature dependence of sintering in perennial isothermal snowpacks. In *Snow Mechanics Symposium*. IAHS Publication 114, pp. 25–41.

Gow, A. J. and T. Williamson (1976). Rheological implications of the internal structure and crystal fabrics of the West Antarctic ice sheet as revealed by deep ice core drilling at Byrd Station. *CRREL Report*, 76-35.

Gow, A. J., D. A. Meese, R. B. Alley *et al.* (1997). Physical and structural properties of the Greeland Ice Sheet Project: A review. *J. Geophys. Res.*, **102** (C12), 26,559–26,575.

Gundestrup, N. S. and B. L. Hansen (1984). Bore-hole survey at Dye 3, South Greenland, *J. Glaciol.*, **30**, 282–288.

Hamann, I., S. Kipfstuhl, S. Faria *et al.* (2009) Subgrain boundaries and related microstructural features in EPICA–Dronning Maud Land (EDML) deep ice core, *J. Glaciol.* (in press).

Hill, R. (1950). *The Mathematic Theory of Plasticity.* New York: Oxford University Press.

Hillert, M. (1965). On the theory of normal and abnormal grain growth. *Acta Metall.*, **13**, 227–238.

Hobbs, P. V. (1974). *Ice Physics*. Oxford: Clarendon Press.

Homer, D. R. and J. W. Glen (1978). The creep activation energies of ice. *J. Glaciol.*, **21**, 429–444.

Hondoh, T. (2000). Nature and behavior of dislocations in ice. In *Physics of Ice Core Records*, ed. T. Hondoh. Sapporo: Hokkaido University Press, pp. 3–24.

Hooke, R. LeB., B. B. Dahlin and M. T. Kauper (1972). Creep of ice containing dispersed fine sand. *J. Glaciol.*, **11**, 327–336.

Hooke, R. LeB. (1981). Flow law for polycrystalline ice in glaciers: comparison of theoretical predictions, laboratory data and field measurements. *Rev. Geophys. Space Phys.*, **19**, 664–672.

Hooke, R. LeB. (1998). *Principles of Glacier Mechanics*. Upper Saddle River, N. J.: Prentice Hall.

Humphreys, F. J. and M. Hatherly (1996). *Recrystallization and Related Annealing Phenomena*. Oxford: Pergamon Press.

Hutchinson, J. W. (1976). Bounds and self-consistent estimates for creep of polycrystalline materials. *Proc. R. Soc. Lond. A*, **348**, 101–127.

Hutchinson, J. W. (1977). Creep and plasticity of hexagonal polycrystals as related to single crystal slip. *Metall. Trans.*, **8A**, 9, 1465–1469.

Hutter, K. (1983). *Theoretical Glaciology.* Dordrecht: Reidel.

Iliescu, D., I. Baker and Hui Chang (2004). Determining the orientations of ice crystals using Electron Backscatter Patterns. *Microsc. Res. Tech.*, **63**, 183–187.

Jacka, T. H. (1984a). The time and strain required for development of minimum strain rates in ice. *Cold Reg. Sci. Technol.*, **8**, 261–268.

Jacka, T. H. (1984b). Laboratory studies on relationship between ice crystal size and flow rate. *Cold Reg. Sci. Technol.*, **10**, 31–42.

Jacka, T. H. (1994). Investigations of discrepancies between laboratory studies on the flow of ice: density, sample shape and size, and grain size. *Ann. Glaciol.*, **19**, 146–154.

Jacka, T. H. and Li Jun (1994). The steady state crystal size of deforming ice. *Ann. Glaciol.*, **20**, 13–18.

Jacka, T. H. and M. Maccagnan (1984). Ice crystallographic and strain rate changes with strain in compression and extension. *Cold Reg. Sci. Technol.*, **8**, 269–286.

Johnston, M. E., K. R. Croasdale and I. J. Jordaan (1998). Localized pressures during ice-structure interaction: relevance to design criteria. *Cold Reg. Sci. Technol.*, **27**, 105–117.

Jones, S. J. and J. G. Brunet (1978). Deformation of ice single crystals close to the melting point. *J. Glaciol.*, **21**, 445–455.

Jones, S. J. and H. A. M. Chew (1981). On the grain-size dependence of secondary creep. *J. Glaciol.*, **27**, 517–518.

Jones, S. J. and H. A. M. Chew (1983a). Effect of sample and grain size on the compressive strength of ice. *Ann. Glaciol.*, **4**, 129–132.

Jones, S.J. and H.A.M. Chew (1983b). Creep of ice as a function of hydrostatic pressure. *J. Phys. Chem.* **87**, 4064–4066.

Jones, S. J., R. E. Gagnon, A. Derradji and A. Bugden (2003). Compressive strength of iceberg ice. *Can. J. Phys.*, **81**, 191–200.

Jordaan, I. J. (2001). Mechanics of ice-structure interaction. *Eng. Fract. Mech.*, **68**, 1923–1960.

Kalifa, P. (1988). Contribution à l'étude de la fissuration dans la glace polycristalline en compression. Thèse de l'Université Joseph Fourier, Grenoble, France.

Kamb, W. B. (1972). Experimental recrystallization of ice under stress. American Geophysical Union, *Geophysical Monograph*, **16**, 211–241.

Kirby, S. H., W. B. Durham, M. L. Beeman, H. C. Heard and M. A. Daley (1987). Inelastic properties of ice Ih at low temperatures and high pressures. *J. Physique Coll.*, **48**, C1-227–C1-232.

Kohlstedt, D. L. and M. E. Zimmerman (1996). Rheology of partially molten mantle rocks, *Ann. Rev. Earth Planet. Sci.*, **24**, 41–62.

Langdon, T. G. (1973). Creep mechanisms in Ice. In *Physics and Chemistry of Ice*, eds. E. Whalley, S. J. Jones and L. W. Gold. Ottawa: Royal Society of Canada, pp. 356–361.

Langdon, T. G. (1993). The role of grain boundaries in high temperature deformation. *Mater. Sci. Eng.*, **A166**, 67–79.

Langdon, T. G. (1994). A unified approach to grain boundary sliding in creep and superplasticity. *Acta Metal. Mater.*, **42**, 2437–2443.

Langdon, M. and Yavari, P. (1982). An investigation of Harper-Dorn creep. II. The flow process. *Acta Metall.*, **30**, 881–887.

Lawson, W. (1996). The relative strength of debris-laden basal ice and clean glacier ice: some evidence from Taylor Glacier, Antarctica. *Ann. Glaciol.*, **23**, 270–276.

Lebensohn, R. A., C. N. Tomé and P. Ponte Castaneda (2007). Self-consistent modeling of the mechanical behavior of viscoplastic polycrystals incorporating intragranular field fluctuations. *Phil. Mag.*, **7**, 4287–4322.

Legrand, M. and P. Mayewski (1997). Glaciochemistry of polar ice cores: a review. *Rev. Geophys.*, **35**, 219–243.

Li Jun, T. H. Jacka and W. F. Budd (1996). Deformation rates in combined compression and shear for ice which is initially isotropic and after the development of strong anisotropy. *Ann. Glaciol.*, **23**, 247–252.

Li Jun, T. H. Jacka and V. Morgan (1998). Crystal size and microparticle record in the ice core from Dome Summit South, Law Dome, East Antarctica. *Ann. Glaciol.*, **27**, 243–348.

Lipenkov, V., N. I. Barkov, P. Duval and P. Pimienta (1989). Crystalline structure of the 2083 m ice core at Vostok Station, Antarctica. *J. Glaciol.*, **35**, 392–398.

Lipenkov, V., A. Salamatin and P. Duval (1997). Bubbly-ice densification in ice sheets: applications. *J. Glaciol.*, **43**, 397–407.

Liu, F., I. Baker and M. Dudley (1995). Dislocation-grain boundary interactions in ice crystals. *Phil. Mag.*, **A71**, 15–42.

Lliboutry, L. (1969). The dynamics of temperate glaciers from the detailed viewpoint. *J. Glaciol.*, **8**, 185–205.

Lliboutry, L. (1971). Permeability, brine content and temperature of a temperate glacier. *J. Glaciol.*, **10**, 15–29.

Lliboutry, L. and P. Duval (1985). Various isotropic and anisotropic ices found in glaciers and polar ice caps and their corresponding rheologies. *Ann. Geophys.*, **3**, 207–224.

Louchet, F., J. Weiss and Th. Richeton (2006). Hall-Petch law revisited in terms of collective dislocation dynamics. *Phys. Rev. Lett.*, **97**, 075504-1–075504-4.

Mangeney, A., F. Califano and K. Hutter (1997). A numerical study of anisotropic, low Reynolds number, free surface flow of ice-sheet modeling. *J. Geophys. Res.*, **102**, B10, 22,749–22,764.

Mangold, N., P. Allemand, P. Duval, Y. Geraud and P. Thomas (2002). Experimental and theoretical deformation of ice-rock mixtures: implications on rheology and ice content of Martian permafrost. *Planet. Space Sci.*, **50**, 385–401.

Marshall, H. P., J. T. Harper, W. T. Pfeffer and N. F. Humphrey (2002). Depth-varying constitutive properties observed in an isothermal glacier. *Geophys. Res. Lett.*, **29**, doi: 10.1029/2002GL015412.

Mellor, M. (1980). Mechanical properties of polycrystalline ice. In *Physics and Mechanics of Ice*, ed. P. Tryde. Berlin: Springer-Verlag.

Mellor, M. and D. M. Cole (1982). Deformation and failure of ice under constant stress or constant strain-rate. *Cold Reg. Sci. Technol.*, **5**, 201–219.

Mellor, M. and R. Testa (1969). Creep of ice under low stress. *J. Glaciol.*, **8**, 147–152.

Melton, J. S. and E. M. Schulson (1998). Ductile compressive failure of columnar saline ice under triaxial loading. *J. Geophys. Res.*, **103** (C10), 21,759–21,766.

Meyssonnier, J. and A. Goubert (1994). Transient creep of polycrystalline ice under uniaxial compression: an assessment of internal state variables models. *Ann. Glaciol.*, **19**, 55–62.

Michel, B. (1978). *Ice Mechanics*. Québec: Laval University Press.

Mohamed, F. A. (2005). On the origin of superplastic flow at very low stresses. *Mat. Sci. Eng.*, **A410–411**, 89–94.

Montagnat, M. and P. Duval (2000). Rate controlling processes in the creep of polar ice: influence of grain boundary migration associated with recrystallization. *Earth Planet. Sci. Lett.*, **183**, 179–186.

Montagnat M. and P. Duval (2004). The viscoplastic behavior of ice in polar ice sheets: experimental results and modelling. *C. R. Physique*, **5**, 699–708.

Montagnat, M., P. Duval, P. Bastie, B. Hamelin and V. Ya Lipenkov. (2003). Lattice distortion in ice crystals from the Vostok core (Antarctica) revealed by hard X-ray diffraction: implication in the deformation of ice at low stresses. *Earth Planet. Sci. Lett.*, **214**, 369–378.

Morgan, V. I. (1991). High-temperature ice creep tests. *Cold Reg. Sci. Technol.*, **19**, 295–300.

Muguruma, J., S. Mae and A. Higashi (1966). Void formation by non-basal glide in ice single crystals. *Phil. Mag.*, **13**, 625–629.

Mukherjee, A. K. (2002). An examination of the constitutive equation for elevated temperature plasticity. *Mater. Sci. Eng.*, **A322**, 1–22.

Nabarro, F. R. N. (2002). Creep at very low stresses. *Metall. Mater. Trans.*, **33A**, 213–218.

Obbard, R., I. Baker and K. Sieg (2006). Using electron backscatter diffraction patterns to examine recrystallization in polar ice sheets. *J. Glaciol.*, **52**, 546–557.

Paterson, W. S. B. (1991). Why is ice-age ice sometimes "soft"? *Cold Reg. Sci. Technol.*, **20**, 75–98.

Paterson, W. S. B. (1994). *The Physics of Glaciers*, 3rd edn. Oxford: Elsevier Science Ltd.

Peltier, W. R., D. L. Goldsby, D. L. Kohlstedt and Lev Tarasov (2000). Ice-age ice sheet rheology: constraints from the Last Glacial Maximum from the Laurentide ice sheet. *Ann. Glaciol.*, **30**, 163–176.

Petrenko, V. F. and R. W. Whitworth (1999). *Physics of Ice*. Oxford: Oxford University Press.

Pettit, A. C. (2006). Ice flow at low deviatoric stresses: Siple Dome, West Antarctica. In *Glacier Science and Environmental Change*, ed. P. G. Knight. Oxford: Blackwell Publishing, pp. 300–303.

Pettit, E. C. and Waddington, E. D. (2003). Ice flow at low deviatoric stress. *J. Glaciol.*, **49**, 359–369.

Pettit, E. C., Th. Thorsteinsson, H. P. Jacobson and E. D. Waddington (2007). The role of crystal fabric in flow near an ice divide. *J. Glaciol.*, **53**, 277–288.

Pimienta, P. and P. Duval (1987). Rate controlling processes in the creep of polar glacier ice. *J. Physique Coll.*, **48**, C1-243–C1-248.

Plé, O. and J. Meyssonnier (1997). Preparation and preliminary study of structure-controlled S2 columnar ice. *J. Phys. Chem.*, **101**, 6118–6122.

Poirier, J. P. (1985). *Creep of Crystals: High Temperature Deformation Processes in Metals, Ceramics and Minerals*. Cambridge: Cambridge University Press.

Raj, R. and M. F. Ashby (1971). On grain boundary sliding and diffusional creep. *Metall. Trans.*, **2**, 1113–1128.

Ramseier, R. O. (1972). Growth and mechanical properties of river and lake ice. Ph.D. Thesis, Laval University, Canada.

Read, W. T. (1953). *Dislocations in Crystals*. New York: McGraw-Hill Book Company.

Richeton, Th., J. Weiss and F. Louchet (2005). Dislocation avalanches: role of temperature, grain size and strain hardening. *Acta Mater.*, **53**, 4463–4471.

Rigsby, G. P. (1951). Crystal fabric studies on Emmons Glacier, Mount Rainier, Washington. *J. Geol.*, **59**, 590–598.

Ruano, O. A., J. Wolfenstime, J. Wadsworth and O. D. Sherby (1991). Harper-Dorn and power law creep in uranium dioxide. *Acta Metall. Mater.*, **39**, 661–668.

Russell-Head, D. S. and W. F. Budd (1979). Ice-flow properties derived from bore-hole shear measurements combined with ice-core studies. *J. Glaciol.*, **24**, 117–130.

Samyn, D., A. Svensson, S. J. Fitzsimons and R. D. Lorrain (2005). Ice crystal properties of amber ice and strain enhancement at the base of cold Antarctic glaciers. *Ann. Glaciol.*, **40**, 185–190.

Sanderson, T. J. O. (1988). *Ice Mechanics: Risks to Offshore Structures*. London: Graham and Trotman Limited.

Schmid, E. and W. Boas (1936). *Kristallplastizität*. Berlin: Springer.

Schulson, E. M. (2001). Brittle failure of ice. *Eng. Fract. Mech.*, **68**, 1839–1887.

Schulson, E. M. and S. E. Buck (1995). The ductile-to-brittle transition and ductile failure envelopes of orthotropic ice under biaxial compression. *Acta Metall. Mater.*, **43**, 3661–3668.

Schulson, E. M. and Nickolayev, O. Y. (1995). Failure of columnar saline ice under biaxial compression: failure envelopes and the brittle-to-ductile transition. *J. Geophys. Res.*, **100** (B11), 22,383–22,400.

Sinha, N. K. (1978). Rheology of columnar-grained ice. *Exp. Mech.*, **18**, 464–470.

Smith, C. S. (1948). Grains, phases, and interfaces: an interpretation of microstructure. *Trans. Metall. Soc. AIME*, **175**, 15–51.

Song, M., D. M. Cole and I. Baker (2005a). Creep of granular ice with and without dispersed particles. *J. Glaciol.*, **51**, 210–218.

Song, M., I. Baker and D. M. Cole (2005b). The effect of particles on dynamic recrystallization and fabric development of granular ice during creep. *J. Glaciol.*, **51**, 377–382.

Song, M., D. M. Cole and I. Baker (2006). Investigation of Newtonian creep in polycrystalline ice. *Phil. Mag. Lett.*, **86**, 763–771.

Steffensen, J. P. (1997). The size distribution of microparticles from selected segments of the GRIP ice core representing different climatic periods. *J. Geophys. Res.*, **102** (C12), 26,755–26,763.

Steinemann, S. (1958). Experimentelle Untersuchungen Zur Plastizität von Eis. *Beitr. Geol. Schweiz, Hydrologie*, **10**, 1–72.

Sunder, S. S. and M. S. Wu (1990). On the constitutive modeling of transient creep of polycrystalline ice. *Cold Reg. Sci. Technol.*, **18**, 267–294.

Tatibouet, J., J. Perez and R. Vassoille (1986). High-temperature internal friction and dislocations in ice Ih. *J. Physique*, **47**, 51–60.

Taupin, V., S. Varadhan, J. Chevy *et al.* (2007). Effects of size on the dynamics of dislocations in ice single crystals. *Phys. Rev. Lett.*, **99**, 155507-1–155507-4.

Thorsteinsson, Th. (2002). Fabric development with nearest-neighbor interaction and dynamic recrystallization. *J. Geophys. Res.*, **107** (B1), 1–13.

Thorsteinsson, Th., J. Kipfstuhl and H. Miller (1997). Textures and fabrics in the GRIP ice core. *J. Geophys. Res.*, **102** (C12), 26,583–26,599.

Van der Veen, C. J. and I. M. Whillans (1990). Flow laws for glacier ice: comparison of numerical predictions and field measurements. *J. Glaciol.*, **36**, 324–339.

Van der Veen, C. J. and I. M. Whillans (1994). Development of fabric in ice. *Cold Reg. Sci. Technol.*, **22**, 171–195.

Wakahama, G. (1967). On the plastic deformation of single crystal of ice. In *Physics of Snow and Ice*, ed. H. Oura. Sapporo: Hokkaido University Press, 291–311.

Weertman, J. (1973). Creep of ice. In *Physics and Chemistry of ice*, eds. E. Whalley, S. J. Jones and L. W. Gold. Ottawa: Royal Society of Canada.

Weertman, J. (1983). Creep deformation of ice. *Ann. Rev. Earth Planet. Sci.*, **11**, 215–240.

Weiss J., J. Vidot, M. Gay *et al.* (2002). Dome Concordia ice microstructure impurities effect on grain growth, *Ann. Glaciol.*, **33**, 552–558.

Wilson, C. J. L. (1986). Deformation induced recrystallization of ice: the application of in-situ experiments. American Geophysical Union, *Geophysical Monograph*, **36**, 213–232.

Wilson, C. J. L. and D. S. Russell-Head (1982). Steady state preferred orientation of ice deformed in plane strain at −1 °C. *J. Glaciol.*, **28**, 145–159.

Wilson, C. J. L. and Zhang, Y. (1996). Development of microstructure in the high-temperature deformation of ice. *Ann. Glaciol.*, **23**, 293–302.

Wolff, E. W. and 27 co-authors (2006). Southern Ocean sea-ice extent, productivity and iron flux over the past eight glacial cycles. *Nature*, **440**, 491–496.

7

Modeling the ductile behavior of isotropic and anisotropic polycrystalline ice

7.1 Introduction

During the gravity-driven flow of glaciers and ice sheets, isotropic ice at the surface progressively becomes anisotropic with the development of textures. Strain-induced textures, combined with the strong anisotropy of the ice crystal, make the polycrystal anisotropic. A polycrystal of ice with most of its c-axes oriented in the same direction deforms at least ten times faster than an isotropic polycrystal, when it is sheared parallel to the basal planes. Depending on the flow conditions, this anisotropy varies from place to place. To construct ice-sheet flow models for the dating of deep ice cores, this evolving viscoplastic anisotropy must be taken into account. Computation with isotropic and anisotropic flow models predicts at depth an age of ice that can differ by several thousand years. From Mangeney *et al.* (1997), anisotropic ice could be more than 10% younger above the bumps of the bedrock and could be older by more than 100% within hollows. An adequate constitutive relationship must be also incorporated within large-scale flow models to simulate the variation of polar ice sheets with climate.

Various models have been proposed to simulate the evolution of the anisotropy and the behavior of such ices. The increasing numerical capability of computers and the advances in theories that link materials' microstructures and properties have enabled the development of new concepts and algorithms that constitute the so-called *multiscale* approach for the modeling of material behavior. In this chapter, we restrict our analysis to physically based micro-macro models using a self-consistent approach. The full-field approach based on the fast Fourier transform (Lebensohn, 2001) is introduced. This model has been used to obtain the intracrystalline mechanical fields that develop in 2D columnar ice and remain inaccessible to homogenization techniques.

Results obtained with other types of models can be found in: Lliboutry (1993); Azuma (1994); Svendsen and Hutter (1996); Gödert and Hutter (1998); Morland and Staroszczyk (1998); Gagliardini and Meyssonnier (1999); Staroszczyk and

Morland (2000); Meyssonnier *et al.* (2001); Thorsteinsson (2001, 2002); Faria (2006); Gillet-Chaulet (2006). A review of theories developed to simulate the behavior of anisotropic ice in the context of the flow of ice sheets has been given by Hutter (1983), Placidi *et al.* (2006) and Gagliardini *et al.* (in press).

7.2 Modeling the behavior of isotropic and anisotropic ice

It is widely accepted that the anisotropy of glacier ice comes from the lattice preferential orientations of grains (i.e., crystallographic textures). Initially isotropic ice near the surface of polar ice sheets progressively becomes anisotropic with lattice rotation by intracrystalline slip. Hence, a directional variation of the viscosity must be considered. Very large variations of strain rates with the applied stress direction have been found with both laboratory and *in-situ* measurements (Russell-Head and Budd, 1979; Duval and Le Gac, 1982; Gundestrup and Hansen, 1984; Budd and Jacka, 1989). The directional viscosity of a given anisotropic polycrystal can differ by more than one order of magnitude depending on the direction of the applied stress. For the same effective shear stress and for ice with most of its *c*-axes oriented near the vertical direction, the minimum strain rate for *horizontal simple shear* is about 40 times that found when this ice is deformed in uniaxial compression along the vertical direction (Duval and Le Gac, 1982). A larger difference in strain rate was found by Shoji and Langway (1988).

The computation of the mechanical behavior of isotropic and anisotropic polycrystalline materials using self-consistent models is nowadays a standard approach. The one-site viscoplastic self-consistent (VPSC) theory (Molinari *et al.* 1987) has been adapted by Lebensohn and Tomé (1993) to fully account for polycrystal anisotropy. This VPSC approach, which uses a discrete description of the texture, was extensively used to simulate the plastic deformation of metallic and geological materials. It has been used to simulate the behavior of isotropic and anisotropic ice and the development of textures (Castelnau *et al.*, 1996a, 1997). Limitations of the tangent approximation, made to linearize the non-linear local mechanical behavior, have been identified (Lebensohn *et al.*, 2004a) and a new self-consistent approach (the second-order method), which leads to estimates that are exact to second order in the heterogeneity contrast, has been proposed to take into account intragranular fluctuations of the mechanical fields (Liu and Ponte Castaneda, 2004).

The VPSC model of Lebensohn and Tomé (1993) is described here and applied to the simulation of texture development in polar ice sheets. It is shown that the second-order (SO) model yields results on *anisotropic* ice in better agreement with experimental evidence than the VPSC model. On the other hand, this model gives information on intragranular field fluctuations. The full-field formulation based on the fast Fourier transform (FFT) is shown to be very efficient in characterizing the intracrystalline fields. As a first step, this model was applied to 2D columnar ice to use higher resolution with more discretization points and to visualize results in 2D representation (Lebensohn *et al.*, in press). Finally, the transient creep of isotropic polycrystalline ice is well reproduced by using a self-consistent approach, which deals with the *elasto-viscoplastic* coupling that appears in the transient regime.

7.2.1 The viscoplastic self-consistent (VPSC) tangent approach

The computation of the mechanical behavior and texture evolution of polycrystalline materials using a *self-consistent approach* is nowadays a standard approach in materials science. The behavior of the polycrystal is inferred from the assumed knowledge of the grain behavior. The usual requirement of polycrystal models is that the loading and displacement conditions at the boundary of the polycrystal are uniform. The polycrystal of interest is replaced by a fictitious homogeneous body with the same global behavior. As part of this approach, the concept of a representative volume element (RVE) is introduced, where the RVE must include enough grains to represent the behavior of the polycrystal as a whole. The average of stress and strain rate over all grains must coincide with the overall stress, strain and strain rate at the boundary.

The viscoplastic response of the polycrystal essentially depends on the current distribution of the grain orientations. For highly anisotropic grains, strong deformation gradients develop within them because of the contrast in properties between neighbors. The treatment of these materials is greatly improved by introducing into the calculations intragranular fluctuations of deformation and stress fields via higher-order statistical moments.

The more commonly used homogenization procedure is the uniform strain rate approximation. This approximation ensures compatibility and provides an *upper bound* to the overall stress, just as this procedure gives an upper bound to the elastic modulus of ice (Chapter 4). Grains of different orientations deform at the same rate. This Taylor model (1938), which leads to multiple slip and requires the activation of five independent slip systems, is not well adapted to strongly anisotropic material (Castelnau *et al.*, 1996a).

Another approach considers that the stress on each grain is equal to the macroscopic stress applied to the polycrystal (Sachs, 1928). This *static* approximation leads to a *lower bound* for the stress on both local and global scales, just as it leads to a lower bound on the elastic modulus. The equilibrium condition within the polycrystal is fully respected and only the easiest slip systems are activated. With this assumption, grains poorly oriented for basal slip may not deform at all. This model gives a rather weak estimate for the polycrystal behavior (Castelnau *et al.*, 1996a) and underestimates the overall anisotropy for textured ices (Castelnau *et al.*, 1997).

On the other hand, the deformation is necessarily incompatible. Several models developed to describe the anisotropic behavior of ice are based on this assumption (Lliboutry, 1993; Van der Veen and Whillans, 1994; Gödert and Hutter, 1998; Gagliardini and Meyssonnier, 1999; Thorsteinsson *et al.*, 1999).

A rigorous approach for the modeling of anisotropic ice is given by self-consistent methods. Accordingly, the overall response of the polycrystal is deduced from the known behavior of the crystal and from an assumption concerning the interaction of each grain with its environment. The viscoplastic self-consistent (VPSC) theory was developed by Hutchinson (1976) and was formulated in a general framework by Molinari *et al.* (1987). This VPSC approach was adapted to anisotropic materials by Lebensohn and Tomé (1993). Elasticity is not taken into account with this approach. The plastic anisotropy of both the grains and the polycrystal is taken into account. This model was used to simulate the behavior of anisotropic ice and the development of textures (Castelnau *et al.*, 1996a, 1997). Stress and strain rate fields are calculated in the polycrystal by solving stress equilibrium and incompressibility equations. This model regards each grain of the polycrystal as an inclusion embedded in an infinite homogeneous equivalent medium (HEM). The behavior of the HEM, which represents that of the polycrystal, is not known in advance.

The overall response of the polycrystal is approximated using a linearization procedure, which is assumed to represent the behavior within a certain range of stress and strain rate. Figure 7.1 gives a schematic representation of such a linearization in the vicinity of the overall strain rate $\bar{D}$ and overall stress $\bar{S}$. The "tangent" approach developed by Molinari *et al.* (1987) makes a *first-order* expansion around the point of interest. Other assumptions are also shown.

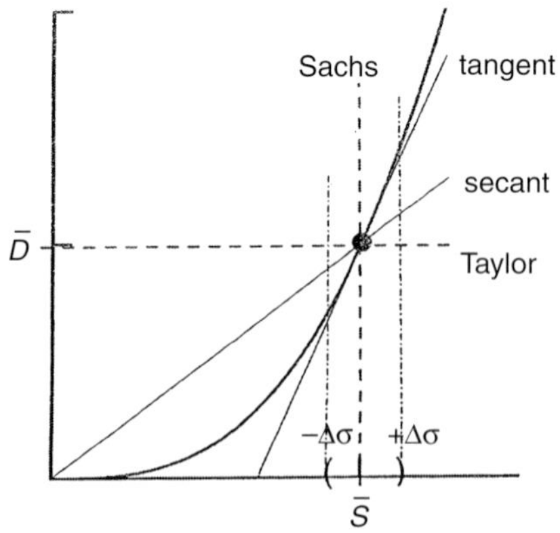

Figure 7.1. Schematic one-dimensional representation of the linearization of the stress against strain rate in the vicinity of the overall equivalent stress and strain rate for different types of interaction assumptions.

The deformation of the ice crystal is assumed to be produced by slip on basal, prismatic and pyramidal planes (Castelnau *et al.*, 1996a). The shear rate associated with the slip system s is a function of the local deviatoric Cauchy stress $\boldsymbol{s}$:

$$\dot{\gamma}^s = \dot{\gamma}_0 \left| \frac{r^s : s}{\tau_0^s} \right|^{n^s - 1} \frac{r^s : s}{\tau_0^s} \tag{7.1}$$

where $\dot{\gamma}_0$ is a reference shear rate, n^s and τ_0^s are, respectively, the stress exponent and the reference resolved shear stress (RRSS) for the system s. The Schmid tensor r^s expresses the orientation of the slip system s relative to the macroscopic axes:

$$r^s = \frac{1}{2}(n_s \otimes b_s + b_s \otimes n_s) \tag{7.2}$$

where n_s and b_s denote the unit vector normal to the slip plane and the unit vector parallel to the Burgers' vector, respectively. A slip system is defined by both plane and direction.

The constitutive law for the single crystal, which expresses the strain rate d^{g} is given by a sum over all slip systems:

$$d^{\mathrm{g}} = \sum_s \dot{\gamma}^s r^s. \tag{7.3}$$

The slip systems introduced for the deformation of the ice crystal were given in Table 6.1. From Equations (7.1) to (7.3), the components of the strain rates in each grain are given by:

$$d_{ij}^{\mathrm{g}} = M_{ijkl}^{\mathrm{g}} s_{kl}^{\mathrm{g}} \tag{7.4}$$

where M_{ijkl}^{g} represents the compliance that gives the instantaneous relation between strain rates and deviatoric stresses for each grain.

The constitutive relation for the polycrystal is given by:

$$D_{ij} = M_{ijkl} S_{kl} \tag{7.5}$$

where M_{ijkl} represents the compliance of the polycrystal.

The interaction equation that relates microscopic and macroscopic strain rates and stresses is defined by:

$$d_{ij}^{\mathrm{g}} - D_{ij} = -\tilde{M}_{ijkl}(s_{kl}^{\mathrm{g}} - S_{kl}) \tag{7.6}$$

where $\tilde{M}$ is the interaction tensor.

This treatment allows one to calculate the mechanical response of an isotropic and an anisotropic polycrystal. The macroscopic stress and strain rate are defined as the volume average of the microscopic states:

$$\left\langle d_{ij}^{g} \right\rangle = D_{ij} \quad \text{and} \quad \left\langle s_{ij}^{g} \right\rangle = S_{ij}. \tag{7.7}$$

Calculations based on this VPSC model show that the behavior of isotropic polycrystalline ice corresponding to secondary creep with $n=3$ can be obtained by assuming that $\tau_b/\tau_a = \tau_c/\tau_a = 70$ (Castelnau *et al.*, 1996a). The activity of basal slip systems is about 90% whereas the activity of pyramidal slip systems is less than 2%. The occurrence of such non-basal slip, however, has never been shown experimentally.

It is worth noting that the viscoplasticity of polycrystalline ice can be described by invoking only four slip systems (Hutchinson, 1977). Calculations based on the VPSC method show that the viscoplasticity of isotropic polycrystalline ice can be described by invoking only basal and prismatic slip, i.e., with $\tau_c/\tau_a = \infty$, but with a strength a little higher than that corresponding to secondary creep (Castelnau *et al.*, 1996a). Overall compatibility would be possible therefore without axial deformation along the c-axis, since only pyramidal slip can produce such deformation. A maximum in strain rate for a given equivalent stress is found when the c-axis is at 45° from the compression axis with the static (lower bound) and VPSC models. But, strain rate in grains oriented with the c-axis along the compression axis can be higher than 10% of the macroscopic counterpart. In these orientations, one expects that pyramidal slip may be activated. From Lebensohn *et al.* (2007), strain is mostly accommodated by basal slip, even in grains exhibiting a vanishing Schmid factor. This shows the importance of the redistribution of stresses within the polycrystal.

This VPSC tangent approach yields improvements over the Taylor and static methods. However, it exhibits limitations for high anisotropy of the grains and for textured ices. Figure 7.2 shows the predictions of several self-consistent models for the overall reference stress as a function of the grain anisotropy.

The tangent model, while very accurate for the lower value of the anisotropy factor τ_c/τ_a, underestimates the macroscopic behavior for the larger values of this factor. This model is based on a linearization scheme at local level that makes use of information on field averages only, disregarding higher-order statistical information within grains. For these anisotropic materials, strong deformation gradients develop inside grains to accommodate the mismatch of slip at grain boundaries (Montagnat *et al.*, 2003). From Lebensohn *et al.* (2004a), the tangent approach underestimates the behavior of stress fluctuations within the polycrystal for large values of the grain anisotropy factor. Considering these normalized results and the relatively low equivalent flow stress given by the tangent model, the non-normalized values of stress fluctuations for the tangent approach would be low compared with those given by other models. As shown in Figure 7.2, the tangent estimates obtained with a large grain anisotropy are close to the static estimates.

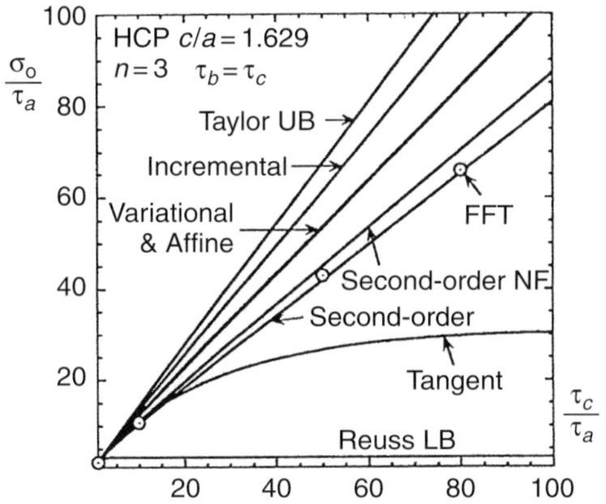

Figure 7.2. Plots of various self-consistent estimates and FFT simulation for the equivalent flow stress σ_0, normalized by the reference stress τ_a for isotropic ice polycrystals undergoing uniaxial tension as a function of the grain anisotropy τ_c/τ_a; three points were obtained with the FFT model. (From Lebensohn *et al.*, 2004a.)

Another limitation of the tangent approach is that it is a one-site model. It considers only the interaction between the inclusion (a grain) and a homogeneous equivalent medium whose mechanical behavior is that of the polycrystal. Several simple models, which take into account the interaction between each grain and the nearest neighbors (two-sites models), have been adapted with success to the behavior of ice (Azuma, 1994, 1995; Thorsteinsson, 2002). But only basal slip is considered in these models. On the other hand, as discussed below, by a rigorous treatment of the interaction between the inclusion and the HEM that incorporates information on field fluctuations within the grains, the *one-site* second-order self-consistent method also gives accurate estimates for the behavior of isotropic ice.

7.2.2 The second-order self-consistent and fast Fourier transform (FFT) models

The *second-order self-consistent* model (SO model) of Liu and Ponte Castaneda (2004) has the advantage that it incorporates information not only on the average stress and strain rate fields in the grains, but also on the second moments of the stress and strain rate over the grain *r*. These second moments are given by $\langle\sigma\otimes\sigma\rangle^{(r)}$ and $\langle\dot{\varepsilon}\otimes\dot{\varepsilon}\rangle^{(r)}$. Intragranular fluctuations serve to characterize the heterogeneities of the mechanical fields within the polycrystal, which are also useful in estimating the microstructural evolution during external loading. This model leads to estimates that are exact to second order in the heterogeneity contrast. This method gives the best agreement by comparison with the fast Fourier transform (FFT) model established for periodic heterogeneous materials (Fig. 7.2). This full-field model solves the

Stokes equations as properly as the finite element method, but with fewer calculations. In this approach, each crystal is decomposed into many elements, allowing one to infer the stress and strain rate heterogeneity at the microscopic scale.

7.3 Flow laws for secondary creep of isotropic ice

From laboratory creep tests and by assuming that the third invariant of the stress tensor has no effect on the material response, the flow law for secondary creep, as described in Chapter 6, is given by:

$$\dot{\varepsilon}_{ij} = \frac{B}{2}\tau^{n-1}\sigma'_{ij}. \tag{7.8}$$

From (7.8), the components of stress deviator and strain rate tensor are proportional. This was verified by Duval (1976) by performing tests in torsion, uniaxial compression and torsion-compression over the stress range $0.1 < \tau < 0.4$ MPa. The effect of high confining pressure is discussed in Chapter 8, in which the rheology of high-pressure ices is discussed.

For deformation conditions excluding cracking, the stress exponent is about $n = 3$ for an equivalent stress higher than 0.1 MPa and lower than 2 at low stresses (see Section 6.2.1). A polynomial flow law with $n = 1$ and 3.5 has been suggested by Lipenkov *et al.* (1997) from densification data of firn and bubbly ice in polar ice sheets for $0.01 < \tau < 0.8$ MPa. The effect of temperature on the coefficient B is discussed in Chapter 6.

7.4 Modeling transient creep

The affine self-consistent model of Masson and Zaoui (1999), which deals with the *elasto-viscoplastic* response of the polycrystal, is applied to the transient creep of isotropic ice. Results from this model are compared with data on the transient creep compiled by Ashby and Duval (1985) and given in Figure 6.4. To describe the transient creep, both the elastic and the viscoplastic response of ice need to be accounted for. The total strain rate $\dot{\varepsilon}(x, t)$ at any spatial position x and time t is given by:

$$\dot{\varepsilon}(x, t) = \dot{\varepsilon}^{\mathrm{el}}(x, t) + \dot{\varepsilon}^{\mathrm{vp}}(x, t). \tag{7.9}$$

The viscoplastic deformation is assumed to occur by dislocation glide on the basal, prismatic and pyramidal slip systems and the three reference stresses τ_a, τ_b and τ_c are assumed to evolve during transient creep as the dislocation density evolves. The stress exponent for the three slip systems is assumed to be different from that for the polycrystal, which is taken as $n = 3$. The stress exponent n_a and the stationary

Table 7.1. *Slip systems, stress exponents and reference shear stresses for the basal, prismatic and pyramidal slip*

	Slip systems	Stress exponent n	Initial reference shear stress (MPa)	Final reference shear stress (MPa)
Basal	$\{0001\}\langle 11\bar{2}0\rangle$	2.00	0.125	0.035
Prismatic	$\{01\bar{1}0\}\langle 2\bar{1}\bar{1}0\rangle$	2.85	0.21	1.30
Pyramidal	$\{11\bar{2}2\}\langle 11\bar{2}\bar{3}\rangle$	4.00	3.0	3.10

Calculations were made for a temperature of −10 °C.

reference stress τ_a^{sta} for the basal slip system have been determined at −10 °C from experimental data on basal slip in single crystals given in Figure 5.8. The stress exponent n_a is found equal to 2. The initial value of τ_a^{init} has been identified from the stress–strain curve given in Figure 5.2 for single crystals deformed for basal slip. Values of τ_a^{init} and τ_a^{sta} are given in Table 7.1. The decrease of τ_a with strain is directly related to the softening of basal slip.

The reference shear stress τ_b^{sta} and τ_c^{sta} for secondary creep and the stress exponents n_b and n_c have been identified to reproduce the viscoplastic response of isotropic ice with $n=3$. The obtained values are given in Table 7.1. As expected, pyramidal slip is found to be more difficult than prismatic slip. The ratio τ_b/τ_a of about 37 can be compared with the value of 70 found with the VPSC tangent model. Again, pyramidal slip contributes less than 2% of the total slip activity. Yet this slip system, even with a minor contribution, is of major importance since it allows one to obtain the value of $n=3$ for the polycrystal.

With data given in Table 7.1, the affine self-consistent model is now applied to simulate primary creep. Model results are shown in Figure 7.3 and compared with the data of Ashby and Duval (1985), expressed for the same loading conditions (uniaxial compression at −10 °C under a stress of 1 MPa). This model well reproduces the transient creep of isotropic ice by introducing a significant increasing resistance to shear of the prismatic slip systems. As shown in Table 7.1, the ratio $\tau_b^{\text{sta}}/\tau_b^{\text{init}}$ is more than 6. This *hardening* is shown to be important to reproduce the kinetics of the decrease of strain rate during primary creep (Castelnau *et al.*, 2008). The activity of pyramidal slip during transient creep is still less than 2%. As discussed above, in spite of its minor contribution to the overall deformation, this non-basal system is required to reproduce the macroscopic behavior with the value of the stress exponent equal to 3.

In conclusion, this self-consistent model, which simulates the behavior of elasto-viscoplastic polycrystals, well reproduces primary creep of isotropic polycrystalline

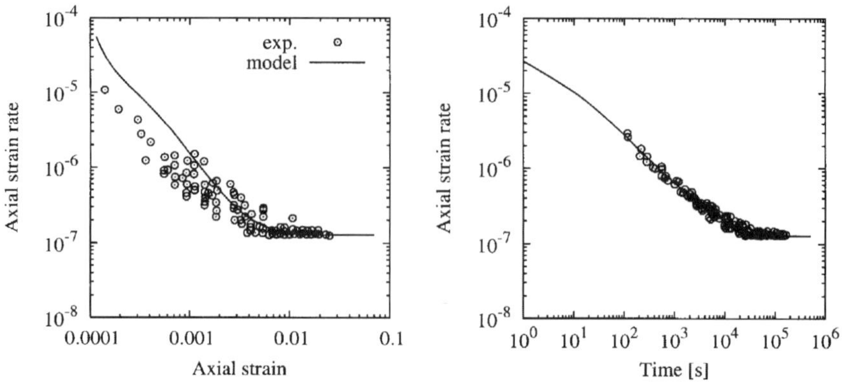

Figure 7.3. Transient creep response of isotropic ice under a uniaxial compressive stress of 1 MPa. Model results are compared to the data of Ashby and Duval (1985), expressed with the same loading conditions. (From Castelnau *et al.*, 2008.)

ice. As stated above, this self-consistent approach does not allow on analysis of transient creep in terms of criticality.

7.5 Modeling viscoplastic behavior of columnar ice with the FFT model

The full-field formulation based on the fast Fourier transform (FFT) has been used to characterize the intracrystalline fields (strain rate and stress) and visualize the results in a 2D representation on S2 columnar ice. An advantage in simulating the behavior of S2 columnar ice is that each crystallographic orientation is characterized by only one angular parameter; namely, the *c*-axis orientation with respect to the compression axis. The simulation was made on specimens tested in uniaxial compression, with the compression axis normal to the columns. The application of the FFT method requires the generation of a periodic unit cell. Figure 7.4 shows the square domain, constructed in such a way that it contained the cross sections of 200 columnar grains and generated by Voronoi tessellation. This unit cell was discretized using a $1024 \times 1024 \times 1$ grid of regularly spaced Fourier points, resulting in an average of around 5250 Fourier points per grain.

Three orientations were selected to analyze their intracrystalline fields with the *c*-axis at 0°, 45° and 90° with respect to the compression axis. Only results on the 45° grains and surrounding grains are given here. Calculation was made with the anisotropy factor $\tau_c/\tau_a = 30$ and $\tau_b = \tau_c$. Figure 7.5 shows the predicted fields of normalized equivalent strain rate, equivalent stress in unit of τ_{basal} and relative basal activity in the vicinity of the 45° grain. The imposed compression strain rate was $1 \times 10^{-8}\ s^{-1}$ and results are given for a finite strain of about 2%. Therefore, only the

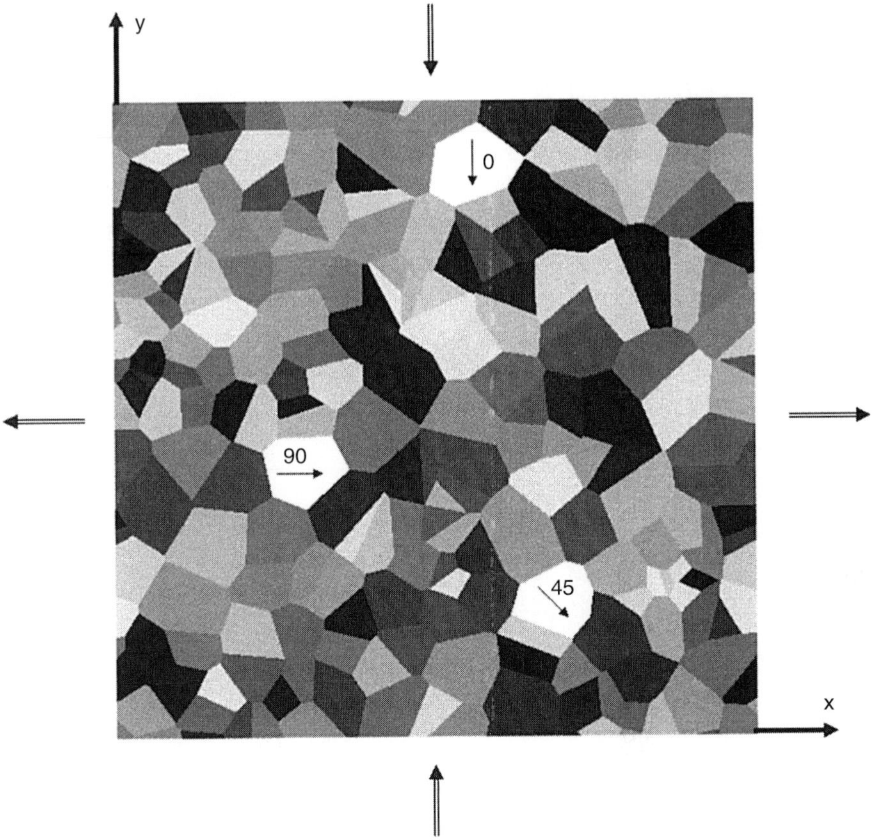

Figure 7.4. Unit cell constructed in such way that it contains the cross sections of 200 columnar grains. Three grains with particular orientations are shown. Note the periodic repetition of this unit cell along the *x* and *y* directions and infinitely long grains along the *z* direction. The analysis was restricted to the deformation behavior in the vicinity of the 45° grain.

instantaneous viscoplastic behavior is investigated. Two intense (i.e., local strain rates higher than 10 times the macroscopic strain rate) bands parallel to the *c*-axis are seen in the 45° grain, connected by several less intense basal shear bands. Both bands go through triple and quadruple points formed by the central grain and neighbor grains. The main *c*-axis band propagates up, following two grains (the −122.6 and −144.2 grains) with the same orientation. From experimental results obtained in the same conditions by Mansuy (2001), these bands are probably kink bands, which delimit two zones of the crystals with the same basal plane orientation. The basal activity in the 45° grain is very high, although some regions of higher non-basal activity can be observed between basal shear bands and outside the kink bands. As expected, low equivalent stress is associated with relatively high basal slip activity (Fig. 7.5). The average strain rate of each grain not only depends on its

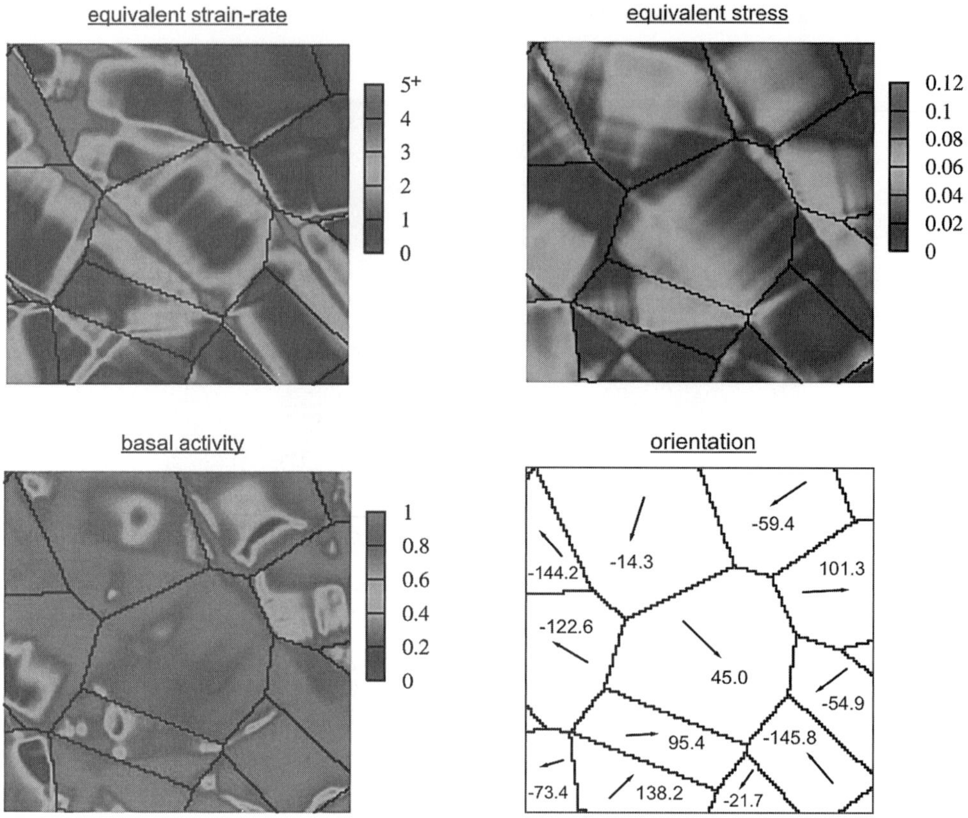

Figure 7.5. Intracrystalline fields predicted by the FFT model in S2 columnar ice deformed in compression across the columns.

orientation, but also depends on the orientation of its neighbors. It is worth noting that all grains deform heterogeneously. Some regions of grains deform up to more than 10 times the macroscopic strain rate, whereas other regions do not deform at all (Fig. 7.5).

The formation of kink bands in grains well oriented for basal slip can be understood by considering the internal stress field induced by the mismatch of slip at grain boundaries. The density of basal edge dislocations increases with deformation and a small component of compressive stress along basal planes can initiate such kink bands. From Nishikawa and Takeshita (1999), the formation of such kink bands (type II kink bands) does not require the high stresses found when the compression axis is parallel to the easy slip plane (type I kink bands). These two kink bands were observed by Wilson *et al.* (1986) during the deformation of layered polycrystalline ice.

In conclusion, under the assumption of creep deformation accommodated exclusively by dislocation glide, the deformation heterogeneity predicted by the FFT model is compatible with the predominance of basal slip. Both stress and strain rate are very heterogeneously distributed. The intragranular strain rate heterogeneity appears to be of the same level as the intergranular heterogeneity. The formation of basal shear bands is in agreement with the observation of slip bands in deformed single crystals (Montagnat *et al.*, 2006). Kink bands parallel to the *c*-axis, associated with basal edge dislocations, are compatible with observations on 2D ice (Mansuy, 2001). Such kink bands were also observed on S1 columnar ice by Manley and Schulson (1997). As expected, the FFT model does not reproduce the formation of type I kink bands by compression along the basal plane, which could form under high stresses. Incipient type I kink bands in ice, when loaded in compression parallel to the *c*-axis by a spherical indenter, could form at relatively high stresses (Zhou *et al.*, 2008). But, the formation of cracks could prevent the formation of such kink bands. It is worth noting that these results obtained on columnar ice cannot be extrapolated to very low stresses since several accommodation processes of basal slip can occur: grain boundary migration, recrystallization and grain boundary sliding (Chapter 6). One of the issues remaining in the modeling of polycrystalline ice is the introduction in models of some accommodation processes of basal slip.

7.6 Modeling texture development with the VPSC tangent model

7.6.1 Description and strength of the texture

The description of texture depends on the choice made for the micro-macro model that is used to describe the mechanical behavior of the aggregate. The first method considers the polycrystal to be an aggregate of a finite number of grains. The description of the texture is given by the Euler angles, which allow one to locate the grain reference frame ${}^{g}R$ with respect to the reference frame R attached to the polycrystal (Fig. 7.6). The number of Euler angles per grain depends on the constitutive model adopted to describe the grain. Three angles are necessary to give the directions of the *c*-axes and *a*-axes of the grain. But only two angles are needed since the polycrystal is considered to be transversely isotropic to any applied shear force acting in the basal plane (Kamb, 1961). Then, only the direction of the *c*-axes is considered (Fig. 7.6).

The other method considers the polycrystal to be made of an infinite number of grains, therefore allowing a continuous description of the texture by means of an orientation distribution function (ODF). When using the transversely isotropic model of a grain, the ODF is a function $f(\theta, \varphi)$ which gives the density of *c*-axes with orientation (θ, φ). The relative number of grains whose orientations lie in the interval $(\theta, \theta + d\theta; \varphi, \varphi + d\varphi)$ is then:

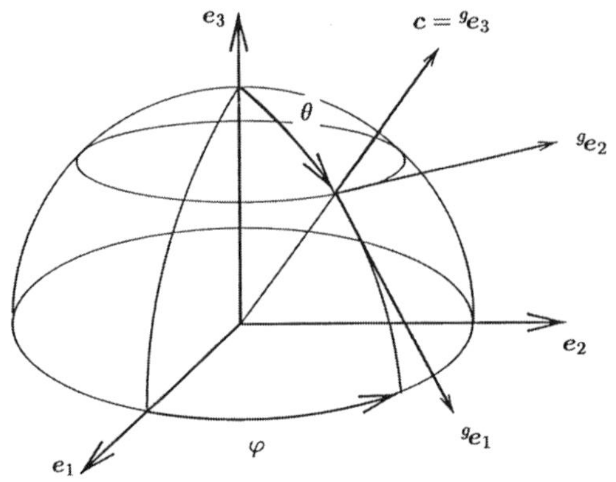

Figure 7.6. Reference frames $R(e_1, e_2, e_3)$ and ${}^gR({}^ge_1, {}^ge_2, {}^ge_3)$ attached to the polycrystal and to the grain, respectively.

$$\mathrm{d}N = \frac{1}{2\pi} f(\theta, \varphi) \sin\theta \, \mathrm{d}\theta \, \mathrm{d}\varphi. \tag{7.10}$$

Assuming that there is no correlation between the volume of a grain and its orientation, dN is also a volume fraction. The ODF fulfils the volume conservation, that is to say:

$$\frac{1}{2\pi} \int_0^{2\pi} \int_0^{\pi/2} f(\theta, \varphi) \sin\theta \, \mathrm{d}\theta \, \mathrm{d}\varphi. \tag{7.11}$$

For a transversely isotropic polycrystal, f does not depend on φ, i.e., $f(\theta, \varphi) \equiv f(\theta)$.

The strength of c-axes textures can be characterized by several techniques. The equal-area Schmidt projection with contour plots was the tool generally used to describe glacier ice textures (Rigsby, 1951). Another parameter used for describing texture strength is the statistical parameter R defined by:

$$R(\%) = \frac{100}{N} \left(2 \left[\sum_{g=1}^{N} c^g \right] - N \right) \tag{7.12}$$

where N is the total number of grains in the thin section (Thorsteinsson *et al.*, 1997). The c-axis of each grain g is treated as a unit vector c^g. R is equal to 0% for a randomly oriented texture and takes a theoretical maximal value of 100% when all c-axes are exactly parallel. In recent years, data have been collected using automatic methods and a thousand measured crystal orientations are common (Russell-Head and Wilson, 2001; Hansen and Wilen, 2002).

A good way to characterize textures in polar ice sheets is to use the second-order orientation tensor defined as:

$$a^{(2)} = \sum_{k=1}^{N} c^k \otimes c^k = \langle c \otimes c \rangle \tag{7.13}$$

where c^k is the unit vector along the c-axis. The tensor $a^{(2)}$ is symmetric and it is possible to find a reference frame in which $a^{(2)}$ is diagonal. The three corresponding eigenvalues $a_i^{(2)}$ $(i = 1, 2, 3)$ represent three orthogonal unit vectors along the axes of an ellipsoid which fits the distribution of c-axes (Durand *et al.* 2006). After normalization, the three eigenvalues follow the relationships:

$$\begin{aligned} &a_1^{(2)} + a_2^{(2)} + a_3^{(2)} = 1, \\ &0 \leq a_3^{(2)} \leq a_2^{(2)} \leq a_1^{(2)} \leq 1. \end{aligned} \tag{7.14}$$

The three values a_1^2, a_2^2 and a_3^2 are equal to $1/3$ for isotropic ice.

$a_2^2 \approx a_3^2 < 1/3$ for a single maximum texture.

Figure 7.7 shows the evolution with depth of the degree of orientation R, the normalized eigenvalues a_1^2, a_2^2 and a_3^2 and c-axes textures for the GRIP ice core

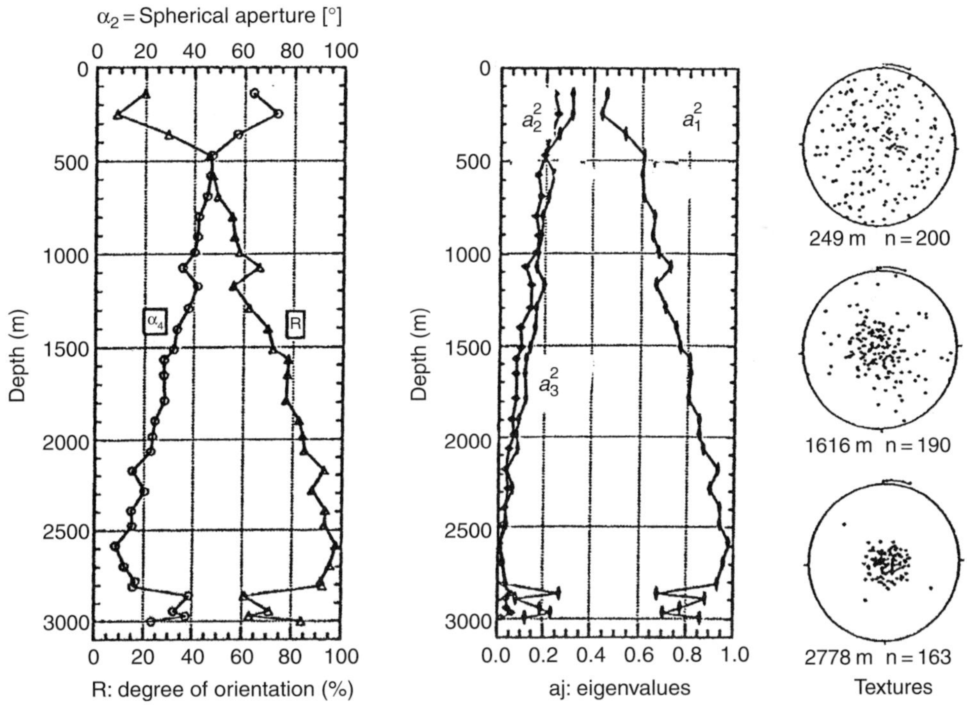

Figure 7.7. Degree of orientation $R(\%)$, spherical aperture α_s, eigenvalues a_i and textures along the GRIP ice core. The spherical aperture α_s gives the radius of a circle, whose center coincides with the position of the tip of the average c-axis. The point distribution is uniform within this circle. (From Thorsteinsson *et al.*, 1997.)

(from Thorsteinsson *et al.*, 1997). This figure shows the gradual evolution of texture with depth. Changes below 2800 m could be related to the occurrence of migration recrystallization (Thorsteinsson *et al.*, 1997).

7.6.2 Texture development

The change in the crystallographic lattice orientation of the grains during deformation determines the evolution of the texture. In a large part of ice sheets, lattice rotation by slip is the major mechanism for texture development, as noted in Chapter 6. Dynamic recrystallization also occurs, but is not taken into account in this analysis.

The lattice rotation rate of each grain is given by:

$$\dot{\omega}_{ij} = \dot{\Omega}_{ij} + \dot{\omega}^{\mathrm{e}}_{ij} - \dot{\omega}^{\mathrm{p}}_{ij} \tag{7.15}$$

where $\dot{\Omega}$ is the antisymmetric component of the macroscopic distortion rate and $\dot{\omega}^{\mathrm{e}}_{ij}$ represents the local rotation rate of the inclusion (Lebensohn and Tomé, 1993). The last term is the antisymmetric component of the plastic rotation rate, given by:

$$\dot{\omega}^{\mathrm{p}}_{ij} = \sum_{s} \frac{1}{2}(b_i n_j - b_j n_i)^s \dot{\gamma}^s. \tag{7.16}$$

Figure 7.8 displays textures obtained in uniaxial tension, uniaxial compression and simple shear after 0.4 equivalent strain with the tangent VPSC model. Textures predicted by the static and Taylor models are given for comparison. Calculations were made with $\tau_b = \tau_c = 70\tau_a$. The evolution of the ellipsoidal grain shape with deformation was taken into account.

Qualitatively, the predicted textures are in agreement with those obtained by assuming only basal slip (Azuma and Higashi, 1985; Lipenkov *et al.*, 1989; Alley, 1992). In uniaxial extension, a concentration of *c*-axes around the plane normal to the direction of extension is found. In uniaxial compression, a single maximum develops, where *c*-axes rotate toward the direction of the applied compression load. In simple shear, the single maximum lies between the normal to the macroscopic shear plane and the principal direction of compression. Large differences between all models appear for the same equivalent strain. The texture strength increases from the Taylor to the static assumptions. Predictions from the tangent VPSC model lie between the two bounds.

A more detailed analysis of these models can be made by considering the relative activity of all slip systems, for orientation changes are directly linked to such activity (Castelnau *et al.*, 1996a). Figure 7.9 shows the activity of all slip systems given by the Taylor and VPSC models. Diagrams from the static model are not plotted since the relative contribution of non-basal slip systems is always less than

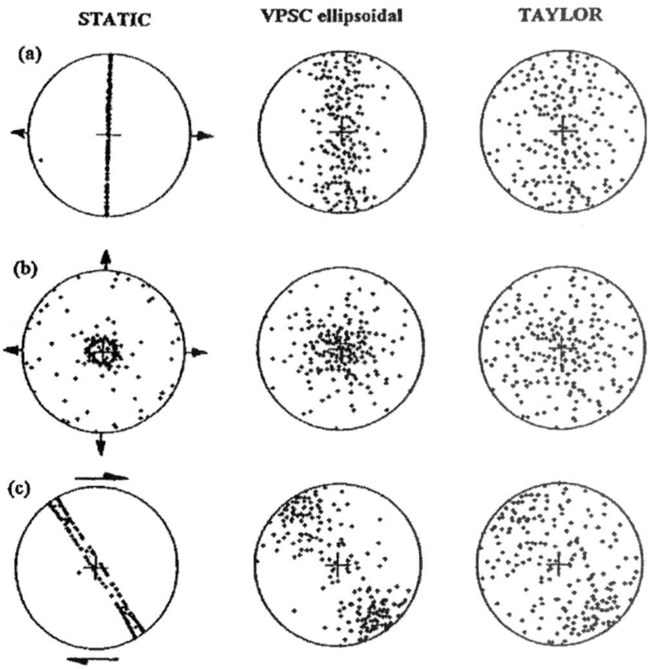

Figure 7.8. c-axes textures simulated with the static, VPSC and Taylor models. Calculations from the VPSC model were made with an evolutive ellipsoidal grain shape. (a) Uniaxial tension; (b) uniaxial compression; (c) simple shear. $\varepsilon_{eq}=0.4$; the arrows indicate the deformation directions. (From Castelnau *et al.*, 1996a.)

1%. The Taylor model predicts rather high activity (>0.2) on the pyramidal and the prismatic slip systems, whereas the basal activity is always less than 0.4 whatever the strain. Such low activity of the basal slip system clearly indicates that the uniform strain field assumption cannot simulate the behavior of the ice polycrystal. It is worth noting that the increasing non-basal activity slows down the rotation rate of c-axes, but does not influence the character of textures.

Results obtained with the tangent model are strongly dependent on the deformation state. In uniaxial extension, the basal activity continuously decreases whereas the prismatic activity increases. A similar tendency is found in simple shear. In uniaxial compression, the basal activity exceeds 0.9 up to a strain of 1. In comparison, the pyramidal activity starts to increase from a strain of 0.8. These results question the tangent assumption for the description of the behavior of textured ices.

It is worth noting that in simple shear two mechanisms compete in texture development: rotation of the c-axes toward the principal direction of compression and dragging of the texture maximum by the macroscopic rotation. Textures are not stable for a strain of 0.4. Indeed, the maximum continuously turns toward

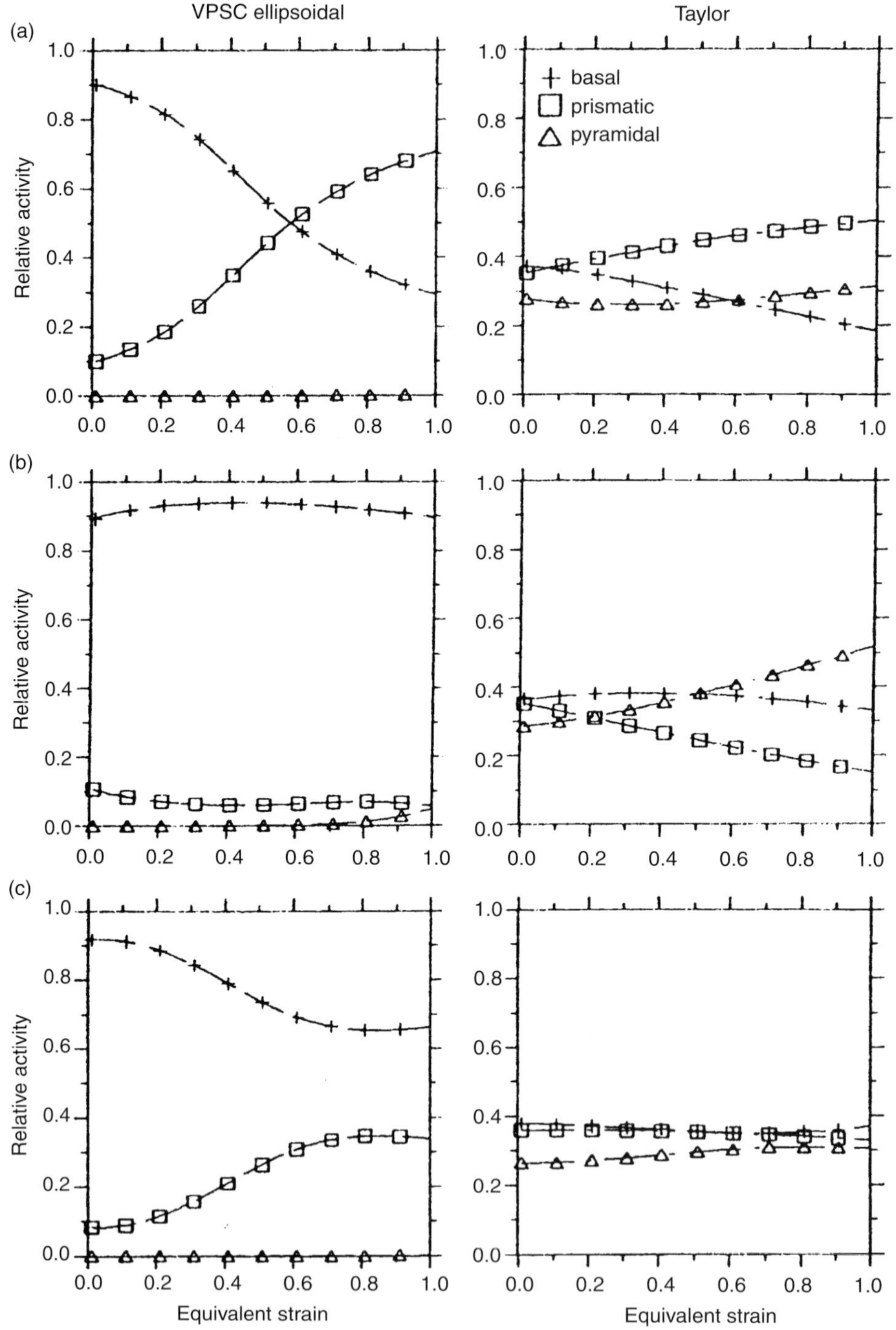

Figure 7.9. Relative activity of basal, prismatic and pyramidal slip systems from VPSC and Taylor models. The non-basal activity is always lower than 1% with the static model. (a) Uniaxial tension; (b) uniaxial compression; (c) simple shear. (From Castelnau *et al.*, 1996a.)

the normal to the macroscopic shear plane with increasing strain (Castelnau *et al.*, 1996a). The macroscopic rotation rate predominates until a stable texture is reached.

Comparison with textures in polar ice sheets

Textures observed along deep ice cores are interpreted by the rotation of the lattice through dislocation slip (Azuma and Higashi, 1985; Alley, 1992; de La Chapelle *et al.*, 1998). But grain growth and rotation recrystallization, which extensively occur in polar ice sheets (Chapter 6), can slow down the lattice rotation rate (Castelnau *et al.*, 1996b). Recrystallization textures associated with migration recrystallization (Chapter 6) are outside this discussion.

Figure 7.10 shows the GRIP *c*-axes texture observed at 991.1 m compared to that found with static, VPSC and Taylor models for a vertical deformation of about 0.4, corresponding to that estimated at this depth. The distribution of *c*-axes around the vertical direction predicted by each model appears to be different. A pronounced texture is found with the static model, and a more open one, with the Taylor model. The self-consistent estimate gives an intermediate result, but closer to that of the static model. The measured texture lies between the predictions of the VPSC and Taylor models. The texture strength therefore probably depends on the strength of the interaction between grains. The rotation rate of grains given by the tangent VPSC and the static models appears too high.

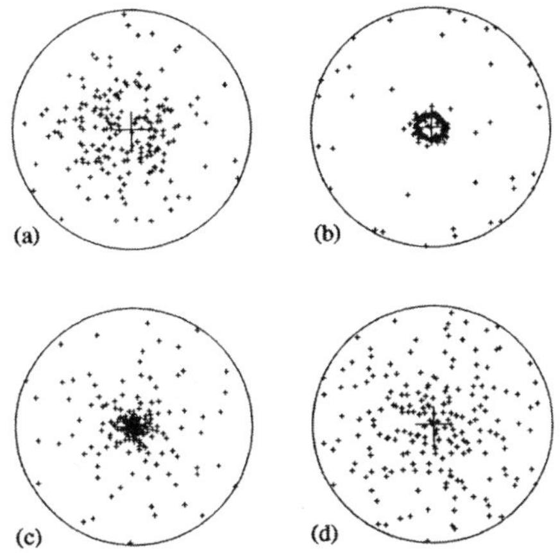

Figure 7.10. Comparison of (a) the texture pattern of the GRIP ice for a depth of 991.1 m (from Thorsteinsson *et al.*, 1997), with textures simulated in uniaxial compression with (b) the static, (c) the VPSC and (d) the Taylor model for a vertical strain of 0.4, corresponding to the estimated strain at this depth. The center of the diagrams indicates the *in-situ* vertical direction and the compression direction. (From Castelnau *et al.*, 1996a.)

A significant stress redistribution within the polycrystal or a more homogeneous strain within the polycrystal would be required to reproduce the observed textures. The model used by Thorsteinsson (2002), which takes into account the effects of nearest-neighbor interaction (NNI), provides a good estimate of the development of textures. It is worth noting that this approach leads to a more homogeneous strain.

The slower texture development, as compared with the VPSC estimate, could also be attributed to the effects of rotation recrystallization (Castelnau *et al.*, 1996b). The formation of new grains by the progressive misorientation of sub-boundaries should slow down the development of textures by slip. With this in mind, Lebensohn *et al.* (2007) obtained a good estimation of texture development with the second-order approach by introducing intragranular stresses and strain rate fluctuations. This self-consistent model predicts a quasi-persistent accommodation of deformation by basal slip, even when the *c*-axes become strongly aligned with the compression axis.

7.7 Modeling mechanical behavior of anisotropic ice

For most deformation models used to describe the behavior of anisotropic ice, each grain is assumed to deform only by basal slip (Lliboutry, 1993; Van der Veen and Whillans, 1994; Azuma, 1994; Gödert and Hutter, 1998). Azuma's model appears to be well adapted to simulate the behavior of textured ices. The stress applied on each grain is related to its Schmid factor and to a mean local Schmid factor that takes into account the first neighbors. Variations of strain-rate enhancement as a function of strain in uniaxial compression, uniaxial extension and simple shear are shown in Figure 7.11. The enhancement factor is defined by $E = \dot{\varepsilon}_{\xi\xi}/\dot{\varepsilon}_{\xi\xi 0}$, where $\dot{\varepsilon}_{\xi\xi 0}$ denotes the strain rate for isotropic ice. This model well reproduces the enhancement factors determined in the laboratory by Pimienta *et al.* (1987), even though the very small value of E in compression and tension from a strain higher than 0.6 is not realistic, since strain homogeneities within grains and/or rotation recrystallization should prevent the development of such strongly textured ices. These results clearly show that the behavior of anisotropic ice cannot be characterized by introducing one enhancement factor in Equation (7.8) established for isotropic ice.

Another approach to model the behavior of anisotropic ice is to obtain the instantaneous response of the ice polycrystal as a function of the texture described by the orientation distribution function (Gillet-Chaulet *et al.*, 2005; Gillet-Chaulet, 2006). Polar ice is considered as a linearly viscous orthotropic material (see Boehler, 1987). This assumption is questionable since there is no evidence of linear behavior in polar ice (Chapter 6). However, this approach has the major advantage of being easy to implement into an ice-sheet flow model (Gillet-Chaulet *et al.*, 2006). A

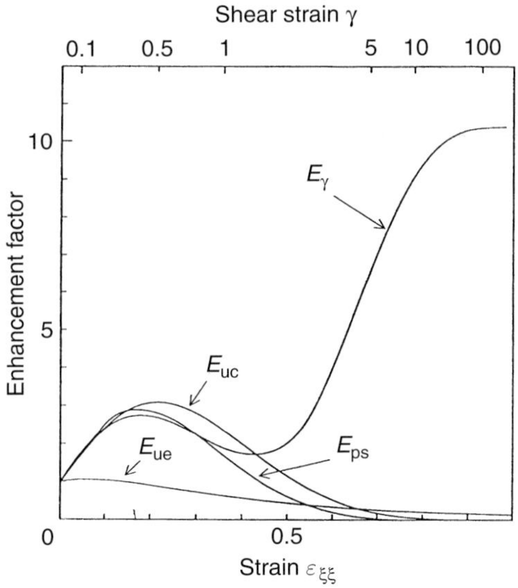

Figure 7.11. Variation of the strain rate enhancement factor ($E = \dot{\varepsilon}_{\xi\xi}/\dot{\varepsilon}_{\xi\xi 0}$) with strain for several stress states. E_{uc} (uniaxial compression), E_{ue} (uniaxial tension), E_{ps} (pure shear). The enhancement factor for simple shear E_γ is given as the ratio $\dot{\gamma}/\dot{\gamma}_0$. (From Azuma, 1994.)

simple, but informative, result has been given by Gillet-Chaulet (2006), which shows the extreme sensitivity of the behavior of ice with a single maximum texture, when ice is deformed in uniaxial compression along the direction of the *c*-axes. A small deviation of the orientation of this maximum ($<7°$) with respect to the direction of the compression can induce a shear higher than the deformation along the compression axis (Gillet-Chaulet, 2006). This effect is expected to be stronger when taking into account a non-linear flow law. Other applications of such continuum models can be found in Gillet-Chaulet (2006).

7.8 Modeling with self-consistent models

We have applied the VPSC tangent model to predict the mechanical behavior of polycrystalline ice samples exhibiting a single maximum texture. The *instantaneous* polycrystal response is compared with experimental results. Table 7.2 compares the enhancement factors obtained from mechanical tests with those given by the VPSC model. The agreement appears to be very good. But, as shown by Castelnau *et al.* (1996a), the progressive decrease of the activity of the basal systems with accumulated strain is not compatible with the deformation behavior of polycrystalline ice.

Predictions with the VPSC and SO models on the evolution of the mechanical behavior of polycrystalline ice deformed in compression are shown in Figure 7.12.

Table 7.2. *Enhancement factors obtained in uniaxial compression and torsion on a natural sample exhibiting a single maximum texture*

Sample	R(%)	E_c^{exp}	E_t^{exp}	E_c^{VPSC}	E_t^{VPSC}
BHC 354	89	0.25	10	0.171	11

E_c is the enhancement factor for uniaxial compression along the symmetry axis of the texture and E_t the enhancement factor in torsion.
The sample BHC 354 is from Law Dome in Antarctica at the depth of 354 m (Russell-Head and Budd, 1979).

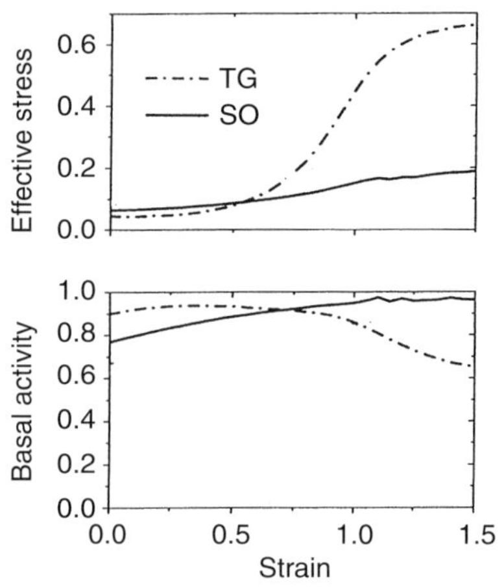

Figure 7.12. Simulation of the behavior of an ice polycrystal under uniaxial compression with the tangent VPSC and the second-order models. Calculations were made with $n=3$, $\tau_b=20\tau_a$ and $\tau_c=200\tau_a$ up to a compressive strain of 1.5. (From Lebensohn *et al.*, 2007; reprinted by permission of Taylor & Francis Ltd, www.infoworld.com.)

As shown in Figure 7.9, there is a decrease of the basal activity from a strain of about 0.8 with the tangent approach together with a rapid increase in the effective stress. This behavior indicates that strain accommodation requires the activation of hard pyramidal slip systems (Castelnau *et al.*, 1996a). The basal activity by itself is not enough to accommodate the deformation when *c*-axes become strongly aligned with the compression axis. Different behavior with greater activity of the basal slip systems is found with the SO model. The basal activity and the effective stress slowly increase with the alignment of *c*-axes along the compression axis (Fig. 7.12). But, as shown by Lebensohn *et al.* (2007), the stress becomes more uniform and the strain rate more heterogeneous as the *c*-axes concentrate along the compression axis. Then, the SO model shows the fundamental role of field fluctuations in grains for the deformation of the ice polycrystal. This model appears to give the best overall predictions for the effective behavior in power-law polycrystals that exhibit very high grain anisotropy (Lebensohn *et al.*, 2004a).

In conclusion, the mechanical behavior of anisotropic ice can be reasonably well described by a self-consistent approach provided that field fluctuations within the grains are introduced. For example, the second-order method developed by Liu and Ponte Castaneda (2004) for polycrystals yields results that are in agreement with full-field solutions based on the FFT algorithm of Lebensohn *et al.* (2004a, b). It is important to point out that the continuously decreasing numerical costs and the high numerical performance of the FFT and SO models do not restrict them to the simulation of the deformation behavior of laboratory samples. Depending on applications in glaciology, a good strategy could be to use these polycrystal models as an input to the development of large-scale ice sheet flow models. Another priority for the simulation of the deformation behavior of glacier ice would be to introduce dynamic recrystallization in these models, since it is very active in glaciers and ice sheets (Chapter 6).

References

Alley, R. B. (1992). Flow-law hypotheses for ice-sheet modeling. *J. Glaciol.*, **38**, 245–256.

Ashby, M. F. and P. Duval (1985). The creep of polycrystalline ice. *Cold Reg. Sci. Technol.*, **11**, 285–300.

Azuma, N. A. (1994). A flow law for anisotropic ice and its application to ice sheets. *Earth Planet. Sci. Lett.*, **128**, 601–614.

Azuma, N. A. (1995). A flow law for anisotropic polycrystalline ice under uniaxial compressive deformation. *Cold Reg. Sci. Technol.*, **23**, 137–147.

Azuma, N. and A. Higashi (1985). Formation processes of ice fabric pattern in ice sheets. *Ann. Glaciol.*, **6**, 130–134.

Boehler, J. P. (1987). Representations for isotropic and anisotropic non-polynomial tensor functions. In *Applications of Tensor Functions in Solid Mechanics*, ed. J. P. Boehler. Berlin: Springer-Verlag, pp. 31–53.

Budd, W. F. and T. H. Jacka (1989). A review of ice rheology for ice sheet modelling. *Cold Reg. Sci. Technol.*, **16**, 107–144.

Castelnau, O., P. Duval, R. A. Lebensohn and G. R. Canova (1996a). Viscoplastic modelling of texture development in polycrystalline ice with a self-consistent approach: comparison with bound estimates. *J. Geophys. Res.*, **101** (B6), 13,851–13,868.

Castelnau, O., Th. Thorsteinsson, J. Kipfstuhl, P. Duval and G. R. Canova (1996b). Modelling fabric development along the GRIP ice core, central Greenland. *Ann. Glaciol.*, **23**, 194–201.

Castelnau, O., G. R. Canova, R. A. Lebensohn and P. Duval (1997). Modelling viscoplastic behavior of anisotropic polycrystalline ice with a self-consistent approach. *Acta Mater.*, **45**, 4823–4834.

Castelnau, O., P. Duval, M. Montagnat and R. Brenner (2008). Elastoviscoplastic micromechanical modelling of the transient creep of ice. *J. Geophys. Res.*, **113**, B11203.

de La Chapelle, S., O. Castelnau, V. Lipenkov and P. Duval (1998). Dynamic recrystallization and texture development in ice as revealed by the study of deep ice cores in Antarctica and Greenland. *J. Geophys. Res.*, **103** (B3), 5091–5105.

Durand, G. and 10 others (2006). Effect of impurities on grain growth in cold ice sheets. *J. Geophys. Res.*, **111**, FO1015, 1–18.
Duval, P. (1976). Lois de fluage transitoire ou permanent de la glace polycristalline pour divers états de contrainte. *Ann. Geophys.*, **32**, 335–350.
Duval, P. and H. Le Gac (1982). Mechanical behavior of Antarctic ice. *Ann. Glaciol.*, **3**, 92–95.
Faria, S. H. (2006). Creep and recrystallization of large polycrystalline masses. I. General continuum theory. *Proc. R. Soc. A*, **462**, 1493–1514.
Gagliardini, O. and J. Meyssonnier (1999). Analytical derivations for the behaviour and fabric evolution of a linear orthotropic ice polycrystal. *J. Geophys. Res.*, **104** (B8), 17,797–17,809.
Gagliardini, O., F. Gillet-Chauvet and M. Montagnat (in press). A review of anisotropic polar ice models: from crystal to ice-sheet flow models. In *Physics of Ice Core Records* 11, ed. T. Hondoh. Sapporo: Hokkaido University Press.
Gillet-Chaulet, F. (2006). Modélisation de l'écoulement de la glace polaire anisotrope et premières applications au forage de Dôme C. Thèse de l'Université Joseph Fourier, Grenoble, France.
Gillet-Chaulet, F., O. Gagliardini, J. Meyssonnier, M. Montagnat and O. Castelnau (2005). A user-friendly anisotropic flow law for ice-sheet modelling. *J. Glaciol.*, **51**, 1–14.
Gillet-Chaulet, F., O. Gagliardini, J. Meyssonnier, T. Zwinger and J. Ruokolainen (2006). Flow-induced anisotropy in polar ice and related ice-sheet flow modeling. *J. Non-Newtonian Fluid Mech.*, **134**, 33–43.
Gödert, G. and K. Hutter (1998). Induced anisotropy in large ice shields: theory and its homogenization. *Continuum Mech. Thermodyn.*, **10**, 293–318.
Gundestrup, N. S. and B. L. Hansen (1984). Bore-hole survey at Dye 3, South Greenland. *J. Glaciol.*, **30**, 282–288.
Hansen, D. P. and L. A. Wilen (2002). Performance and applications of an automated *c*-axis fabric analyser. *J. Glaciol.*, **48**, 159–170.
Hutchinson, J. W. (1976). Bounds and self-consistent estimates for creep of polycrystalline materials. *Proc. R. Soc. Lond. A*, **348**, 101–127.
Hutchinson, J. W. (1977). Creep and plasticity of hexagonal polycrystals as related to single crystal slip. *Metall. Trans.* **8A** (9), 1465–1469.
Hutter, K. (1983). *Theoretical Glaciology*. Dordrecht: Reidel Publishing Company.
Kamb, W. B. (1961). The glide direction in ice. *J. Glaciol.*, **3**, 1097–1106.
Lebensohn, R. A. (2001). N-site modeling of a 3D viscoplastic polycrystal using Fast Fourier Transform. *Acta Mater.*, **49**, 2723–2737.
Lebensohn, R. A. and C. N. Tomé (1993). A self-consistent anisotropic approach for the simulation of plastic deformation and texture development of polycrystals: application to zirconium alloys. *Acta Metall.*, **41**, 2611–2624.
Lebensohn, R. A., Yi Liu and P. Ponte Castaneda (2004a). On the accuracy of the self-consistent approximation for polycrystals: comparison with full-field numerical simulations. *Acta Mater.*, **52**, 5347–5361.
Lebensohn, R. A., Yi Liu and P. Ponte Castaneda (2004b). Macroscopic properties and field fluctuations in model power-law polycrystals: full-field solutions versus self-consistent estimates. *Proc. R. Soc. Lond. A*, **460**, 1381–1405.
Lebensohn, R. A., C. N. Tomé and P. Ponte Castaneda (2007). Self-consistent modeling of the mechanical behavior of viscoplastic polycrystals incorporating intragranular field fluctuations. *Phil. Mag.*, **87**, 4287–4322.
Lebensohn, R. A., M. Montagnat, P. Mansuy *et al.* (2009). Modeling viscoplastic behavior and heterogeneous intracrystalline deformation of columnar ice polycrystals, *Acta Mater.* (in press).

Lipenkov, V., N. I. Barkov, P. Duval and P. Pimienta (1989). Crystalline structure of the 2083 m ice core at Vostok Station, Antarctica. *J. Glaciol.*, **35**, 392–398.

Lipenkov, V., A. Salamatin and P. Duval (1997). Bubbly-ice densification in ice sheets: applications. *J. Glaciol.*, **43**, 397–407.

Liu, Yi and P. Ponte Castaneda (2004). Second-order theory for the effective behavior and field fluctuations in viscoplastic polycrystals. *J. Mech. Phys. Solids*, **52**, 467–495.

Lliboutry, L. (1993). Anisotropic, transversely isotropic nonlinear viscosity of rock ice and rheological parameters inferred from homogenization. *Int. J. Plast.*, **9**, 619–632.

Mangeney, A., F. Califano and K. Hutter (1997). A numerical study of anisotropic, low Reynolds number, free surface flow of ice-sheet modeling. *J. Geophys. Res.*, **102** (B10), 22,749–22,764.

Manley, M. E. and E. M. Schulson (1997). Kinks bands and cracks in S1 ice under across-column compression. *Phil. Mag.*, **75**, 83–90.

Mansuy, Ph. (2001). Contribution à l'étude du comportement viscoplastique d'un multicristal de glace: hétérogénéité de la déformation et localisation, expériences et modèles. Thèse de l'Université Joseph Fourier, Grenoble, France.

Masson, R. and A. Zaoui (1999). Self-consistent estimates for the rate-dependent elasto-plastic behaviour of polycrystalline materials. *J. Mech. Phys. Sol.*, **47**, 1543–1568.

Meyssonnier, J., P. Duval, O. Gagliardini and A. Philip (2001). Constitutive modeling and flow simulation of anisotropic polar ice. In *Continuum Mechanics and Applications in Geophysics and the Environment*. Berlin: Springer-Verlag.

Molinari, A., G. R. Canova and S. Ahzy (1987). A self-consistent approach of the large deformation polycrystal viscoplasticity. *Acta Metall.*, **35**, 2983–2994.

Montagnat, M., P. Duval, P. Bastie, B. Hamelin and V. Ya. Lipenkov (2003). Lattice distortion in ice crystals from the Vostok core (Antarctica) revealed by hard X-ray diffraction: implication in the deformation of ice at low stresses. *Earth Planet. Sci. Lett.*, **214**, 369–378.

Montagnat, M., J. Weiss, P. Duval *et al.* (2006). The heterogeneous nature of slip in ice single crystals deformed under torsion. *Phil. Mag.*, **86**, 4259–4270.

Morland, L. and R. Staroszczyk (1998). Viscous response of polar ice with evolving fabric. *Continuum Mech. Thermodyn.*, **10**, 135–152.

Nishikawa, O. and T. Takeshita (1999). Dynamic analysis and two types of kink bands in quartz veins deformed under subgreenschist conditions. *Tectonophysics*, **301**, 21–34.

Pimienta, P., P. Duval and V. Ya. Lipenkov (1987). Mechanical behavior of anisotropic polar ice. In *The Physical Basis of Ice Sheet Modeling*, eds. E. D. Waddington and J. S. Walder. Vancouver: IAHS Publication 170, pp. 57–66.

Placidi, L., K. Hutter and S. H. Faria (2006). A critical review of the mechanics of polycrystalline polar ice. *GAMM-Mitteilungen*, **29**, 80–117.

Rigsby, G. P. (1951). Crystal fabric studies on Emmons Glacier, Mount Rainier, Washington. *J. Geol.*, **59**, 590–598.

Russell-Head, D. S. and W. F. Budd (1979). Ice-flow properties derived from bore-hole shear measurement combined with ice-core studies. *J. Glaciol.*, **24**, 117–130.

Russell-Head, D. S. and C. J. L. Wilson (2001). Automated fabric analyzer system for quartz and ice. *Geol. Soc. Austral. Abstr.*, **64**, 159.

Sachs, G. (1928). Zur Ableitung einer Fliessbedingung. *Z. Ver. Dtsch. Ing.*, **72**, 734–736.

Shoji, H. and C. C. Langway Jr. (1988). Flow-law parameters of the Dye 3, Greenland, deep ice core. *Ann. Glaciol.*, **10**, 146–150.

Staroszczyk, R. and L. W. Morland (2000). Plane ice-sheet flow with evolving orthotropic fabric. *Ann. Glaciol.*, **30**, 93–101.

Svendsen, B. and K. Hutter (1996). A continuum approach for modeling induced anisotropy in glaciers and ice sheets. *Ann. Glaciol.*, **23**, 262–269.

Taylor, G. I. (1938). Plastic strain in metals. *J. Inst. Met.*, **62**, 307–324.

Thorsteinsson, Th. (2001). An analytical approach to deformation of anisotropic ice-crystal aggregates. *J. Glaciol.*, **47**, 507–516.

Thorsteinsson, Th. (2002). Fabric development with nearest-neighbor interaction and dynamic recrystallization. *J. Geophys. Res.*, **107** (B1), 1–13.

Thorsteinsson, Th., J. Kipfstuhl and H. Miller (1997). Textures and fabrics in the GRIP ice core. *J. Geophys. Res.*, **102** (C12), 26,583–26,599.

Thorsteinsson, Th., E. D. Waddington, K. C. Taylor, R. B. Alley and D. D. Blankenship (1999). Strain-rate enhancement at Dye 3, Greenland. *J. Glaciol.*, **45**, 338–345.

Van der Veen, C. J. and I. M. Whillans (1994). Development of fabric in ice. *Cold Reg. Sci. Technol.*, **22**, 171–195.

Wilson, C. J. L., J. P. Burg and J. C. Mitchell (1986). The origin of kinks in polycrystalline ice. *Tectonophysics*, **127**, 27–48.

Zhou, A. G., S. Basu and M. W. Barsoum (2008). Kinking nonlinear elasticity, damping and microyielding of hexagonal close-packed metals. *Acta Mater.*, **56**, 60–67.

8

Rheology of high-pressure and planetary ices

8.1 Introduction

When subjected to high pressures and varying temperatures, ice can form in 12 known ordered phases. On Earth, only ice Ih is present because polar ice sheets are too thin to reach the critical conditions for the formation of ice II and ice III. The situation is very different for the large icy satellites of the outer planets. Temperature and pressure are such that current models of the internal structure of the principal moons of the outer planets suggest that a thick icy shell containing several high-pressure phases of ice surrounds the silicate core. The occurrence and properties of such ices are the subject of numerous studies to understand the tectonics and dynamics of these ices. A review of the main physical properties of high-pressure ices can be found in Klinger *et al.* (1985), Schmitt *et al.* (1998) and Petrenko and Whitworth (1999).

The regions of stability of ice crystalline phases on the pressure–temperature diagram are shown in Figure 8.1. Several phases are not mentioned in this figure because they are not stable or not present in icy satellites. In this chapter, the analysis of the mechanical properties of high-pressure ices is restricted to ices II, III, V and VI, the only ices studied in the laboratory. Crystalline structure, density and shear modulus of several phases of water ice are given in Table 8.1.

Due to the difficulty in performing deformation experiments under high pressure and low temperature, few data are available and results concern a narrow range of strain rates, roughly between 10^{-7} and $10^{-4}\,s^{-1}$. Extrapolation of these results to lower strain rates is always a challenge since deformation mechanisms that occur at high (with respect to creep) laboratory strain rates will not necessarily occur at lower strain rates. Considering the decrease of the stress exponent when the equivalent stress decreases, extrapolation of laboratory measurements to lower strain rates should provide an upper bound on the ductile strength.

A review of most results obtained on the rheology of ice at conditions applicable to icy satellites will be made with reference to the work of Durham *et al.* described

Table 8.1. *Structure and physical properties of several phases of water ice*

	Crystal system	Space group	T (K)	P (GPa)	Density ($\mathrm{kg\,m^{-3}}$)	Shear modulus (GPa)	Reference
Ice I	Hex.	$P6_3/mmc$	250	0	920	3.3[a]	(1)
			250	0.21	941	3.3	(1)
Ice II	Rhomb.	$R\bar{3}$	238	0.283	1193	6.2	(2)
Ice III	Tetra.	$P4_12_12_1$	246	0.276	1166	4.6	(2)
Ice V	Monocl.	A2/a	237.5	0.48	1267	6.1	(2)
Ice VI	Tetra.	$P4_2/nmc$	237.5	0.777	1360	7.5	(2)

[a] A higher value of 3.52 GPa was given by Gammon *et al.* (1983) at 257 K (Chapter 4).
(1) Shaw (1986); (2) Gagnon *et al.* (1987)

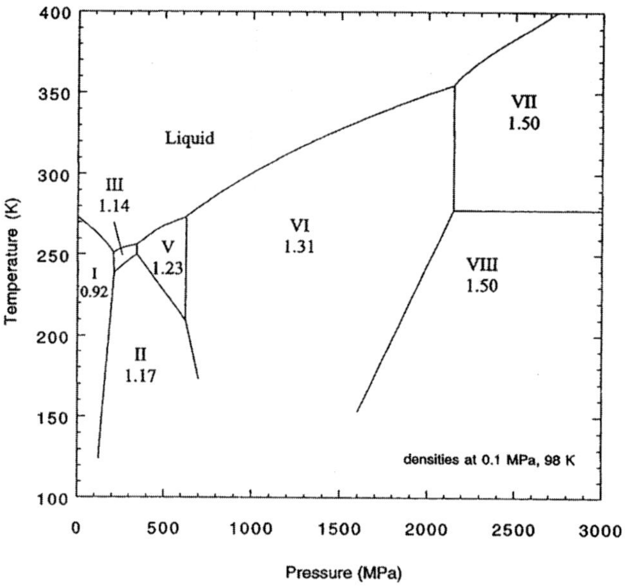

Figure 8.1. Phase diagram of water ice. Data are from Bridgman (1912) and Fletcher (1970). (From Durham and Stern, 2001.)

in several papers and summarized in Durham *et al.* (1997) and Durham and Stern (2001). Gas hydrates exist at moderate and high pressure on Earth and in icy satellites of giant planets. Carbon dioxide hydrate is present on the Mars south polar cap. Gas clathrate hydrates may impact the dynamics of icy satellites by their particular physical properties. The rheology of methane clathrate hydrate under relatively low confining pressure (< 100 MPa) is analyzed on the basis of results given by Durham *et al.* (2003a,b). Finally, ammonia has been proposed as a major constituent of several satellites of the outer Solar System. For a mole fraction x of NH_3 smaller

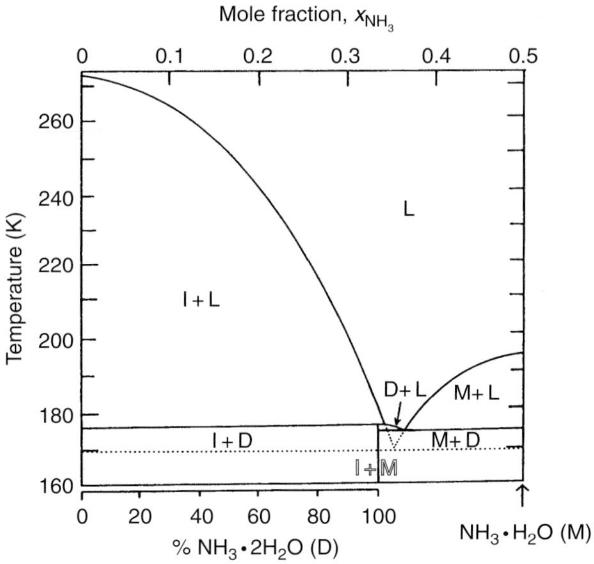

Figure 8.2. Phase diagram for the system NH_3–H_2O at about 0.1 MPa for ammonia mole fraction <0.5. The metastable ice–monohydrate eutectic is shown by the dotted lines. (From Durham *et al.*, 1993.)

than 0.33, water ice and the ammonia dihydrate are present at a temperature lower than 176 K, the peritectic temperature, and at the pressure of 0.1 MPa (Fig. 8.2). A short analysis of the flow of the ammonia–water system will be made.

8.2 Experimental methods

Most of the mechanical tests described here were performed with a conventional triaxial high-pressure apparatus surrounded by a cryostatic tank (Durham *et al.*, 1983). The capacity of the rig is about 1 GPa, making possible the study of ice V and VI. Samples were generally deformed in compression at constant displacement rate. The strain rate range of this system is 3.5×10^{-4} to 3.5×10^{-7} s^{-1}. Results obtained from experiments performed at constant stress are also given. Test samples were typically circular cylinders of 63.5 mm length and 25.4 mm in diameter. The rheology of high-pressure ices V and VI was determined with smaller samples, 32 mm in length and 12.7 mm in diameter. Ice samples were prepared in the laboratory from distilled water and sieved seed ice. After freezing and grain growth, a grain size of 250 ± 50 μm was obtained (Stern *et al.*, 1997). For the study of grain-size sensitive rheology, a pressure-release treatment procedure was developed (Stern *et al.*, 1997). Standard polycrystalline ice samples are pressurized to about 300 MPa to convert ice I to ice II. After the complete transformation, the confining pressure is rapidly reduced to a pressure lower than 50 MPa. Nucleation of ice grains in underpressurized ice II produces fine grains. Polycrystalline ice with grain size smaller than 10 μm can be produced with this technique on the condition of a pressure release at a temperature lower than 220 K.

Conversion of ice I to the higher-pressure phases is simply accomplished by pressurizing samples. Phase identification is generally based on the observed pressure anomalies during transformation. X-ray diffraction measurements on high-pressures samples, metastably recovered at room temperature, can confirm the inferred phase.

Some mechanical experiments are performed by using a diamond anvil cell (Poirier *et al.*, 1981; Sotin *et al.*, 1985; Sotin and Poirier, 1987). The hydrostatic pressure gradient in the cell can induce plastic deformation. The cell is installed under an optical microscope to follow displacement markers initially embedded in ice. Samples used with this technique are discs with a thickness of about 200 μm and a diameter of about 1.5 mm. Identification of phases, small size of samples, stress determination and strain measurements can be limiting factors for the determination of the rheology of such ices.

8.3 Viscoplasticity of ice II and III

Deformation experiments on ices II and III were performed by Durham *et al.* (1988) under the following conditions: $178 < T < 257$ K, $220 \leq P \leq 500$ MPa and $3.7 \times 10^{-6} < \dot{\varepsilon} < 5.2 \times 10^{-4}\ \mathrm{s}^{-1}$. Grain size was about 1 mm. Stress–strain curves are comparable to those obtained with the same conditions on ice I (Durham *et al.*, 1983). Following a quasi-linear response, a strength maximum is reached. A steady-state stress is achieved after a strain generally higher than 4%. Steady state is described by using the empirical law:

$$\dot{\varepsilon} = A\sigma^n \exp[-(E^* + PV^*)/RT] \tag{8.1}$$

where A is a flow constant, n the stress exponent, E^* the activation energy, V^* the activation volume and R the gas constant.

Values of these parameters are given in Table 8.2. Depending on temperature, the results give two flow laws with a boundary at 220 K for ice II and 230 K for ice III. From Durham *et al.* (1988), the flow of ice II is well characterized, whereas the reproducibility is poor for ice III. The value of the activation volume could not be determined for ice III from experimental data. A value of $V^* = 0$ is therefore given by the authors in the table. Ice II appears to be harder than ice I and ice III. The greatest contrast between ice I and ice II is in the activation volume, being negative for ice I and positive for ice II (Durham *et al.*, 1987). It is worth noting that the negative value of V^* for ice I is generally explained by the effect of pressure on the melting temperature and accommodation processes of basal slip (Jones and Chew, 1983).

Table 8.2. *Rheological constants for different ices used in the equation:* $\dot{\varepsilon} = A\sigma^n \exp[-(E^* + PV^*)/RT]$

	T (K)	$\log A$ ($MPa^{-n} s^{-1}$)	n	E^* ($kJ\,mol^{-1}$)	V^* ($cm^3\,mol^{-1}$)	Reference
Ice I	240–258	11.8 ± 0.4	4 ± 0.6	91 ± 2	−13	(1)
Ice I	195–240	5.1 ± 0.03	4 ± 0.1	61 ± 2	−13	(1)
Ice I	< 195	−1.4	5.6	43		(1)
Ice II	220–250	11.7 ± 1.6	5.2 ± 0.3	98 ± 8	4 ± 4	(2)
Ice II	< 220	1.84 ± 0.27	5.3 ± 0.1	55 ± 2	7 ± 1	(2)
Ice III	230–250	26.4 ± 4.8	4.3 ± 2	151 ± 21		(2)
Ice III	<220	13.3 ± 1.5	6.3 ± 0.4	103 ± 8		(2)
Ice V	209–270	23	6.0 ± 0.7	136 ± 38	29 ± 8	(3)
Ice V	233–263	−1.27	2.7 ± 1	32.4 ± 10	10.1 ± 9	(4)
Ice VI	> 250	6.7	4.5	110 ± 20	11	(3)
Ice VI	< 250	6.7	4.5	66 ± 4	11 ± 6	(3)
Ice VI[a]	253–293	−10.6	1.9	28.5 ± 1	8.1 ± 0.8	(4)
$NH_3.2H_2O$	145–176	21.55	5.81	107.5		(5)

[a] The mole fraction of dihydrate is equal to 1.
(1) Durham *et al.* (1992); (2) Durham *et al.* (1988); (3) Durham *et al.* (1996); (4) Sotin and Poirier (1987); (5) Durham *et al.* (1993)

Results discussed above are in agreement with those found by Echelmeyer and Kamb (1986) from high-pressure extrusion experiments.

Grain-size dependent behavior

A grain-size related weakening effect in ice II was assumed by Stern *et al.* (1997). Creep experiments performed on fine-grained ice II at 200 K and compression stresses between 4 and 15 MPa by Kubo *et al.* (2006) show a clear grain-size effect. A change in the stress exponent from 5 to 2.5 is correlated with a decrease of grain size from about 40 to 6 μm (Fig. 8.3). The value of 2.5 for the stress exponent is found with grain size of about 6 μm and at strain rates lower than $10^{-7}\,s^{-1}$. To take into account this grain size effect, the flow law (8.1) can be written as:

$$\dot{\varepsilon} = Ad^{-p}\sigma^n \exp[-(E^* + PV^*)/RT] \tag{8.2}$$

where d is grain size and p is the grain-size exponent.

The value of $p > 1.5$ estimated by the authors must be taken with caution because results were obtained with only two values of grain size. Unfortunately, no information was given on the microstructure after deformation. It is important to point out that no softening was found when fine-grained ice was deformed at a constant strain rate of about $4 \times 10^{-8}\,s^{-1}$ (Kubo *et al.*, 2006). The occurrence of dynamic recrystallization observed at relatively high strain rates would not be an efficient softening process at low strain rates when the stress exponent is about 2.5.

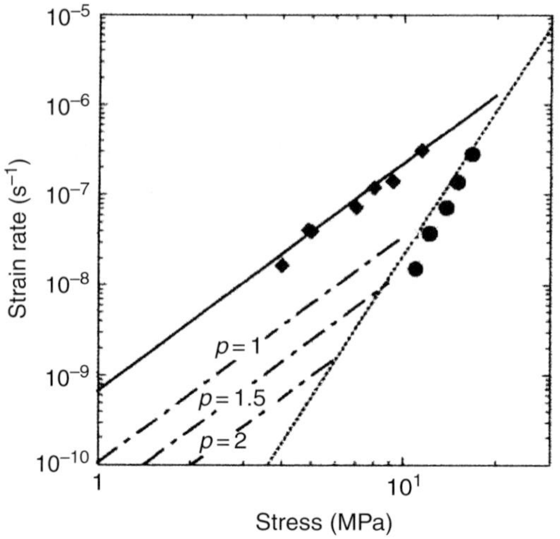

Figure 8.3. Creep data for ice II at 200 K. The grain size is estimated to be 38 ± 14 μm (circles) and 6 ± 2 μm (diamonds). Data are adjusted to a pressure of 200 MPa. The dashed lines show the rheology for a grain size of 38 μm with different values of p (Eq. 8.2). The solid line is obtained with a stress exponent $n = 2.5$. (From Kubo *et al.*, 2006; reprinted with permission from AAAS.)

8.4 Viscoplasticity of ice V and VI

Deformation experiments on ice V and VI have been performed with the sapphire anvil cell by Sotin *et al.* (1985) and Sotin and Poirier (1987) and with the triaxial apparatus by Durham *et al.* (1988, 1996). Values of the flow parameters for the two studies are given in Table 8.2. Large differences in the activation energy and the stress exponent are found. In spite of these discrepancies, there is relatively good agreement on the strength of ice VI. A serious inconsistency between the two studies is in the strength of ice V. For example, strain rate deduced from data given in Table 8.2 for a temperature of 250 K, a pressure of 500 MPa and a stress of 10 MPa is respectively 4×10^{-7} with the sapphire anvil cell and $3.6 \times 10^{-3}\ s^{-1}$ with the triaxial press. Despite discussions between the two laboratories, there is no clear explanation for this large difference. The deformation of small ice samples in the anvil cell could be inhomogeneous at high stresses ($\sigma > 50$ MPa in Sotin and Poirier, 1987).

8.5 Comparison of ices

The rheologies of high-pressure ices are compared to the rheology of ice I in Figure 8.4. Rheology for each phase is shown at representative pressures. Taking into account the pressure dependence (Table 8.2), ice II is the hardest phase, whereas ice III is the weakest. Ice III shows a viscosity contrast of several orders

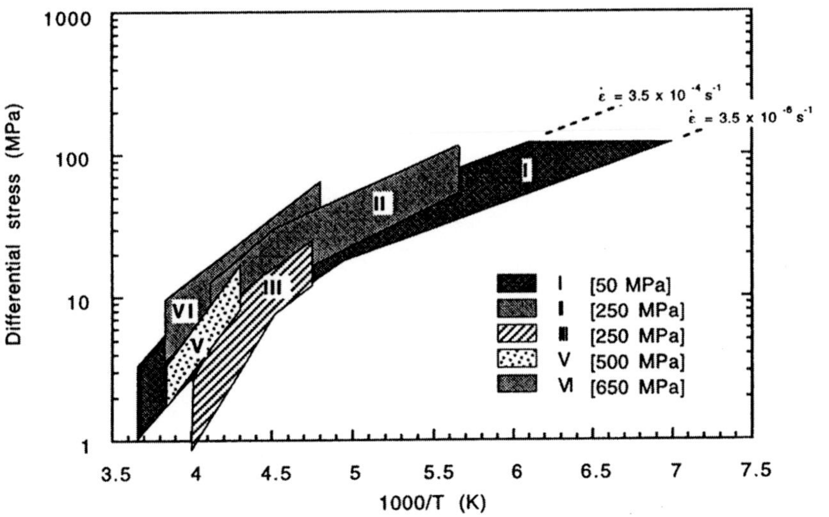

Figure 8.4. Steady-state strength of ices I, II, III, V and VI as a function of temperature. Vertical variations of strength for each ice are related to variations of strain rate. Comparison is made at pressures representative for each phase. (From Durham *et al.*, 1998; with permission of Springer Science and Business Media.)

of magnitude between 210 and 250 K. But, as noted above, the reproducibility of results on ice III is poor (Durham *et al.*, 1988). Ice V and VI are distinct in terms of their activation energy and stress exponent, but their strengths overlap at laboratory conditions.

8.6 Flow of methane clathrate hydrates

The ductile behavior of methane hydrates with structure 1, which belongs to cubic Pm3n (Sloan, 1998), was analyzed by Durham *et al.* (2003a,b) by conducting compression creep tests over the temperature range 269–287 K and confining pressure of 50–100 MPa. Cylindrical samples were synthesized from granular water seed ice held under a pressure of methane gas higher than 27 MPa and heated to a temperature close to the pure ice melting point. The polycrystalline aggregates contained about 27% porosity after full reaction and prior to testing. Samples consisted of clusters of equant grains of about 250 μm. The porosity was eliminated by compaction of samples at about 175 K.

Samples were tested in the triaxial apparatus described by Durham *et al.* (1983), under creep conditions at differential stresses between 5.7 and 51.7 MPa. Most of the samples were shortened by about 20% and steady state was considered. Figure 8.5 shows the strength of methane hydrate samples at the strain rate of $3.5 \times 10^{-7}\,s^{-1}$ and for a confining pressure of 50 MPa. Methane hydrate appears

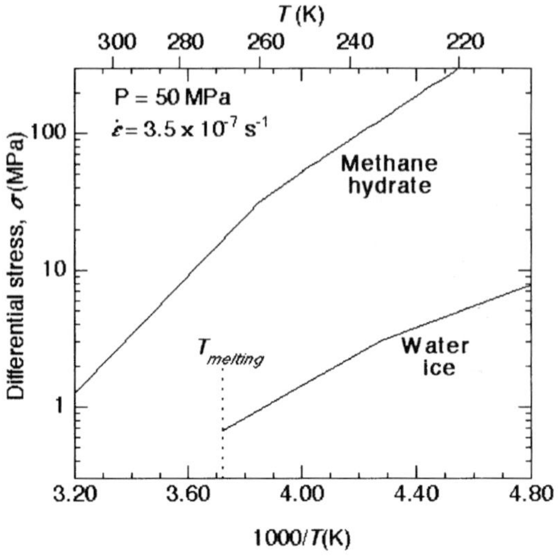

Figure 8.5. Ductile flow behavior of methane hydrate in comparison with water ice. (From Durham *et al.*, 2003b; with permission of NRC Research Press.)

to be very strong compared with ice Ih (Fig. 8.5). This difference could be related to the nature and the mobility of dislocations in such a material. Deformation produced by molecular diffusion is expected to be very slow in gas hydrates (Sloan, 1998). The high strength of methane clathrate hydrate could be extrapolated to other hydrates with significant implications for the convective flow in icy satellites and in the development of landforms on Mars with the presence of carbon dioxide hydrate. On Earth, the mass movement of hydrate-bearing sediments triggered by gas hydrate decomposition could be amplified by the drastic loss of strength of sediments.

8.7 Flow of ice in the ammonia–water system

This analysis is based on results obtained by Durham *et al.* (1993) with a mole fraction of ammonia x_{NH_3} lower than 0.33, which gives for temperatures lower than 176 K (the peritectic temperature of this ammonia–water system) a mixture of ice I and ammonia dihydrate crystals. Polycrystalline samples with controlled grain size and various composition x_{NH_3} were prepared by packing seed ice into moulds, evacuating the moulds, backfilling with ammonia–water liquid to saturate pores and, finally, putting the moulds at a temperature lower than 176 K (Durham *et al.*, 1993).

Constant strain rate tests were generally carried out with the triaxial press below 176 K. Some experiments were performed between 176 and 220 K on samples containing a liquid phase. Results are summarized as follows:

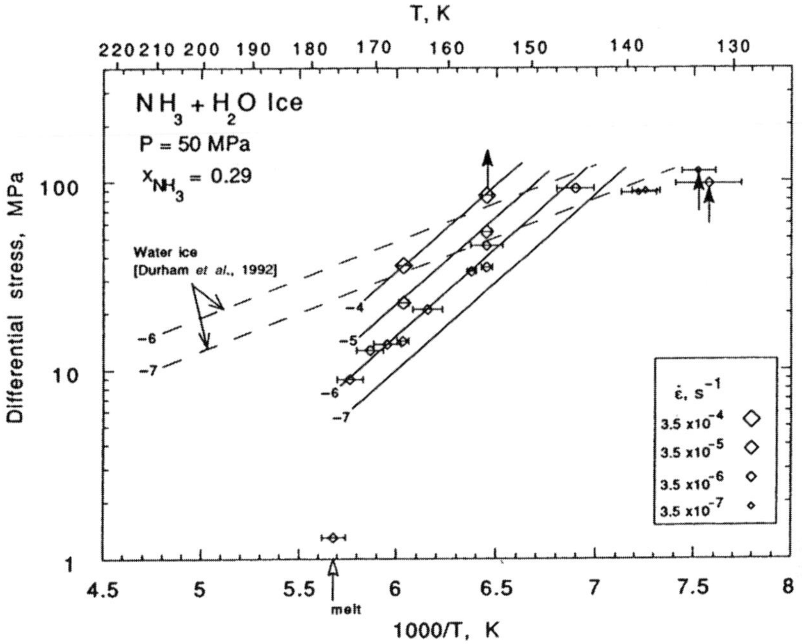

Figure 8.6. Steady-state strength of nearly pure ammonia dihydrate ($x_{NH_3} = 0.295$ and a mole fraction of dihydrate equal to 0.9). Dashed lines correspond to the behavior of ice I at the strain rates of 3.5×10^{-6} and $3.5 \times 10^{-7}\,s^{-1}$. (From Durham *et al.*, 1993.)

- At $T > 145$ K, samples of composition $x_{NH_3} = 0.295$, corresponding to a mole fraction of dihydrate of about 0.9, are weaker than pure water-ice for strain rates corresponding to the pure ductile behavior and show a greater temperature sensitivity (Fig. 8.6). At $T < 140$ K, the dihydrate is as strong as or stronger than water-ice.
- At $T > 176$ K, samples with $x_{NH_3} > 0.15$ have negligible strength. The narrow melting interval of the dihydrate would cause this weakening (Durham *et al.*, 1993).
- At $T > 176$ K, samples with $x_{NH_3} < 0.05$ show significant strength in spite of the increasing liquid phase.

Considering the flow law (8.1), for $145 < T < 176$ K, the stress exponent is about 5.8 for samples of composition $x_{NH_3} = 0.295$, to be compared with the value of about 6 for pure ice. On the other hand, the activation energy, of about 100 kJ mol^{-1}, is much higher than that for pure ice at the same conditions (of about 40 kJ mol^{-1}).

8.8 Conclusions

Laboratory experiments performed on high-pressure ices are needed to characterize the strength of water ice at the conditions of the interiors of the large icy moons of the outer Solar System. A good constraint on the constitutive laws for the steady-state flow of ice phases I, II, III, V and VI has been obtained in spite of the difficulty of performing such experiments. It can be significant to point out that the rheology of planetary ices is far better known than that of other materials under geological conditions because the relevant temperatures and pressures are much easier to reach in the laboratory. Also, pure ice samples or ices containing impurities or particles can be easily produced in the laboratory. A remaining difficulty is to extrapolate results obtained at relatively high strain rates to planetary conditions. More studies are needed to quantify the grain-size-dependent flow law at low strain rates. The difficulty of following the evolution of the ice microstructure during deformation does not facilitate the understanding of the deformation behavior of such ices.

References

Bridgman, P. W. (1912). Water, in the liquid and five solid forms, under pressure. *Proc. Am. Acad. Arts Sci.*, **47**, 441–558.

Durham, W. B. and L. A. Stern (2001). Rheological properties of water ice – applications to satellites of the outer planets. *Annu. Rev. Earth Planet. Sci.*, **29**, 295–330.

Durham, W. B., H. C. Heard and S. H. Kirby (1983). Experimental deformation of polycrystalline H_2O ice at high pressure and low temperature: preliminary results. *J. Geophys. Res.*, **88**, B377–B392.

Durham, W. B., S. H. Kirby, H. C. Heard and L. A. Stern (1987). Inelastic properties of several high pressure crystalline phases of H_2O: ices II, III and V. *J. Physique*, **48**, C1-221–C1-226.

Durham, W. B., S. H. Kirby, H. C. Heard, L. A. Stern and C. O. Boro (1988). Water ice phases II, III and V: plastic deformation and phase relationships. *J. Geophys. Res.*, **93** (B9), 10,191–10,208.

Durham, W. B., S. H. Kirby and L. A. Stern (1992). Effects of dispersed particulates on the rheology of water ice at planetary conditions. *J. Geophys. Res.*, **97** (E12), 20,883–20,897.

Durham, W. B., S. H. Kirby and L. A. Stern (1993). Flow of ices in the ammonia-water system. *J. Geophys. Res.*, **98** (B10), 17,667–17,682.

Durham, W. B., L. A. Stern and S. H. Kirby (1996). Rheology of water ices V and VI. *J. Geophys. Res.*, **101** (B2) 2989–3001.

Durham, W. B., S. H. Kirby and L. A. Stern (1997). Creep of water ices at planetary conditions: a compilation. *J. Geophys. Res.*, **102** (E7), 16,293–16,302.

Durham, W. B., S. H. Kirby and L. A. Stern (1998). Rheology of planetary ices. In *Solar System Ices*, eds. B. Schmitt, C. De Bergh and M. Festou. Dordrecht: Kluwer Academic Publishers, pp. 63–78.

Durham, W. B., L. A. Stern and S. H. Kirby (2003a). The strength and rheology of methane clathrate hydrate. *J. Geophys. Res.*, **108** (B4), 2182, ECV2-1–ECV2-11.

Durham, W. B., L. A. Stern and S. H. Kirby (2003b). Ductile flow of methane hydrate. *Can. J. Phys.*, **81**, 373–380.

Echelmeyer, K. and B. Kamb (1986). Rheology of ice II and ice III from high-pressure extrusion. *Geophys. Res. Lett.*, **13**, 693–696.

Fletcher, N. H. (1970). *The Chemical Physics of Ice*. New York: Cambridge University Press.

Gagnon, R. E., H. Kiefte, M. J. Clouter and E. Whalley (1987). Acoustic velocities in ice Ih, II, III, V and VI, by Brillouin Spectroscopy, *J. Physique*, **48**, C1-29–C1-35.

Gammon, P. H., H. Kiefte, M. J. Clouter and W. W. Denner (1983). Elastic constants of artificial and natural ice samples by Brillouin spectroscopy. *J. Glaciol.*, **29**, 433–460.

Jones, S. J. and H. A. M. Chew (1983). Creep of ice as a function of hydrostatic pressure. *J. Phys. Chem.*, **87**, 4064–4066.

Klinger, J., D. Benest, A. Dollfus and R. Smoluchowski (1985). *Ices in the Solar System*. Dordrecht: Reidel Publishing Company.

Kubo, T., W. B. Durham, L. A. Stern and S. H. Kirby (2006). Grain size-sensitive creep in ice II. *Science*, **311**, 1267–1269.

Petrenko, V. F. and R. W. Whitworth (1999). *Physics of Ice*. New York: Oxford University Press.

Poirier, J. P., C. Sotin and J. Peyronneau (1981). Viscosity of high-pressure ice VI and evolution and dynamics of Ganymede. *Nature*, **292**, 225–227.

Schmitt, B., C. De Bergh and M. Festou (1998). *Solar System Ices*. Dordrecht: Kluwer Academic Publishers.

Shaw, G. H. (1986). Elastic properties and equation of state of high pressure ice. *J. Chem. Phys.*, **84**, 5862–5868.

Sloan, E. D. (1998). *Clathrate Hydrates of Natural Gases*, 2nd edn. New York: Dekker.

Sotin, C. and J. P. Poirier (1987). Viscosity of ice V. *J. Physique*, **48**, C1-233–C1-238.

Sotin, C., P. Gillet and J. P. Poirier (1985). Creep of high-pressure ice VI. In *Ices in the Solar System*, eds. Klinger *et al.* Dordrecht: Reidel Publishing Company, pp. 109–117.

Stern, L. A., W. B. Durham and S. H. Kirby (1997). Grain-size induced weakening of H_2O ices I and II and associated anisotropic recrystallization. *J. Geophys. Res.*, **102** (B3), 5313–5325.

9

Fracture toughness of ice

9.1 Introduction

We turn now from lower-rate to higher-rate deformation and thus from creep to fracture. Accordingly, we focus on cracks. They are ubiquitous within natural features, such as the ice cover on the Arctic Ocean (http://psc.apl.washington.edu/Harry/Radarsat/images.html) and the icy crust of Europa (http://solarsystem.nasa.gov/galileo/gallery/europa.cfm), and play a major role in evolution of the bodies. Our interest is in their behavior, particularly in their resistance to propagation. This is expressed in terms of fracture toughness, a property that is fundamental not only to the scenes noted above but also to the calving of icebergs (Nye, 1957; Vaughan and Doake, 1996), to ice forces on engineered structures (Chapter 14) and to the ductile-to-brittle transition (Chapter 13). Fracture toughness is fundamental also to tensile (Chapter 9) and compressive (Chapters 11, 12) failure.

Our objectives in this chapter are to review briefly the principles underlying fracture mechanics, and then to discuss methods of measuring the fracture toughness of ice and factors that affect the property. For comparison, we include a short discussion of lightly consolidated snow.

9.2 Principles of fracture mechanics

The energy dissipated during fast crack propagation through ice is governed to a large degree by the energy required to create new surface. Hence, we base our discussion upon the theory of linear-elastic-fracture mechanics (LEFM). More complete treatments of fracture mechanics may be found in books by Knott (1973), Broek (1982), Atkinson (1989), Lawn (1995) and Anderson (1995).

9.2.1 Griffith's energy-balance concept

LEFM is based upon Griffith's (1921) seminal work. Aware of an analysis by Inglis (1913) of stress concentration about an elliptically shaped hole within a body loaded under tension, Griffith imagined that a crack could be viewed as an infinitesimally narrow elliptical hole. He then invoked equilibrium thermodynamics to determine its stability under load. Accordingly, he considered an isolated crack of length $2c$ embedded within an isotropic, linearly elastic body loaded by an external force, and sought the configuration that minimizes the total energy of the system. The system comprises the body plus the external loading device. In that state, the crack would be in equilibrium and just on the verge of propagation. Taking the total energy U to be the sum of the elastic energy within the body U_{E}, the potential energy within the external loading system U_{P} and the surface energy associated with the crack U_{S}, Griffith proposed that over an increment of crack extension $\mathrm{d}c$ thermodynamic equilibrium prevails when the different energies are balanced. Since $\mathrm{d}U_{\mathrm{S}}/\mathrm{d}c > 0$ as new surface is created, then $\mathrm{d}U_{\mathrm{M}} < 0$ where U_{M} is the mechanical energy and is given by the sum $U_{\mathrm{M}} = U_{\mathrm{E}} + U_{\mathrm{P}}$. This is Griffith's *energy-balance concept*. It leads to the equilibrium requirement:

$$\mathrm{d}U/\mathrm{d}c = 0. \tag{9.1}$$

For a small displacement from its equilibrium length, a crack would either extend or close up depending upon whether the left-hand side of Equation (9.1) is either negative or positive. In other words, crack growth is driven by a reduction in the mechanical energy of the entire system.

To quantify the approach, Griffith employed Inglis' (1913) analysis. Accordingly,

$$U_{\mathrm{E}} = \frac{\pi c^2 \sigma^2}{E'} \tag{9.2}$$

where σ is the applied tensile stress; for thin bodies under a state of plane stress $E' = E$, while for thick bodies under plane strain $E' = E/(1 - \nu^2)$ where E denotes Young's modulus and ν denotes Poisson's ratio (Chapter 4). Also, from elasticity theory (e.g., see Lawn, 1995):

$$U_{\mathrm{P}} = -2U_{\mathrm{E}} \tag{9.3}$$

for a body under constant load during crack formation. Finally, per unit width of crack front:

$$U_{\mathrm{S}} = 4c\gamma_{\mathrm{S}} \tag{9.4}$$

where γ_{S} is the surface energy per unit area of the solid–vapor interface (Chapter 2). The energy for a static crack system may then be expressed by the relationship:

$$U = U_{\mathrm{P}} + U_{\mathrm{E}} + U_{\mathrm{S}} = -U_{\mathrm{E}} + U_{\mathrm{S}} = -\frac{\pi c^2 \sigma^2}{E'} + 4c\gamma_{\mathrm{S}}. \tag{9.5}$$

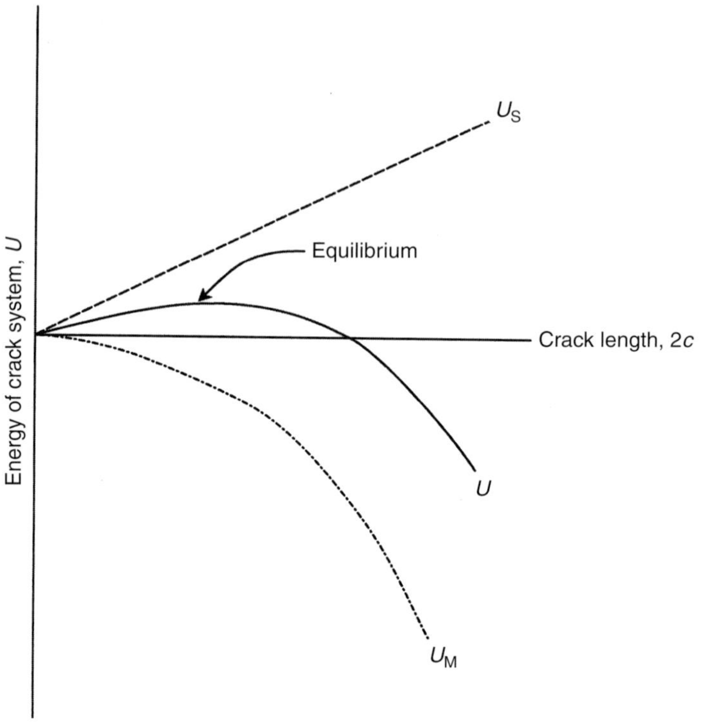

Figure 9.1. Energy U of a Griffith crack system loaded under uniform tension, showing the competition between an increase in surface energy U_S and a decrease in mechanical energy U_M with increasing crack length c. Note the unstable equilibrium where U reaches a maximum.

Thus, upon applying Equation (9.1), the critical stress for crack propagation under constant load is given by:

$$\sigma_P = \left(\frac{2E'\gamma_S}{\pi c}\right)^{1/2} \tag{9.6}$$

where the product $\sigma_P\sqrt{\pi c}$ is a material constant. The energy is a *maximum* at equilibrium, Figure 9.1, and consequently the configuration is unstable. This means that a stress slightly greater than that given by Equation (9.6) will cause the crack to grow in an uncontrolled manner.

The analysis dictates that the critical stress decreases with increasing crack length, scaling as $1/(\text{crack length})^{1/2}$. Griffith (1921) tested this prediction in experiments on glass tubes and found it to be in excellent agreement with observation. The basis of modern fracture mechanics was thus established.

9.2.2 Crack-extension force

In considering crack propagation, it is also instructive to define a *crack-extension force*, G. This is taken to be the reduction in mechanical energy with respect to the increase in crack area. Generally, LEFM theory considers planar cracks of unit thickness. Consequently, the crack extension force is defined with respect to an increase in crack length, through the relationship:

$$G = -\frac{\partial U_{\mathrm{M}}}{\partial c}. \tag{9.7}$$

The parameter G is also termed the *strain energy release rate*, where "rate" is understood with respect to crack length and not time.

9.2.3 *Modes of crack propagation*

Griffith considered crack propagation under tension. However, cracks can grow via three modes, Figure 9.2: crack opening, termed tensile mode or *mode-I*; crack sliding, termed *mode-II*; and crack tearing, termed *mode-III*. Propagation may also occur through a mixture of modes. In practice, sliding and tearing generally offer greater resistance to propagation, and so for most materials, including ice, mode-I loading is usually of greatest importance. That point is evident from the observation that a cylindrical bar, when twisted, fractures not along a plane of maximum shear stress, but along a spiral surface oriented at $\sim 45°$ to the axis of the cylinder (see Figure 6 of Butkovich, 1954); i.e., along a plane on which the tensile stress reaches a maximum. Even "shear crevasses" that form at an angle to the wall of valley glaciers lie perpendicular to the local highest principal stress (Rist *et al.*, 1999). Thus, we consider below only the resistance to crack propagation under mode-I loading.

9.2.4 *Crack-tip stress state under mode-I loading*

For a body loaded uniformly under a far-field tensile stress σ, the stress state near the tip of a through-thickness, sharp crack of radius ρ and length $2c$ is given by the stress

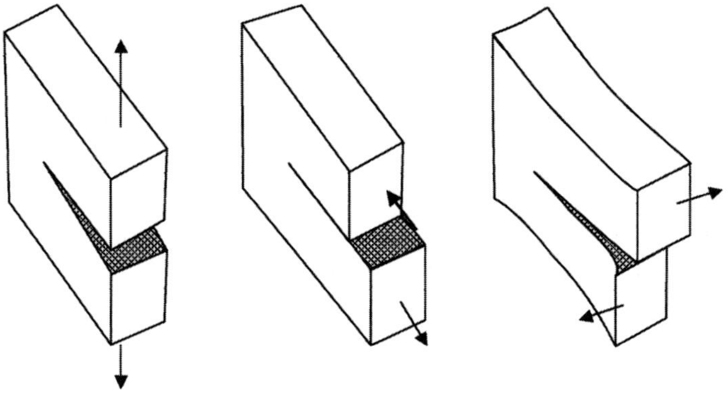

(a) (b) (c)

Figure 9.2. Three modes of crack propagation: (a) I: crack opening or tensile mode; (b) II: crack sliding mode; and (c) III: crack tearing mode.

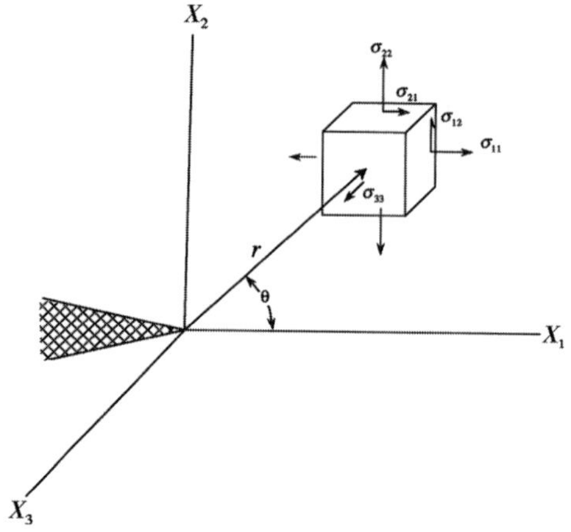

Figure 9.3. Stress field within the vicinity of the tip of a crack loaded uniformly in mode-I.

components (given with respect to the coordinate system defined in Figure 9.3, where σ acts along direction X_2):

$$\sigma_{ij} = \frac{K_{\mathrm{I}}}{(2\pi r)^{1/2}} f_{ij}(\theta) \tag{9.8}$$

for $\rho < r < c$, where r is the distance from the crack tip; K_{I} is termed the mode-I *stress intensity factor*, defined as $K_{\mathrm{I}} = \sigma\sqrt{\pi c}$, and $f_{ij}(\theta)$ is a geometrical factor given by (Paris and Sih, 1965):

$$f_{11} = \cos(\theta/2)[1 - \sin(\theta/2)\sin(3\theta/2)], \tag{9.9a}$$

$$f_{22} = \cos(\theta/2)[1 + \sin(\theta/2)\sin(3\theta/2)], \tag{9.9b}$$

$$f_{12} = \sin(\theta/2)\cos(\theta/2)\cos(3\theta/2). \tag{9.9c}$$

On the plane of the crack ($\theta = 0$) $\sigma_{12} = 0$ and $\sigma_{11} = \sigma_{22}$. For a thin body (plane stress) the other components of the stress tensor $\sigma_{33} = \sigma_{13} = \sigma_{23} = 0$. For a thick body (plane strain) $\sigma_{33} = \upsilon(\sigma_{11} + \sigma_{22})$. The K_{I} parameter depends only on the applied stress and on crack geometry and so determines the intensity of the local stress field. This means that a system defined by a long crack and a low stress is in the same state as one defined by a short crack and a high stress. K_{I} is intimately related to the condition for crack propagation (more below). For crack systems loaded uniformly in which the geometry is different from the geometry of the through-thickness crack, the stress intensity factor is defined by $K_{\mathrm{I}} = Y\sigma\sqrt{\pi c}$ where Y is a geometrical factor of order unity. For a buried, penny-shaped crack, for instance, $Y = \sqrt{2/\pi}$. Tada (1973) lists Y-values for a variety of geometries.

To determine the strain energy release rate, the components of the displacement field are also needed. These terms were derived by Paris and Sih (1965) and are listed in the books noted in the introduction to this chapter. Suffice it to say that from the sum of the product of the displacement and the appropriate component of the stress tensor, integrated over the crack-tip region, it can be shown that for mode-I loading:

$$G = \frac{K_I^2}{E'}. \tag{9.10}$$

This equation gives the driving force for the crack. For crack systems loaded non-uniformly, such as the double-cantilevered beam and the notched beam under three-point loading, the driving force may be found in the books cited above.

9.2.5 Fracture toughness and toughness

The parameters K_I and G define the state of a loaded crack system. In other words, the parameters by themselves do not constitute a criterion for crack propagation. However, when coupled with Griffith's concept that cracks propagate when the reduction in mechanical energy within the entire system exceeds the increase in surface energy, K_I or G may be viewed as a measure of how close to propagation is a loaded crack. When the stress intensity factor reaches a critical level for the material at hand, i.e., $K_I = K_{Ic}$, or when the strain energy release rate reaches a critical rate, i.e., $G = G_c$, the crack propagates. The parameters K_{Ic} and G_c are termed *fracture toughness* and *toughness*, respectively, and are properties of the material. More specifically, K_{Ic} is termed the *plane-strain fracture toughness* for mode-I loading and implies that in both its measurement and its application the material at the crack tip is constrained to deform in plane strain. The two materials properties are related. From Equation (9.10):

$$K_{Ic} = (G_c E')^{1/2}. \tag{9.11}$$

Of the two properties, K_{Ic} is the one more commonly employed in ice mechanics.

9.2.6 Stable versus unstable crack propagation

Once a crack begins to propagate, the question is the manner in which it propagates. Under far-field tensile loads applied uniformly or homogeneously, propagation occurs uncontrollably, as already noted. In other scenarios, propagation can occur in a controlled manner. An example of stable growth is the lengthening of a crack between a thin flake of mica and its parent upon inserting a wedge between the two (Obreimoff, 1930): the crack tip advances no more rapidly than the speed at which the wedge advances, owing to a *minimum* in the energy of the system at a certain crack length. Another example, particularly relevant to compressive failure of ice

(Chapters 11, 12), is the out-of-plane growth of extensile or wing cracks from the tips of an inclined, parent crack loaded under far-field compression (Griffith, 1924; Ashby and Hallam, 1986): the crack lengthens only when the load is increased.

The governing principle may be expressed in terms of the crack extension force G and the resistance to growth, denoted R. For purely brittle fracture $R = G_c = 2\gamma_S$. Two conditions must be met for crack growth. For *stable crack growth*:

$$G = R \tag{9.12}$$

and

$$\mathrm{d}G/\mathrm{d}c < \mathrm{d}R/\mathrm{d}c. \tag{9.13}$$

For *unstable growth* the first condition is the same, while the second condition becomes:

$$\mathrm{d}G/\mathrm{d}c > \mathrm{d}R/\mathrm{d}c. \tag{9.14}$$

The derivative $\mathrm{d}R/\mathrm{d}c$ is a materials characteristic and is expected to be close to zero when surface energy plays a major role in the resistance to crack propagation, as appears to be the case for ice (Section 9.5). The other derivative $\mathrm{d}G/\mathrm{d}c$ depends upon the configuration of the system and has positive values under far-field tension and negative values under far-field compression.

9.3 Measurement of fracture toughness

Methods have generally been employed in which cracks grow in an uncontrolled manner. Gold (1963) appears to be the first person to have reported measurements. He employed thermal shocking and obtained values that ranged from $K_{\mathrm{Ic}} = 50$ to 150 kPa $\mathrm{m}^{1/2}$ under ambient conditions. More modern methods are now employed, although they generate similar values.

9.3.1 *Valid measurement of* K_{Ic}

Three conditions must be satisfied to obtain a valid measure of plane strain fracture toughness:

- the size of the inelastic zone within the material at the tip of the crack (or at the tip of a notch used to simulate a crack) must be sufficiently small in relation to the crack length and to any other dimension of the body to impart plane strain deformation;
- the notch tip must be sharp; and
- all dimensions should exceed about 10 to 15 grain diameters to ensure polycrystalline behavior. Otherwise, as Andrews (1985) and Dempsey (1991) noted, measurements are more typical of multiple crystals.

Loading rate Perhaps the easiest condition to violate is the first one. When the loading rate is too low, energy that flows towards the crack mainly from the strain energy stored within the body is expended not only in the creation of new surface, but also in plastic work. Consequently, the applied load must reach a higher level to induce propagation. As a result, the apparent fracture toughness exceeds the plane strain value.

How low is too low? To answer this question Nixon and Schulson (1987, 1988) calculated and then measured the critical loading rate $\dot{K}_I^*$ below which plane strain is not achieved. By assuming that the requirement for plane-strain deformation at the notch tip is violated when the notch-tip radius exceeds $1/50$ of the depth of the notch, following Brown and Srawley (1966) on steels and aluminum alloys, and by applying the model of Riedel and Rice (1980) to determine crack-tip creep, they calculated that for granular polycrystals of fresh-water ice $\dot{K}_I^*$ increased with increasing temperature, from $\dot{K}_I^* = 0.005\,\text{kPa}\,\text{m}^{1/2}\,\text{s}^{-1}$ at $-50\,°\text{C}$ to $\dot{K}_I^* = 200\,\text{kPa}\,\text{m}^{1/2}\,\text{s}^{-1}$ at $-2\,°\text{C}$. Measurements using circumferentially notched tensile bars (91 mm diameter, notch depth 9.2 mm) of manufactured ice of 2–3 mm grain size confirmed the calculations. Upon increasing the loading rate from $\dot{K}_I = 10^{-2}\,\text{kPa}\,\text{m}^{1/2}\,\text{s}^{-1}$ to $\dot{K}_I = 10^5\,\text{kPa}\,\text{m}^{1/2}\,\text{s}^{-1}$, the apparent fracture toughness first decreased and then remained more or less constant once the rate reached the temperature-dependent critical rate. Table 9.1 summarizes the calculations and the experimental results. Bentley *et al.* (1989) and Weber and Nixon (1996a,b) confirmed the effect.

The results in Table 9.1 were obtained for fresh-water ice. Salt-water ice creeps more rapidly, by about a factor of ten for salinities typical of first-year sea ice (de La Chapelle *et al.*, 1995; Cole, 1998). Consequently, the critical loading rate for sea ice is expected to be about an order of magnitude greater than the values listed.

Not apparent from the table is the apparent fracture toughness when $\dot{K}_I < \dot{K}_I^*$. Values of about two to four times the measured plane strain value are obtained at

Table 9.1. *Critical crack-tip loading rate vs. temperature for granular fresh-water ice of 2–3 mm grain size*

Temperature (°C)	Calculated, $\dot{K}_I^*$ $(\text{kPa}\,\text{m}^{1/2}\,\text{s}^{-1})$	Observed, $\dot{K}_I^*$ $(\text{kPa}\,\text{m}^{1/2}\,\text{s}^{-1})$
−50	0.005	<0.01
−20	0.5	0.3–10
−10	2	0.3–10
−5	100	10–1000
−2	200	10–1000

From Nixon and Schulson (1987)

loading rates lower than the critical rate by an order of magnitude or two (Liu and Miller, 1979; Nixon and Schulson, 1987; Bentley *et al.*, 1989).

Notch-tip acuity It is not clear whether, as in high-strength steel (Knott, 1973), there is a notch-tip radius below which the apparent fracture toughness remains constant. The best practice is to sharpen the notch immediately prior to loading (to reduce blunting through sublimation) using a fresh razor-blade. Fatigue sharpening, although possible in principle, is difficult to apply owing to the relatively low toughness of ice (more below).

Polycrystalline behavior The other points to note are the depth of the notch relative to the grain size and the number of grains across the loaded section. On the first point, Nixon (1988) found from measurements using circumferentially notched, 91 mm diameter tensile bars of granular ice (d = 3.4 mm and 7.3 mm) loaded at −10 °C that, as long as the depth exceeded about two grain diameters, little effect of the ratio of grain size to notch depth could be detected. This result agreed with an earlier observation by Andrews (1985) who measured the fracture toughness of coarsely grained (d = 9.6 mm) glacier ice using internally pressurized notched rings. On the number of grains across the load-bearing section, Rist *et al.* (1996) found that experimental scatter is reduced when the number of grains across the diameter of a notched cylinder exceeds about 15. This number is reminiscent of Elvin's (1996) requirement (Chapter 4) of 230 grains within a loaded area for homogeneous elastic behavior and it is similar to the number recommended by Jones and Chew (1981) when examining the tensile and compression behavior of granular ice.

9.3.2 Methods employed

Provided that the above conditions are satisfied and that care is taken to avoid mixed-mode loading, good measurements of the mode-I critical stress intensity factor can be made using a variety of modern methods. Examples include three- and four-point loaded notched beams (Goodman and Tabor, 1978; Timco and Frederking, 1982; Wei *et al.*, 1991; Weber and Nixon, 1996a,b; Xu *et al.*, 2004), compact tensile specimens (Liu and Miller, 1979), double-cantilever beams (Bentley *et al.*, 1989), circumferentially notched tensile bars (Nixon and Schulson, 1987, 1988) and short rods with chevron notches (Stehn *et al.*, 1994; Sammonds *et al.*, 1998; Rist *et al.*, 1999, 2002). Timco and Frederking (1986) reported no significant difference between three- and four-point loading configurations, provided that care is taken to avoid shear stresses across the notched section. Pyramidal and conical indentation (Goodman and Tabor, 1978) and a modified ring test

(Fischer *et al.*, 1995) have been used as well, but appear to suffer from complications related to induced stresses.

Dempsey (1991) questioned the use of circumferentially notched tensile bars. He argued that the geometry employed by Nixon and Schulson (1987, 1988) – namely, 91 mm diameter bars notched to a depth of 8 mm – resulted in sub-sized specimens that were notch insensitive. The interesting point is that the values obtained fall well within the range of measurements obtained using other methods (more below).

No method is perfect and none dominates, as is the case in the study of ceramics (Rice, 2000). We caution that, in applying any of them, the results obtained should be strictly viewed as estimates of the true fracture toughness of ice.

9.4 Measured values of K_{Ic}

9.4.1 *Temperature*

Figure 9.4 shows K_{Ic} versus temperature (−2 °C to −50 °C) for polycrystalline ice. The data were obtained from several studies, cited in the key. In each case the

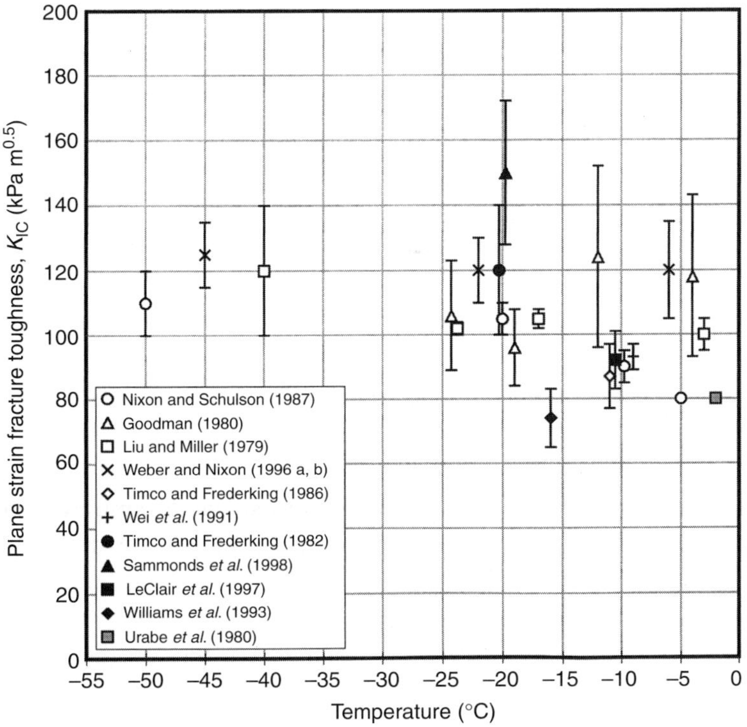

Figure 9.4. Graph of plane-strain fracture toughness vs. temperature for both fresh-water ice (open symbols plus × and +) and first-year sea ice (closed and shaded symbols). Some data obtained at −10 °C are plotted at −9 °C and −11 °C, to avoid overlap.

loading rate exceeded the critical rate. Included are data from both granular ice and S2 fresh-water ice[1] of a variety of grain sizes, prepared in the laboratory (open symbols, ×'s and +'s), from S2 first-year sea ice harvested from the Beaufort Sea (solid symbols), and from first-year sea ice harvested from an inland sea in Japan (light grey square). The data cluster around the value 100 kPa $m^{1/2}$. However, they exhibit considerable scatter, and so it is difficult to detect a clear effect of temperature.

The strongest statement that can be made is that a linear relationship through the data exhibits a slope of −0.5 kPa $m^{1/2}$ $°C^{-1}$ with a regression coefficient of $r^2 = 0.22$. This suggests perhaps a slight increase in fracture toughness with decreasing temperature over the range explored. Neither the surface energy nor the stiffness of ice increases much with decreasing temperature (Chapter 4), and so from Equation (9.11) it is perhaps not surprising that the fracture toughness increases very little.

Over the temperature range examined, it is difficult to distinguish the fracture toughness of first-year sea ice from fresh-water ice, given the scatter in the data. This contrasts with crack velocity, which on average appears to be significantly lower within salt-water ice, namely ∼10 m s^{-1} vs. ∼200–1300 m s^{-1} at temperatures around −10 °C (Petrenko and Gluschenkov, 1996). One might expect from the effect of porosity (Section 9.4.4) that, as temperature increases to the point (> −10 °C) where the brine volume begins to increase significantly, the fracture toughness of sea ice would begin to decrease. There is perhaps a slight indication from Figure 9.4 that that is the case. However, the data are too few and too scattered to draw any meaningful conclusion.

9.4.2 Grain size

The literature is interesting on the role of grain size. Upon comparing measurements on finely grained (1–2 mm diameter) fresh-water S2 columnar ice with values compiled by Urabe and Yoshitake (1981) for more coarsely grained (5–10 mm diameter) columnar ice, Timco and Frederking (1986) suggested, albeit from rather limited data, that K_{Ic} increases with increasing grain size, from K_{Ic} (1–2 mm) = 120 kPa $m^{1/2}$ to K_{Ic}(5–10 mm) = 148 kPa $m^{1/2}$. Nixon and Schulson (1988) reported the opposite effect. From the results of systematic experiments at −10 °C on bubble-free, granular fresh-water ice manufactured in the laboratory, whose grain size was varied from $d = 1.6$ mm to 9.3 mm (measured using the method of linear

[1] Recall from Chapter 3 that granular ice is defined as an aggregate of equiaxed and randomly oriented grains. S2 ice is defined as a polycrystal composed of columnar-shaped grains that possesses the S2 growth texture in which the *c*-axes are randomly oriented in a plane perpendicular to the growth direction.

intercepts), they found that fracture toughness decreased slightly with increasing grain size, through the relationship:

$$K_{\mathrm{Ic}} = K_{\mathrm{Io}} + \phi d^{-1/2} \tag{9.15}$$

where, for d in millimeters, $K_{\mathrm{Io}} = 58.3$ kPa m$^{1/2}$ and $\phi = 42.4$ kPa mm$^{1/2}$; the correlation coefficient has the value $r^2 = 0.60$. Finally, Rist *et al.* (2002) observed no effect at all of grain size on the fracture toughness of granular ice harvested from different depths within the Ronne Ice Shelf. The size ranged from around 1 mm at a depth of 29.8 m to 9 mm at a depth of 134.7 m. However, shelf ice is porous and the porosity is higher near the surface where the grains are smaller (Rist *et al.*, 2002). This suggests that an embrittling effect of porosity, described below, may have countered a toughening effect of grain refinement.

Our sense is that grain refinement toughens ice. The effect, however, is a relatively small one, reminiscent of that in ceramics (Rice, 2000). Its origin could reside in an increasing density of crack-tip deformation damage with decreasing grain size, as discussed by Nixon and Schulson (1987).

9.4.3 Growth texture

The role of texture is not clear. Timco and Frederking (1986) found that cracks in S2 fresh-water ice, when propagating in a plane parallel to the long axis of columnar grains, experience greater resistance when the front moves outward in a radial direction (with respect to the parent ice cover) than when it moves in a longitudinal direction parallel to the columns; i.e., $K_{\mathrm{Ic(radial)}} = 120$ kPa m$^{1/2}$ and $K_{\mathrm{Ic(longitudinal)}} = 87$ kPa m$^{1/2}$. This behavior remains to be confirmed.

9.4.4 Porosity

The fracture toughness of fresh-water granular ice decreases with increasing porosity, as evident from several studies. Smith *et al.* (1990) introduced gas bubbles by freezing carbonated water. They found that K_{Ic}, although scattered, decreased by about 25% upon increasing the volume fraction of porosity from $p \sim 0$ to $p \sim 0.15$. Similarly, Fischer *et al.* (1995) found that the fracture toughness of Greenland firn of $p \sim 0.34$ was about 30% lower than that of manufactured ice with $p \sim 0.03$. (Although the ring method employed in this instance is questionable, for the reason noted above, the qualitative effect is not.) From a more thorough study, Rist *et al.* (1999) observed that the fracture toughness of ice from the Ronne Ice Shelf, although scattered, decreased by about 50–60% as the porosity increased to $p \sim 0.4$, Figure 9.5. If one ignores the possibility of a threshold below which there

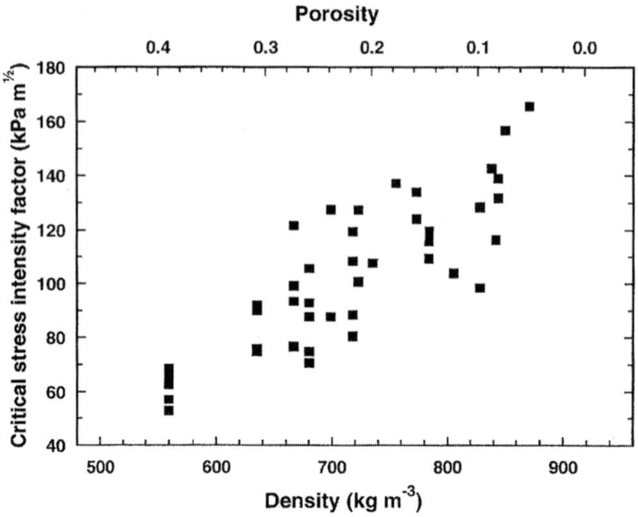

Figure 9.5. Graph of plane-strain fracture toughness vs. porosity and density for fresh-water ice from the Ronne Ice Shelf. (From Rist *et al.*, 1999.)

is little influence, then the effect observed by Rist *et al.* (2002) may be described to a first approximation by the linear relationship:

$$K_{\mathrm{Ic}}(p) = K_{\mathrm{Ic}}(1 - p) \tag{9.16}$$

for $p < 0.4$. Porosity reduces the stiffness of ice (Chapter 4) and this may account, through Equation (9.11), for its effect on fracture toughness.

9.4.5 Particulates and fibers

The role of hard, second phases is an open question. Preliminary work (Smith *et al.*, 1990) in which particles of kaolinite (up to 3% by vol.) were added to fresh-water granular ice revealed little effect, owing perhaps to a liquid-like, ice–clay interface that inhibits bonding. Fibers, on the other hand, seem to toughen ice, perhaps through crack bridging in the way they toughen ceramics. For instance, during World War II, Geoffrey Pyke suggested that the addition of a little wood pulp might raise the toughness-controlled tensile strength (Chapter 10) to the point that icebergs could serve as aircraft carriers. Subsequent research proved the concept and showed that the addition of about 14% Canadian Spruce increased shock resistance to the level of acceptability. The material so created is termed *pykrete* and is described by Perutz (1948) and Gold (1990). More recently, Nixon and Smith (1987) reported that wood derivatives (newspaper, pulp, sawdust, blotting paper and bark) at volume fractions up to 0.1–0.2 raise the apparent fracture toughness at −22 °C by about an order of magnitude, to as high as $K_{I,apparent} = 1500\,\text{kPa}\,\text{m}^{1/2}$. They attributed the toughening to additional energy consumed in de-bonding and in pulling out fibers from the matrix. Unfortunately, they did not report a value for fiber-free ice. Had one been obtained using their experimental technique and method of calculating K_I, the accuracy of the apparent improvement would have been easier to access.

9.4.6 Damage

There is some evidence that deformation damage reduces the fracture toughness of ice. Timco and Frederking (1986) performed a series of notched-beam experiments at −10 °C on fresh-water S2 columnar ice of 1–4 mm column diameter. A field of through-thickness cracks had been introduced by pre-straining within the ductile regime. The crack number density varied from $n \sim 1/\text{cm}^2$ to $n \sim 7/\text{cm}^2$, which (for grain-sized cracks) corresponds to damage $D = n(d/2)^2 < 0.3$. Over this range, the fracture toughness, although again scattered, decreased from $\sim 127\,\text{kPa}\,\text{m}^{1/2}$ to $\sim 100\,\text{kPa}\,\text{m}^{1/2}$. The effect is probably caused by a crack-induced reduction in stiffness (Chapter 4).

9.4.7 External water

The presence of water appears not to be a significant factor. This we base mainly upon the results of a series of tensile tests on doubly notched cylindrical bars of fresh-water bubbly ice loaded at −2 °C (Sabol and Schulson, 1989). In each test, one of the notches was immersed in water of 35 ppt salinity, while the other was exposed to laboratory air. From a total of 20 experiments, 12 specimens fractured in the wet notch and 8 failed in the dry one. Earlier work by Liu and Miller (1979) in which specimens were fractured in a mixture of water (80%) and anti-freeze (20%) was inconclusive, owing to scatter in the data.

9.4.8 Size

Whether size affects fracture toughness is not clear. Motivated by an early call by Weeks and Assur (1972) for measurements of mechanical properties on scales larger than the typical size of a laboratory specimen, Dempsey *et al.* (1999a,b) carried out a field study on floating ice sheets. They performed two series of measurements, the first at Spray Lakes Reservoir, Alberta, and the second at Nunavut (then North West Territories). The cover on Spray Lake was ~0.5 m thick and was composed of warm ice (>-5 °C at the surface) and was very coarsely grained (~150 mm). Also, it possessed the S1 growth texture.[2] The cover on Resolute Bay was ~1.8 m thick and was composed of colder (−17 to −12 °C at the surface), coarsely grained (~15 mm) material that possessed the S2 growth texture and a melt-water salinity of 5 ± 2 ppt. Plate-shaped specimens were cut from each sheet, notched through the thickness and then, after ensuring that newly formed ice was removed from around the cuts, were split apart using flat-jacks. Generally, the specimens were subjected to multiple loadings during which elastic and creep properties were obtained as well. In each set of tests, the principal variable was the length/width of the test specimen. It ranged from 0.34 m to 28.64 m for the lake ice and from 0.5 m to 80 m for the first-year sea ice. For comparison, smaller specimens were harvested from the same parent ice sheets, notched and then loaded monotonically in the laboratory.

In reporting the results, Dempsey *et al.* were careful to describe them not as values of plane strain fracture toughness, but as values of *apparent fracture toughness*. The reason for caution is that given the limited time and resources that were available, it was not possible to examine the role of loading rate and geometry, factors that are probably important, particularly within ice so near to its melting point.

What they found is interesting. In both scenarios, the apparent fracture toughness, although scattered, increased with increasing size. For the lake ice it increased from $K_{\mathrm{I,apparent}} \sim 100\,\mathrm{kPa\,m^{1/2}}$ for the smallest specimens (0.34 m) to $\sim 300\,\mathrm{kPa\,m^{1/2}}$ for the largest ones (28.64 m). Similarly, for the sea ice the apparent fracture toughness increased from $\sim 100\,\mathrm{kPa\,m^{1/2}}$ for the smallest specimens (0.5 m) to $\sim 250\,\mathrm{kPa\,m^{1/2}}$ for the largest ones (80 m). From these results they concluded that the apparent fracture toughness of ice exhibits a size effect.

The key word here is "apparent". While the results speak for themselves, it is not clear that the trend exhibited in both series of tests is caused by size per se. Instead, other factors may have played a role. The primary one is loading rate. For the lake ice the loading rate was not constant but varied in a non-systematic manner from $\dot{K}_{\mathrm{I}} \sim 5\,\mathrm{kPa\,m^{1/2}\,s^{-1}}$ for the smallest specimens to $\sim 60\,\mathrm{kPa\,m^{1/2}\,s^{-1}}$ for the largest. For the sea ice it ranged from $\dot{K}_{\mathrm{I}} \sim 200\,\mathrm{kPa\,m^{1/2}\,s^{-1}}$ for the smallest specimens to $0.17\,\mathrm{kPa\,m^{1/2}\,s^{-1}}$ for the largest. Judging from the discussion in Section 9.3.1, and assuming that the loading rates listed in Table 9.1 are reasonably applicable to both S1 and S2 ice (with the appropriate amplification by a factor of ten for the higher creep rate of salt-water ice), the ice in the field was probably loaded too slowly to impede creep. Plastic work would then have accompanied crack propagation, thereby raising the fracture toughness. Exacerbating this effect may have been the

[2] Recall that the S1 growth texture means that the crystallographic *c*-axes of the individual grains are oriented more or less parallel to the growth direction, i.e., perpendicular to the horizontal plane of the parent ice sheet (Chapter 3).

multiple/cyclic loads that were applied before the specimens were fractured (see Section 9.4.9). The ice cover may also have been too thin to impart plane strain deformation, particularly of the larger plates.

That said, Dempsey's results should not be underestimated. From engineering and geophysical viewpoints, the values obtained offer a measure of the resistance to crack propagation *in situ* when a floating ice cover is loaded rather slowly. However, the results should not be taken as evidence that the fracture toughness of ice is size dependent.

9.4.9 Cyclic loading

One can imagine a situation in nature where cracked ice is loaded cyclically, for example, through the action of waves on a floating ice sheet. The question is whether loading in that way affects the resistance to crack propagation. Were the material to behave in a purely elastic manner, the answer clearly would be no. However, ice creeps even under very low stresses (Chapter 5) and that process, as discussed in Section 9.3.1, can lead to an apparent resistance greater than the plane strain value. In mind, in particular, is creep during down-loading. We note in Chapter 11 that cyclic loading can increase both the compressive and the tensile strength of ice, two properties that are intrinsically related to fracture toughness. We wonder, therefore, whether cyclic loading might also increase the resistance to crack propagation.

To our knowledge, this question has not been examined systematically, although a few measurements suggest an effect. In particular, Rist *et al.* (1996) performed fracture toughness tests on notched, cylindrical bars of laboratory-grown, fresh-water, granular ice of 2–8 mm grain size subjected to three-point loading at rates from 5 to 20 kPa $m^{1/2}$ s^{-1} at temperatures from −10 to −25 °C. At these temperatures and loading rates we would expect little creep during monotonic loading. However, the bars were loaded cyclically at about 0.05 Hz and generated an apparent fracture toughness of ~200 kPa $m^{1/2}$. This value is greater than any of the monotonic measurements shown in Figure 9.4. Whether this difference is truly an effect of cyclic loading remains to be seen.

9.5 The role of surface energy

Returning to the probable grain size effect, it is informative to consider the magnitude of K_{Io} of Equation (9.15). This parameter may be interpreted to be the fracture toughness of non-textured ice of very large grain size. When expressed in terms of the critical strain energy release rate or toughness through Equation (9.11), we obtain for the same very coarsely grained material the value $G_{co} = 320$ mJ m^{-2}, using

$E' = 10.5$ GPa (Chapter 4). This value implies that the intrinsic toughness of ice is greater than twice the surface energy ($2\gamma_S = 138$ to $218\,\text{mJ m}^{-2}$, Chapter 2). The difference between G_{co} and $2\gamma_S$ is probably related to inelastic deformation accompanying crack propagation, the nature of which remains to be determined. In comparison, the creation of new surface accounts for a much lower fraction of the toughness of ceramics and rocks ($\sim 1/25$) and for a very much lower fraction of the toughness of ductile metals ($<1/1000$). In other words, the creation of new surface plays a major role in the fracture toughness of ice.

9.6 Fracture toughness of snow

When lightly consolidated, snow is a material that can be considered to be foam of ice (Kirchner *et al.*, 2001). Although its fracture toughness has been recognized for many years as an important parameter in the mechanics of avalanches (McClung, 1981), only recently have values been reported. Experiments have been performed under predominantly mode-I loading (Kirchner *et al.*, 2000; Schweizer *et al.*, 2004; Faillettaz *et al.*, 2002), under predominantly mode-II loading (Kirchner *et al.*, 2002a; Schweizer *et al.*, 2004) and under mixed mode-I and mode-II loading (Kirchner *et al.*, 2002b). The method employed in all of the experiments, which have been performed both in the field and in the laboratory, was essentially the same. Specifically, into a cantilevered slab of material (width = 10 cm, height = 20 cm, length = variable) that had been harvested from a natural snow-fall, a saw-cut was introduced, either vertically or horizontally, to a depth that caused the beam to break off, either under its own weight or under additional loading. From beam mechanics and from the depth of the notch (determined from post-test examination of the fracture surface), the fracture toughness was calculated. The results suggested that fracture toughness in shear, K_{IIc}, has about the same value and fracture toughness in tension, K_{Ic} (Kirchner *et al.*, 2002a,b). Again, we consider only K_{Ic}.

Its value is extremely low. It falls within the range $K_{Ic} \sim 50$–$2000\,\text{Pa m}^{1/2}$ (or 0.05–$2.0\,\text{kPa m}^{1/2}$) and depends principally upon density. Over the range $0.05 < \rho_{snow}/\rho_{ice} < 0.5$, it may be described by the relationship:

$$K_{Ic} = A(\rho_{snow}/\rho_{ice})^s. \tag{9.17}$$

Different values of the coefficient A and the exponent s have been reported, from $A = 7.8\,\text{kPa m}^{1/2}$ and $s = 2.3$ for very warm snow at -1 °C (Kirchner *et al.*, 2000) to $A = 21.6\,\text{kPa m}^{1/2}$ and $s = 2.1$ for colder snow at about -10 °C (Schweizer *et al.*, 2004). While the difference may be related to the difference in temperature, for there is some evidence that the fracture toughness of snow increases slightly with decreasing temperature (Schweizer *et al.*, 2004) as does the fracture toughness of ice, it might also be related to differences in the microstructure of the material. There is also a report that K_{Ic} increases with the length of the cantilever beam, by about a factor of four as the length increases from 10 cm to 30 cm (Faillettaz *et al.*, 2002). This result has been attributed to an effect of size, but we wonder whether densification under the compressive stress of the beam below its neutral axis might have played a role.

These values should best be viewed as order-of-magnitude estimates, for two reasons. The material was not loaded purely in tension, for the weight of the beam near the notch created a shear load and this load appears to have influenced the fracture path which, in at least one set of experiments (Schweizer *et al.*, 2004), was not perfectly perpendicular to the long axis of the beam. There is also a question about the application of beam mechanics and whether the relationships used to compute fracture toughness are exactly appropriate. Nevertheless, the results show that the fracture toughness of snow is about two orders of magnitude lower than the fracture toughness of solid ice, and is possibly lower than that of any natural material.

To understand this result, Kirchner *et al.* (2000) viewed snow as an open, cellular material. Accordingly, from the mechanics of such materials (Gibson and Ashby, 1997) and from the assumption that Young's modulus E scales with density (Chapter 4) through the relationship $E_{snow}/E_{ice} = (\rho_{snow}/\rho_{ice})^2$ they postulated that:

$$\frac{K_{Ic,snow}}{K_{Ic,ice}} = \left(\frac{\rho_{snow}}{\rho_{ice}}\right)^{1/2}\left(\frac{E_{snow}}{E_{ice}}\right)^{1/2} = \left(\frac{\rho_{snow}}{\rho_{ice}}\right)^{3/2}. \tag{9.18}$$

In reality, the modulus appears to depend on density to a power >2, as noted in Chapter 4, and this greater sensitivity could account for the observation that in Equation (9.17) $s \sim 2$ instead of $s = 3/2$. It is important to note that Equation (9.18) does not extrapolate well to ice, even to the highly porous material whose fracture toughness is shown in Figure 9.5, nor is it intended to: once the relative density reaches $\sim$0.5, snow transforms to firn; at that point it can no longer be viewed as an open-celled material.

9.7 Comparison with other materials

Figure 9.6 compares the fracture toughness of a variety of materials, including ice. When snow is considered as well (not plotted), the range is large, from < 1 kPa m$^{1/2}$ to $>10^5$ kPa m$^{1/2}$. Snow and ice fall near the lower end of the range, along with other brittle materials. There, linear elastic fracture mechanics works well, and so the fracture toughness is a well-defined property.

Why is its value so low? The reason, as already noted, is that the toughness of ice is related largely to the energy consumed in creating two new surfaces. In turn, surface energy is governed by the interatomic spacing, a, and by Young's modulus through the relationship $\gamma_S \sim Ea/20$ (Ashby, 1989). Thus, as an approximation:

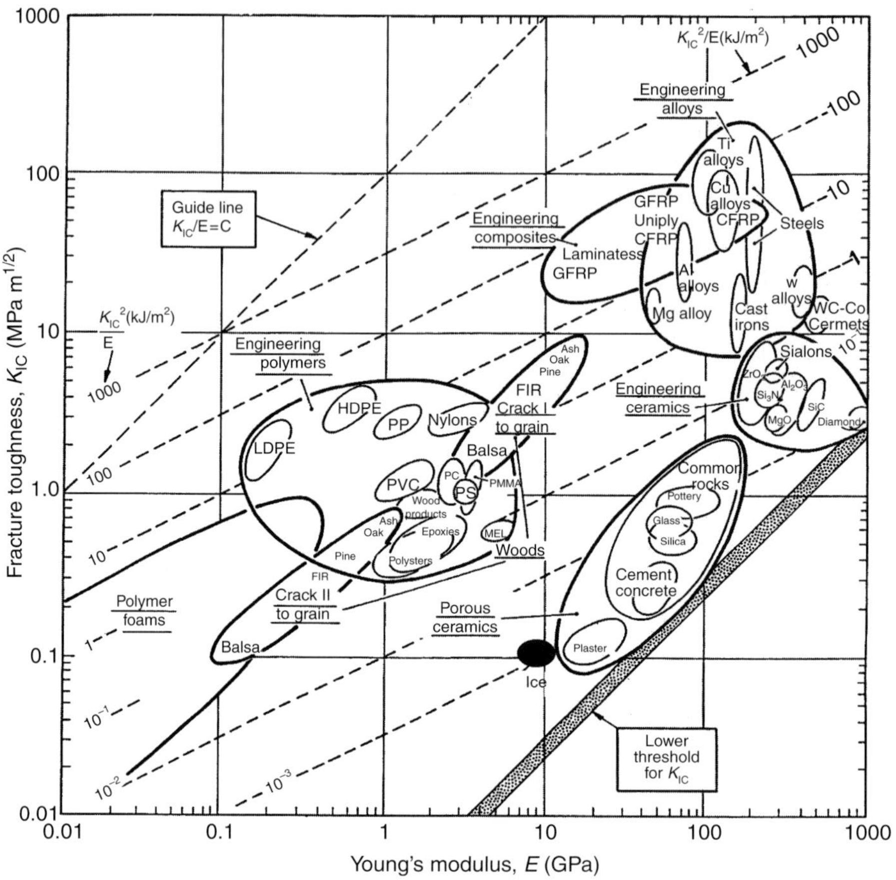

Figure 9.6. Fracture toughness vs. density for a variety of materials, including ice. (From Ashby, 1989.)

$$K_{\mathrm{Ic,approx}} = \sqrt{G_c E'} = \sqrt{2\gamma_S E'} \approx E' \sqrt{a/10}. \tag{9.19}$$

Taking $E' = 10.5$ GPa (Chapter 4) and $a = 0.45$ nm (Chapter 2), then $K_{\mathrm{Ic,approx}} = 70$ kPa m$^{1/2}$. In other words, the reason that the fracture toughness of ice is so low is that the stiffness itself is relatively low, owing to the weak intermolecular hydrogen bonds that hold the water molecules together in the solid state.

References

Anderson, T. L. (1995). *Fracture Mechanics: Fundamentals and Applications*, 2nd edn. Boca Raton: CRC Press.

Andrews, R. M. (1985). Measurement of the fracture toughness of glacier ice. *J. Glaciol.*, **31**, 171–176.

Ashby, M. F. (1989). Materials selection in conceptual design. *Mater. Sci. Technol.*, **5**, 517–525.

Ashby, M. F. and S. D. Hallam (1986). The failure of brittle solids containing small cracks under compressive stress states. *Acta. Metall.*, **34**, 497–510.

Atkinson, B. K., Ed. (1989). *Fracture Mechanics of Rock*. London: Academic Press.

Bentley, D. L., J. P. Dempsey and D. S. Sodhi (1989). Fracture toughness of columnar freshwater ice from large scale dcb tests. *Cold Reg. Sci. Technol.*, **17**, 7–20.

Broek, D. (1982). *Elementary Engineering Fracture Mechanics*, 4th edn. Boston: Springer.

Brown, W. F. and J. E. Srawley (1966). Plane strain crack toughness testing of high strength metallic materials. *STP 410*, ASTM.

Butkovich, T. R. (1954). Ultimate strength of ice. *U.S. Snow, Ice and Permafrost Research Establishment*, Research Paper, 15.

Cole, D. M. (1998). Modeling the cyclic loading response of sea ice. *Int. J. Sol. Struct.*, **35**, 4067–4075.

de La Chapelle, S., P. Duval and B. Baudelet (1995). Compressive creep of polycrystalline ice containing a liquid phase. *Scr. Metall. Mater.*, **33**, 447–450.

Dempsey, J. P. (1991). Fracture toughness of ice. In *Ice Structure Interactions*, ed. S. J. Jones. Berlin: Springer-Verlag, pp. 109–125.

Dempsey, J. P., R. M. Adamson and S. V. Mulmule (1999a). Scale effects on the in-situ tensile strength and fracture of ice. Part I: Large grained freshwater ice at Spray Lakes Reservoir, Alberta. *Int. J. Fract.*, **95**, 325–345.

Dempsey, J. P., S. J. DeFranco, R. M. Adamson and S. V. Mulmule (1999b). Scale effects on the in-situ tensile strength and fracture of ice. Part II: First-year sea ice at Resolute, N.W.T. *Int. J. Fract.*, **95**, 347–366.

Elvin, A. A. (1996). Number of grains required to homogenize elastic properties of polycrystalline ice. *Mech. Mater.*, **22**, 51–64.

Faillettaz, J., D. Daudon, D. Bonjean and F. Louchet (2002). Snow toughness measurements and possible applications to avalanche triggering. *International Snow Science Workshop 2002*, Penticton, B.C., Canada.

Fischer, M. P., R. B. Alley and T. Engelder (1995). Fracture toughness of ice and firn determined from the modified ring test. *J. Glaciol.*, **41**, 383–394.

Gibson, L. G. and M. F. Ashby (1997). *Cellular Solids*, 2nd edn. Cambridge: Cambridge University Press.

Gold, L. W. (1963). Deformation mechanisms of ice. In *Ice and Snow*, ed. W. D. Kingery. Cambridge, Mass.: MIT Press, pp. 8–27.

Gold, L. W. (1990). *The Canadian Habbakuk Project*. International Glaciological Society.

Goodman, D. J. (1980). Critical stress intensity factor (K_{IC}) measurements at high loading rates for polycrystalline ice. In *Physics and Mechanics of Ice*, ed. P. Tryde. IUTAM Symposium, Copenhagen. Berlin: Springer-Verlag, pp. 129–146.

Goodman, D. J. and D. Tabor (1978). Fracture toughness of ice: A preliminary account of some new experiments. *J. Glaciol.*, **21**, 651–660.

Griffith, A. A. (1921). The phenomena of rupture and flow in solids. *Phil. Trans. R. Soc. Lond. A*, **221**, 163–198.

Griffith, A. A. (1924). The theory of rupture. In *Proc. First Internat. Congr. Appl. Mech.*, eds. C. B. Biezeno and J. M. Burgers. Delft: J. Waltman Jr., 55–63.

Inglis, C. E. (1913). Stresses in a plate due to the presence of cracks and sharp corners. *Trans. Inst. Naval Architects*, **55**, 219–230.

Jones, S. J. and H. A. M. Chew (1981). On the grain-size dependence of secondary creep. *J. Glaciol.*, **27**, 517–518.

Kirchner, H. O. K., G. Michot and J. Schweizer (2000). Fracture toughness of snow in tension. *Phil. Mag. A*, **80**, 1265–1272.

Kirchner, H. O. K., G. Michot, N. Narita and T. Suzuki (2001). Snow as a foam of ice: plasticity, fracture and the brittle-to-ductile transition. *Phil. Mag. A*, **81**, 2161–2181.

Kirchner, H. O. K., G. Michot and J. Schweizer (2002a). Fracture toughness of snow in shear under friction. *Phys. Rev.*, **66**, 027103.

Kirchner, H. O. K., G. Michot and J. Schweizer (2002b). Fracture toughness of snow in shear and tension. *Scr. Mater.*, **46**, 425–429.

Knott, J. F. (1973). *Fundamentals of Fracture Mechanics*. New York: John Wiley and Sons.

Lawn, B. (1995). *Fracture of Brittle Solids*, 2nd edn. Cambridge: Cambridge University Press.

LeClair, E. S., R. M. Adamson and J. P. Dempsey (1997). Core-based fracture of aligned first-year sea ice (Phase I). *J. Cold Reg. Eng.*, ASCE, **11**, 45–58.

Liu, H. W. and K. J. Miller (1979). Fracture toughness of fresh-water ice. *J. Glaciol.*, **22**, 135–143.

McClung, D. M. (1981). Fracture mechanical models of dry slab avalanche release. *J. Geophys. Res.*, **86**, 783–790.

Nixon, W. A. (1988). The effect of notch depth on the fracture toughness of freshwater ice. *Cold Reg. Sci. Technol.*, **15**, 75–78.

Nixon, W. A. and E. M. Schulson (1987). A micromechanical view of the fracture toughness of ice. *J. Physique*, **48**, 313–319.

Nixon, W. A. and E. M. Schulson (1988). Fracture toughness of ice over a range of grain sizes. *J. Offshore Mech. Arctic Eng.*, **110**, 192–196.

Nixon, W. A. and R. A. Smith (1987). The fracture toughness of some wood-ice composites *Cold Reg. Sci. Technol.*, **14**, 139–145.

Nye, J. F. (1957). The distribution of stress and velocity in glaciers and icesheets. *Proc. R. Soc. Lond., Ser. A*, **239**, 113–133.

Obreimoff, J. W. (1930). The splitting strength of mica. *Proc. R. Soc. A*, **127**, 290–297.

Paris, P. C. and G. C. Sih (1965). Stress analysis of cracks. In *Symposium on Fracture Toughness Testing: ASTM 381*, 30–77.

Perutz, M. F. (1948). Description of the iceberg aircraft carrier and the bearing of the mechnical properties of frozen wood pulp upon some problems of glacier flow. *J. Glaciol.*, **1**, 95–104.

Petrenko, V. F. and O. Gluschenkov (1996). Crack velocities in freshwater and saline ice. *J. Geophys. Res.*, **101**, 11,541–11,551.

Rice, R. W. (2000). *Mechanical Properties of Ceramics and Composites: Grain and Particle Effects*. New York: CRC.

Riedel, H. and J. R. Rice (1980). Tensile cracks in creeping solids. *ASTM-STP*, **7700**, 112–130.

Rist, M. A., P. R. Sammonds, S. A. F. Murrell *et al.* (1996). Experimental fracture and mechanical properties of Antarctic ice: preliminary results. *Ann. Glaciol.*, **23**, 284–292.

Rist, M. A., P. R. Sammonds, S. A. F. Murrell *et al.* (1999). Experimental and theoretical fracture mechanics applied to Antarctic ice fracture and surface crevassing. *J. Geophys. Res.*, **104**, 2973–2987.

Rist, M. A., P. Sammonds, H. Oerter and C. S. M. Doake (2002). Fracture of Antarctic shelf ice. *J. Geophy. Res. Solid Earth*, **107** (B1), 2002, doi:10.1029/2000JB000058.

Sabol, S. A. and E. M. Schulson (1989). The fracture toughness of ice in contact with salt water. *J. Glaciol.*, **35**, 191–192.

Sammonds, P. R., S. A. F. Murrell and M. A. Rist (1998). Fracture of multi-year sea ice. *J. Geophys. Res.*, **103**, 21,795–21,815.

Schweizer, J., G. Michot and H. O. K. Kirchner (2004). On the fracture toughness of snow. *Ann. Glaciol.*, **38**, 1–8.

Smith, T. R., M. E. Schulson and E. M. Schulson (1990). The fracture toughness of porous ice with and without particles. *9th International Conference on Offshore Mechanics and Arctic Engineering.*

Stehn, L. M., S. J. DeFranco and J. P. Dempsey (1994). Fracture resistance determination of freshwater ice using a chevron notched tension specimen. *Int. J. Fract.*, **65**, 313–328.

Tada, H. (1973). *The Stress Analysis of Cracks Handbook.* Hellertown, Pa.: Del Research Corporation.

Timco, G. W. and R. M. W. Frederking (1982). Flexural strength and fracture toughness of sea ice. *Cold Reg. Sci. Technol.*, **8**, 35–41.

Timco, G. W. and R. M. W. Frederking (1986). The effects of anisotropy and microcracks on the fracture toughness of freshwater ice. *Proceedings of Fifth International Offshore Mechanics and Arctic Engineering (OMAE) Symposium*, Tokyo. Vol. 4, eds. V. J. Lunardini, Y. S. Wang, O. A. Ayorinde and D. V. Sodhi. New York: American Society of Mechanical Engineers, pp. 341–348.

Urabe, N. and A. Yoshitake (1981). Strain rate dependent fracture toughness (K_{Ic}) of pure ice and sea ice. *IAHR Ice Symposium*, 410–420.

Urabe, N., T. Iwasaki and A. Yoshitake (1980). Fracture toughness of sea ice. *Cold Reg. Sci. Technol.*, **3**, 29–37.

Vaughan, D. G. and C. S. M. Doake (1996). Recent atmospheric warming and retreat of ice shelves on the Antarctic Peninsula, *Nature*, **379**, 328–331.

Weber, L. J. and W. A. Nixon (1996a). Fracture toughness of freshwater ice – Part I: Experimental technique and results. *J. Offshore Mech. Arctic Eng.*, **118**, 135–140.

Weber, L. J. and W. A. Nixon (1996b). Fracture toughness of freshwater ice – Part II: Analysis and micrography. *J. Offshore Mech. Arctic Eng.*, **118**, 141–147.

Weeks, W. F. and A. Assur (1972). Fracture of lake and sea ice. In *Fracture*, ed. H. Leibowitz. New York: Academic Press, pp. 879–978.

Wei, Y., S. J. DeFranco and J. P. Dempsey (1991). Crack-fabrication techniques and their effects on the fracture toughness and CTOD for fresh-water columnar ice. *J. Glaciol.*, **37**, 270–280.

Williams, F. M., C. Kirby and T. Slade (1993). *Strength and Fracture Toughness of First-year Arctic Sea Ice.* Report No. Tr-1993–12. Institute For Marine Dynamics, NRC-Canada, St. John's, Nfld.

Xu, X., G. Jeronimidia, A. G. Atkins and P. A. Trusty (2004). Rate-dependent fracture toughness of pure polycrystalline ice. *J. Mater. Sci.*, **39**, 225–233.

10

Brittle failure of ice under tension

10.1 Introduction

Ice fractures under tension in a number of engineering and geophysical situations. Examples include ice breaking by ships (Michel, 1978), the bending of floating ice sheets against offshore structures (Riska and Tuhkuri, 1995), the formation of thermal cracks (Evans and Untersteiner, 1971) and the building of pressure ridges (Hopkins *et al.*, 1999) within the sea ice cover on the Arctic Ocean. Other examples include the fracture of pancake ice within the southern Atlantic Ocean (Dai *et al.*, 2004), the calving of icebergs (Nye, 1957; Nath and Vaughan, 2003) and the crevassing of ice shelves (Rist *et al.*, 1999, 2002; Weiss, 2004). Extra-terrestrial tensile failures include the initiation of polygonal features within the ground ice on Mars (Mellon, 1997; Mangold, 2005) and the formation of long lineaments within the icy crust of Europa (e.g., Greenberg *et al.*, 1998; Greeley *et al.*, 2000). In many cases, fast crack propagation is at play, which is to say that crack growth occurs so rapidly that the dissipation of mechanical energy through creep deformation is not a major consideration. Linear elastic fracture mechanics is then a valid method of analysis. In other cases, such as the formation of crevasses and the slow propagation of cracks within ice shelves, analysis based upon non-linear processes and/or sub-critical crack growth may be more useful (Weiss, 2004). In still other cases, more in the laboratory than in the field, tensile strength is limited by crack nucleation, as will become apparent.

Our objectives in this chapter are to describe the basic factors that affect tensile failure, and then to consider the underlying mechanisms. We focus on observations from the laboratory, in hopes that behavior there will help to enlighten tensile failure in the field.

Ice, incidentally, is unusual in exhibiting brittle behavior right up to the melting point, at deformation rates that are well below dynamic ($< 1\ s^{-1}$) rates. Single crystals as well as polycrystals exhibit this behavior, implying that it originates not through the restriction to slip imposed either by an insufficient number of independent slip systems (Chapter 5) or by grain boundaries (Chapter 6), but by a more fundamental constraint. At root is a combination of sluggish dislocation kinetics (Chapter 2) and low resistance to crack propagation (Chapter 9).

10.2 Methods of measurement

Both indirect and direct methods have been employed to measure the tensile strength. During the early years of ice research, the *ring test* was used by several investigators (Assur, 1958; Butkovich, 1958; Weeks, 1962). Adopted from rock mechanics, the test consists of applying a compressive load across the diameter of a short cylindrical specimen through which a hole has been drilled along the axis. Fracture initiates at the top and bottom of the hole, where the induced tensile stress is a maximum, and then propagates diametrically in a direction perpendicular to the tensile stress. Although easy to apply and useful as an index in the qualitative exploration of effects such as porosity weakening (more below), the measure obtained is difficult to interpret because the stress state is a biaxial one and is not uniform throughout the specimen (Jaegar and Cook, 1979).

Related to the ring test is the *Brazil test*. The specimen in this case is a disc or short cylinder without a hole, again compressed diametrically. Although as easy to perform as the ring test, the Brazil test also suffers from a spatially varying stress state.

In a systematic study, Mellor and Hawkes (1971) found that the strength of finely grained ($d \sim 0.7$ mm), bubbly, fresh-water granular ice determined using both the Brazil test and the ring test underestimates the strength obtained directly from uniaxial loading experiments. In this regard, ice seems to be different from rock (Jaegar and Cook, 1979) where the two indirect methods generally overestimate tensile strength.

The other indirect method is the *bend test*. Again the stress varies spatially. However, in relation to problems such as ice-breaking, flexural strength may be a more useful parameter than uniaxial tensile strength. With that point in mind, Timco and O'Brien (1994) reviewed over two thousand measurements that were obtained from both fresh-water ice and sea ice under ambient conditions (more below). Upon extrapolating to zero porosity (1.76 MPa), the bend strength of sea ice was almost identical to that of fresh-water ice (1.73 MPa). This result lends confidence to structure–strength relationships that one can explore using this approach.

Incidentally, one might think that the bend strength of brittle materials should be equal to the tensile strength. However, it is larger, typically by a factor of ~ 1.7 (Ashby and Jones, 2005). For instance, the bend strength of atmospheric ice of ~ 0.5 mm average grain size strained at $10^{-3}\ s^{-1}$ at $-10\ °C$ is ~ 2.7 MPa (Kermani *et al.*, 2008), compared with an expected tensile strength of ~ 1.7 MPa for similar material loaded under the same conditions (Section 10.4.3). The reason is that the largest flaw, which governs the strength, is generally not near the surface of bent material.

When the uniaxial strength is the point of interest, as is the case in mechanistic studies, the best method is the *direct tensile test*, for it yields measures that are easier to interpret. Although more challenging owing to gripping and alignment, the method has been successfully applied in a number of studies on both fresh-water ice (Butkovich, 1954; Carter, 1971; Hawkes and Mellor, 1972; Michel, 1978; Currier and Schulson, 1982; Schulson *et al.*, 1984, 1989; Lee and Schulson, 1988; Cole, 1990) and salt-water ice (Dykins, 1970; Kuehn *et al.*, 1990; Richter-Menge and Jones, 1993; Sammonds *et al.*, 1998; Weiss, 2001). Unless otherwise noted, we focus below on measurements obtained using the direct method.

The number of grains across the load-bearing section is once again an important consideration. As noted earlier (Chapters 4 and 9), more than 15 grains across any linear dimension is desirable when probing the behavior of granular ice, although as few as 10 may suffice. However, when measuring the strength of columnar-grained ice by loading in a direction perpendicular to the long axis of the grains (the so-called across-column direction), this constraint can be relaxed in the along-column direction, for when loaded in this direction columnar ice tends not to deform inelastically along the columns.

10.3 Single crystals

A thorough study of the brittle behavior of single crystals under uniaxial loading was made by Carter (1971). He performed over 600 tests on cylindrical-shaped (5 cm diameter × 15 cm length) crystals of fresh-water ice loaded under either tension or compression at strain rates from $4 \times 10^{-7}\,s^{-1}$ to $2.5 \times 10^{-1}\,s^{-1}$ at temperatures from −30 °C to 0 °C. He also varied the orientation of the crystallographic *c*-axis, from $\alpha = 0°$ to 90° with respect to the tensile axis. When deformed at strain rates below about $10^{-4}\,s^{-1}$ single crystals are ductile, while at rates above about $10^{-3}\,s^{-1}$ they are brittle. Between these limits they exhibit transitional behavior.

The brittle tensile strength is essentially independent of strain rate and temperature, but depends strongly upon the orientation of the *c*-axis, Figure 10.1. It ranges between ~4 MPa ($\alpha = 0$) and 6 MPa ($\alpha = 90°$) for the two extremes of orientation and reaches a minimum of ~2 MPa when the basal planes are inclined by $\alpha = 45°$ to the loading direction. Fracture occurs via cleavage on macroscopic planes that run parallel to the basal planes when the angle between them and the tensile load exceeds about 45°. For steeper orientations, cleavage occurs on planes that are oriented parallel to the prismatic planes. The crystallographic character of fracture is consistent with calculations by Parameswaran (1982), which show that cracks form

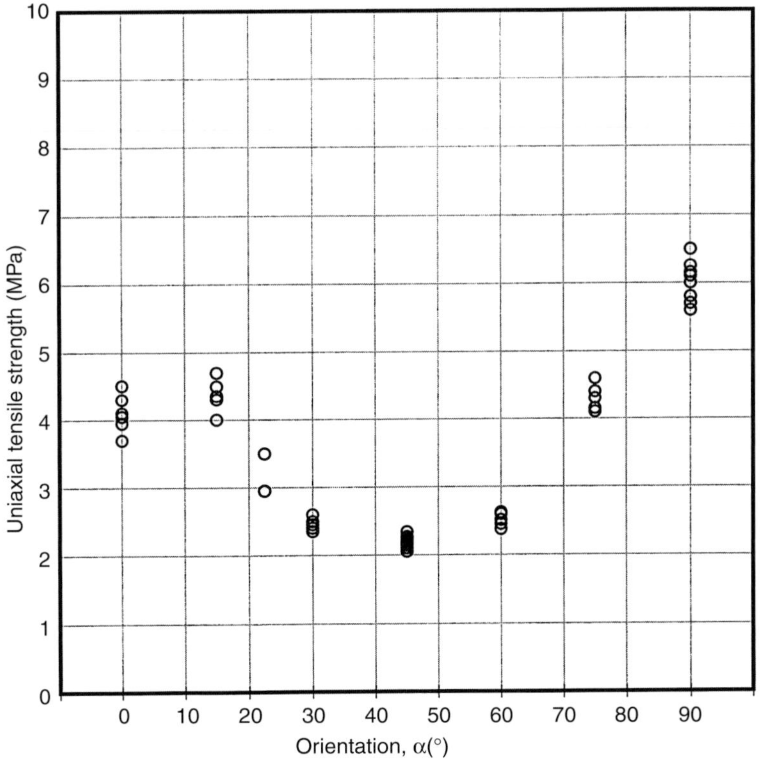

Figure 10.1. Tensile strength of single crystals of ice at −10 °C vs. orientation α of the basal plane, where α denotes the angle between the c-axis and the tensile load. (Data from Carter, 1971.)

as easily on prismatic planes as on basal planes. The character is consistent also with Gold's (1963) observation that, when thermally shocked, ice cracks preferentially on either basal or prismatic planes.

Other measurements of the tensile fracture stress of single crystals were reported by Jones and Glen (1969), following a study of the creep of colder (−90 °C to −50 °C) ice of variable orientation. The nominal tensile strength (load at fracture divided by original cross-sectional area) increased with decreasing temperature, from 1.25 ± 0.06 MPa at −50 °C to 2.1 MPa at −80 °C and then to 3.2 ± 0.2 MPa at −90 °C. Whether this behavior is truly characteristic of brittle fracture, however, is not clear, for some of the crystals deformed plastically (elongation up to 40% in two cases at −50 °C).

10.4 Polycrystals

Polycrystals of typical grain size (1–10 mm) are weaker than single crystals when deformed within the brittle regime. They fail between about 0.5 MPa and 2 MPa, compared with 2–6 MPa for single crystals. This contrasts with ductile deformation where polycrystals are significantly stronger (Chapter 6). Another difference is the strain rate that marks the ductile-to-brittle transition: it is lower for polycrystals, of the order $10^{-7}\,s^{-1}$ at −10 °C for aggregates of ∼1 mm grain size. However, fracture again occurs via cleavage, in this case across the grains in a transgranular mode. In examining the tensile failure of polycrystalline ice, several factors have been considered. We address each in turn.

10.4.1 Temperature

The tensile strength of fresh-water ice increases slightly with decreasing temperature (Butkovich, 1954; Carter, 1971; Schulson *et al.*, 1984). Over the range from 0 °C to −30 °C the strength of columnar-grained ice increases by about 10%, from around 1 MPa to ∼1.1 MPa, Figure 10.2. This behavior is similar to the behavior of fracture toughness (Chapter 9) which, as we show below, governs the strength under many conditions.

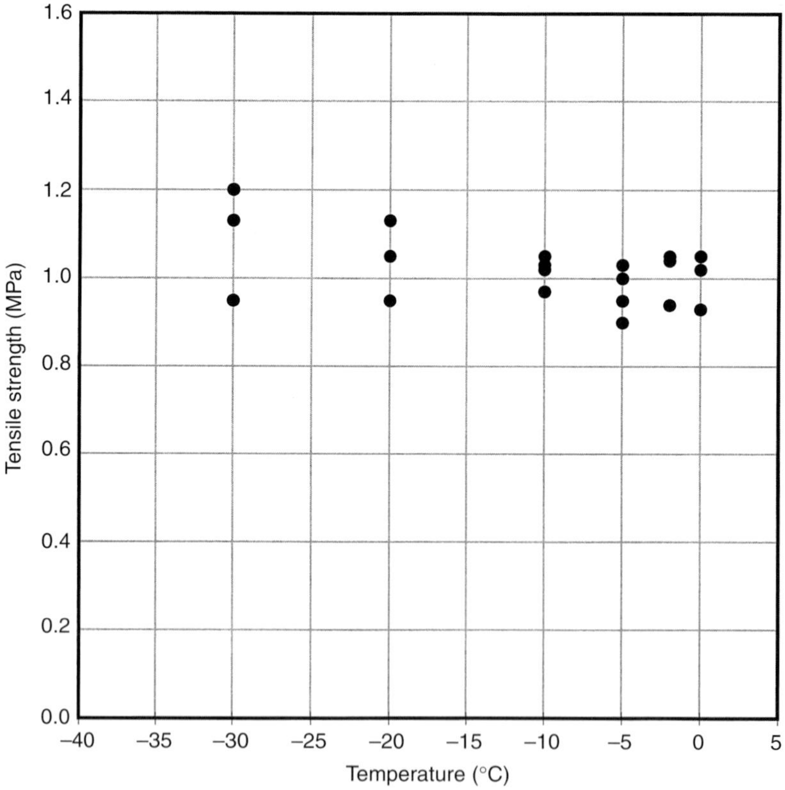

Figure 10.2. Along-column tensile strength of S2 columnar-grained fresh-water ice of ∼6 mm column diameter vs. temperature. (Data from Carter, 1971.)

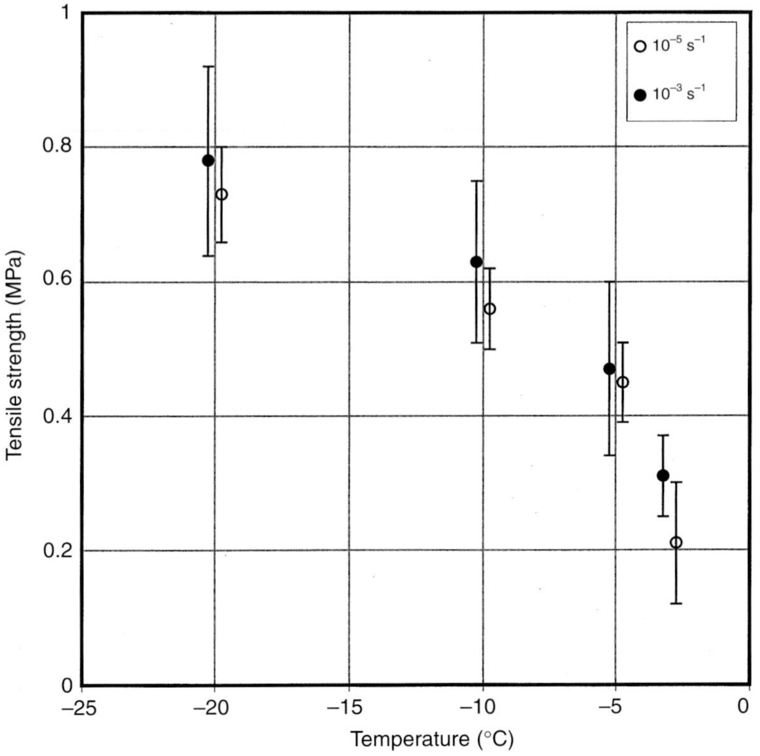

Figure 10.3. The tensile strength of columnar-grained first-year sea ice loaded uniaxially across the columns vs. temperature, at strain rates of $10^{-5}\,s^{-1}$ and $10^{-3}\,s^{-1}$. (Data from Richter-Menge and Jones, 1993.)

Salt-water ice exhibits greater thermal sensitivity (Weeks, 1962). Over similar temperatures, the across-column strength of columnar-grained material of ~5 ppt melt-water salinity increases by about a factor of four (Richter-Menge and Jones, 1993), from 0.2–0.3 MPa at −2.5 °C to 0.7–0.9 MPa at −20 °C, Figure 10.3. The difference in thermal sensitivity is related not to a fundamental change in the deformation process, but to porosity/brine pockets within salt-water ice. As discussed in Chapter 3, the equilibrium brine fraction decreases with decreasing temperature until the NaCl–H_2O eutectic is reached at −21.2 °C (Chapter 3). At temperatures above the eutectic, salt-water ice is weaker than fresh-water ice, owing to the stress-concentrating effect of the pores (more below). At lower temperatures the strengths of the two kinds of ice are similar, at least to −35 °C (Weeks, 1962).

10.4.2 Strain rate

When deformed "quasi-statically" – i.e., at strain rates over the range $10^{-7}\,s^{-1}$ to $10^{-1}\,s^{-1}$ – polycrystals fail under a tensile stress that is essentially independent of strain rate (Hawkes and Mellor, 1972; Michel, 1978; Schulson *et al.*, 1984; Druez *et al.*, 1987). On the other hand, when deformed dynamically they appear to exhibit *strain-rate hardening*. For instance, when impacted at temperatures between about −40 °C and −20 °C at a strain rate around $10^4\,s^{-1}$, polycrystalline plates of 6 mm thickness appear to support a tensile stress of ~17 MPa (Lange and Ahrens, 1983). Similarly, thin (0.7 mm) polycrystalline coatings grown onto a thin (0.8 mm) substrate of aluminum appear to support a tensile stress of ~47 MPa when subjected to a short (3 ns) pulse from a Nd:YAG laser during which the strain rate is estimated to reach $\sim 10^7\,s^{-1}$ (Gupta and Archer, 1999). Impact and laser spallation experiments are difficult to perform, and so some caution is appropriate in interpreting these results.

10.4.3 Grain size

The tensile strength σ_t decreases with increasing grain size *d*, Figure 10.4. This is evident from a number of studies that include granular ice prepared in the laboratory (Hawkes and Mellor, 1972; Currier and Schulson, 1982; Schulson *et al.*, 1984;

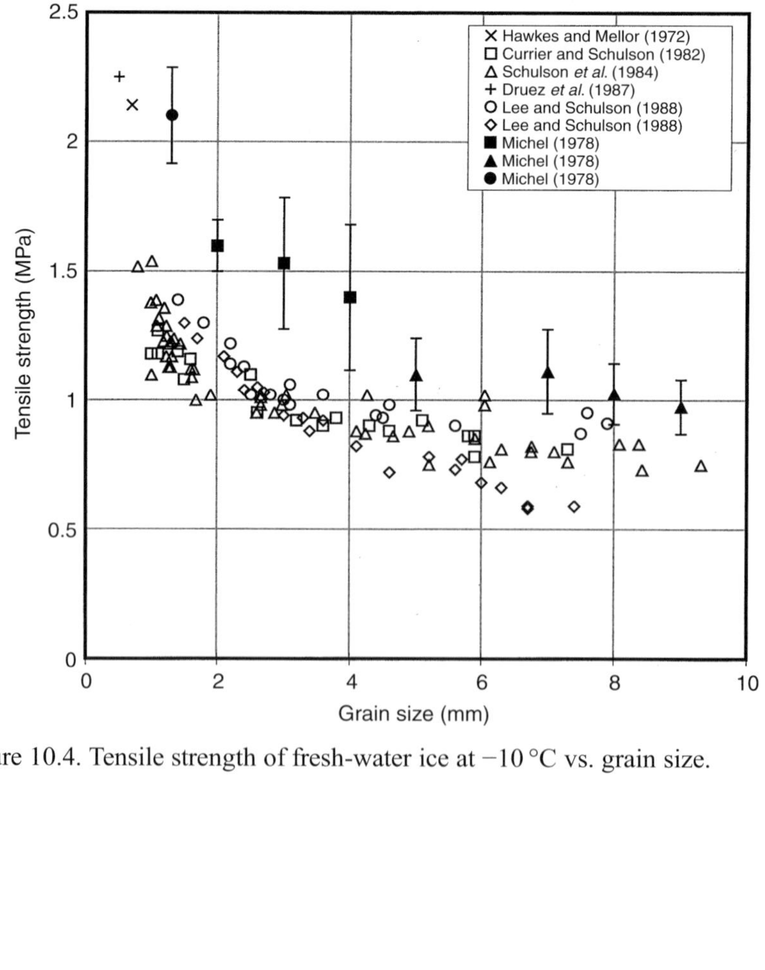

Figure 10.4. Tensile strength of fresh-water ice at −10 °C vs. grain size.

Lee and Schulson, 1988), S2 columnar ice harvested from the cover on the St. Lawrence river (Michel, 1978) and atmospheric ice (Druez *et al.*, 1987).

The functional relationship is of the form $\sigma_t \propto d^{-1/2}$. This functionality was established by Currier and Schulson (1982), by Schulson *et al.* (1984) and by Lee and Schulson (1988), from systematic measurements on cylindrical-shaped specimens (10 cm diameter × 25 cm length) of granular, fresh-water ice. The grain size was measured using the method of linear intercepts[1] and was varied from 1.3 to 8.4 mm, similar to the range that occurs naturally. Although a size range of less than one decade is often insufficient to establish with confidence functional relationships, in this case the data exhibit sufficiently high regression coefficients r^2 to lend confidence to the analysis. At low strain rates, e.g., at $10^{-7}\,\mathrm{s}^{-1}$, the function takes the Orowan form (Orowan, 1954), Figure 10.5a:

$$\sigma_t = Kd^{-0.5} \tag{10.1}$$

where K is a materials constant that has the value $K = 0.052\,\mathrm{MPa\,m}^{1/2}$ ($r^2 = 0.97$) at −10 °C. Correspondingly, the ice exhibits a small amount of inelastic deformation, 0.1–0.4%, that increases with decreasing grain size, Figure 10.6a. The inelasticity originates in both dislocation creep as well as cracking. The latter contribution is suggested by the presence of a low concentration (~0.01 crack/cm^3) of remnant cracks within the two parts of a broken specimen (Schulson, 1987; Lee and Schulson, 1988). At higher strain rates, e.g., at $10^{-3}\,\mathrm{s}^{-1}$, the function takes the Hall–Petch form, Figure 10.5b:

$$\sigma_t = \sigma_o + k_t d^{-0.5} \tag{10.2}$$

where σ_o and k_t are other materials constants that have the values $\sigma_o = 0.52$ MPa and $k_t = 0.030\,\mathrm{MPa\,m}^{1/2}$ ($r^2 = 0.84$) at −10 °C. Correspondingly, the ice exhibits much less inelastic deformation, about 0.01%, Figure 10.6b, and the two parts of the broken specimen are free from cracks. As we show below, Orowan strengthening can be explained in terms of *crack propagation* and Hall–Petch strengthing, in terms of *crack nucleation*. In both scenarios, crack size is limited by grain size (Lee and Schulson, 1988).

We caution against an indiscriminate use of Equations (10.1) and (10.2). Although the two relationships are expected to be general, the parametric values are probably limited to the kind of ice from which they were obtained; namely, bubble-free, equiaxed and randomly oriented aggregates of fresh-water ice initially free from cracks. Atmospheric ice, for instance, although generally more

[1] The method of linear intercepts offers an estimate of grain size that is lower by about a factor of 0.7 than that offered by counting the number of grains per unit area, N, and then by assuming that grains are spherical in shape, which gives $d = \sqrt{\frac{6}{\pi N}}$.

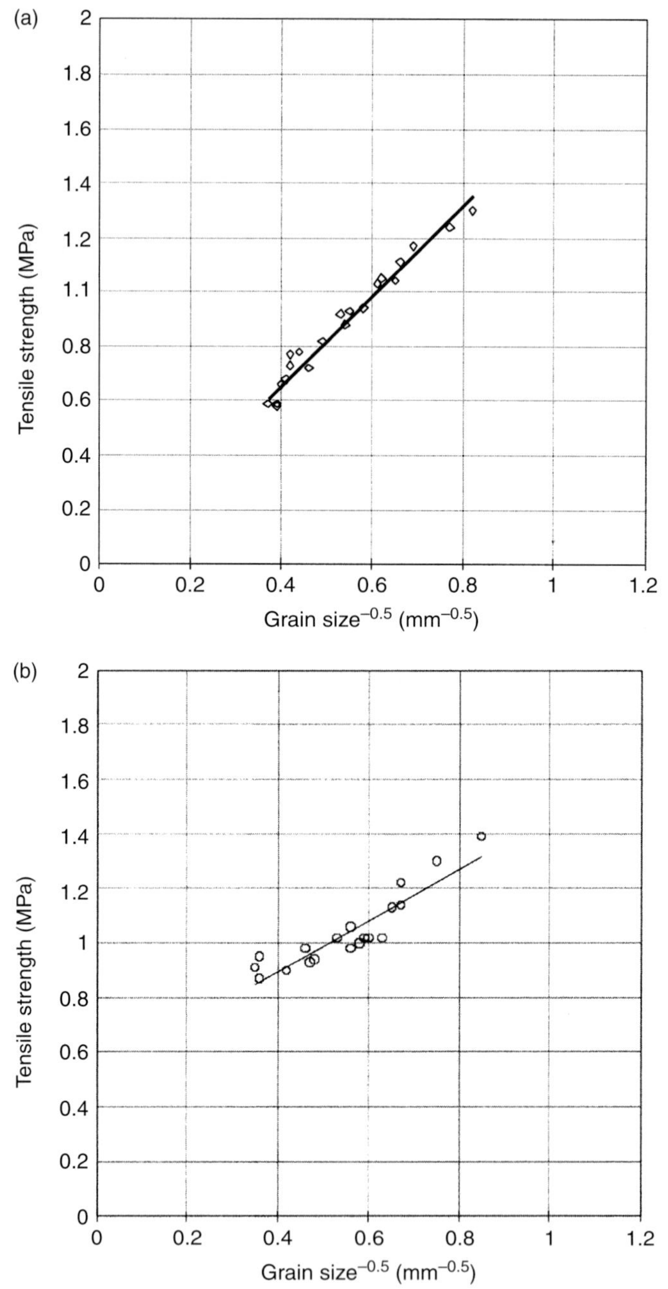

Figure 10.5. Tensile strength of fresh-water granular ice vs. (grain size)$^{-0.5}$ at -10 °C: (a) at a strain rate of 10^{-7} s^{-1} and (b) at 10^{-3} s^{-1}. Note that the strength at the lower strain rate extrapolates through the origin. (Data from Schulson *et al.*, 1984; Lee and Schulson, 1988.)

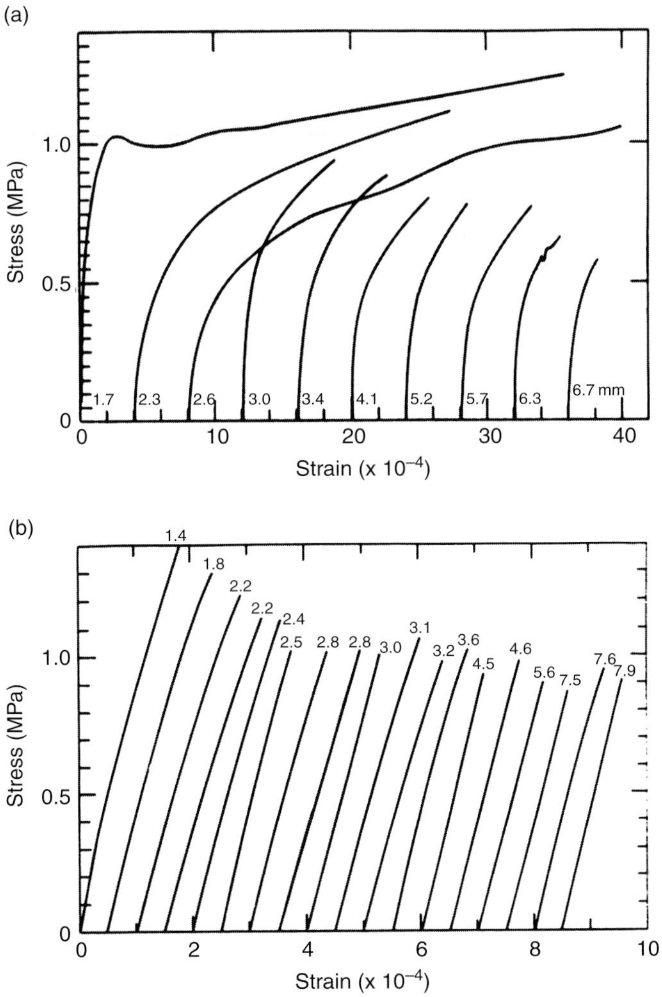

Figure 10.6. Typical stress–strain curves for fresh-water granular ice at −10 °C vs. grain size (in mm) from which were obtained some of the data in Figure 10.5: (a) at a strain rate of $10^{-7}\,s^{-1}$ and (b) $10^{-3}\,s^{-1}$. (From Lee and Schulson, 1988.)

finely grained ($d < 1$ mm), usually contains a significant amount of porosity which may weaken the material by more than fine grains strengthen it (more below). Atmospheric ice may also possess a growth texture. Columnar ice definitely possesses a growth texture (Chapter 3) that, although unlikely to significantly affect the Orowan constant K owing to its relationship to fracture toughness (more below), may well affect the Hall–Petch parameters. In other words, the material parameters should be evaluated for each kind of polycrystal.

An implication of the two different dependencies upon grain size is the concept of a *critical grain size* d_c. Following Cottrell in steels, Schulson (1979) suggested that polycrystalline aggregates of ice can be expected to exhibit a small amount of ductility, provided that $d < d_c$. The idea is that within such finely grained material, cracks of grain-size dimensions are stable upon nucleation; i.e., $K_I < K_{Ic}$ (Chapter 9). Consequently, the ice can support an applied stress greater than the stress required to nucleate the first crack. The higher stress imparts additional inelastic deformation that increases with decreasing grain/crack size. On the other hand, within more coarsely grained aggregates ($d > d_c$) cracks are unstable upon nucleation, and so fracture ensues immediately. The critical grain size is obtained by equating Equations (10.1) and (10.2). Thus:

$$d_c = \left(\frac{K - k_t}{\sigma_o}\right)^2. \tag{10.3}$$

The parameters k_t and σ_o are expected to increase slightly with increasing strain rate and with decreasing temperature, and so the critical grain size is expected to vary with loading conditions as well. When viewed in this light, Orowan strengthening operates when the size of the grains within even the most coarsely grained aggregates is smaller than the critical size under the conditions of the experiment, Figure 10.7. (The critical size is defined by the crossover point.) Hall–Petch strengthening, on the other hand, operates when the size of the grains within even the most finely grained aggregates is larger than the critical size.

10.4.4 Damage

Pre-existing cracks generally act as Griffith flaws (Chapter 9). Consequently, they weaken ice. However, there are two exceptions (Schulson *et al.*, 1989). Material that contains a low number density of optimally oriented short cracks of shorter length than the critical grain size (under the conditions of interest) is not weakened. Similarly, material that contains long cracks that have been blunted by creep deformation is not weakened. In both cases, new cracks must nucleate before the ice fails.

10.4.5 Growth texture

Growth texture is a significant factor. For instance, columnar-grained, fresh-water ice of ~6 mm column diameter that possesses the S2 growth texture is stronger by about 25% when loaded in a direction parallel to the columns than when loaded in a direction perpendicular to them (Carter, 1971). We attribute this difference not to the greater stiffness along the columns. That difference is too small: $E_{pll} = 9.71$ GPa vs. $E_{perp} = 9.33$ GPa (Chapter 4). Instead, we propose that within relatively coarsely grained material, the column diameter is greater than the critical

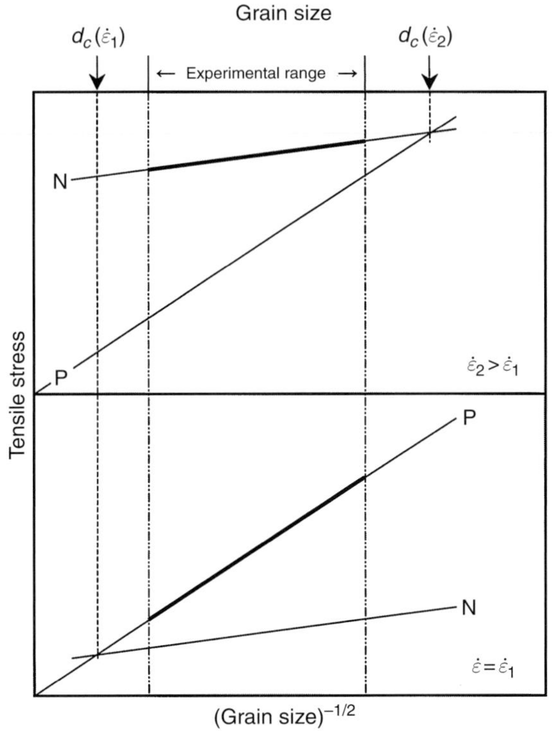

Figure 10.7. Schematic sketch illustrating the critical grain size d_c for two different strain rates. At the lower strain rate, d_c is larger than the grain size of the most coarsely grained aggregate within the experimental range and so the tensile strength is controlled by crack propagation (P); correspondingly the ice behaves inelastically. At the higher strain rate, d_c is smaller than the grain size of the most finely grained aggregate within the experimental range and so the tensile strength is controlled by crack nucleation (N) and the ice behaves elastically.

grain size, in which case crack nucleation governs the strength. Whether the along-column nucleation stress is in fact greater than the across-column nucleation stress remains to be seen.

10.4.6 Brine pockets, brine-drainage channels and porosity

When loaded across the columns, columnar-grained S2 salt-water ice is weaker than S2 fresh-water ice, but is not significantly more ductile. For instance, and further to the points noted in Section 10.4.1, when loaded across the columns within the brittle regime, salt-water specimens of 3 to 6 ppt salinity possess across-column tensile strength of ~0.4 to 0.6 MPa and ductility of 0.01–0.02% at −10 °C (Dykins, 1970;

Richter-Menge and Jones, 1993). This compares with an across-column strength of ~1 MPa for brine-free, fresh-water ice (Carter, 1971). However, when loaded along the columns, the strength of saline ice is about three times greater than its across-column strength and is close to the along-column strength of brine-free fresh-water ice (Dykins, 1970; Kuehn *et al.*, 1990). This behavior reflects the effect of brine pockets and their anisotropic distribution. As discussed in Chapter 3, the pockets form as platelet-like arrays aligned in the growth (i.e., along-column) direction and are preferentially oriented on basal planes. The arrays concentrate stress when the ice is loaded across the columns and thus weaken the material (Assur, 1958; Anderson and Weeks, 1958). The effect of microstructural anisotropy is evident as well within beams of ice harvested from a first-year sea ice cover (Shapiro and Weeks, 1993). The bend strength reaches a minimum when beams are loaded in a direction perpendicular to the platelets. Brine drainage channels (Chapter 3) weaken sea ice still further (Shapiro and Weeks, 1995), for the same reason.

The effects just mentioned are part of a larger picture, where porosity is the governing factor. From a compilation of over 2000 measurements on sea ice, Figure 10.8, Timco and O'Brien (1994) proposed that the bend strength of that material decreases exponentially with increasing porosity v_b. Although scattered, it may be described by the relationship:

$$\sigma_f(\text{MPa}) = 1.76 \exp(-5.88\sqrt{v_b}) \tag{10.4}$$

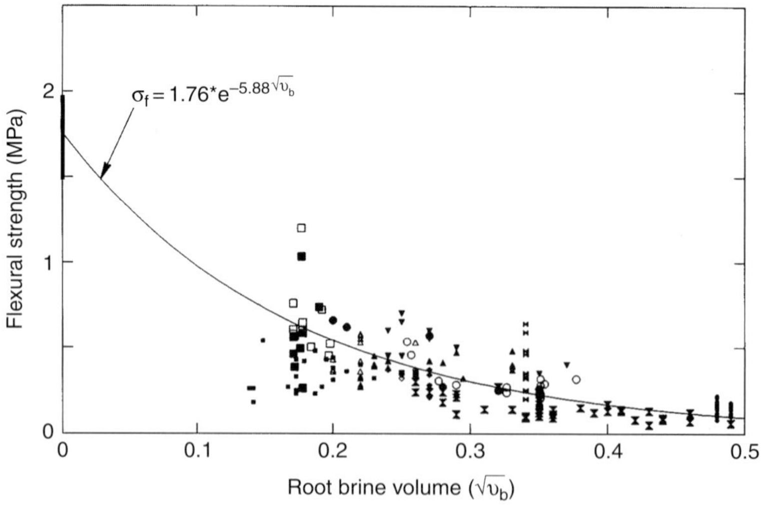

Figure 10.8. The flexural strength of saline ice vs. square root of brine volume where volume is expressed as volume fraction. (From Timco and O'Brien, 1994.)

for $\nu_b < 0.3$. Porosity is expressed as a volume fraction and is defined as the sum of the volume fractions of brine pockets and air bubbles (although the latter is usually much lower than the former). This relationship (i.e., the first term in its expansion) is consistent with one suggested earlier (Weeks and Assur, 1969) in which the strength scaled as $\sigma_f \propto (1 - b\nu_b^{1/2})$, where b is a materials constant. The effect of porosity is a large one, evident from the ~70% reduction in strength for $\nu_b = 0.05$, the porosity that is typical of first-year arctic sea ice. When the bend strength is divided by 1.7, the factor by which the bend strength of brittle material typically overestimates the tensile strength (Ashby and Jones, 2005), the value is consistent with the uniaxial strength given above. The scatter could be caused by variations in the shape and distribution of the pores.

10.4.7 Size

The results described above were generally obtained from tests on specimens whose volume ranged from about 0.1 to 2 liters and whose load-bearing area contained a sufficient number of grains to impart polycrystalline behavior. To determine whether specimen size per se is a factor, Parsons *et al.* (1992) performed flexural tests on beams of different sizes, harvested from a sheet of first-year sea ice, from an iceberg and from S2 fresh-water ice grown in the laboratory. The specimen shape was held constant by setting the width equal to the thickness and by setting both dimensions to 10% of the length. The length of the sea ice beams was varied from 2 m to 0.2 m, of the lab ice beams from 1.3 m to 0.3 m, and of the iceberg ice from 0.6 m to 0.3 m. Thus, the volume was varied by up to a factor of 10^3. The sea ice beams were prepared so that the future tensile surface was defined by the top surface of the parent ice sheet. For each beam, the strain rate of the outer fibers was maintained constant at $10^{-3}\ s^{-1}$, which was high enough to impart brittle behavior. The temperature of the ice was not specified, but judging from the context, was in the range −10 °C to −20 °C. The strength of the fresh-water ice exhibited little statistically significant dependence on specimen volume. The strength of the sea ice, on the other hand, appeared to decrease slightly with increasing volume, scaling as $(\text{volume})^{-1/12}$.

Although this effect was well supported by a large amount of data, we question whether it is related to specimen size per se. Instead, it is probably related to differences in microstructure. Judging from the fact that finely grained, randomly oriented granular ice formed the top ~2 cm of the parent ice cover, from the preservation of that surface in all test specimens large and small, and from the dimensions of the test beams, the next to smallest specimens contained ~50% granular ice. The smallest specimens contained ~100% granular ice. In comparison, columnar S2 ice dominated the microstructure of the larger beams. In other words, an increase in grain size and a change in both grain shape and texture may have contributed more to the reduction in strength than did the increase in specimen size per se.

10.4.8 Pre-strain

Interestingly, both fresh-water and salt-water ice exhibit an elongation of 5–10% or greater – in one case, over 54% – with little loss in strength when pre-strained by a small percentage under compression and then loaded in tension under conditions where it usually fractures before an elongation of $<0.3\%$ (Schulson and Kuehn, 1993; Kuehn and Schulson, 1994). The pre-strain introduces a network of short stable cracks oriented along the loading direction, as well as pockets of recrystallized grains whose size is <0.5 mm. The fracture surface of the ductile ice is somewhat porous in appearance, more granular than transgranular. The processes underlying this behavior are not known, but may include grain boundary sliding in view of the small size of the recrystallized grains and of the topography of the fracture surface. One wonders whether this is a kind of *superplasticity* and whether processing of the kind described here could be mapped onto engineered ceramics to improve their ductility.

10.5 Ductile-to-brittle transition

Returning to the behavior of virgin material, a phenomenological criterion for the ductile-to-brittle transition may now be obtained by equating the relationships for ductile (creep) and brittle (fracture) strengths. Accordingly, upon taking the brittle strength to be governed by crack propagation (Eq. 10.1) and then relating the Orowan constant to fracture toughness K_{Ic} as discussed below (Section 10.6.2), and by letting the ductile strength be governed by power-law or secondary creep (Eq. 6.4 or $\dot{\varepsilon} = B\sigma^n$ where B is a temperature-dependent materials parameter), we obtain for the transition strain rate under tension the expression:

$$\dot{\varepsilon}_{\mathrm{tt}} = \frac{BK_{\mathrm{Ic}}^n}{(\lambda d)^{n/2}} \tag{10.5}$$

where λ is a numerical factor that relates crack size to grain size. For fresh-water polycrystals of $\sim$1 mm grain size loaded at -10 °C, we obtain from Figure 6.7 the parametric values $B = 1 \times 10^{-7}$ MPa^{-3} s^{-1} and $n = 3$. Upon taking $K_{\mathrm{Ic}} = 0.1$ MPa m$^{1/2}$ (Chapter 9) and $\lambda = 3.7$ ((Lee and Schulson, 1988) we obtain $\dot{\varepsilon}_{\mathrm{tt}} \sim 4 \times 10^{-7}$ s^{-1}, in reasonable agreement with experimental observations.

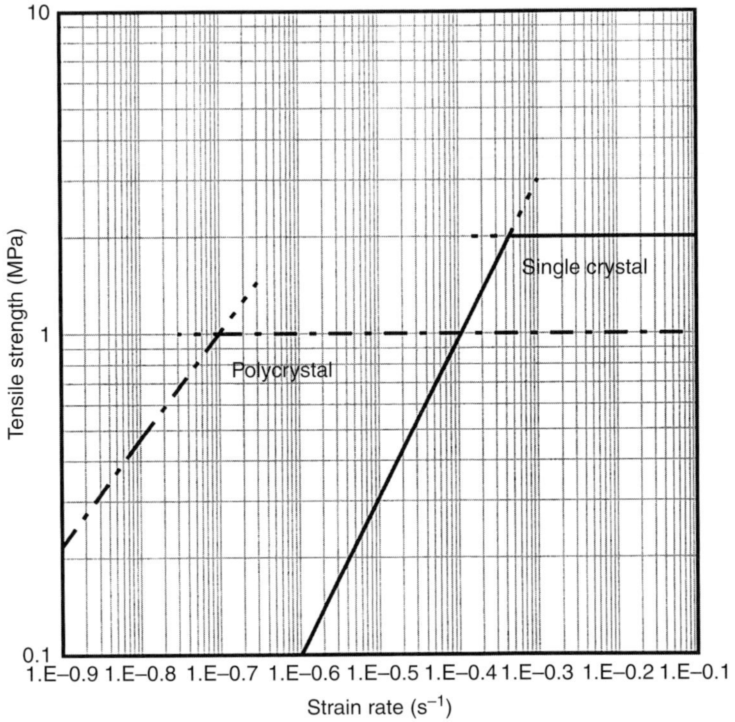

Figure 10.9. Tensile strength at −10 °C vs. strain rate for fresh-water single crystals undergoing basal slip and for fresh-water granular polycrystals of ~1 mm grain size. The intersection of the strain-rate hardening (ductile) and the strain-rate neutral (brittle) regimes marks the ductile-to-brittle transition. The curves for ductile behavior were obtained from data given by Duval *et al.* (1983) that are shown in Figure 6.7. The curves for brittle behavior were obtained from data given in this chapter.

Equation (10.5) is strictly limited to polycrystals. The underlying principle applies to single crystals as well. At a given temperature the creep strength of single crystals is lower than the creep strength of polycrystals, while the fracture strength is higher. As a result, the strain rate at which the two strengths are equal is higher in single crystals than in polycrystals, as shown in Figure 10.9. Correspondingly, the transition strain rate is higher.

10.6 Strength-limiting mechanisms

Two criteria must be satisfied before ice, initially free from cracks, fractures under tension:

- the applied stress must be sufficient to nucleate cracks; and
- once nucleated, the applied stress must be sufficient to propagate them.

10.6.1 Single crystals

A striking characteristic of single crystals is the minimum in the tensile strength when basal planes are inclined at ~45° to the tensile axis. This suggests that in some way basal slip contributes to the failure process. In considering crack nucleation, Carter (1971) invoked a mechanism proposed by Orowan (1954) in which slip dislocations interact with a subgrain tilt boundary composed of edge-dislocation walls. The interaction displaces part of the wall, creating at the free end a tensile stress high enough to overcome the cohesive strength of the material. To model the nucleation stress, Carter introduced both Stroh's (1957) analysis and Griffith's (Chapter 9) analysis of crack mechanics and obtained for sub-boundaries of an assumed height of around 0.01 m a result that was several orders of magnitude lower than the measured tensile strength. On that basis, crack nucleation, at least via the Orowan mechanism, appears not to govern the tensile strength of single crystals.

On crack-propagation control, if we assume that the length $2c$ of slip-cum-sub-boundary induced cracks is of the order of the height of the sub-boundaries, as Friedel (1964) imagined, then (from Chapter 9) the crack-propagation stress is expected to be around $K_{\mathrm{Ic}}/\sqrt{\pi c} \sim 0.5\,\mathrm{MPa}$. This is of the order of the observed strength. The fracture toughness of ice is almost independent of temperature and is independent of strain rate (Chapter 9), and so this kind of explanation would appear to account also for the insensitivity of the strength to these factors. Caution, however, is appropriate. The analysis is based upon not only the action of sub-boundaries, but also upon their height, and it is not clear that such features behave as imagined. Nor is it clear that sub-boundaries are a general feature of single crystals, at least of good ones. Consequently, their role as an integral feature in the failure process is, at best, limited to crystals of lower quality. The surface, for instance, might be a more important feature.

The mechanism that limits the tensile strength of single crystals thus remains an open question.

10.6.2 Polycrystals

Polycrystals offer a more definitive picture – one that rests largely upon the effect of grain size.

Crack-propagation control

Crack propagation governs the tensile strength of aggregates whose grain size is smaller than the critical size. This is suggested by the presence of remnant cracks within the two parts of broken specimens and is confirmed by a simple calculation of the materials constant K in Equation (10.1). Assume that the cracks can be modeled as non-interacting penny-shaped flaws buried within the body and oriented perpendicular to the direction of loading. Under an applied tensile stress σ the mode-I stress intensity factor is given by:

$$K_I = \left(\frac{2}{\pi}\right)^{1/2} \sigma\sqrt{\pi c}. \qquad (10.6)$$

The length of the cracks that propagate was found (Lee and Schulson, 1988) to be directly proportional to grain size through the relationship $2c = \lambda d$. Thus, upon setting $K_I = K_{Ic}$ the tensile strength when governed by crack propagation is given by:

$$\sigma_t = \frac{K_{Ic}}{\lambda^{1/2}} d^{-1/2}. \qquad (10.7)$$

Taking $K_{Ic} = 0.1$ MPa m$^{1/2}$ (Chapter 9) and $\lambda = 3.7$ (Lee and Schulson, 1988), the coefficient of $d^{-1/2}$ has the value 0.051 MPa m$^{1/2}$, in excellent agreement with the measured value $K = 0.052$ MPa m$^{1/2}$.

The skeptic might question this analysis. He/she might argue that linear elastic fracture mechanics is inappropriate in this case, given the length of the loading time ($\sim 3 \times 10^{-3}/10^{-7}\ \mathrm{s}^{-1} \sim 3 \times 10^4$ s). However, that argument would require critical cracks to be present at the beginning of the test, which is not the case for virgin material. Instead, cracks that eventually propagate appear to nucleate nearer the end of the loading period, in which case they are probably loaded rapidly enough to prevent crack-tip creep from violating plane strain conditions.

Crack nucleation control

When the grain size exceeds the critical size and the tensile strength is given by Equation (10.2), crack nucleation appears to control the tensile strength. The questions then concern the nucleation mechanism and the process that serves to concentrate the applied stress. Several possibilities have been considered (for review, see Frost 2001), including thermal expansion mismatch at grain boundaries, elastic mismatch and precursor flaws. Dislocation pile-ups and grain boundary sliding also concentrate stress. Thermal mismatch can be immediately ruled out, because ice is thermally isotropic (Petrenko and Whitworth, 1999). Similarly, elastic mismatch appears to be improbable (Frost and Gupta, 1993; Picu and Gupta, 1996), because the elastic modulus of ice varies by no more than ~30% with crystallographic direction (Chapter 4). And the precursors suggested by Sunder and Wu (1990) seem unlikely, for such features would need to be of millimeter dimensions and flaws that size are large enough to be detectable by the unaided eye.

This leaves dislocation pile-ups and grain boundary sliding. Both mechanisms concentrate stress on grain boundaries in proportion to $\sqrt{L}$ where L is either the length of a dislocation pile-up impinging upon boundaries or the distance along which boundaries slide (Zener, 1948; Stroh, 1957; Smith and Barnby, 1967). Length

L is usually considered to scale with grain size, and so both mechanisms dictate that the applied stress to nucleate cracks scales as $1/\sqrt{d}$. Equation (10.2) follows. The parameters σ_o and k_t in that equation may then be viewed, respectively, as the strength of an infinitely large grained polycrystal and the effectiveness with which grain boundaries impede either dislocation slip or sliding. On the pile-up mechanism, Stroh's (1957) analysis of crack nucleation is based upon the principle of a reduction in strain energy accompanying the creation of new surface. It leads to the expression:

$$k_t = \langle m \rangle \left(\frac{3\pi\gamma_S G}{8(1-v)} \right)^{1/2} \tag{10.8}$$

where $\langle m \rangle$ is the Taylor orientation factor and G is the shear modulus. Upon taking $\langle m \rangle \sim 5$, as estimated from hexagonal-close-packed metals for which basal slip dominates, and upon taking $G = 3.52$ GPa (Chapter 4), this relationship gives $k_t = 0.10$ MPa m$^{1/2}$. Similarly, Zener's (1948) analysis, which is based upon the same principle but focuses on grain boundary sliding, leads to the relationship:

$$k_t = \left(\frac{48\gamma_S G}{\pi} \right)^{1/2} \tag{10.9}$$

which gives $k_t = 0.06$ MPa m$^{1/2}$. In other words, both mechanisms lead to grain boundary impediments that are of the same order of magnitude as the measured value $k_t = 0.03$ MPa m$^{1/2}$.

In addition to energetics, however, we must also consider kinetics. On the pile-up model, Cole (1988) argued that owing to the low velocity with which dislocations glide on the basal planes, of the order of $v \sim 10^{-6}$ m s^{-1} MPa^{-1} at the temperature of interest (Ahmad and Whitworth, 1988; Shearwood and Whitworth, 1991), the time to form a pile-up within grains of the order of $d = 1$ mm would be around 10^3 s. This estimate is considerably greater than the time of a typical experiment which, at a strain rate of 10^{-3} s^{-1} and tensile strain of $\sim 10^{-4}$, is between 0.1 and 1 s. On that basis, the dislocation-based mechanism would seem to be unlikely. However, Cole's argument assumes that dislocations in ice move as individuals. In reality, they move collectively and intermittently, as discussed in Chapters 5 and 6. This behavior is a result of mutual elastic coupling in which the motion of one dislocation can trigger the movement of many (Louchet, 2006). Correspondingly, this collective, intermittent movement can lead to the kind of dislocation avalanche heard by Weiss and Grasso (1997), by Miguel *et al.* (2001) and by Richeton *et al.* (2005). It also leads directly to the Hall–Petch relationship (Louchet *et al.*, 2006). As a result, the actual time to form a pile-up is probably lower than the time estimated from the movement of a single dislocation. Thus, the pile-up of dislocations, which in fact has been observed in ice via X-ray topography (Liu *et al.*, 1995), remains a credible crack-nucleation mechanism.

On GBS, the observation that decohesions form along grain boundaries within polycrystals strained at rates as high as $10^{-2}\,s^{-1}$ at $-10\,°C$ (Picu and Gupta, 1995), albeit under compression, would seem to suggest that this mechanism also occurs rapidly enough. Perhaps pre-melting at grain boundaries (for review, see Dash *et al.*, 1995) might assist, at sufficiently high temperatures.

We conclude, therefore, that either dislocation pile-ups or grain boundary sliding can account for the nucleation of cracks and, thus, for the Hall–Petch character of the tensile strength of polycrystals of grain size larger than the critical size.

10.6.3 The role of brine pockets in sea ice

The competition between Orowan and Hall–Petch behavior may not be relevant in sea ice. Crack nucleation may not even be an issue, given the stress-concentrating effect of the resident pores and the progressive freezing and expansion of brine pockets upon cooling. In this case, the tensile strength is probably governed by crack propagation, where the crack size is related more to pore size and to the spacing between them than to grain size. It would be erroneous, however, to associate an individual pore with a critical crack, because the critical crack diameter is estimated to be about $2c \sim (K_{Ic}/\sigma_t)^2 \sim (0.1/0.5)^2 10^3 \sim 40\,mm$. In comparison, the pore diameter is $< 1\,mm$. More likely, the critical flaw consists of a macroscopic defect made from the collection of pores within the planar arrays that characterize the microstructure of sea ice (Chapter 3).

10.6.4 Dynamic strength

In the above discussion we accounted for tensile failure in terms of the propagation of a single crack, either immediately upon nucleation or thereafter. Under dynamic loading a different process appears to operate. This is suggested by the fact that, while "statically" loaded tensile bars break into two pieces, ones loaded dynamically break into a large number of small fragments (Lange and Ahrens, 1983). The implication is that in the latter case failure results from the interaction of many cracks.

That this is reasonable follows from the fact that a single crack travels too slowly. If we assume that the tensile strain to failure under dynamic conditions is similar to that under static ones, namely $\varepsilon_t \leq 10^{-3}$, then at strain rates of the order imposed by Lange and Ahrens (1983) of $10^4\,s^{-1}$, the time to failure is less than 10^{-7} s. Crack velocities cannot exceed the speed of sound, which can be estimated from the relationship $\sqrt{E/\rho}$ where E is Young's modulus ($\sim$10 GPa, Chapter 4) and ρ is density (917 kg m^{-3}) and is thus about 3300 m s^{-1}. At that speed a crack travels less than 0.3 mm in 10^{-7} s. In other words, during dynamic failure it is unlikely that a single crack can propagate across the load-bearing section of anything other than the smallest body. The question, then, which remains to be answered, is: how do multiple cracks interact within a body of ice dynamically loaded under tension?

References

Ahmad, S. and R. W. Whitworth (1988). Dislocation-motion in ice – a study by synchrotron x-ray topography. *Phil. Mag. A*, **57**, 749–766.

Anderson, D. L. and W. F. Weeks (1958). A theoretical analysis of sea ice strength. *Trans. Amer. Geophys. Union*, **39**, 632–640.

Ashby, M. and D. R. H. Jones (2005). *Engineering Materials 2: An Introduction to Microstructures, Processing and Design*, 3rd edn. Oxford: Butterworth-Heinemann.

Assur, A. (1958). Composition of sea ice and its tensile strength. In *Arctic Sea Ice*. Washington, D.C.: U.S. National Academy of Sciences, pp. 106–138.

Butkovich, T. R. (1954). Ultimate strength of ice. *U.S. Snow, Ice and Permafrost Research Establishment*, Research Paper, 15.

Butkovich, T. R. (1958). Recommended standards for small-scale ice strength tests. *Trans. Eng. Inst. Can.*, **2**, 112–115.

Carter, D. (1971). Lois et mechanismes de l'apparente fracture fragile de la glace de rivière et de lac. Ph.D. thesis, University of Laval.

Cole, D. M. (1988). Crack nucleation in polycrystalline ice. *Cold Reg. Sci. Technol.*, **15**, 79–87.

Cole, D. M. (1990). Reversed direct-stress testing of ice: initial experimental results and analysis. *Cold Reg. Sci. Technol.*, **18**, 303–321.

Currier, J. H. and E. M. Schulson (1982). The tensile strength of ice as a function of grain size. *Acta Metall.*, **30**, 1511–1514.

Dai, M. R., H. H. Shen, M. A. Hopkins and S. F. Ackley (2004). Wave rafting and the equilibrium pancake ice cover thickness. *J. Geophys. Res.*, **109**, C07023.

Dash, J. G., F. Haiying and J. S. Wettlaufer (1995). The premelting of ice and its environmental consequences. *Rep. Prog. Phys.*, **58**, 115–167.

Druez, J., J. Cloutier and L. Claveau (1987). Etude comparative de la résistance à la traction et à la compression de la glace atmospherique. *J. Physique Coll.*, **49**, C1-337–C331-343.

Duval, P., M. F. Ashby and I. Anderman (1983). Rate-controlling processes in the creep of polycrystalline ice. *J. Phys. Chem.*, **87**, 4066–4074.

Dykins, J. E. (1970). *Ice Engineering: Tensile Properties of Sea Ice Grown in a Confined System*. Naval Civil Engineering Laboratory.

Evans, R. J. and N. Untersteiner (1971). Thermal cracks in floating ice sheets. *J. Geophys. Res.*, **76**, 694–703.

Friedel, J. (1964). *Dislocations*. Addison-Wesley Series in Metallurgy and Materials. New York: Pergamon Press.

Frost, H. J. (2001). Crack nucleation in ice. *Eng. Fract. Mech.*, **68**, 1823–1837.

Frost, H. J. and V. Gupta (1993). Crack nucleation mechanisms and fracture toughness measurements in freshwater ice. *ASME*, AMD, Vol. 163.

Gold, L. W. (1963). Deformation mechanisms of ice. In *Ice and Snow*, ed. W. D. Kingery. Cambridge, Mass.: MIT Press, pp. 8–27.

Greeley, R., P. H. Figueredo, D. A. Williams *et al.* (2000). Geologic mapping of Europa. *J. Geophys. Res.*, **105** (E9), 22,559–22,578.

Greenberg, R., P. Geissler, G. Hoppa *et al.* (1998). Tectonic processes on Europa: tidal stresses, mechanical response, and visible features. *Icarus*, **135**, 64–78.

Gupta, V. and P. Archer (1999). Measurement of the grain-boundary tensile strength in columnar freshwater ice. *Phil. Mag. Lett.*, **79**, 503–509.

Hawkes, I. and M. Mellor (1972). Deformation and fracture of ice under uniaxial stress. *J. Glaciol.*, **11**, 103–131.

Hopkins, M. A., J. Tuhkuri and M. Lensu (1999). Rafting and ridging of thin ice sheets. *J. Geophys. Res.*, **104**, 13,605–13,613.

Jaeger, J. C. and N. G. W. Cook (1979). *Fundamentals of Rock Mechanics*, 3rd edn. London: Chapman and Hall.

Jones, S. J. and J. W. Glen (1969). The mechanical properties of single crystals of pure ice. *J. Glaciol.*, **8**, 463–473.

Kermani, M., M. Farzaneh and R. Gagnon (2008). Bend strength and effective modulus of atmospheric ice. *Cold Reg. Sci. Technol.*, **53**, 162–169.

Kuehn, G. A. and E. M. Schulson (1994). Ductile saline ice. *J. Glaciol.*, **40**, 566–568.

Kuehn, G. A., R. W. Lee, W. A. Nixon and E. M. Schulson (1990). The structure and tensile behavior of first-year sea ice and laboratory-grown saline ice. *J. Offshore Mech. Arctic Eng.*, **112**, 357–363.

Lange, M. A. and T. J. Ahrens (1983). The dynamic tensile-strength of ice and ice-silicate mixtures. *J. Geophys. Res.* **88** (B2), 1197–1208.

Lee, R. W. and E. M. Schulson (1988). The strength and ductility of ice under tension. *J. Offshore Mech. Arctic Eng.*, **110**, 187–191.

Liu, F., I. Baker and M. Dudley (1995). Dislocation-grain boundary interactions in ice crystals. *Phil. Mag. A*, **71**, 15–42.

Louchet, F. (2006). From individual dislocation motion to collective behaviour. *J. Mater. Sci.*, **41**, 2641–2646.

Louchet, F., J. Weiss and T. Richeton (2006). Hall-Petch law revisited in terms of collective dislocation dynamics. *Phys. Rev. Lett.*, **97**, 0775504.

Mangold, N. (2005). High latitude patterned grounds on Mars: Classification, distribution and climatic control. *Icarus*, **174**, 336–359.

Mellon, M. T. (1997). Small-scale polygonal features on Mars: Seasonal thermal contraction cracks in permafrost. *J. Geophys. Res.*, **102**, 25,617–25,628.

Mellor, M. and I. Hawkes (1971). Measurement of tensile strength by diametral compression of discs and annuli. *Eng. Geol.*, **5**, 194–195.

Michel, B. (1978). *Ice Mechanics*. Quebec: Laval University Press.

Miguel, M. C., A. Vespignani, S. Zapperi, J. Weiss and J. R. Grasso (2001). Intermittent dislocation flow in viscoplastic deformation. *Nature*, **410**, 667–671.

Nath, P. C. and D. G. Vaughan (2003). Subsurface crevasse formation in glaciers and ice sheets. *J. Geophys. Res.*, **108** (B1), 2020, doi :10.1029/2001JB000453.

Nye, J. F. (1957). The distribution of stress and velocity in glaciers and icesheets. *Proc. R. Soc. Lond., Ser. A*, **239**, 113–133.

Orowan, E. (1954). Dislocations and mechanical properties. In *Dislocations in Metals*, ed. M. Cohen. New York: American Institute of Mining and Metallurgical Engineers, pp. 69–195.

Parameswaran, V. R. (1982). Fracture criterion for ice using a dislocation model. *J. Glaciol.*, **28**, 161–169.

Parsons, B. L., M. Lal, F. M. Williams *et al.* (1992). The influence of beam size on the flexural strength of sea ice, fresh-water ice and iceberg ice. *Phil. Mag. A*, **66**, 1017–1036.

Petrenko, V. F. and R. W. Whitworth (1999). *Physics of Ice*. New York: Oxford University Press.

Picu, R. C. and V. Gupta (1995). Crack nucleation in columnar ice due to elastic anistropy and grain boundary sliding. *Acta Metall. Mater.*, **43**, 3783–3789.

Picu, R. C. and V. Gupta (1996). Singularities at grain triple junctions in two-dimensional polycrystals with cubic and orthotropic grains. *J. Appl. Mech. T. ASME*, **63**, 295–300.

Richeton, T., J. Weiss and F. Louchet (2005). Dislocation avalanches: Role of temperature, grain size and strain hardening. *Acta Mater.*, **53**, 4463–4471.

Richter-Menge, J. A. and K. F. Jones (1993). The tensile strength of first-year sea ice. *J. Glaciol.*, **39**, 609–618.

Riska, K. and J. Tuhkuri (1995). Application of ice cover mechanics in design and operations of marine structures. *Sea Ice Mechanics and Arctic Modeling Workshop*, Anchorage, Alaska, Northwest Research Associates, Inc., Bellevue, WA.

Rist, M. A., P. R. Sammonds, S. A. F. Murrell *et al.* (1999). Experimental and theoretical fracture mechanics applied to Antarctic ice fracture and surface crevassing. *J. Geophys. Res.*, **104**, 2973–2987.

Rist, M. A., P. Sammonds, H. Oerter and C. S. M. Doake (2002). Fracture of Antarctic shelf ice. *J. Geophys. Res. Solid Earth*, **107** (B1), 2002, doi:10.1029/2000JB000058.

Sammonds, P. R., S. A. F. Murrell and M. A. Rist (1998). Fracture of multi-year sea ice. *J. Geophys. Res.*, **103**, 21,795–21,815.

Schulson, E. M. (1979). An analysis of the brittle to ductile transition in polycrystalline ice under tension. *Cold Reg. Sci. Technol.*, **1**, 87–91.

Schulson, E. M. (1987). The fracture of ice 1h. *J. Physique*, C1-207–C201-207.

Schulson, E. M. and G. A. Kuehn (1993). Ductile ice. *Phil. Mag. Lett.*, **67**, 151–157.

Schulson, E. M., P. N. Lim and R. W. Lee (1984). A brittle to ductile transition in ice under tension. *Phil. Mag. A*, **49**, 353–363.

Schulson, E. M., S. G. Hoxie and W. A. Nixon (1989). The tensile strength of cracked ice. *Phil. Mag. A*, **59**, 303–311.

Shapiro, L. H. and W. F. Weeks (1993). The influence of crystallographic and structural properties on the flexural strength of small sea ice beam. *Ice Mechanics-1993; 1993 Joint ASME Applied Mechanics and Materials Summer Meeting AMD*, New York, American Society of Mechanical Engineers.

Shapiro, L. H. and W. F. Weeks (1995). Controls on the flexural strength of small plates and beams of first-year sea ice. *Ice Mechanics; 1995 Joint ASME Applied Mechanics and Materials Summer Meeting AMD*, American Society of Mechanical Engineers.

Shearwood, C. and R. W. Whitworth (1991). The velocity of dislocations in ice. *Phil. Mag. A*, **64**, 289–302.

Smith, E. and J. T. Barnby (1967). Crack nucleation in crystalline solids. *Met. Sci. J.*, **1**, 56–64.

Stroh, A. N. (1957). A theory of the fracture of metals. *Phil. Mag. Supp.*, **6**, 418–465.

Sunder, S. S. and M. S. Wu (1990). Crack nucleation due to elastic anisotropy in polycrystalline ice. *Cold Reg. Sci. Technol.*, **18**, 29–47.

Timco, G. W. and S. O'Brien (1994). Flexural strength equation for sea ice. *Cold Reg. Sci. Technol.*, **22**, 285–298.

Weeks, W. F. (1962). Tensile strength of NaCl ice. *J. Glaciol.*, **4**, 25–52.

Weeks, W. F. and A. Assur (1969). Fracture of lake and sea ice. *U.S. Cold Regions Research and Engineering Laboratory*, Research Report, 269.

Weiss, J. (2001). Fracture and fragmentation of ice: a fractal analysis of scale invariance. *Eng. Fract. Mech.*, **68**, 1975–2012.

Weiss, J. (2004). Subcritical crack propagation as a mechanism of crevasse formation and iceberg calving. *J. Glaciol.*, **50**, 109–115.

Weiss, J. and J. R. Grasso (1997). Acoustic emission in single crystals of ice. *J. Phys. Chem. B*, **101**, 6113–6117.

Zener, C. (1948). *Elasticity and Anelasticity of Metals*. Chicago: Chicago University Press.

11

Brittle compressive failure of unconfined ice

11.1 Introduction

Ice fails under compressive loads in a variety of situations. During the exploration and extraction of oil and gas from ice-infested waters, for instance, risers must be protected against impact, necessitating a defense system sufficiently strong to resist both local and global failure. The forces can be large. The 110-meter wide Molikpaq exploration platform, deployed in the Canadian Arctic, experienced on 12 April 1986 interactions with a multi-year hummock 8–12 meters in thickness that led to global loads that were estimated to have reached 420 MN (Frederking and Sudom, 2006) or about 68% of the design load (Klohn-Crippen, 1998). Similarly, the 100-meter diameter Hibernia platform, now in production in the North Atlantic Ocean approximately 315 kilometers southeast of St. John's, Newfoundland, was designed and built to withstand global ice loads of about 1300 MN that could arise through interactions with icebergs (Hoff *et al.*, 1994). In these instances and others like them the limiting load was/is set by brittle compressive failure of the ice (Chapter 14).

Compressive loading threatens not only the integrity of off-shore structures, but also the arctic sea ice cover itself. The cover extends over an area of about 12 million km^2 and plays a significant role in both local and global climate (e.g., Zhang and Walsh, 2006, and references therein). It is loaded under compression by wind and ocean currents (Thorndike and Colony, 1982). When strong enough, this forcing induces stresses that break the ice, through processes we discuss in Chapter 15. Fracture in this instance affects the ice thickness distribution and, hence, the transfer of heat and moisture from the ocean to the atmosphere and the oceanic salt budget (Maykut, 1982; Heil and Hibler, 2002).

Brittle compressive failure is a consideration in extra-terrestrial scenarios as well. For instance, Spaun *et al.* (2001) suggested that certain near-equatorial linear features within the trailing quadrant of Europa may have originated not from tensile fracture as had been thought, but from compressive shear faulting. Similarly,

Schenk and McKinnon (1989) remarked that in the southeastern portion of Europa's wedge crack region sufficient compression may have existed to create a shear fault. And Schulson (2002) analyzed a specific wedge crack within Europa's icy crust on the basis of frictional crack sliding driven by compressive stresses and obtained an estimate of maximum compressive stress that agrees reasonably well with estimates based upon elastic deformation. The stress in this situation is of tidal origin exacerbated by non-synchronous rotation (Greenberg *et al.*, 1998).

In this chapter our objectives are to describe the characteristics of brittle compressive failure, and then to consider the underlying physical mechanisms. We focus mainly on unconfined ice, although we offer a few words on the behavior of confined single crystals. We address confinement of polycrystals in Chapter 12. Again, we emphasize quasi-static failure. Dynamic or high-speed failure, of the kind that occurs between hailstones and jet engines, for instance, or between atmospheric ice and the Space Shuttle (Carney *et al.*, 2006), has been less well studied. Unless otherwise noted, the results to be described were obtained in the laboratory on homogeneous material of known microstructure and of uniform temperature. This contrasts with the situation in the field, particularly within floating sea ice covers, where the stress state is often not known, thermal gradients exist and structural defects of a variety of sizes and origins are generally present.

As will become apparent, brittle compressive failure is a more complicated process than tensile failure. Cracks nucleate as before, but now grow in a stable or quasi-stable manner as the load rises. They interact and this leads to the creation of macroscopic faults of one kind or another. In addition, frictional sliding enters the picture and plays a major role, rendering ice a Coulombic material.

11.2 Measurement of brittle compressive strength

Unconfined compressive strength might appear to be one of the simplest measurements to make. Good values, however, require great care in experimental procedure. Brittle behavior is less forgiving than is ductile behavior. Consequently, attention must be paid not only to the alignment of the loading train, but also to a number of other factors.

Stiffness of loading frame The stiffer the system, the better it is. Brittle failure occurs suddenly and is accompanied by a sudden release of energy stored within both the specimen and the load frame. Stiff systems store less energy than compliant ones, and so promote more controlled failure. Complete control requires feedback of the kind applied to the study of granite (Lockner *et al.*, 1991). The stiffness of the relatively rigid servo-hydraulic controlled multi-axial testing system (MATS) at Dartmouth's Ice Research Laboratory, for instance, is 8 MN/mm, but even that is not sufficiently great to completely suppress unstable failure of liter-sized specimens.

Loading platens Lateral constraint introduced by loading platens is a key factor. Constraint induces a triaxial state of stress, which promotes shearing and fragmentation instead of longitudinal splitting, and leads to ostensibly greater strength. To reduce end constraint, one practice is to cap specimens using a material for which the ratio $(E/v)_{cap} = (E/v)_{ice}$ where E and v denote Young's modulus and Poisson's ratio, respectively. "Synthane", a phenolic-based material strengthened with fibers, meets that requirement quite well. E/v matching, however, is not a complete solution: while it eliminates end constraint within the elastic regime, once ice begins to crack, which it does well before terminal failure (more below), constraint sets in. A better practice is to employ axially stiff and laterally compliant brush platens in which the dimensions of the bristles are chosen on the basis of Euler buckling (Hausler, 1981; Schulson *et al.*, 1989). Potential problems with proud bristles can be eliminated by "creep seating" under a low stress (~0.1 MPa) for a short time (~15–30 minutes). Better, still, would be the use of fluid cushions to reduce boundary constraints altogether, particularly under multi-axial loading. Platens of this kind have been used successfully in the compression testing of mortar and concrete (Gerstle *et al.*, 1980), but have not yet been applied to ice.

Parallelism of ends Ideally, the ends of the specimens should be parallel to each other and perpendicular to the loading axis. Investigators have generally found that good results could be obtained when the ends of liter-sized specimens of length of 100–250 mm, for instance, are parallel to within ±0.025 mm.

Length-to-diameter ratio When testing cylindrically shaped specimens, it is good practice to maintain a *L*:*D* ratio of ~2.5:1, in the interests of promoting a relatively uniform stress state within the central part of the specimen.

Specimen size In addition to the *L*:*D* ratio, the specimen should be large enough to include a sufficient number of grains across the load-bearing section. The comments made with respect to tensile strength (Chapter 10) apply here as well. We consider specimen size per se in Section 11.5.8.

Specimen shape When examining granular ice, there is little to recommend one shape over another. Both cylinders and cubes have been examined successfully. Even when testing columnar ice by loading in the across-column direction, where cylindrically shaped specimens lead to a variation in the length of the columns across the load-bearing area, there is little to suggest that a cylinder is inferior to a rectangular prism.

11.3 Ductile versus brittle behavior: an overview

Whether in the form of a single crystal or a polycrystal, ice under compressive loading exhibits either ductile or brittle behavior, depending upon the loading conditions. In this chapter we distinguish on the basis of macroscopic behavior. Later (Chapter 13) we offer a more fundamental distinction and consider the ductile-to-brittle transition directly, in terms of the factors that affect it and the underlying mechanism. At this juncture, the point to note is that under compressive loading the strain rate $\dot{\varepsilon}_{tc}$ that marks the transition within test specimens is around one to four orders of magnitude greater than it is under tension; namely, $\dot{\varepsilon}_{tc} \sim 10^{-2}\,s^{-1}$ vs. 10^{-4}–$10^{-3}\,s^{-1}$ for single crystals and $\dot{\varepsilon}_{tc} \sim 10^{-4} - 10^{-3}\,s^{-1}$ vs. $\sim 10^{-7}\,s^{-1}$ for polycrystals loaded under ambient conditions. The reason for the higher transition strain rate, within the context of the discussion in the previous chapter (Section 10.5), is that the brittle compressive strength, as we will see, is considerably greater than the tensile strength. A change in any factor that has the effect of increasing the brittle strength by more than the ductile strength also has the effect of increasing the transition strain rate.

The transition, incidentally, occurs gradually rather than abruptly. The reason is that the overall inelastic deformation comprises a mixture of creep via dislocation slip and cracking, the proportion of which varies in the vicinity of the transition. Thus, although we refer to a transition strain rate, the actual change in macroscopic behavior occurs over a range of strain rates that vary by about a factor of three.

Ductile behavior is characterized by a smooth stress–strain curve that rises and then either levels off or, at higher rates within the ductile regime, reaches a maximum followed by descent towards a plateau (Michel, 1978), Figure 11.1. Correspondingly, plastic strain in excess of 0.1 can be imparted without macroscopic failure, even though the material, which initially may be crack-free and completely transparent, becomes riddled with non-propagating, grain-sized microcracks when deformed at higher rates within the ductile regime. Granular ice, for instance, becomes milky-white in appearance, owing to the scattering of light from cracks of a variety of orientations. In addition, ductile behavior is characterized by strain-rate hardening ($\sigma \sim \dot{\varepsilon}^{1/n}$ where $n \sim 3$) and by thermal softening (Michel 1978) and is sensitive to salinity and porosity (Peyton, 1966; Schwarz and Weeks, 1977), but is independent of grain size (Cole, 1987). Also, the maximum shear stress – i.e., one-half the difference between the maximum and minimum principal stress – is independent of pressure, in the manner of ductile metals.

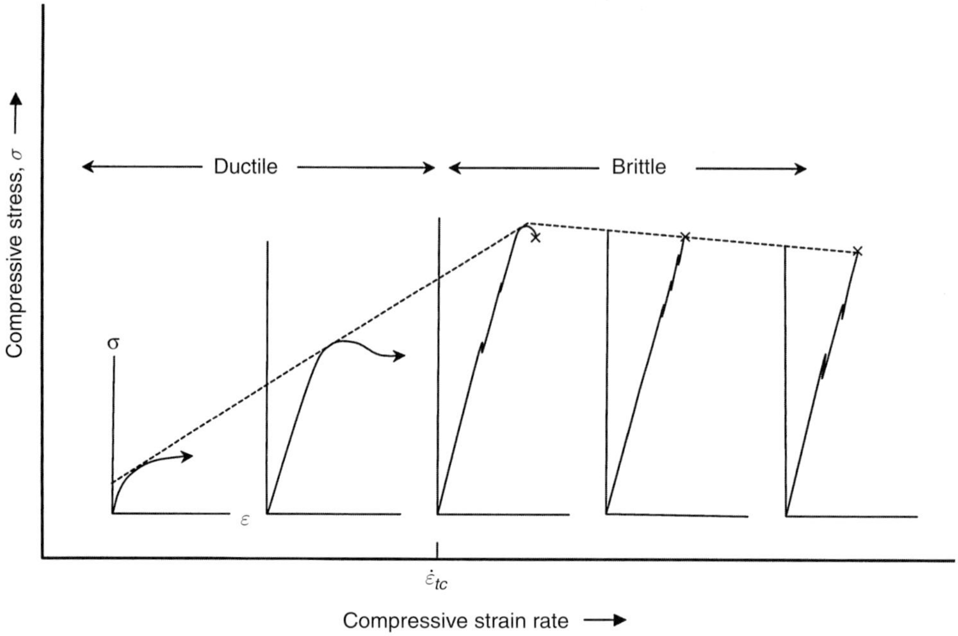

Figure 11.1. Schematic compressive stress–strain curves. At low strain rates ice exhibits ductile behavior accompanied by strain-rate hardening. At high rates its behavior is brittle.

In comparison, brittle behavior is characterized by a stress–strain curve that rises pseudo linearly and then, owing to the development of a mechanical instability, suddenly drops off after a strain of <0.003 with little evidence of roll-over (Schulson, 1990; Rist and Murrell, 1994; Arakawa and Maeno, 1997; Iliescu and Schulson, 2004). The curve is punctuated by small load-drops (Figure 11.1) once the stress reaches around one-third of the terminal failure stress, after which it exhibits slightly negative curvature. The load-drops and the curvature are related to the creation and growth of micro-cracks. Terminal failure is accompanied by an audible report, owing to the sudden release of strain energy, and (at least in lab-sized specimens) is marked by multiple splitting of unconfined material along the direction of loading and by shear faulting and spalling of confined material (more below). In addition, brittle behavior is characterized by moderate strain-rate softening[1] and by strengthening through grain refinement, as described below, but appears to be insensitive to the low level of salinity/porosity encountered in first-year sea ice, at least at −10 °C (Schulson and Gratz, 1999; Schulson *et al.*,

[1] Hints of strain-rate softening were first reported over 60 years ago (e.g., Khomichevskaya, 1940; Korzhavin, 1955). The effect was later ascribed (Hawkes and Mellor, 1972) to an artifact of experimental technique. More recent experiments, discussed in this chapter, have confirmed its existence.

2006). Also, brittle behavior is marked by much higher sensitivity to pressure than is ductile behavior (Chapter 12), in the manner of rocks and minerals (Paterson and Wong, 2005).

A consequence of both strain-rate hardening (ductile regime) and strain-rate softening (brittle regime) is that the compressive strength reaches a maximum at the ductile-to-brittle transition. This behavior is exhibited by both single crystals and polycrystals, and by atmospheric ice (Kermani *et al.*, 2007). This maximum is of considerable importance in relation to ice forces on engineered structures (Chapter 14).

11.4 Single crystals

Several factors affect the strength of single crystals.

Temperature and strain rate Figure 11.2 shows the compressive strength of unconfined single crystals vs. strain rate, at 0 °C, −10 °C and −30 °C (Carter, 1971). The

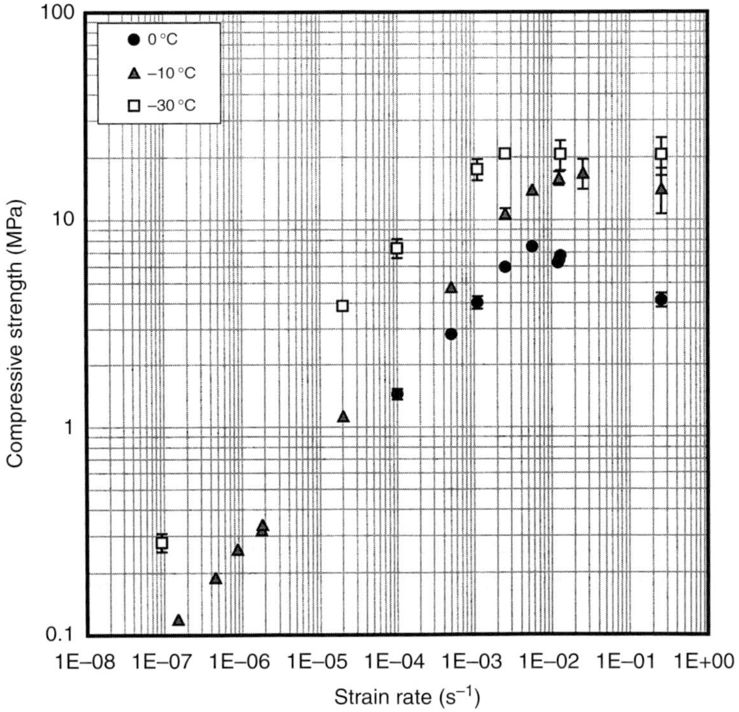

Figure 11.2. Unconfined compressive strength of single crystals of ice vs. strain rate. The basal plane was inclined by 45° to the direction of loading. The ductile-to-brittle transition occurs at a strain rate between about $10^{-3}\,s^{-1}$ and $10^{-2}\,s^{-1}$. (Data from Carter, 1971.)

c-axes are oriented at 45° to the direction of loading and the strain rate spans over six orders of magnitude. The strength is defined as the applied normal stress, and not the shear stress resolved onto the basal planes. At strain rates up to about $10^{-2}\,s^{-1}$ single crystals exhibit ductile behavior, as already noted; correspondingly, the strength increases with increasing strain rate. At higher strain rates crystals exhibit brittle behavior and their strength decreases slightly; correspondingly, terminal failure occurs by splitting along the loading direction. Within both regimes, the strength increases with decreasing temperature. The magnitude of the effect within the brittle regime is $\sim$0.4 MPa $°C^{-1}$.

Orientation of c-axis Figure 11.3 shows that the brittle compressive strength, like the tensile strength, reaches a minimum when the c-axis is oriented at $\alpha \sim 45°$ with respect to the loading direction. The compressive strength is greater than the tensile strength, by about a factor of five to six at the temperature (−10 °C) of measurement.

Confinement The principal effects of confinement are to suppress axial splitting and to increase strength. If sufficient, confinement can also impart a brittle-to-ductile

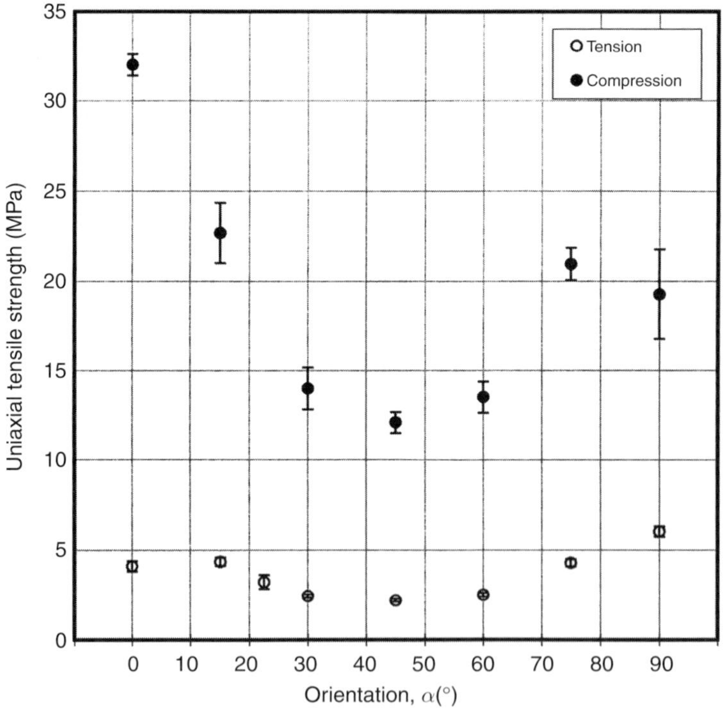

Figure 11.3. Unconfined brittle compressive strength (and tensile strength) of single crystals of ice at −10 °C vs. orientation of the basal plane. The parameter α denotes the angle between the c-axis and the loading direction. (Data from Carter, 1971.)

transition, analogous to that in rocks and minerals (Duba *et al.*, 1990). For instance, Rist (1997) found that single crystals of $\alpha \sim 45°$ fail through localized basal slip under an axial stress of ~34 MPa, when subject to a confining pressure of 10 MPa and deformed at $10^{-2}\,s^{-1}$ at −20 °C. At the same strain rate and temperature, but in the absence of confinement, the crystals fail by splitting, under an axial stress of ~16 MPa (obtained by interpolating Carter's (1971) measurements). Similarly, crystals of $\alpha \sim 0°$ fail via localized prismatic slip under an axial stress of ~50 MPa when confined under a pressure of 10 MPa and deformed at $10^{-2}\,s^{-1}$ at −20 °C; when unconfined, the crystals fail by splitting under an axial stress ~24 MPa, (Carter, 1971). For both orientations ($\alpha \sim 45°$ and $\alpha \sim 0°$), terminal failure under confinement is often accompanied by recrystallization (Rist, 1997) within localized deformation bands. These features imply pressure-induced plasticity, even though the attendant stress–strain curves exhibit a sudden drop (Section 12.3.2).

11.5 Polycrystals

When loaded under compression, polycrystals of typical grain size ($d > 1$ mm) are generally weaker than single crystals, just as they are under tension (Chapter 10). Presumably, internal stress concentrators are again at play. The compressive strength is affected by several factors.

11.5.1 Strain rate

Figure 11.4 shows the unconfined brittle compressive strength of fresh-water granular ice of ~1 mm grain size vs. strain rate. The strength decreases by about 20% to 30% upon increasing strain rate from $10^{-3}\,s^{-1}$ to $10^{-2}\,s^{-1}$ and then levels off at strain rates up to $2.5 \times 10^{-1}\,s^{-1}$, at temperatures from 0 °C to about −20 °C (Carter, 1971; Schulson, 1990). At somewhat lower temperatures (−21.4 °C and −32 °C) the ice exhibits moderate strain-rate softening over the entire range of strain rate noted in the figure (~$10^{-3}\,s^{-1}$ to $2.5 \times 10^{-1}\,s^{-1}$). Atmospheric ice exhibits the same effect (Kermani *et al.*, 2007) when loaded at −3, −10 and −20 °C at strain rates from $10^{-3}\,s^{-1}$ to $3 \times 10^{-2}\,s^{-1}$; at lower rates the ice exhibits strain-rate hardening.

There is some evidence that *dynamic compressive strength* increases moderately with increasing strain rate. For instance, Jones (1997) compressed laboratory-grown, columnar-grained S2 fresh-water ice of ~6 mm column diameter by loading uniaxially along the columns (Jones, personal communication, 2005) at −11 °C, at strain rates from $10^{-1}\,s^{-1}$ to $10\,s^{-1}$. He found that, although scattered, the strength increased slightly with increasing strain rate and could be described by the relationship $\sigma_c = 8.9\dot{\varepsilon}^{0.15}$(MPa) for $\dot{\varepsilon}$ in s^{-1}, with a correlation coefficient of $r^2 = 0.46$. Schulson *et al.* (2005) compressed the same kind of material under similar

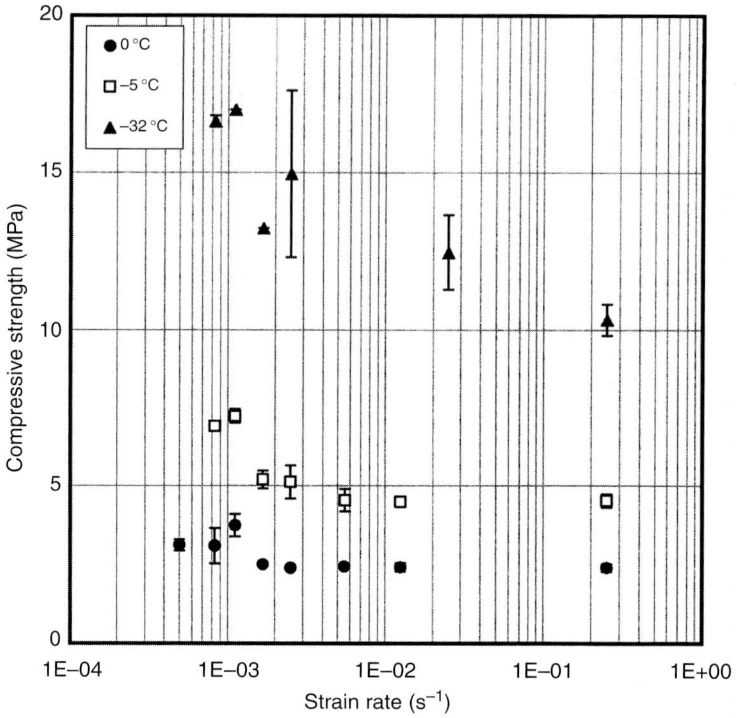

Figure 11.4. Unconfined brittle compressive strength of fresh-water granular ice of ~1 mm grain size vs. strain rate. (Data from Carter, 1971.)

conditions (−10 °C, $10^{-2}\,s^{-1}$ to $1.6\,s^{-1}$). They, too, found that the along-column strength, although again scattered, increased with increasing strain rate and could be described by the relationship $\sigma_c = 13.0\dot{\varepsilon}^{0.16}(\mathrm{MPa})$ with $r^2 = 0.61$. In both sets of experiments the ice failed by splitting into many shards. The other evidence is from Shazly *et al.* (2006). They performed a series of experiments on single crystals of uncontrolled orientation loaded at −10 °C within the dynamic range ($\dot{\varepsilon} = 90\,s^{-1}$ to $1122\,s^{-1}$) using a modified split Hopkinson pressure bar. It was necessary to use relatively thin specimens (~5.5 mm length, ~18.5 mm diameter) to allow the stress to reach equilibrium (Davies and Hunter, 1963). The data were scattered, but once again could be described by the same kind of relationship; namely, $\sigma_c = 10.2\dot{\varepsilon}^{0.16}(\mathrm{MPa})$ with $r^2 = 0.54$. Taken collectively, the measurements indicate that dynamic strength at −10 °C may be described by the relationship $\sigma_c = 9.8\dot{\varepsilon}^{0.14}(\mathrm{MPa})$ with $r^2 = 0.71$, Figure 11.5. (That single crystals and polycrystals seem to obey the same relationship should not be taken as evidence that grain size is not a factor here: the single crystals examined by Shazly *et al.* (2006) contained a relatively large degree of substructure and the polycrystals contained fewer than ten grains across the diameter.) This kind of strain-rate hardening is of different origin from the dislocation-based hardening

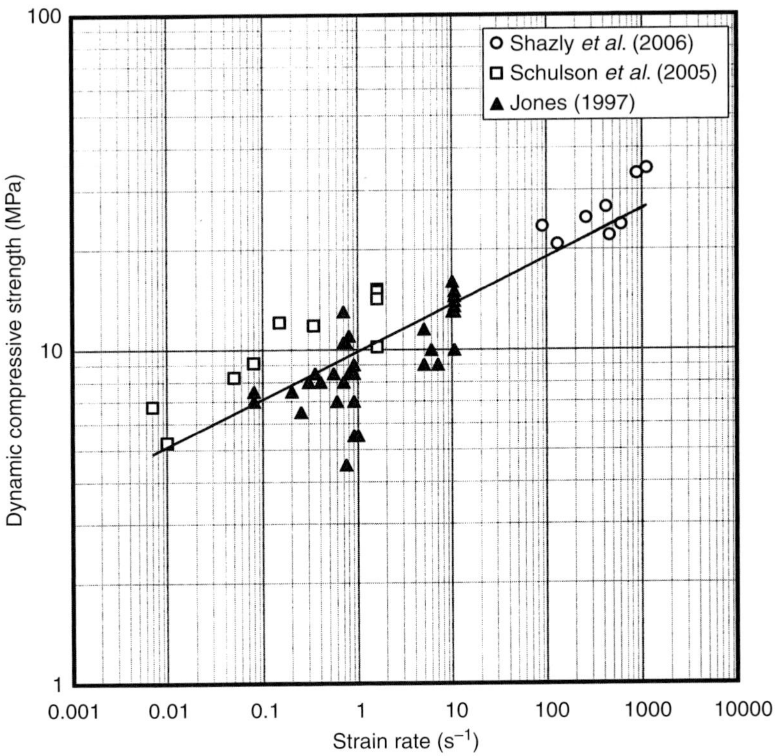

Figure 11.5. The dynamic compressive strength of ice at −10 °C.

within the ductile regime, and is reminiscent of the dynamic behavior of granite (Shan *et al.*, 2000) and silicon carbide (Sarva and Nemat-Nasser, 2001). The nature of the effect remains to be elucidated.

11.5.2 Temperature

Figure 11.6 shows the unconfined brittle compressive strength of fresh-water granular ice of ~1 mm grain size vs. temperature (Carter, 1971; Schulson, 1990; Arakawa and Maeno, 1997). Over the range from 0 °C to −50 °C the strength increases with decreasing temperature, by approximately 0.3 MPa °C^{-1}, similar to the increase within single crystals. At lower temperatures, the thermal sensitivity appears to first increase and then decrease. However, caution is appropriate. Ice from which the low-temperature (<−50 °C) measurements were made was bonded to copper seats. Upon reaching terminal failure, the material shattered instead of splitting (Arakawa and Maeno, 1997), suggesting that end-constraint probably introduced a triaxial state of stress. The low-temperature strengths shown in Figure 11.6 are thus probably somewhat higher than the true unconfined strength.

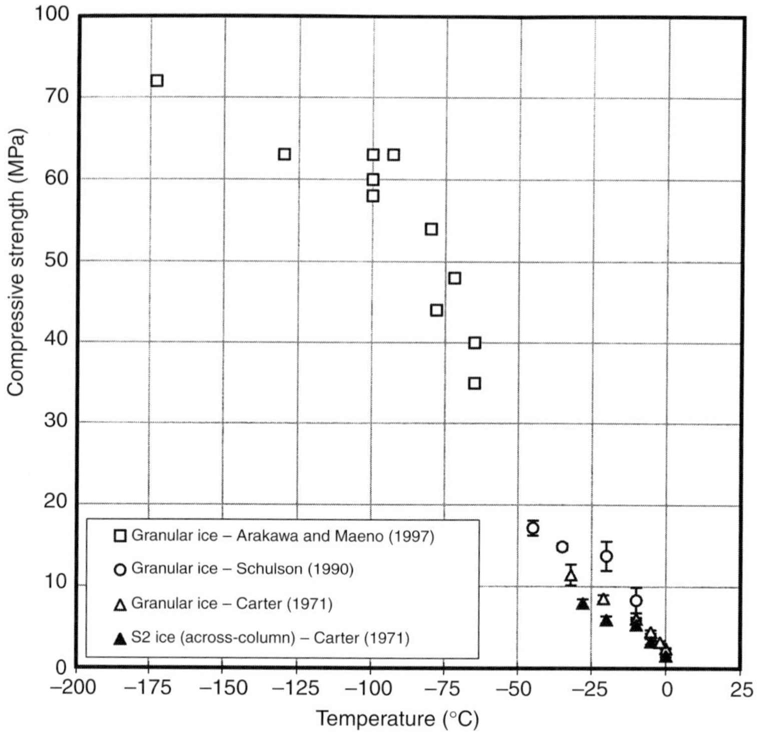

Figure 11.6. Unconfined brittle compressive strength of fresh-water granular ice of ~1 mm grain size vs. temperature. Also shown is the across-column strength of S2 fresh-water ice of ~5 mm column diameter.

11.5.3 Crystallographic growth texture

When compressed across the columns, the brittle compressive strength of columnar-grained ice that possesses the S2 growth texture is essentially indistinguishable from that of granular ice of the same grain size/column diameter. However, when compressed along the columns, the strength is greater by a factor of two to four. More specifically, the ratio of along-column to across-column strength of fresh-water ice is ~3.6 at 0 °C, but decreases to ~2.6 at −10 °C and to ~1.9 at −28 °C (Carter, 1971). For S2 salt-water ice of 4–6 ppm salinity the ratio of along-column to across-column strength is 2.0 ± 0.7 and shows no systematic variation with temperature over a similar range (Kuehn and Schulson, 1994).

Columnar ice that possesses the S3 growth texture (in which the *c*-axes are aligned within a plane perpendicular to the growth direction, Chapter 3) has been less well characterized. There is evidence, however, that S3 ice (unlike S2 ice) is slightly mechanically anisotropic when loaded within the plane perpendicular to the columns. For instance, from a series of unconfined compression tests on aligned first-year sea ice deformed at $-10\,°C$ at $10^{-3}\,s^{-1}$ Richter-Menge (1991) found that specimens with a $c{:}\sigma$ angle of ~0° have somewhat greater strength (7.30 ± 1.91 MPa) than samples with a $c{:}\sigma$ angle of ~90° (6.56 ± 1.88 MPa). She found also that specimens of both orientations have greater strength than ones with a $\sigma{:}c$ angle of ~45° (4.27 ± 0.83 MPa). Interestingly, the orientation dependence is similar to that exhibited by single crystals (Figure 11.3), although polycrystals are weaker.

Columnar ice that possesses the S1 growth texture (where the *c*-axes are preferentially aligned with the growth direction, Chapter 3) remains to be systematically examined. This variant appears to form less frequently than either S2 or S3 ice, and so is less important from a practical perspective.

11.5.4 Grain size

Figure 11.7a shows the unconfined brittle compressive strength of fresh-water granular ice vs. grain size d over the range 1–8 mm (measured using the method of linear intercepts), at $-10\,°C$, $-30\,°C$ and $-50\,°C$ at $10^{-3}\,s^{-1}$. Although scattered, the compressive strength decreases with increasing grain size (Cole, 1987; Schulson, 1990). This behavior is reminiscent of the tensile strength and reminiscent as well of the compressive strength of ceramics (Knudsen, 1959; Rice, 2000) and rock (Fredrich *et al.*, 1990; Olsson, 1990). The functionality, however, is somewhat uncertain. One possibility (Schulson, 1990) is:

$$\sigma_c = \sigma_o + k_c d^{-p} \qquad (11.1)$$

where σ_o and k_c are materials constants. The problem is that analysis on this basis generates rather low correlation coefficients; e.g., $r^2 \leq 0.6$ for $p = 1/2$, Figure 11.7b. Why the compressive strength is more highly scattered than the tensile strength over the same range of grain size is not clear.

In comparing compressive strength to tensile strength of the same grain size (Figure 10.5), we note that the ratio increases with decreasing temperature, from $\sigma_c/\sigma_t \sim 8$ at $-10\,°C$ to ~20 at $-50\,°C$. This is a direct result of the effect of temperature on compressive strength or, more fundamentally, of the increase in the resistance to frictional sliding (more below).

11.5.5 Porosity/salinity

One might expect porosity to decrease compressive strength, based upon its effect on rock (Knudsen, 1959; Scott and Nielsen, 1991) and ceramics (Rice, 1998). About all we can say for ice is that at temperatures near the upper end of the terrestrial scale, porosity at the level of ~5% seems to have little effect. This is based upon a

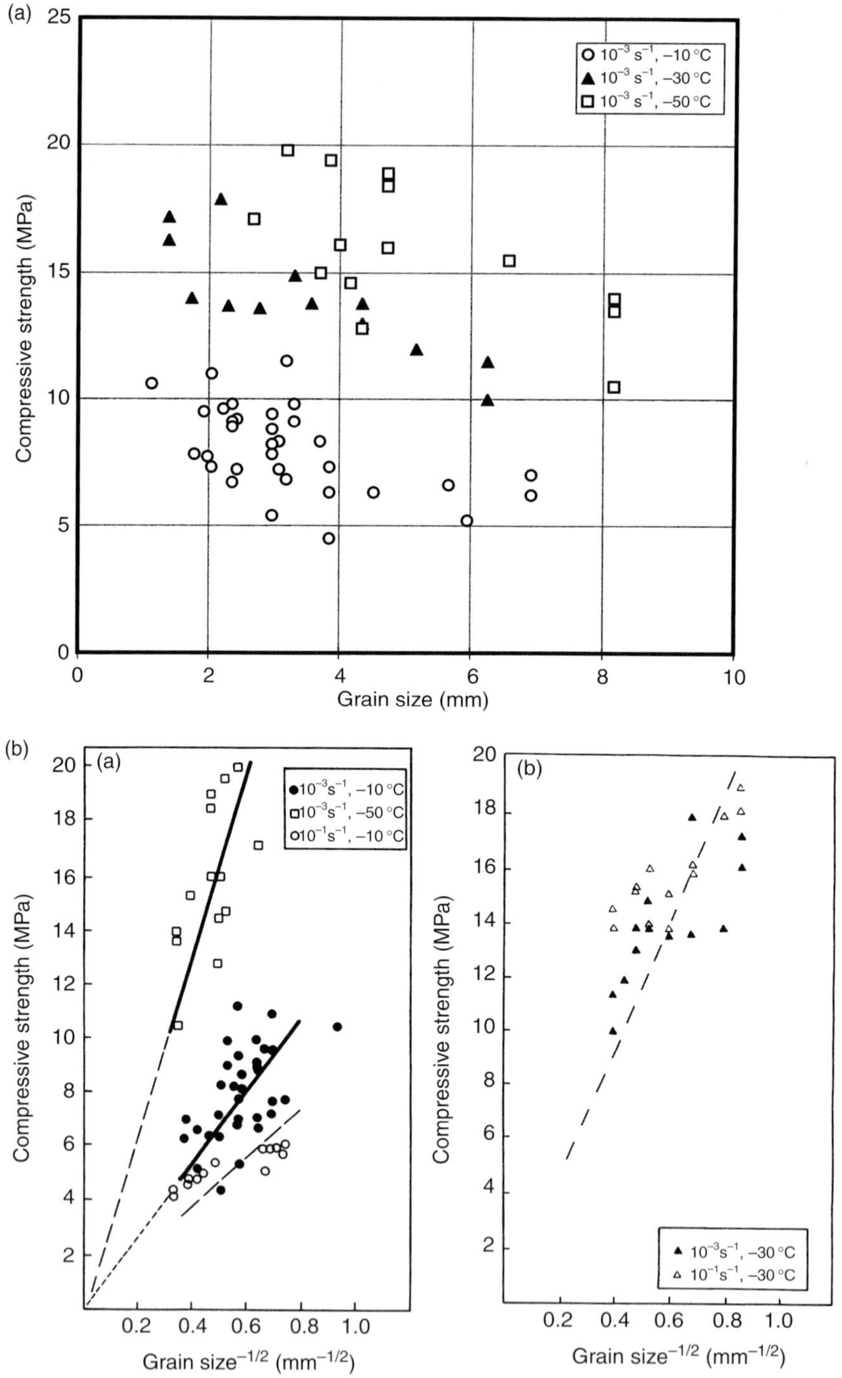

Figure 11.7. Unconfined brittle compressive strength of fresh-water granular ice (a) vs. grain size, and (b) vs. (grain size)$^{-1/2}$. (Data from Schulson, 1990.)

comparison of the across-column strengths of S2 columnar-grained, salt-water ice of ~5 ppt salinity and fresh-water ice of similar column diameter (~5 mm) and growth texture. At −5 °C the brittle compressive strength of unconfined saline ice is 3.0 ± 0.5 MPa (Kuehn and Schulson 1994), compared with 3.2 ± 0.2 MPa for fresh-water ice (Carter, 1971); and at −10 °C the strengths of the salt-water and fresh-water ice are 4.0 ± 1.0 MPa and 4.8 ± 0.3 MPa, respectively. That said, we caution against concluding that sheets of salt-water ice possess the same compressive strength as sheets of fresh-water ice. Sheets of sea ice contain brine drainage channels (Chapter 3), features that may weaken the body. That porosity at the level noted has little effect on the brittle compressive strength is in marked contrast to its weakening effect on the ductile strength (Chapter 6).

11.5.6 Damage

Systematic experiments in which the brittle compressive strength has been measured as a function of crack density are few. A preliminary study was performed by Couture (1992), on columnar-grained plates of S2 fresh-water ice of ~5 mm column diameter loaded across the columns at −10 °C at $10^{-3}\ s^{-1}$ and at $10^{-2}\ s^{-1}$. She introduced through-thickness damage by first straining specimens within the ductile regime at $10^{-5}\ s^{-1}$ at −10 °C until the peak of the stress–strain curve was reached. This treatment produced cracks of grain-size dimensions d distributed uniformly throughout the body, of number density of approximately $n \sim 2/cm^2$. This corresponds to a level of damage $D = n(d/2)^2 \sim 0.1$. Rather surprisingly, her results suggest no significant effect. More recently, Schulson *et al.* (2005) introduced damage in a similar manner into granular fresh-water ice of 3 mm and 9 mm grain size, at the level of $D \sim 0.3$. Again, no significant effect was detected, at least at −10 °C at $10^{-2}\ s^{-1}$.

We hesitate to conclude that damage at the level noted is not important. Both studies included relatively few tests and the data were scattered. Yet these results are reminiscent of ones from an indentation study by Timco (1987). He found that the peak pressure during crushing of a 10 ± 2 mm thick cover of finely grained (1–2 mm) S2 fresh-water ice, when indented at a rate of 200 mm s^{-1} by a cylindrical indenter 60 mm in diameter, was little affected by damage below the level $D \sim 0.2$. At higher levels, the peak pressure decreased. Perhaps, as Timco suggests, ice possesses a *damage threshold* above which the brittle compressive strength decreases.

11.5.7 Cyclic loading

The brittle compressive strength appears to increase under moderate cyclic loading. For instance, from a series of experiments on S2 fresh-water columnar ice of

4–8 mm column diameter proportionally loaded across the columns under a small degree of biaxial confinement ($\sigma_2/\sigma_1 \sim 0.07$, see Chapter 12 for procedure), Iliescu and Schulson (2002) found that the strength at −10 °C at 4×10^{-3} s^{-1} increased by a factor of about 1.5 upon subjecting the material to 10 cycles of up–down loading. During each test, the applied stress was progressively increased until terminal failure. Gupta *et al.* (1998) had previously reported a similar effect. An important point is the rate of unloading: if too low, then significant creep occurs and attendant creep damage develops to the point of suppressing the effect and perhaps even of lowering the strength. Cyclic hardening has also been observed under tensile loading (Cole, 1990).

On the *fatigue* of ice per se, less is known. Nixon and Smith (1984, 1987) performed cyclic bending tests and obtained preliminary evidence that the fatigue strength *S* decreases with increasing number of cycles *N*, typical of the so-called *S–N* behavior exhibited by metallic materials. Neither the nature of the behavior nor its reproducibility is known.

11.5.8 Size and boundary conditions

There is no evidence that sample size per se affects the compressive strength of ice that is initially free from structural defects. For instance, Kuehn *et al.* (1993) performed a series of uniaxial compression tests at −10 °C on cubes, 10 mm to 150 mm on edge, of fresh-water granular ice of ~1 mm grain size. The results from 183 experiments showed that within the ductile regime ($\dot{\varepsilon} = 10^{-5}$ s^{-1}), size does not affect strength at all. Nor does size affect the brittle compressive strength, provided that care is taken to eliminate effects of boundary conditions. For instance, when the compressive strength within the brittle regime was measured using polished and ground platens, no effect was detected. However, when measured using platens that were somewhat concave and roughened, the strength decreased with increasing size, from 7.5 ± 1.5 MPa for 10 mm cubes to 3.8 ± 0.6 MPa for 150 mm cubes (Kuehn *et al.*, 1993). In a related study on S2 fresh-water ice loaded across the columns, Fortt and Schulson (2007) found that over the range investigated (25 mm to 150 mm) the thickness of the specimen in the direction of the columns had no systematic effect on the brittle compressive strength.

Ice covers are different from laboratory specimens. Ice in the field contains a variety of structural defects (cracks, brine drainage channels) that can weaken the cover, particularly as the size of the loaded area increases. In addition, boundaries against which ice is pushed may also play a role. As a result, field strengths (when expressed in units of stress) may well depend upon size. We return to this point in Chapter 14.

11.6 Failure process

To elucidate the process of compressive failure of polycrystals, Cannon *et al.* (1990) performed a series of step-loaded compression tests on unconfined columnar-grained S2 fresh-water ice, initially free from cracks and of ~8 mm column diameter, loaded across the columns at −20 °C at $\dot{\sigma} = 200$ MPa s^{-1} or $\dot{\varepsilon} = 2 \times 10^{-2}$ s^{-1}. Upon loading to ~30% of the stress at terminal failure, the ice began to crack. The first crack that formed (detected by eye) was inclined by ~45° to the direction of loading and was oriented on a plane that ran parallel to the long axis of the grains. It appeared to have formed either along or near a grain boundary and was similar in length to the column diameter. Upon increasing the load, out-of-plane extensions or secondary cracks developed from the tips of the primary cracks, oriented in a direction sub-parallel to the direction of loading and parallel to the long axis of the grains, Figure 11.8a. The extensions are termed *wing cracks*. Upon raising the load further, the wings grew predominantly trans-granularly (Cannon *et al.*, 1990) and began to interact, Figure 11.8b. Finally, at a level of stress close to the terminal failure stress, the wings lengthened further and propagated all the way to the ends of the specimen and formed a longitudinal split, Figure 11.8c, a prelude to material

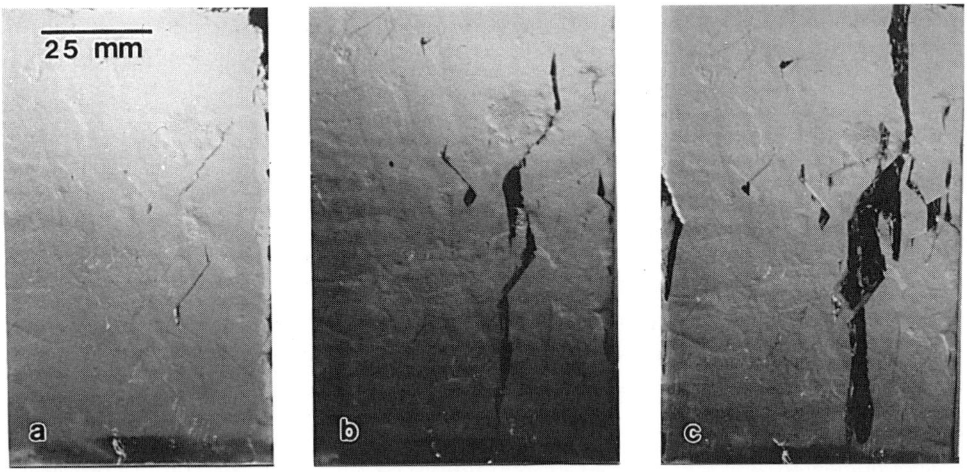

Figure 11.8. Photographs of wing cracks in a specimen of columnar-grained S2 fresh-water ice of ~8 mm column diameter, pulse-loaded across the columns at 200 MPa s^{-1} at −20 °C, as viewed along the columns. (a) Loaded to 5.2 MPa. Note wing cracks extending from the tips of two inclined primary cracks. (b) Loaded to 6.2 MPa. Note that the wing cracks of (a) have lengthened and a new set has formed on another primary crack. (c) Loaded to 6.6 MPa, near terminal failure. Note that the original wing cracks have extended all the way to the ends of the specimens and created an axial split and that new ones have formed. (From Cannon *et al.*, 1990.)

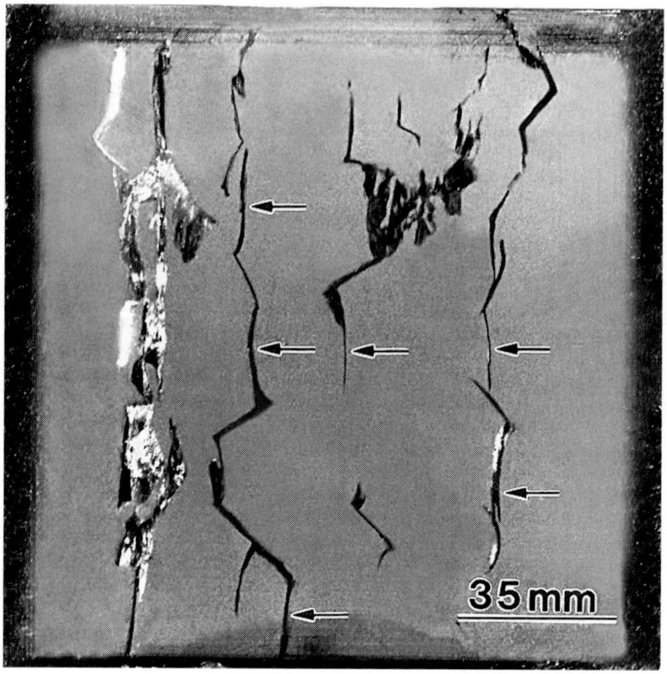

Figure 11.9. Photograph of wing cracks (arrowed) creating, through interaction, longitudinal splits in columnar-grained S2 fresh-water ice of ~6 mm column diameter loaded uniaxially (vertical in image) across the columns at −10 °C at $4 \times 10^{-3}\,s^{-1}$, as viewed along the columns. (From Iliescu and Schulson, 2004.)

collapse. Figure 11.9 shows another example of wing cracks. In this case, they lead to multiple splits.

Schulson *et al.* (1991) confirmed these observations and expanded upon them, using high-speed photography (1000 frames/second). The specimens were again made from S2 columnar-grained, fresh-water ice of ~8 mm column diameter and were loaded across the columns at −10 °C at $2 \times 10^{-2}\,s^{-1}$. In these experiments the load was not stepped, but was allowed to increase monotonically until terminal failure. Figure 11.10 shows a typical example of a whole specimen as viewed along the columns; also shown is the corresponding stress–strain curve. Again, the first cracks nucleated at ~30% of the terminal failure stress, Figure 11.10a. As the stress increased, other cracks nucleated throughout the body, and wing cracks initiated from the tips of an inclined parent crack (crack A, Figure 11.10d). The wings lengthened in a rather jerky manner as the stress increased further, Figure 11.10f, and reached around twice the length of the parent crack as the stress approached the terminal level. Figure 11.11 quantifies the length–stress relationship. Along the path to failure, other wing cracks initiated from the tips of other inclined primary cracks, labeled B and C in Figure 11.10g. Of particular note is crack C. It was located closer

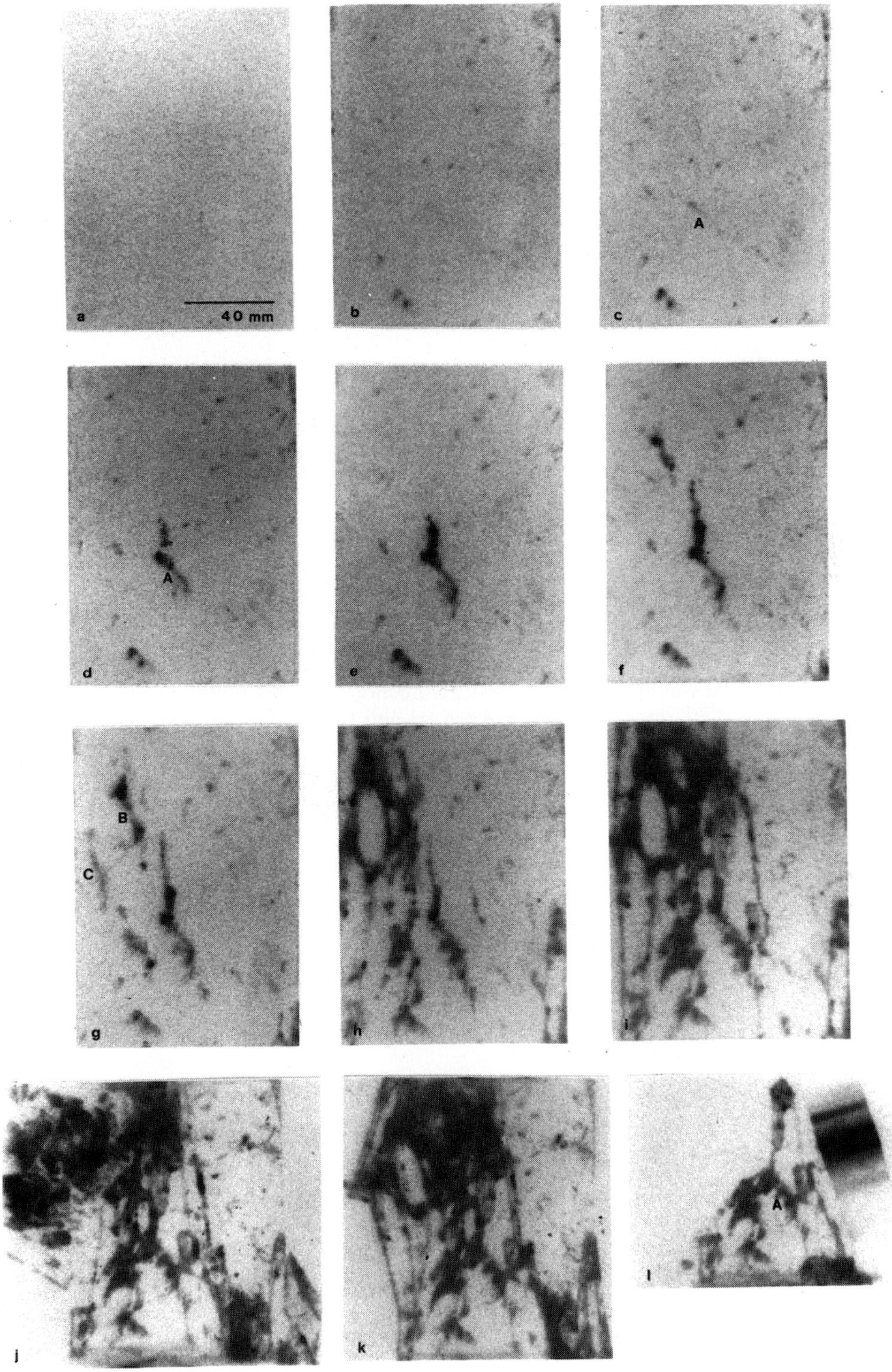

Figure 11.10.

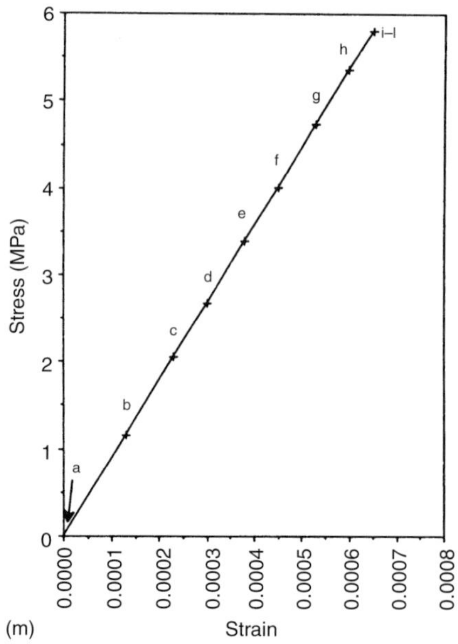

(m)

Figure 11.10. High-speed photographs (a)–(l) illustrating the development of deformation damage and of wing cracks within columnar-grained S2 fresh-water ice of ~8 mm column diameter compressed across the columns to terminal failure at −10 °C at $2 \times 10^{-2}\,s^{-1}$, as viewed along the columns. In the photographs shown, the load was applied in the vertical direction. Note the initiation away from the ice–platen interface of wing cracks A, B and C. (m) The stress–strain curve corresponding to images (a)–(l). (From Schulson *et al.*, 1991.)

to the free surface and grew very rapidly, Figure 11.10h, creating in the process a long, thin column of ice, Figure 11.10i. The column appears to have buckled and then to have blown outward, Figure 11.10j, and thus to have triggered a mechanical instability. Beyond that point the individual steps in the process are more difficult to discern, but probably involve repeated episodes of near-surface wing-crack growth and blowing out of material.

The sequence of events described above characterizes granular ice as well (Schulson *et al.*, 1989; Schulson, 1990). Figure 11.12, for instance, shows a wing crack within a shard that broke away from a granular specimen of ~6 mm grain size that was loaded to terminal failure at −20 °C at $2 \times 10^{-2}\,s^{-1}$. The shape of the grains, however, affects the orientation of the wing cracks and, hence, the deformation of the body. Although still aligned more or less with the direction of loading, the plane of the crack appears to be randomly oriented within the zone of the applied stress. As a result, lateral displacement is symmetric about the loading direction and this imparts macroscopically three-dimensional inelastic deformation. Correspondingly,

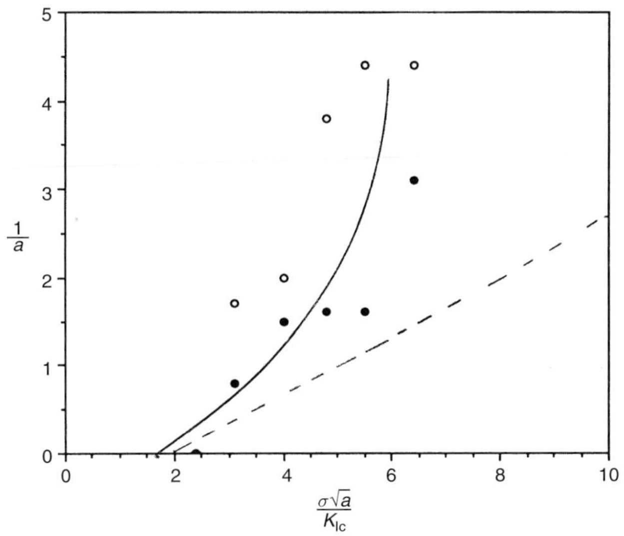

Figure 11.11. Showing the length, l, of the upper (open points) and the lower (closed points) wing cracks A of Figure 11.10 vs. applied compressive stress. The length is normalized with respect to one-half of the length of the primary, parent crack. The stress is normalized with respect to the parent-crack half-length and fracture toughness. The solid and broken curves, respectively, were calculated with and without consideration of crack interaction with the free surface. (From Schulson *et al.*, 1991.)

axial splits can form on any plane parallel to the loading direction, leading to the creation of shards. In comparison, within columnar ice loaded uniaxially across the columns, the wing-crack plane is restricted by the microstructure and is oriented more or less parallel to the long axis of the columnar grains. This restriction leads to two-dimensional inelastic deformation, confined to the plane perpendicular to the columns. As a result, axial splits form only on macroscopic planes parallel to the columns.

Wing cracks, incidentally, are more difficult to detect within granular material. Within columnar ice loaded across the columns, both the primary and the secondary wing cracks are parallel to the long axis of the grains. The primary cracks initiate either near or on the free surface penetrated by the columns, and then quickly deepen by growing inward along weak paths provided by grain boundaries. The secondary cracks also deepen as they lengthen along the loading direction. Because the length of columnar grains is often greater than the thickness of a test specimen, there is little opportunity for cracks to shield each other when the ice is viewed along the columns. In comparison, within granular ice, cracks do shield each other. Primary cracks often nucleate within the interior of the body on randomly oriented grain

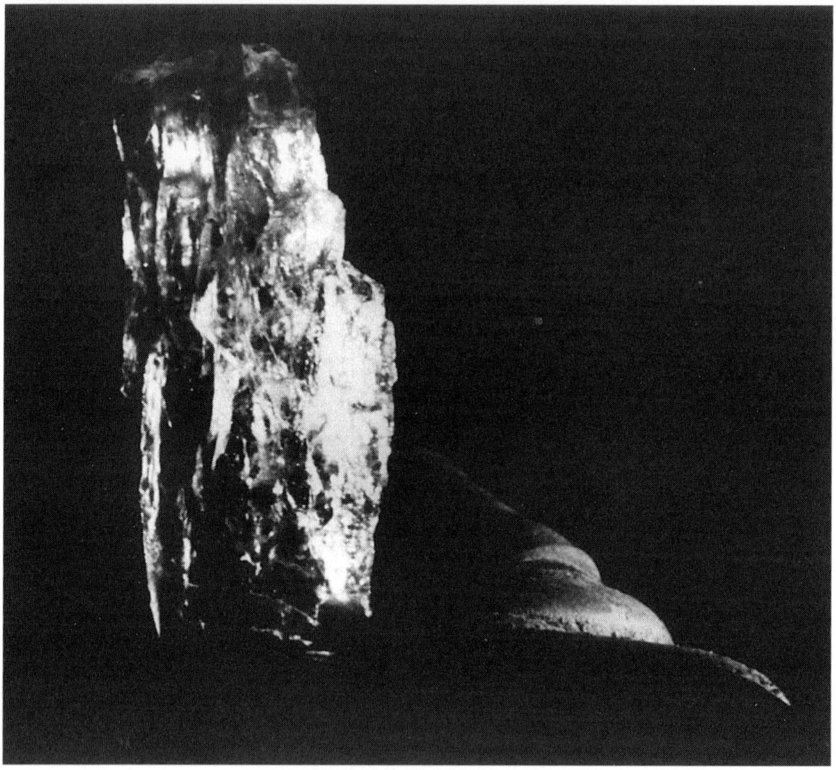

Figure 11.12. Photograph of a wing crack in a hand-held shard of fresh-water ice released from unconfined granular material of 6 mm grain size that had been loaded to terminal failure at −20 °C at 200 MPa s^{-1}. The load was applied along the length of the fragment. Note the internal cracks (white) oriented along the loading direction, and note near the center of the shard the inclined crack from the ends of which wing cracks extend in the loading direction. (From Schulson, 1990.)

Figure 11.13. Schematic models for the development of wing cracks from an initial, inclined crack in columnar and in granular ice. (a) Wing cracks in a thin plate, roughly described as a two-dimensional model. The initial inclined crack covers an entire grain boundary facet. (b) Wing cracks in a thick plate, such as S2 columnar ice, that undergoes two-dimensional deformation. The grain boundary provides a weak path, which allows the initial, inclined crack to extend through the thickness. (c) Wing crack in granular material. The initial, semicircular, inclined crack does not grow. Instead, axial extensions develop around it through combined mode-II and mode-III loading and then move into the depth of the specimen on approximately the same plane that is parallel to the loading direction. (From Cannon *et al.*, 1990.)

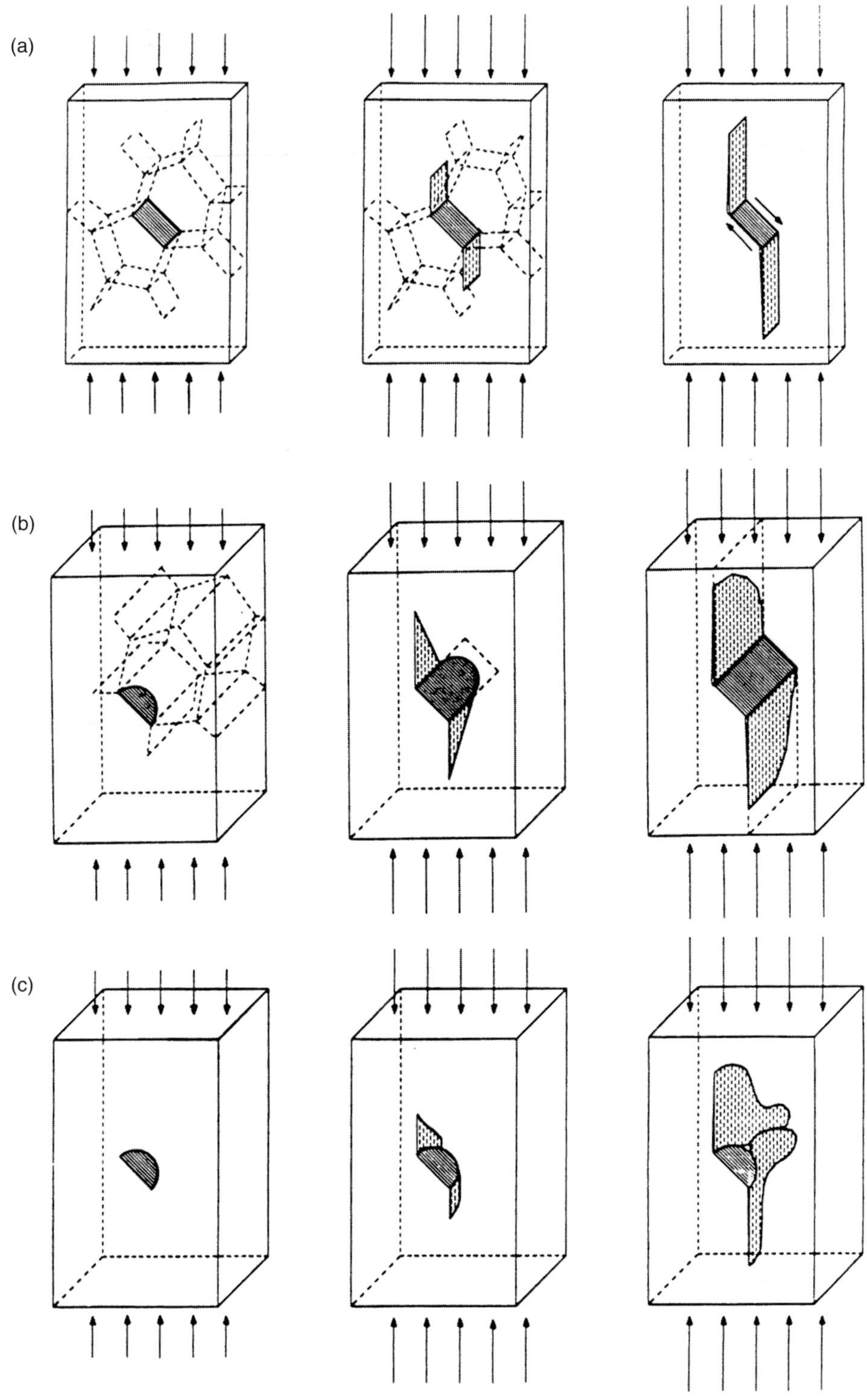

Figure 11.13.

boundaries, but do not deepen as secondary cracks sprout from their tips. Wing-crack growth then occurs by lengthening along the direction of loading as well as by wrapping around the periphery of the parent crack and then expanding both inward and outward, on a plane approximately parallel to the plane of lengthening, as sketched in Figure 11.13. This means that the characteristic inclined step that marks the primary crack is more difficult to discern within granular ice, particularly within thin sections that do not capture the parent crack.

The same can be said of rocks and minerals. Wing cracks are rarely seen in those materials (Paterson and Wong, 2005), and when they are seen they are attributed more to the preparation of the specimen than to the deformation of the body from which the specimen was made. For that reason, failure mechanisms that invoke wing cracks appear to be regarded with some suspicion by the rock mechanics community, even though the concept was first proposed to explain brittle compressive failure within that class of materials (more below).

Thus, the picture emerges that brittle compressive failure is a process. It begins relatively early in the deformation history as the first primary cracks nucleate; it develops continuously as the load rises and new cracks nucleate while older ones develop wings that grow and interact; and the process terminates when the degree of crack interaction with each other and with free surfaces (which include both the original external surface as well as newly formed splits) is sufficient to destabilize the body.

11.7 Wing-crack mechanics

Wing cracks originate through sliding along inclined cracks, as sketched in Figure 11.14. The concept was first proposed by Griffith (1924), advanced by McClintock and Walsh (1962) and by Brace and Bombolakis (1963) who recognized the role of friction, and then developed further by Kachanov (1982), by Horii and Nemat-Nasser (1985) and by Ashby and Hallam (1986). In a little known study, Goetze (1965) was the first person to invoke the concept to explain the fracture of ice. In essence, the mechanics operate as follows.

Imagine a body that contains a set of randomly oriented cracks of length determined by the grain size, loaded monotonically. Once the shear traction induced by the far-field stress is great enough to overcome frictional resistance, opposing faces of the most favorably oriented cracks begin to slide. At that point, tensile stresses develop within the material adjacent to the crack tips and increase in magnitude as sliding continues. Provided that creep relaxation occurs slowly enough (Chapter 13), the tensile stress increases to the level that a pair of extensile cracks subsequently nucleates, forming out-of-plane extensions or wings. Stress is then redistributed. Within elastically isotropic material, wings are expected to be

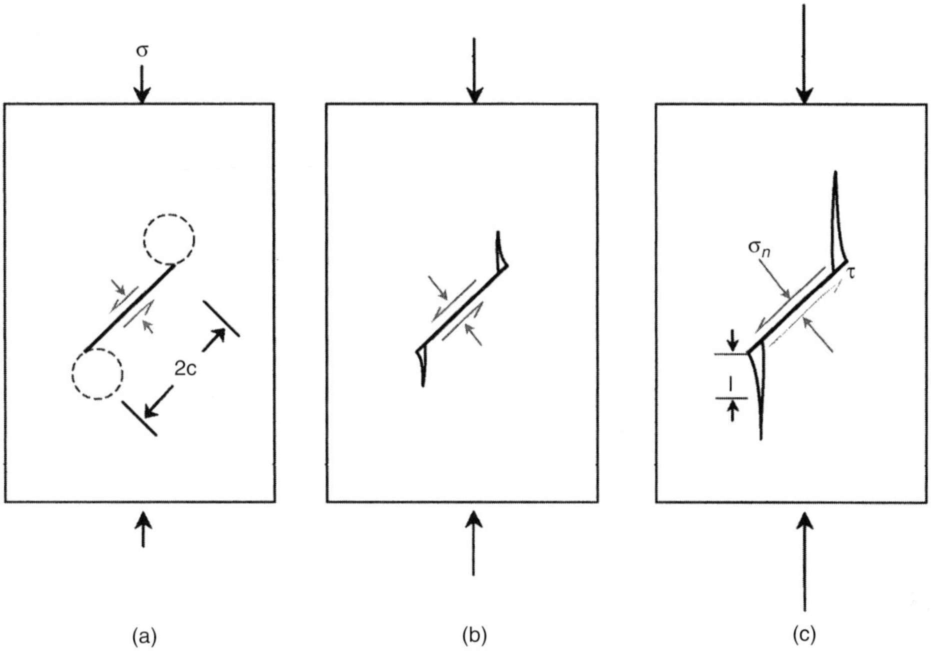

Figure 11.14. Schematic sketch showing the early stages in the initiation of wing cracks under an applied compressive loading. (a) Under low stresses tensile zones develop at the tips of inclined cracks. (b) Under greater stress, wing cracks initiate within the tensile zones, through frictional sliding across the parent, inclined crack. (c) Under still greater applied stress, the wing cracks lengthen as their mouths are pried open by further sliding.

initially oriented at 70.5° to the in-plane extension of the parent crack (i.e., along the trajectory of maximum tensile stress, see Ashby and Hallam, 1986). There is some suggestion of that here (Figure 11.8a), for the elastic behavior of ice is not highly anisotropic (Chapter 4). As the load increases, sliding continues and this pries open the wing-crack mouths. The wings then tend toward the loading direction and continue to lengthen as long as the attendant mode-I stress intensity factor is maintained at the critical level; i.e., $K_{\mathrm{I}} = K_{\mathrm{Ic}}$. Growth occurs in a macroscopically stable manner because $\mathrm{d}G/\mathrm{d}a < \mathrm{d}R/\mathrm{d}a$, as discussed in Chapter 9, a characteristic that distinguishes brittle compressive fracture from tensile fracture.

A complete three-dimensional treatment of wing-crack growth appears not to have been developed, although progress has been made (Germanovich *et al.*, 1994). Instead, two-dimensional analyses have been made of crack growth within isotropic material (Horii and Nemat-Nasser, 1985; Ashby and Hallam, 1986), a situation not unlike the case of crack growth within S2 columnar ice loaded across the columns. Underlying the models is Coulomb's failure criterion (see Jaeger and Cook, 1979):

$$|\tau| - \mu\sigma_n = \tau_o \tag{11.2}$$

where $|\tau|$ denotes the value of the shear stress generated by the applied stress, μ the coefficient of friction, σ_n the normal stress across the sliding plane and τ_o the cohesive strength. The left-hand side of Equation (11.2) defines the excess shear stress that is effective in causing sliding. The excess or effective shear stress is concentrated by a factor that is proportional to the square root of the length of the sliding crack $2c$ and this leads to a singular stress field that is characterized by the product $(\tau - \mu\sigma_n)\sqrt{2c}$. On the plane of the crack, the stress field is predominantly of shear character, but on planes inclined to the crack tip the field contains a tensile component. The mode-I stress intensity factor is obtained by finding the plane on which the tensile stress is a maximum and then by finding the orientation of the plane on which the excess shear stress is a maximum. In ice where cracks did not exist prior to loading but were created on grain boundaries inclined by ~45° to the loading direction, the orientation of the crack plane is fixed. In this case, it can be shown that the mode-I stress intensity factor generated by through-thickness sliding is given by the relationship (Ashby and Hallam, 1986):

$$K_{\mathrm{I},\infty} = \frac{\sigma\sqrt{\pi c}(1-\mu)}{\sqrt{3}} \left[\frac{\left(\beta L + \dfrac{1}{(1+L)^{1/2}} \right)}{(1+L)^{3/2}} \right] \quad \text{for } \mu < 1 \tag{11.3}$$

where σ denotes the applied stress, $L = l/c$ where l denotes the wing-crack length and $\beta \sim 0.4$ is a geometrical factor. Within a finite plate of width $2t$ an additional term must be added to take account of outward bending of the column created by the wing, given by (Ashby and Hallam, 1986):

$$K_{\mathrm{I,B}} = \frac{\sigma\sqrt{\pi c}}{2} \left[\frac{\dfrac{1}{\sqrt{2\pi}}\left(\dfrac{c}{t}\right)^{1/2}}{1 + \dfrac{12}{\pi^2}\left(\dfrac{c}{t}\right)^2 \left(L + \dfrac{1}{\sqrt{2}}\right)^2 \dfrac{\sigma}{E}} \right] \tag{11.4}$$

where E signifies Young's modulus. Indeed, it was the effect of this outward bending that caused wing crack C of Figure 11.10 to grow more rapidly than wing crack A. The criterion for crack growth is then given by:

$$K_\mathrm{I} = K_{\mathrm{I},\infty} + K_{\mathrm{I,B}} = K_{\mathrm{Ic}}. \tag{11.5}$$

The curves shown in Figure 11.11 compare the dictates of the model with the measured length of wing crack A of Figure 11.10, for both the infinite plate (broken curve) and the finite one (continuous curve). The calculation was made by setting $2c = 16$ mm and $2t = 100$ mm and by using the parametric values $K_{\mathrm{Ic}} = 0.1$ MPa m$^{1/2}$

(Chapter 9) and $E = 9.3$ GPa (Chapter 4); the friction coefficient was set at $\mu = 0.5$ and was chosen from Figure 4.6 for a temperature of sliding of -10 °C and a sliding speed of $v_s \sim 10^{-4}\,\mathrm{m\,s^{-1}}$, which was estimated from the relationship $v_s \sim 2\dot{\varepsilon}d$. Without including the effect of bending, the model over-predicts the stress to drive the crack, by about a factor of two. Upon including this effect, theory and experiment agree within the scatter in the data. Ice, in other words, appears to obey the dictates of wing-crack mechanics rather well.

11.8 Strength-limiting mechanism

If we assume that frictional sliding and stable crack growth continue to be important steps in the failure process right up to the point of material collapse, and if we assume further that the terminal failure stress scales as $\sigma_c \propto K_{Ic} c^{-1/2}/(1-\mu)$, as dictated by Equations (11.3)–(11.5) and as appears to be the case (Schulson, 1990), then a number of the characteristics of the unconfined compressive strength can be understood.

Consider its dependence upon grain size. From the crack-sliding analysis and from the fact that $2c \propto d$ (Cole, 1987; Iliescu, 2000), we would expect the compressive strength to scale as $d^{-1/2}$. The experimental results, although not definitive on this point, are at least consistent with it (Figure 11.7b).

On the effects of strain rate under quasi-static loading (Figure 11.4) and of temperature (Figure 11.6), we rule out variations in fracture toughness: temperature has little effect on this parameter (Chapter 9) and the crack tip loading rate is of the order of $10\,\mathrm{kPa\,m^{1/2}\,s^{-1}}$ or greater, which in fresh-water ice is high enough to suppress contributions from creep deformation, as discussed in Chapter 9. The kinetic coefficient of friction, on the other hand, depends upon both temperature and sliding speed (Chapter 4) and this parameter can account for the effects observed. The kinetic coefficient of friction for smooth (roughness $\sim 3 \times 10^{-6}$ m) surfaces of ice on ice ranges from $\mu = 0.8$ to $\mu = 0.05$ over the range of temperature from -40 °C to -3 °C and over the range of sliding speeds from $v_s = 10^{-6}\,\mathrm{m\,s^{-1}}$ to $10^{-2}\,\mathrm{m\,s^{-1}}$. If we assume that the friction coefficient for crack sliding is similar to that for sliding across a smooth interface, then at -10 °C the value decreases from $\mu = 0.45$ at the lower strain rate to $\mu = 0.15$ at the higher rate. Through the apparent dependence of the failure stress on the relationship $1/(1-\mu)$ for $\mu < 1$, this reduction is expected to impart a reduction in strength of ~35% as strain rate increases from $\dot{\varepsilon} = 10^{-3}$ to $10^{-1}\,\mathrm{s^{-1}}$, in reasonable agreement with observation. Similarly, the friction coefficient at the appropriate sliding speed of $v_s \sim 10^{-4}\,\mathrm{m\,s^{-1}}$ (Section 11.7) increases from $\mu = 0.30$ at -3 °C to $\mu = 0.75$ at -40 °C (Figure 4.6). This increase is expected to raise the strength by ~200%, in fair agreement with observation (Figure 11.6).

This kind of reasoning also helps to explain the absence of an effect of salinity/porosity at the levels examined, which were ~5 ppt ~ 5%. At these levels, as discussed earlier, neither the fracture toughness (Chapter 9) nor the friction coefficient (Chapter 4) depends significantly upon porosity/salinity.

The logic is less successful in explaining the apparent absence of an effect of low-level ($D \sim 0.2$) damage. One might have expected slight weakening, given the effect of damage on fracture toughness (Chapter 9). However, given the paucity of data and the scatter within them, the apparent absence of weakening is little basis for discounting the role of crack growth and frictional sliding in terminal failure.

Further insight into the strength-limiting step must await the results of additional studies that exploit higher-speed and higher-resolution photography. At this juncture, we conclude that frictional crack sliding and wing-crack growth are important steps.

References

Arakawa, M. and N. Maeno (1997). Mechanical strength of polycrystalline ice under uniaxial compression. *Cold Reg. Sci. Technol.*, **26**, 215–229.

Ashby, M. F. and S. D. Hallam (1986). The failure of brittle solids containing small cracks under compressive stress states. *Acta Metall.*, **34**, 497–510.

Brace, W. F. and E. G. Bombolakis (1963). A note of brittle crack growth in compression. *J. Geophys. Res.*, **68**, 3709.

Cannon, N. P., E. M. Schulson, T. R. Smith and H. J. Frost (1990). Wing cracks and brittle compressive fracture. *Acta Metall. Mater.*, **38**, 1955–1962.

Carney, K. S., D. J. Benson, P. DuBois and R. Lee (2006). A phenomenological high strain rate model with failure for ice. *Int. J. Sol. Struct.*, **43**, 7820–7839.

Carter, D. (1971). Lois et mechanismes de l'apparente fracture fragile de la glace de riviere et de lac. Ph.D. thesis, University of Laval.

Cole, D. M. (1987). Strain rate and grain size effects in ice. *J. Glaciol.*, **33**, 274–280.

Cole, D. M. (1990). Reversed direct-stress testing of ice: initial experimental results and analysis. *Cold Reg. Sci. Technol.*, **18**, 303–321.

Couture, M. L. (1992). Do Cracks Strengthen Ice? A.B. honors thesis, Thayer School of Engineering, Hanover, Dartmouth College.

Davies, E. D. H. and S. C. Hunter (1963). The dynamic compression testing of solids by the method of the split Hopkinson pressure bar. *J. Mech. Phys. Sol.*, **11**, 155–179.

Duba, A. G., W. B. Durham, J. W. Handin and H. F. Wang, Eds. (1990). *The Brittle-Ductile Transition in Rocks: The Heard Volume*. Geophysical Monograph 56. Washington, D.C.: AGU Geophysical Monograph Board.

Fortt, A. L. and E. M. Schulson (2007). The resistance to sliding along coulombic shear faults in ice. *Acta Mater.*, **55**, 2253–2264.

Frederking, R. and D. Sudom (2006). Maximum ice force on the Molikpaq during the April 12, 1986 event. *Cold Reg. Sci. Technol.*, **46**, 147–166.

Fredrich, J. T., B. Evans and T.-F. Wong (1990). Effect of grain size on brittle and semibrittle strength: implications for micromechanical modelling of failure in compression. *J. Geophy. Res.*, **95**, 10,907–10,920.

Germanovich, L. N., R. L. Salganik, A. V. Dyskin and K. K. Lee (1994). Mechanisms of brittle-fracture of rock with preexisting cracks in compression. *Pure Appl. Geophys.*, **143**, 117–149.

Gerstle, K. H., R. Belloti, P. Bertacchi *et al.* (1980). Behavior of concrete under multiaxial stress states. *J. Eng. Mech. Div.-ASCE*, **106**, 1383–1403.

Goetze, C. G. (1965). A study of brittle fracture as applied to ice. U.S. Army CRREL, 63 pages (technical note, unpublished).

Greenberg, R., P. Geissler, G. Hoppa *et al.* (1998). Tectonic processes on Europa: Tidal stresses, mechanical response, and visible features. *Icarus*, **135**, 64–78.

Griffith, A. A. (1924). The theory of rupture. In *Proceedings of the First International Congress on Applied Mechanics*, eds. C. B. Biezeno and J. M. Burgers. Delft: J. Waltman Jr., pp. 55–63.

Gupta, V., J. Bergstrom and C. R. Picu (1998). Effect of step-loading history and related grain-boundary fatigue in freshwater columnar ice in the brittle deformation regime. *Phil. Mag. Lett.*, **77**, 241–247.

Hausler, F. U. (1981). Multiaxial compressive strength tests on saline ice using brush-type loading platens. *IAHR Ice Symposium*, Quebec, Canada.

Hawkes, I. and M. Mellor (1972). Deformation and fracture of ice under uniaxial stress. *J. Glaciol.*, **11**, 103–131.

Heil, P. and W. D. Hibler (2002). Modelling the high-frequency component of arctic sea-ice drift and deformation. *J. Phys. Oceanogr.*, **32**, 3039–3057.

Hoff, G. C., R. C. Johnson, D. C. Luther, H. R. Woodhead and W. Abel (1994). The Hibernia Platform. *Fourth (1994) International Offshore and Polar Engineering Conference, April 10–15, 1994*, Osaka, Japan, The International Society of Offshore and Polar Engineering.

Horii, H. and S. Nemat-Nasser (1985). Compression-induced microcrack growth in brittle solids: axial splitting and shear failure. *J. Geophys. Res.*, **90**, 3105–3125.

Iliescu, D. (2000). Contributions to brittle compressive failure of ice. Ph.D. thesis, Thayer School of Engineering, Dartmouth College.

Iliescu, D. and E. M. Schulson (2002). Brittle compressive failure of ice: monotonic versus cyclic loading. *Acta Mater.*, **50**, 2163–2172.

Iliescu, D. and E. M. Schulson (2004). The brittle compressive failure of fresh-water columnar ice loaded biaxially. *Acta Mater.*, **52**, 5723–5735.

Jaeger, J. C. and N. G. W. Cook (1979). *Fundamentals of Rock Mechanics*, 3rd edn. London: Chapman and Hall.

Jones, S. J. (1997). High strain-rate compression tests on ice. *J. Phys. Chem. B*, **101**, 6099–6101.

Kachanov, M. L. (1982). A microcrack model of rock inelasticity, part II: propagation of microcracks. *Mech. Mater.*, **1**, 29–41.

Kermani, M., M. Farzaneh and R. Gagnon (2007). Compressive strength of atmospheric ice. *Cold Reg. Sci. Technol.*, **49**, 195–205.

Khomichevskaya, I. S. (1940). O vremennom soprotivlenii szhatiyu vechnomerzlykh gruntov i l'da yestestvennoy struktury. (The ultimate compressive strength of permafrost and ice in their natural states.) *Tr. Kom. Vechnoy Merzlote*, **10**, 37–83.

Klohn-Crippen (1998). DynaMAC: Molikpaq ice loading experience. *PERD/CHC Report 14–62*. Calgary, Canada, National Research Council of Canada, 96.

Knudsen, F. P. (1959). Dependence of mechanical strength of brittle polycrystalline specimens on porosity and grain size. *J. Amer. Ceram. Soc.*, **42**, 376–387.

Korzhavin, K. N. (1955). Vliyaniye skorosti deformirovaniya na velichinu predela prochnosti rechnogo l'da priodnoosnom szhatii. (The effect of the speed of deformation on the ultimate strength of river ice subject to uniaxial compression.) *Tr. Novosib. Inst. Inzh. Zheleznodorozhn. Transp.*, **11**, 205–216.

Kuehn, G. A. and E. M. Schulson (1994). The mechanical properties of saline ice under uniaxial compression. *Ann. Glaciol.*, **19**, 39–48.

Kuehn, G. A., E. M. Schulson, D. E. Jones and J. Zhang (1993). The compressive strength of ice cubes of different sizes. *J. Offshore Mech. Arctic Eng. (ASME)*, **12**, 142–148.

Lockner, D. A., J. D. Byerlee, V. Kuksenko, A. Pnomarev and A. Sidorin (1991). Quasi-static fault growth and shear fracture energy in granite. *Nature*, **350**, 39–42.

Maykut, G. A. (1982). Large-scale heat exchange and ice production in the central arctic. *J. Geophys. Res.*, **87**, 7971–7984.

McClintock, F. A. and J. B. Walsh (1962). Friction of Griffith cracks in rock under pressure. *Proceedings 4th U.S. National Congress on Applied Mechanics 1962*, New York, pp. 1015–1021.

Michel, B. (1978). *Ice Mechanics*. Quebec: Laval University Press.

Nixon, W. A. and R. A. Smith (1984). Preliminary results on the fatigue behavior of polycrystalline freshwater ice. *Cold Reg. Sci. Technol.*, **9**, 267–269.

Nixon, W. A. and R. A. Smith (1987). The fatigue behavior of freshwater ice. *J. Physique*, **C1**, 329–336.

Olsson, W. A. (1990). Grain size dependence of yield stress in marble. *J. Geophys. Res.*, **79**, 4859–4862.

Paterson, M. S. and T.-F. Wong (2005). *Experimental Rock Deformation: The Brittle Field*, 2nd edn. New York: Springer-Verlag.

Peyton, H. R. (1966). *Sea Ice Strength*. University of Alaska, 86.

Rice, R. W. (1998). *Porosity of Ceramics*. New York: Marcel Dekker, Inc.

Rice, R. W. (2000). *Mechanical Properties of Ceramics and Composites: Grain and Particle Effects*. New York: CRC.

Richter-Menge, J. (1991). Confined compressive strength of horizontal first-year sea ice samples. *J. Offshore Mech. Arctic Eng.*, **113**, 344–351.

Rist, M. A. (1997). High stress ice fracture and friction. *J. Phys. Chem. B*, **101**, 6263–6266.

Rist, M. A. and S. A. F. Murrell (1994). Ice triaxial deformation and fracture. *J. Glaciol.*, **40**, 305–318.

Sarva, S. and S. Nemat-Nasser (2001). Dynamic compressive strength of silicon carbide under uniaxial compression. *Mater. Sci. Eng. A*, **317**, 140–144.

Schenk, P. M., and W. B. McKinnon (1989). Fault offsets and lateral crustal movement on Europa – evidence for a mobile ice shell. *Icarus*, **79**, 75–100.

Schulson, E. M. (1990). The brittle compressive fracture of ice. *Acta Metall. Mater.*, **38**, 1963–1976.

Schulson, E. M. (2001). Brittle failure of ice. *Eng. Fract. Mech.*, **68**, 1839–1887.

Schulson, E. M. (2002). On the origin of a wedge crack within the icy crust of Europa. *J. Geophys. Res.*, **107**, doi:10.1029/2001JE001586.

Schulson, E. M. and E. T. Gratz (1999). The brittle compressive failure of orthotropic ice under triaxial loading. *Acta Mater.*, **47**, 745–755.

Schulson, E. M., M. C. Geis, G. J. Lasonde and W. A. Nixon (1989). The effect of the specimen-platen interface on the internal cracking and brittle fracture of ice under compression: high-speed photography. *J. Glaciol.*, **35**, 378–382.

Schulson, E. M., G. A. Kuehn, D. E. Jones and D. A. Fifolt (1991). The growth of wing cracks and the brittle compressive failure of ice. *Acta Metall. Mater.*, **39**, 2651–2655.

Schulson, E. M., D. Iliescu and A. Fortt (2005). Characterization of ice for return-to-flight of the Space Shuttle 1: Hard ice. *Glenn Research Center Technical Reports*. Glenn Research Center, NASA. NASA CR-2005-213643, 95.

Schulson, E. M., A. Fortt, D. Iliescu and C. E. Renshaw (2006). Failure envelope of first-year Arctic sea ice: The role of friction in compressive fracture. *J. Geophys. Res.*, **111**, doi: 10.1029/2005JC003234186.

Schwarz, J. and W. F. Weeks (1977). Engineering properties of sea ice. *J. Glaciol.*, **19**, 499–531.

Scott, T. E. and K. C. Nielsen (1991). The effects of porosity on the brittle-ductile transition in sandstones. *J. Geophys. Res.*, **96**, 405–414.

Shan, R., Y. Jiang and B. Li (2000). Obtaining dynamic complete stress-strain curves for rock using the split Hopkinson pressure bar technique. *Int. J. Rock Mech. Min. Sci.*, **37**, 983–992.

Shazly, M., V. Prakash and B. A. Lerch (2006). High-strain-rate compression testing of ice. N.S.P. Office, NASA, 2006-213966, 92.

Spaun, N. A., R. T. Pappalardo, J. W. Head and N. D. Sherman (2001). Characteristics of the training equatorial quadrant of Europa from Galileo imaging data: evidence for shear failure in forming lineae. *Lunar and Planetary Science Conference*, XXXII, CD ROM 1228, Houston, Tex.

Thorndike, A. S. and R. Colony (1982). Sea ice motion in response to geostrophic winds. *J. Geophys. Res.*, **87**, 5845–5852.

Timco, G. W. (1987). Ice structure interaction tests with ice containing flaws. *J. Glaciol.*, **33**, 186–194.

Weiss, J. and E. M. Schulson (1995). The failure of fresh-water granular ice under multiaxial compressive loading. *Acta Metall. Mater.*, **43**, 2303–2315.

Zhang, X. and J. E. Walsh (2006). Towards a seasonally ice-covered Arctic Ocean: Scenarios from the IPCC AR4 model simulations. *J. Climate*, **19**, 1730–1747.

12

Brittle compressive failure of confined ice

12.1 Introduction

Compressive failure more often than not occurs under a multi-axial state of stress. For example, during the interaction between a floating ice feature and an engineered structure, material within the contact zone is compressed not only along the direction of impact, but also in orthogonal directions, owing to constraint imposed by surrounding material. The confinement induces biaxial (thin feature, wide structure) and triaxial (thick feature, narrow structure) stress states, which, as we show below, have a large effect on the strength of the ice and on its mode of failure.

This is not surprising. Based upon the failure of unconfined material and on the importance to that process of frictional crack sliding cum the development of secondary cracks (Chapter 11), confinement plays two roles: it lessens the excess or effective shear stress that drives sliding; and it lowers the mode-I stress intensity factor that drives crack growth. Higher applied stresses are thus required to activate the mechanism. Also, in holding the ice together, confinement promotes the development of shear faults. Indeed, very little confinement is required (Wachter *et al.*, 2008), to the effect that from a practical perspective faulting and not axial splitting is the more important failure mode.

In this chapter we review the observations and their interpretation. We again begin with a short discussion of experimental methods, and then quantify the effects of confinement on the behavior of both granular and columnar polycrystalline ice. We present further observations on the evolution of deformation damage, and introduce another kind of secondary crack, termed the comb crack. This feature plays a major role in the development of Coulombic shear faults which, as will become apparent, limit strength under moderate confinement. Under higher confinement, different failure modes are activated, the character of which depends upon both the structure of the ice (granular vs. columnar) and the multi-axiality of the stress state (biaxial vs. triaxial). We limit our discussion to the quasi-static failure of homogeneous material deformed under homogeneous stress states. For completeness, we end with a discussion of post-terminal failure.

12.2 Experimental methods

Triaxial loading Confinement can be imposed in the laboratory using three different methods. The most common is to subject ice to a pressurized fluid and then to apply an additional axial load while holding constant the fluid pressure (Jones, 1982; Durham *et al.*, 1983; Rist and Murrell, 1994; Gagnon and Gammon, 1995; Sammonds *et al.*, 1998; Melanson *et al.*, 1999). This is the so-called triaxial test that is frequently used in the study of soil and rock. The pressure p and axial stress σ_a induce a triaxial state of stress in which the second and third principal stresses, σ_2 and σ_3, are equal and the first principal stress σ_1 is equal to the sum of the pressure and the applied stress, Figure 12.1a; i.e., $\sigma_1 > \sigma_2 = \sigma_3$ where σ_1 is taken to be the most compressive stress and all compressive stresses are taken to be positive quantities. More explicitly, $\sigma_1 = p + \sigma_a$ and $\sigma_2 = \sigma_3 = p$. A variation of the method is to set the pressure to be a fixed proportion of the axial load (Richter-Menge, 1991), thereby achieving proportional loading. With this method, the ice must be protected to prevent entry of the fluid into either pre-existing cracks or ones that are generated as load is applied. Otherwise, the invading fluid would tend to counteract the confining pressure and the failure stress would tend to be lower, judging from the behavior of rock (Paterson and Wong, 2005). Protection of ice has been achieved by using jackets made from either indium (Durham *et al.*, 1983; Rist and Murrell, 1994; Sammonds *et al.*, 1998) or rubber (Richter-Menge, 1991; Gagnon and Gammon, 1995; Melanson *et al.*, 1999). The advantages of the pressure-cell method are that it is based on many years of experience and only the ends of the specimen are subject to the kind of additional confinement that platens can impose, as discussed in the previous chapter. The disadvantage is that only axisymmetric, triaxial states of stress can be achieved.

Passive proportional loading Another method is to confine the ice between walls. Frederking (1977) applied this method, taking care to reduce friction at the ice–platen interfaces with thin sheets of polyethylene. He placed prismatic-shaped specimens between a pair of parallel, rigid walls and then applied an axial load, thereby inducing a biaxial state of stress in which the minor stress was a fixed proportion of the major stress. The constant of proportionality R during the elastic stage is given by Poisson's ratio, i.e., $R = \sigma_2/\sigma_1 = \nu$; during the inelastic stage of deformation, $R < \nu < 0.5$. In a derivative of this method, Wachter *et al.* (2008) varied the compliance of the confining walls and were able to impose very low levels of confinement. We term this kind of loading passive proportional loading.

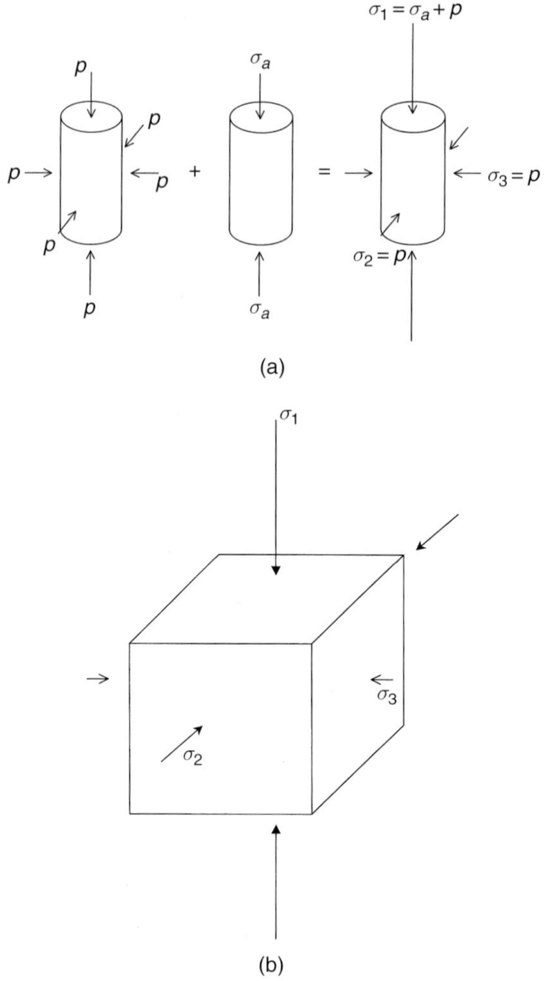

Figure 12.1. Schematic sketches of (a) triaxial loading, and (b) true triaxial loading, where p denotes confining pressure and σ_1, σ_2 and σ_3 denote the first, second and third principal stresses, respectively, where σ_1 is taken as the most compressive stress.

True triaxial loading Another method is to use actuators, where the confining loads are slaved to the driving load. In this way, Hausler (1981, 1989), Schulson *et al.* (1991), Weiss and Schulson (1995), Gratz and Schulson (1997), Melton and Schulson (1998) and Schulson and Gratz (1999) applied triaxial stress states to cube-shaped specimens in which $\sigma_1 \neq \sigma_2 \neq \sigma_3$, Figure 12.1b. This kind of experiment is termed true triaxial loading, because the three principal stresses can be varied independently and all loading paths in principal stress space can be explored, including biaxial ones. The loading paths are defined by the ratio $(\sigma_1{:}\sigma_2{:}\sigma_3) = (1{:}R_{21}{:}R_{31})\sigma_1$ where $R_{21} = \sigma_2/\sigma_1 \leq 1$ and $R = \sigma_3/\sigma_1 \leq 1$. The ability to vary principal stresses independently has proven to be invaluable in elucidating failure mechanisms (more below). The problem is that, unless the edges of the specimen are strengthened, failure can occur prematurely (Weiss and Schulson, 1995), particularly under high degrees of confinement. In the absence of premature failure, terminal failure under true triaxial loading and under under constant pressure sets in at essentially the same maximum principal stress and via the same mode (Weiss and Schulson, 1995), suggesting that the brittle compressive strength of ice under confinement depends only upon the stress state and not upon the path taken to get there.

Indentation The other method of applying confinement is to use an indenter. This is the easiest method to apply, and in some sense relates directly to the practical problem of ice loads on engineered structures (Chapter 14). However, interpretation is more difficult: owing to the stress and strain gradients within the contact zone, more than a single mechanism operates.

Whichever method is applied, *pressure melting* must be kept in mind. The effect is quantified by the Clausius–Clapeyron relationship $\mathrm{d}T/\mathrm{d}P = \mathrm{d}V/\mathrm{d}S$ where $\mathrm{d}T$ is reduction in the equilibrium melting point, $\mathrm{d}V$ is the difference in molar volume between the liquid and the solid ($\mathrm{d}V < 0$) and $\mathrm{d}S$ is the difference in entropy between the liquid and the solid ($\mathrm{d}S > 0$); $\mathrm{d}P$ is the increase in pressure P, which in triaxial loading experiments is given by $\mathrm{d}P = (\sigma_1 + \sigma_2 + \sigma_3)/3 = (\sigma_1 + 2p)/3 > p$ since $\sigma_1 > p$. Both theory and experiment yield $\mathrm{d}T/\mathrm{d}P \sim -0.074$ °C MPa^{-1} at temperatures near the unconfined melting point. Thus, for instance, the equilbrium melting temperature of a body uniformly confined under a pressure $p = 70$ MPa and then loaded in one direction by an axial stress $\sigma_a = 15$ MPa is reduced by $\Delta T = [(70 + 70 + 85)/3] \times 0.074 = 5.6$ °C, implying that under these conditions ice whose temperature is −5.6 °C would be in equilibrium with water. Of course, even if the thermodynamics is favorable, the kinetics of heat (of fusion) transfer could prevent melting should insufficient time be available.

An example of pressure bringing ice closer to its melting temperature and thereby reducing its strength can be found in the pressure-cell experiments of Jones (1982). Upon loading finely grained ($d \sim 1$ mm) fresh-water ice at an ambient temperature of −11.5 °C at 5×10^{-3} s^{-1}, he found that upon increasing the confining pressure from $p = 50$ MPa to $p = 85$ MPa the maximum stress difference decreased from $(\sigma_1 - \sigma_3) \sim 20$ MPa to $(\sigma_1 - \sigma_3) \sim 12$ MPa. The higher pressure effected a reduction of ~7 °C in the equilibrium melting point which brought the ice to within 4 degrees of the phase transition. In similar experiments, but on colder ice at −40 °C, Rist and Murrell (1994) reported no significant effect on strength, for in this case the pressure-induced reduction in the melting point did not bring the ice significantly closer to its melting point to make a difference.

12.3 Granular ice

Of the two forms of polycrystalline ice – granular ice and textured columnar ice – granular material is the easier to understand. Although anisotropic on the scale of individual grains, on the scale of the aggregate as a whole the material deforms isotropically.

12.3.1 Biaxial loading

From proportional loading experiments along the paths $(1{:}R_{21}{:}0)\sigma_1$ using a true multi-axial loading system, Weiss and Schulson (1995) found that the biaxial compressive strength of granular ice is essentially indistinguishable from the strength of unconfined material. Specifically, over the range of confinement from $R_{21}=0$ to 1.0, the biaxial strength of fresh-water material of ~7 mm grain size deformed at −40 °C at $10^{-3}\ s^{-1}$ along the direction of shortening was 7 ± 1 MPa at $R_{21}=0$ and 8 ± 1 MPa at $R_{21}=1.0$. The only difference was a minor change in the mode of failure, from multiple longitudinal splitting and the formation of shards under uniaxial loading to longitudinal splitting and the formation of slabs under biaxial loading, where the normal to the slabs was parallel to the no-load direction. This behavior is not surprising, for the process of frictional crack sliding and wing crack growth that operates under uniaxial loading still operates, under essentially the same value of the maximum principal stress. The difference is that fewer primary, inclined cracks participate; namely, the ones that are oriented in a plane parallel to the confining stress. Others, should they exist, experience a lower effective shear stress (as defined by Equation (11.2)) and hence a lower mode-I stress intensity factor, and so are much less likely to contribute to terminal failure.

12.3.2 Triaxial loading

Under triaxial loading the confining stress acts upon all inclined primary cracks, regardless of orientation. Consequently, it raises the maximum principal stress to effect sliding and crack growth, thereby increasing strength. For instance, when loaded along the path $(1{:}0.2{:}0.2)\sigma_1$, the material described in the preceding paragraph supports a stress that is higher by a factor of about four than under either uniaxial or biaxial loading (Weiss and Schulson, 1995). The triaxial strength appears to be independent of the intermediate principal stress, implying that it is governed by the maximum stress difference $(\sigma_1 - \sigma_3)_{max}$. This difference is sometimes termed the *differential stress* at failure.

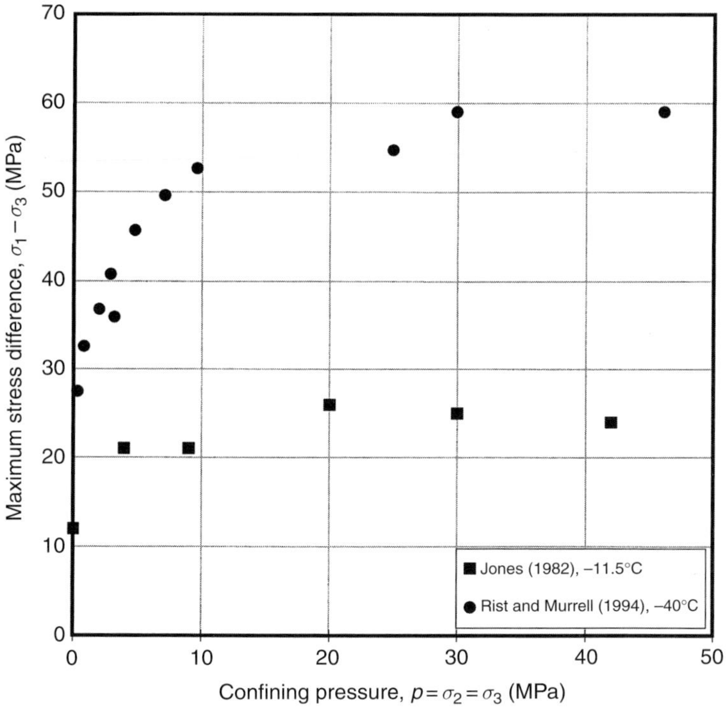

Figure 12.2. Maximum stress difference (or differential stress) vs. confining pressure for fresh-water granular ice of ~1 mm grain size deformed at −11.5 °C and −40 °C at $\dot{\varepsilon}_1 = 1 \times 10^{-2}\ \mathrm{s}^{-1}$. (The data were taken from Jones (1982) and Rist and Murrell (1994).)

Figure 12.2 shows the effect of confining pressure $p = \sigma_2 = \sigma_3$ on the maximum stress difference. The data were obtained from triaxial tests on manufactured fresh-water granular ice of ~1 mm grain size at −11.5 °C and −40 °C (Jones, 1982; Rist and Murrell, 1994). Two regions are apparent. Under lower confinement, the stress difference increases with increasing pressure, the effect being more marked at the lower temperature. Under higher confinement, the stress difference is independent of pressure. The same behavior is exhibited by glacier ice and iceberg ice triaxially loaded under similar conditions (Gagnon and Gammon, 1995) and by much colder (−196 °C and −160 °C) and more slowly loaded ($\dot{\varepsilon}_1 = 3 \times 10^{-6}\ \mathrm{s}^{-1}$) finely grained ($d \sim$ 1–2 mm) manufactured ice (Durham *et al.*, 1983). The behavior is characteristic of ice and not of the method of loading, evident from the fact that Taylor (2005) found it as well through true triaxial proportional loading experiments on similar material. Pressure hardening is a manifestation of frictional sliding: it is greater at lower temperatures because the friction coefficient is greater (Chapter 4).

The absence of the effect implies a failure process based upon volume-conserving plastic[1] flow (more below).

Except for loading along paths of very low triaxial confinement, such as $(1{:}0.02{:}0.02)\sigma_1$ where terminal failure still occurs by splitting (Weiss and Schulson, 1995), loading along all other triaxial paths, whether of low or of high confinement, is terminated by shear faulting. The faults are inclined to the direction of maximum principal stress and, as shown by Taylor (2005) from proportional loading experiments in which $\sigma_1 \neq \sigma_2 \neq \sigma_3$, tend to lie parallel to the direction of the intermediate principal stress. They form occasionally in conjugate sets. There are, in fact, two kinds of shear fault. Lower-confinement faults are inclined by 25°–30° to σ_1 and are made up of a relatively high concentration of deformation damage (more below). In comparison, higher-confinement faults are inclined by $\sim 45 \pm 5°$ to σ_1 (Durham *et al.*, 1983; Rist and Murrell, 1994; Gagnon and Gammon, 1995; Meglis *et al.*, 1999); i.e., they are oriented approximately parallel to planes on which the applied shear stress is a maximum. They are narrower (i.e., generally less than one grain diameter) than the lower-confinement faults and are accompanied by fewer cracks. Also, in the few cases examined to date, higher-confinement faults contain fine, recrystallized grains (Durham *et al.*, 1983; Meglis *et al.*, 1999), indicative of dislocation-based, microstructural restoration (Chapter 6). The two kinds are illustrated schematically in Figure 12.3. Examples are shown below (Figures 12.6a and 12.12b) within the discussion on columnar ice. Lower-confinement faults are reminiscent of features commonly found in rocks and minerals (Jaeger and Cook, 1979; Paterson and Wong, 2005; Haimson, 2006), which also exhibit pressure hardening under moderate confinement. We term them *frictional* or *Coulombic faults*. We term higher-confinement faults *non-frictional* or *plastic faults* and ascribe them to localized plastic flow. As we show below, both kinds form in sea ice as well when confined triaxially and deformed rapidly (Sammonds *et al.*, 1989, 1998; Schulson and Gratz, 1999).

We consider the nature of the faults in Sections 12.7 and 12.8. At this juncture, suffice it to say that the plastic faults develop under conditions where confinement is sufficiently great to suppress frictional sliding. We show in Section 12.6 that an upper limit on "sufficiently great" is defined by the relationship (Schulson, 2002):

$$R \geq \frac{(1+\mu^2)^{1/2} - \mu}{(1+\mu^2)^{1/2} + \mu} \tag{12.1}$$

[1] By the term "plastic flow" we mean volume-conserving, inelastic deformation. Accordingly, we include plasticity that exhibits relatively low dependence upon time/temperature, as well as plasticity that exhibits relatively high dependence upon time/temperature. The magnitude of the dependence depends upon the rate controlling mechanism (Chapter 6).

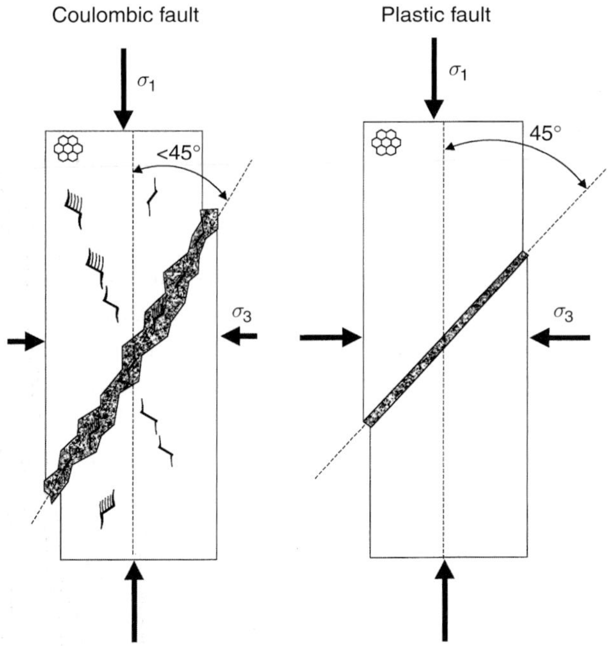

Figure 12.3. Schematic sketch of two kinds of compressive shear fault. Plastic faults form under higher degrees of triaxial confinement and are inclined by ~45° to the direction of maximum principal stress, σ_1. Coulombic faults form under lower degrees of confinement and are inclined by 25°–30° to σ_1. (From Schulson, 2002.)

where $R = (\sigma_3/\sigma_1)$ and μ is the same friction coefficient incorporated in Equation (11.2). For instance, $\mu \sim 0.5$ at $-10\,°C$ and $\mu \sim 0.8$ at $-40\,°C$ (Figure 4.6), implying that $R(-10) > 0.4$ and $R(-40) > 0.2$. In other words, confinement need not be unusually high to suppress frictional sliding and the attendant characteristics of pressure hardening and Coulombic faulting, particularly at lower temperatures.

12.4 Columnar ice

When considering columnar ice, it is useful to define the applied stress state in terms of a coordinate system that reflects the microstructure of the material. Accordingly, and in keeping with the system defined in Section 4.3.2, directions X_1 and X_2 are taken to be perpendicular to the long axis of the grains and X_3, to be parallel to the long axis. For ice that possesses the S2 growth texture, direction X_1 may be chosen randomly within the plane normal to the columns (i.e., within the horizontal plane of a floating sheet of first-year sea ice, for instance), for within this plane the inelastic behavior is isotropic. For ice that possesses the S3 growth texture, we recommend that X_1 be taken parallel to the direction of strongest alignment. It is also useful to describe applied normal stresses by using the double subscript notation. Accordingly, σ_{11} is the stress applied in the X_1 direction to the plane perpendicular to direction X_1, etc. Although this may seem redundant, the value lies in avoiding confusion with principal stresses. For instance, under biaxial loading where the higher compressive stress is applied along the columns in direction X_3 and the lower compressive stress is applied across the columns in direction X_1, the principal stresses are defined by $\sigma_1 = \sigma_{33}$, $\sigma_2 = \sigma_{11}$ and $\sigma_3 = \sigma_{22} = 0$.

We focus the discussion on S2 ice. Of the three columnar variants described in Chapter 3, S2 is the one that has received by far the greatest attention, owing to its frequency of occurrence within floating ice covers on rivers and oceans. We only consider loads applied either across or along the columns. Loads applied along other directions have received relatively little attention.

Readers familiar with the multi-axial loading of columnar ice will note the absence from our discussion of early work on the subject by Frederking (1977), Hausler (1981) and by Timco and Frederking (1986). The reason is that that work was performed at relatively low strain rates ($< 10^{-3}\ s^{-1}$) and thus within the regime of ductile behavior. Within that regime S2 ice, whether loaded biaxially or triaxially, obeys Hill's (1950) criterion for the yielding of plastically orthotropic material (Schulson and Nickolayev, 1995; Melton and Schulson, 1998), as discussed in Chapter 6 and mentioned again below.

12.4.1 Biaxial loading

There are three biaxial loading paths to consider: $(\bar{\sigma}_{11}{:}0{:}\sigma_{33})$, $(\sigma_{11}{:}0{:}\bar{\sigma}_{33})$ and $(\bar{\sigma}_{11}{:}\sigma_{22}{:}0)$. The bar denotes the major stress.

$(\bar{\sigma}_{11}{:}0{:}\sigma_{33})$ From proportional loading experiments on both laboratory-grown fresh-water ice and salt-water ice of 4.5 ± 0.3 ppt salinity deformed at $-10\ °C$ at $\dot{\varepsilon}_{11} = 6 \times 10^{-3}\ s^{-1}$, Schulson and Gratz (1999) found that the strength, taken as the value of the across-column stress at terminal failure, is independent of the along-column stress. Failure occurs through the development of a series of plates that are oriented more or less parallel to the $X_1 - X_3$ plane, just as it does under uniaxial across-column loading. This behavior can be understood within the context of the frictional sliding-wing-crack mechanism discussed in the previous chapter. Accordingly, in this instance the primary or parent cracks that serve as sliding surfaces form preferentially on grain boundaries inclined by $\sim 45°$ to direction X_1. Wings initiate from the tips and then lengthen in the X_1 direction. The along-column stress does not contribute to the shear and normal components of stress acting on the sliding crack, and so does not affect the mechanics of the process.

$(\sigma_{11}{:}0{:}\bar{\sigma}_{33})$ Ice has not been subjected to systematic biaxial loading in which the major stress is applied along the columns. However, we expect the strength (taken in this case to be the value of σ_{33} at failure) not to be affected by low levels of confinement, but to be greater than in the case just considered. We expect greater strength because the along-column strength under uniaxial loading is greater than the uniaxial, across-column strength (Chapter 11). Also, we expect failure to occur by splitting along the X_3 direction, leading to the formation of thin plates oriented perpendicular to the no-load direction. Under higher levels of confinement where the level of σ_{11} reaches the across-column strength of unconfined ice, the strength is expected to decrease.

$(\bar{\sigma}_{11}:\sigma_{22}:0)$ The brittle compressive strength of S2 ice loaded biaxially across the columns has been studied extensively (Smith and Schulson, 1993, 1994; Schulson and Nickolayev, 1995; Schulson and Buck, 1995; Iliescu and Schulson, 2004; Schulson *et al.*, 2006a,b), owing to the importance of across-column loading to the failure of floating sea ice sheets when pushed against themselves or against off-shore structures. Whether measured through proportional loading or under a constant confining stress – i.e., along either path OB or path OAB, Figure 12.4 – the strength is essentially the same (Fortt and Schulson, 2007a). It is also the same when measured under proportional straining (Schulson and Iliescu, 2006). The path independence is a reflection of significant elastic deformation within the brittle regime and of the reversibility of elasticity.

Figure 12.5 shows the compressive-compressive part of the biaxial, across-column failure envelope at −10 °C. The data are plotted twice (mirrored w.r.t. $\sigma_{11}=\sigma_{22}$) to reflect the isotropy of S2 ice within the X_1–X_2 loading plane, although in the subsequent discussion we refer only to the part of the envelope where

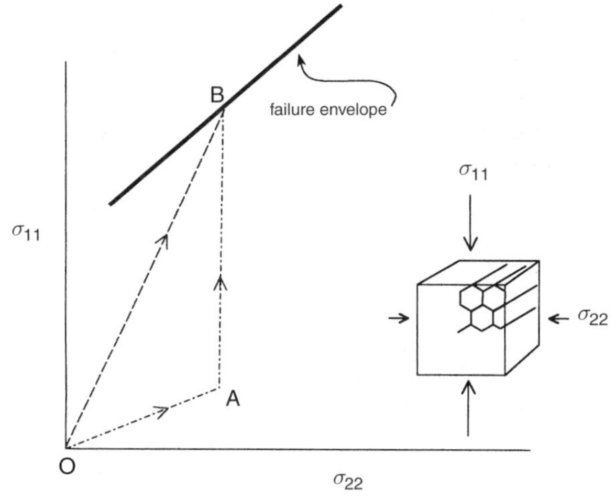

Figure 12.4. Schematic sketch showing two loading paths, proportional loading and constant confinement.

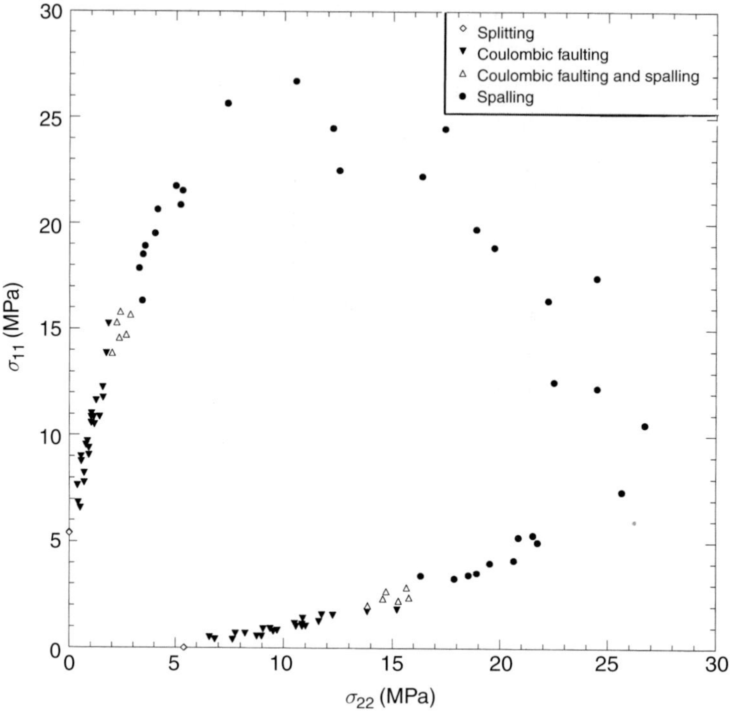

Figure 12.5. Brittle compressive failure envelope for columnar-grained S2 fresh-water ice of 6 ± 2 mm column diameter, biaxially loaded across the columns at $-10\,^{\circ}$C at $\dot{\varepsilon}_{11} = 4 \times 10^{-3}\,s^{-1}$. The data were obtained from measurements on plates ($150 \times 150 \times 25\,mm^3$) where the smallest dimension was parallel to the long axis of the columns. The failure mode is denoted by the shape of the points. (From Iliescu and Schulson, 2004.)

$\sigma_{11} > \sigma_{22}$. The envelope possesses both an ascending or *confinement-strengthening* branch and a descending or a *confinement-weakening* branch. Along the lower segment ($\sigma_{22} \lessapprox 0.1\sigma_{11}$) of the rising branch, the strength (defined as the maximum value of σ_{11}) is limited by the development of a Coulombic shear fault, Figure 12.6a. The fault is inclined by $27 \pm 3°$ to the direction of maximum principal stress (Iliescu and Schulson, 2004) and is decorated by fragments of gouge-like material (Smith and Schulson, 1993). Although Figure 12.6a shows a conjugate set, more often than not single faults form. Unlike the case within granular ice loaded triaxially, the fault within S2 ice loaded biaxially across the columns is not parallel to the direction of the intermediate principal stress. Instead, it is parallel to the direction of the third principal stress. This is not a violation of the Rudnicki–Rice (1975) tenet that the plane of a fault must be parallel to the intermediate principal stress, but a reflection of the mechanical anisotropy of S2 ice in the $X_1(X_2) - X_3$ loading plane. The envelope expands/contracts with decreasing/increasing temperature (Smith

(a)

(b)

Figure 12.6. (a) Coulombic shear faults in columnar-grained S2 fresh-water ice biaxially loaded across the columns to terminal failure under moderate confinement, under the conditions described in the caption to Figure 12.5, as viewed along the columns. (b) A fully developed set of across-column cracks in the same kind of ice biaxially loaded across the columns under higher confinement, showing a set of thin plates that eventually collapse, leading to spalling out of the loading plane, as viewed across the columns. Scale: smallest division = 1 mm. In each image, the maximum stress was applied in the vertical direction. (From Iliescu and Schulson, 2004.)

and Schulson, 1993; Schulson *et al.*, 2006b) and expands slightly with decreasing grain size (Iliescu and Schulson, 2004).

Along the descending branch of the envelope ($\sigma_{22} \gtrsim 0.2\sigma_{11}$) the strength is limited by failure through the thickness of the plate, out of the loading plane. An example of this through-thickness mode is shown Figure 12.6b. What seems to happen is that under these higher confinements frictional sliding along directions within the loading plane is suppressed, thereby preventing the formation of Coulombic faults. Instead, another mode of failure sets in. Unlike the high-confinement plastic mode that develops within triaxially loaded granular ice, the mode that becomes activated within biaxially loaded columnar ice is still a result of brittle failure, but different in origin from Coulombic faulting. The process involves the development of near-parallel sets of across-column cleavage cracks that extend across many grains and then link up, culminating in the creation of thin plates that subsequently collapse in an Euler-like manner. We term this mode of failure *spalling*. Spalling and Coulombic faulting operate together over the intermediate range of confinement $0.1 \lesssim \sigma_{22}/\sigma_{11} \lesssim 0.2$ (i.e., on the upper segment of the rising branch). Incidentally, the faulting-to-spalling transition can be defined by Equation (12.1) where for the present case of across-column loading $R = \sigma_{22}/\sigma_{11}$.

Sea ice exhibits the same truncated Coulombic behavior. Figure 12.7 shows the failure envelope for first-year S2 sea ice ($880\,\mathrm{kg\,m^{-3}}$ density, 6 ± 1 ppt salinity, 3.9 ± 0.4 mm column diameter), proportionally loaded and biaxially compressed across the columns at $-10\,°\mathrm{C}$ at $\dot{\varepsilon}_{11} = 1.5 \times 10^{-2}\,\mathrm{s^{-1}}$ (Schulson *et al.*, 2006a). Included are tensile strengths obtained by Richter-Menge and Jones (1993). The all-compressive part is indistinguishable from the envelope for fresh-water ice, at least under the conditions noted. So, too, are the failure modes.

Figure 12.8 summarizes schematically the failure envelope and the failure modes. For completeness, we include the tensile quadrants. (Note: The tensile strength, points a and a′, is lower than the intercept one would obtain by extrapolating the lower-confinement segment c(c′)–d(d′) of the all-compressive quadrant to the uniaxial tensile loading path. Such extrapolation is prohibited because the failure mechanism is expected to change at some point within the compressive-tensile quadrant, from one based upon frictional crack sliding and attendant quasi-stable growth of secondary cracks, segment b(b′)–c(c′), to one based upon the unstable propagation of a single primary crack, segment a(a′)–b(b′).) The ascending/Coulombic branch of the all-compressive quadrant may be described by the relationship:

$$\sigma_{11} = \sigma_{\mathrm{u}} + q\sigma_{22} \tag{12.2}$$

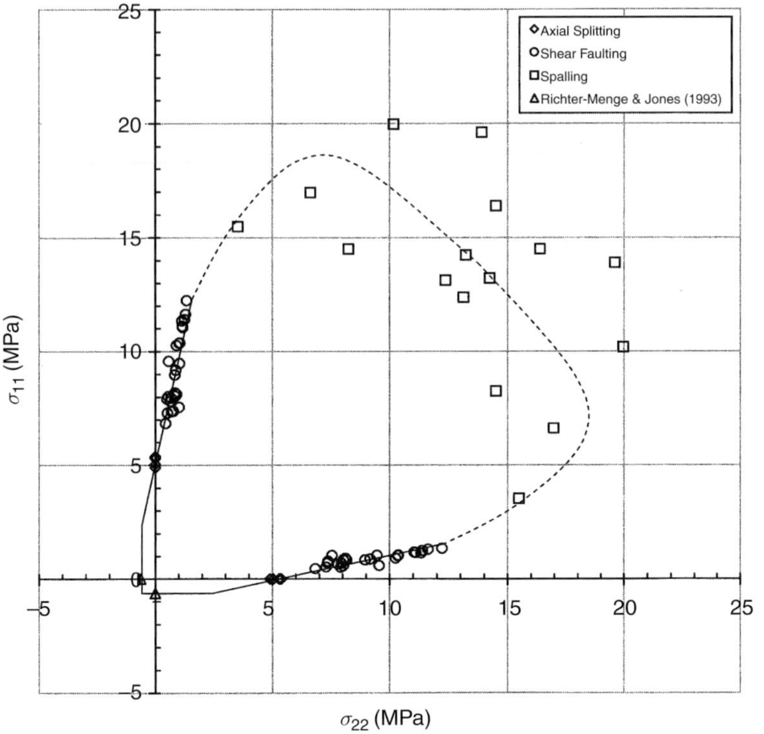

Figure 12.7. Brittle compressive failure envelope for columnar-grained S2 first-year arctic sea ice of 880 kg m^{-3} density, 6 ± 1 ppt salinity, and 3.9 ± 0.4 mm column diameter, biaxially compressed across the columns at −10 °C at 1.5×10^{-2} s^{-1}. The tensile data were obtained from Richter-Menge and Jones (1993). (From Schulson *et al.*, 2006b.)

where σ_u is the unconfined, across-column strength and q is the slope which is governed by the internal friction coefficient[2] (Section 12.6); Table 12.1 lists parametric values. We describe the phenomenological basis for this relationship in Section 12.6 and the physical basis in Section 12.8. The descending branch may be described by:

$$\sigma_{11} + \sigma_{22} = C \tag{12.3}$$

where C is a constant, also listed in Table 12.1. In this case linearity is a poorer description, given the scatter in the data. In Section 12.9 we present a model of the higher-confinement failure mechanics that is based upon a combination of two processes, across-column cleavage-crack growth and Euler buckling: both

[2] Internal friction within the present context is different from internal friction within the context of anelasticity and dislocation motion that was discussed in Chapter 5 on creep of single crystals. Here, we refer to the resistance to frictional sliding near the point of terminal failure.

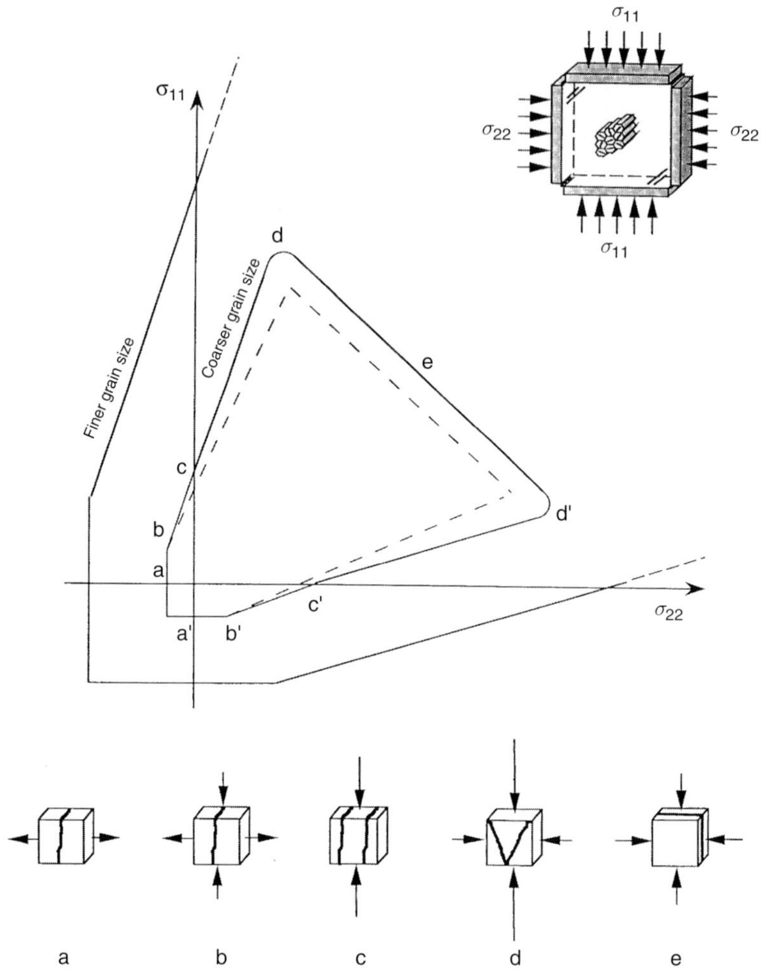

Figure 12.8. Schematic sketches of the brittle failure envelope for columnar-grained S2 ice loaded across the columns and of the failure modes. The dashed envelope denotes strain-rate softening. Compressive stresses are positive: (a–a′) uniaxial tension; (b–b′) biaxial tension and compression; (c–c′) uniaxial compression; (d–d′) biaxial compression, lower confinement; (e–e′) biaxial compression, higher confinement. (From Schulson, 2001.)

processes are driven by the two normal components of the applied stress tensor, thus accounting for confinement weakening.

12.4.2 Triaxial loading

The triaxial compressive strength of both fresh-water and salt-water S2 ice exhibits three regimes of behavior, evident from the results of true triaxial loading

Table 12.1. *Values of the parameters in Equations (12.2) and (12.3) for the brittle compressive failure of S2 columnar ice biaxially loaded across the columns*

Ice	Column dia. (mm)	Temp. (°C)	Strain rate, $\dot{\varepsilon}_{11}\,(s^{-1})$	σ_u (MPa)	q	C (MPa)	References
Fresh-water[a]	6 ± 2	−3	$4 \pm 2 \times 10^{-3}$	~ 2.5	3.5 ± 0.2	20 ± 2	Schulson *et al.* (2006b)
Fresh-water[a]	6 ± 2	−10	$4 \pm 1 \times 10^{-3}$	4.6 ± 0.2	5.7 ± 0.3	32 ± 3	Iliescu and Schulson (2004); Schulson *et al.* (2006b)
Fresh-water[a]	6 ± 1	−40	1×10^{-2}	9.5 ± 0.5	~ 6	—	Smith and Schulson (1993)
First-year arctic sea ice[b]	3.9 ± 0.4	−10	1.5×10^{-2}	5.0 ± 0.2	5.2 ± 0.7	28 ± 4	Schulson *et al.* (2006a)

[a] Laboratory-grown ice.
[b] Salinity = 6 ± 1 parts/thousand; density = $880 \pm 20\ kg\,m^{-3}$.

experiments at $-10\,°C$ at $\dot{\varepsilon}_{11} = 6 \times 10^{-3}\,s^{-1}$ (Gratz and Schulson, 1997; Schulson and Gratz, 1999):

Regime 1 ($\bar{\sigma}_{11}:\sigma_{22} \lessapprox 0.2\sigma_{11}:\sigma_{33}$) Under low across-column confinement, the triaxial compressive strength, when taken as the value of the greater across-column stress at terminal failure, increases approximately linearly with across-column confining stress, but is independent of the along-column confining stress, Figure 12.9a. Correspondingly, terminal failure occurs via the development of Coulombic shear faults that form on macroscopic planes parallel to X_3 and are inclined by $31 \pm 6°$ to σ_{11}. Again, frictional sliding can account for the observed behavior. Within this regime, the along-column confining stress does not act on primary and secondary cracks, owing to their orientation being parallel to X_3. Hence, σ_{33} does not affect the strength.

Regime 2 ($\bar{\sigma}_{11}:\sigma_{22} > 0.2\sigma_{11}:\sigma_{33}$) The situation is different under higher across-column confinement. Here, frictional sliding along directions in the X_1–X_2 plane is suppressed and across-column cracking sets in. Along these loading paths, the triaxial strength increases with increasing along-column stress, Figure 12.9b. Effectively, σ_{33} impedes long-range crack propagation. Correspondingly, failure occurs via across-column faulting on planes inclined by $33 \pm 13°$ to X_3.

Regime 2 actually consists of two sub-regimes, 2a and 2b. Sub-regime 2a develops under lower levels of along-column confining stress where the strength increases markedly with increasing σ_{33}; correspondingly, the fracture planes are generally less highly inclined with respect to X_3. Sub-regime 2b develops under higher levels ($\sigma_{33} \gtrapprox 0.15\sigma_{11}$) of along-column confinement where the strength increases less markedly; correspondingly, the faults are more steeply inclined to X_3, generally by $41 \pm 3°$. Post-test examination of both fresh-water and salt-water ice deformed within sub-regime 2b revealed that the faults, like the plastic faults that form within granular ice when highly confined (Section 12.3.1), are decorated with recrystallized grains (Schulson and Gratz, 1999; Golding, 2009), Figure 12.10.

Regime 3 ($\sigma_{11}:\sigma_{22} < \sigma_{11}:\bar{\sigma}_{33}$) Under loading mainly along the columns, the compressive strength (now taken as the value of the along-column stress at terminal failure) increases linearly in proportion to the lower of the two across-column confining stresses, Figure 12.9c. Correspondingly, failure occurs via Coulombic faulting, on planes that are inclined by $22 \pm 9°$ to the long axis of the columns and approximately parallel to the greater of the two across-column confining stresses.

Within each of the three regimes the triaxial strength may be described by relationships similar to Equation (12.2); namely:

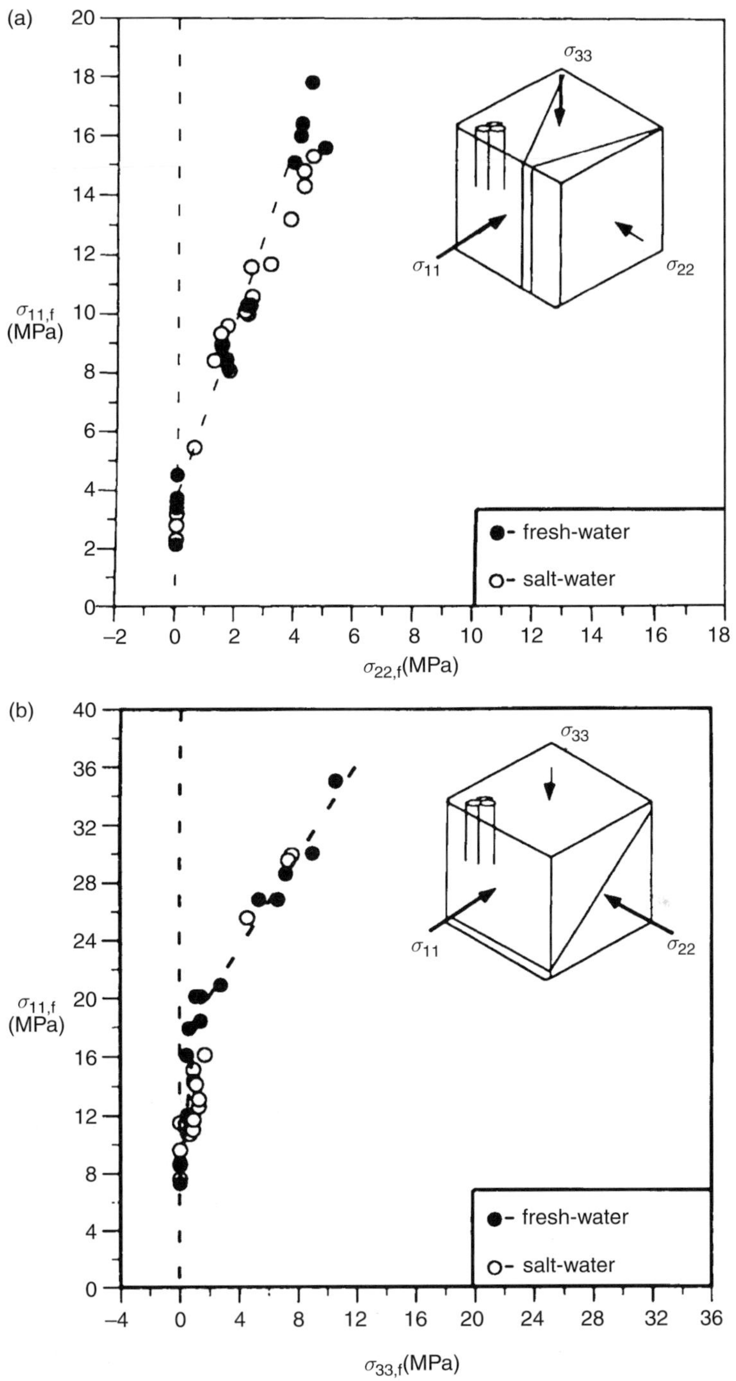

Figure 12.9. Triaxial compressive strength of S2 fresh-water ice and salt-water ice at −10 °C. (a) Regime 1, (b) regime 2 and (c) regime 3. (From Schulson and Gratz, 1999.)

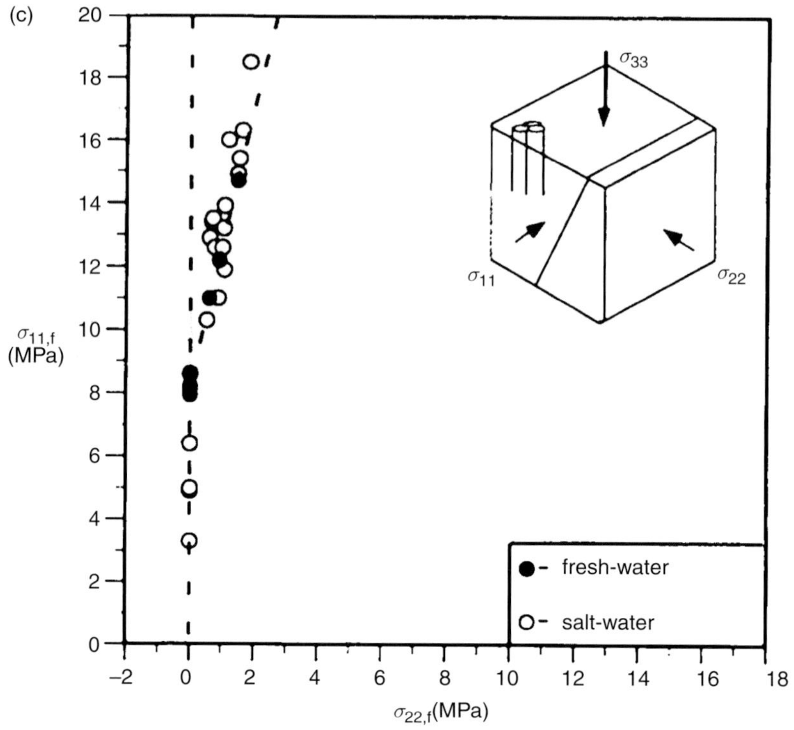

Figure 12.9. (cont.)

$$\sigma_f = \sigma_o + k''\sigma_c \quad (12.4)$$

where the first term denotes the unconfined strength, the second denotes the strengthening imparted by the appropriate component of confining stress σ_c, and k'' is a dimensionless constant governed by the coefficient of internal friction (Section 12.6). Table 12.2 lists the parametric values. The value of k'' of the triaxial tests appears to be lower than the value of q (Table 12.1) of the biaxial tests, for essentially the same conditions, a difference that may be more a reflection of fewer tests in the former case than of some fundamental difference in the underlying friction coefficient that governs these parameters (more below). Note that once again, under the conditions given in Table 12.2, the strength of salt-water ice is practically indistinguishable from the strength of fresh-water ice.

Behavior under higher confinement In the true triaxial experiments just described, the ice was confined to a relatively moderate degree. In experiments using pressure cells, it has been more highly confined. The behavior, however, is similar. For instance, Richter-Menge (1991) harvested cylindrical specimens of first-year S3 columnar ice from the ice cover on the Arctic Ocean. Their long axis was oriented

Table 12.2. *Experimental constants describing the triaxial compressive strength of fresh-water S2 ice and salt-water S2 ice at $-10\,°C$ at $\dot{\varepsilon}_{11}=6\times10^{-3}\ s^{-1}$*

	Definition		Fresh-water ice		Salt-water ice[a]	
Regime	σ_f	σ_c	σ_o (MPa)	k''	σ_o (MPa)	k''
1	$\sigma_{11,f}$	$\sigma_{22,f}$	3.8	3.1	3.5 ± 0.7	3.1
2a	$\sigma_{11,f}$	$\sigma_{33,f}$		10 ± 5		4.5 ± 0.4
2b	$\sigma_{11,f}$	$\sigma_{33,f}$		1.6 ± 0.1 ($R_{21}=1$)[b]	Insufficient data	Insufficient data
				1.3 ± 0.1 ($R_{21}=0.5$)		
3	$\sigma_{33,f}$	$\sigma_{22,f}$		4.4 ± 0.1	5.6 ± 1.7	7.2 ± 0.4

[a] The salt-water ice was characterized by density ($910\pm3\ kg\,m^{-3}$), salinity ($4.5\pm\ 0.3$ ppt), porosity (3.9%), brine pocket spacing (0.6 ± 0.2 mm) and pore diameter (~0.5 mm). These values are similar to those of first-year arctic sea ice.
[b] $R_{21}=\sigma_{22}/\sigma_{11}$.

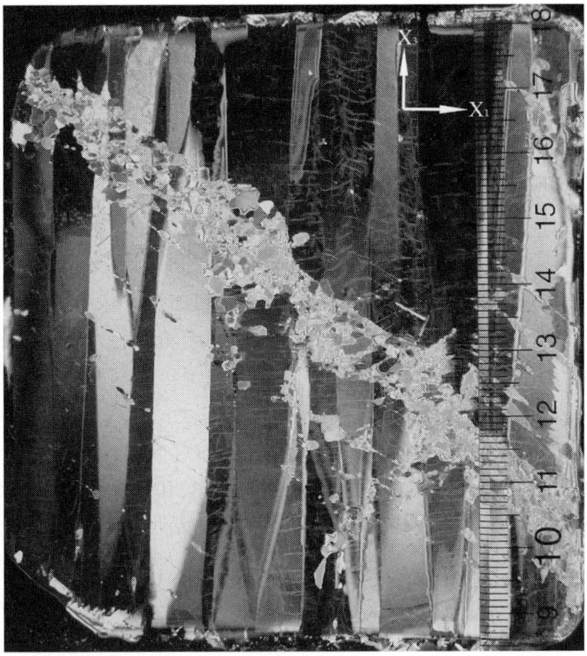

Figure 12.10. Photograph of a thin section from S2 columnar fresh-water ice, as observed through polarizing filters. The parent specimen was proportionally loaded within sub-regime 2b to just beyond terminal failure (to $\varepsilon_{11}=0.024$) along the path (σ_{11}: σ_{22}: σ_{33} = 1:0.5:0.2) at $-10\,°C$ at $\dot{\varepsilon}_{11}=6\times10^{-3}\ s^{-1}$. Note the plastic fault and the recrystallized grains within. The short, white lines are reflections from non-propagating, across-column cracks. Scale: smallest division = 1 mm. (From Golding, 2009.)

within the horizontal plane and was either parallel to, perpendicular to, or at 45° to the direction of *c*-axis alignment. She loaded the ice proportionally at −10 °C at two strain rates, $\dot{\varepsilon}_{11} = 10^{-5}\,\mathrm{s}^{-1}$ and $10^{-3}\,\mathrm{s}^{-1}$. In effect, the most compressive stress was applied across the columns and the confining stress was applied equally both across and along the columns, giving loading paths $(\bar{\sigma}_{11}\colon R\sigma_{11}\colon R\sigma_{11})$ where in this scenario $R = \sigma_{22}/\sigma_{11} = \sigma_{33}/\sigma_{11}$ and $0 \leq R \leq 0.75$. The ice exhibited brittle behavior only when unconfined and strained at the higher rate. Under all other conditions, the confinement was sufficient to impart macroscopically ductile behavior. Correspondingly, the triaxial strength increased with increasing confining stress, but at a rate more in keeping with sub-regime 2b than with sub-regime 2a. Similarly, Sammonds *et al.* (1998) observed pressure-induced plasticity within multi-year sea ice deformed at temperatures from −3 °C to −40 °C at $\dot{\varepsilon}_{11} = 10^{-3}\,\mathrm{s}^{-1}$ and $10^{-2}\,\mathrm{s}^{-1}$, manifested in the form of shear faults oriented at 45 ± 3° to the direction of shortening. The pressure-cell observations are thus consistent with the results from true triaxial loading.

12.5 Failure surfaces

In the interests of summarizing the foregoing observations, Figure 12.11 shows sketches of proposed failure surfaces for both granular ice and S2 columnar ice loaded within the all-compressive octant of stress space. (We retain the double subscript notation to distinguish principal stresses from applied normal stresses, which are specified in terms of the columnar-grained microstructure.) For both variants, brittle failure limits strength under lower confinement, while ductile failure, albeit localized within plastic faults under the higher deformation rates relevant to this discussion, limits strength under higher confinement.

The ductile strength of the plastically isotropic granular material is governed by the well-known von Mises (1928) criterion. Of the plastically orthotropic columnar material, the strength is governed by the Hill (1950) criterion (Melton and Schulson, 1998), as defined in Chapter 6 by Equation (6.5). Although not shown, under very high pressure the von Mises/Hill surfaces are expected to narrow to a point on the hydrostatic axis owing to the pressure-induced reduction in the equilibrium melting point, as discussed by Nadreau and Michel (1986). Note that the projection of the S2 surface onto the σ_{11}–σ_{22} plane creates the S2 failure envelope shown in Figure 6.14. Both failure criteria dictate smooth surfaces which, when projected on the π-plane or viewed along the hydrostatic axis ($\sigma_1 = \sigma_2 = \sigma_3$), are seen as either a circle (von Mises) or an ellipse (Hill). In comparison, the surface that defines brittle failure is faceted and opens in a cone-like manner with increasing confining stress until it intersects the ductile failure surface. The surface defining ductile failure contracts with decreasing strain rate, eventually dominating over the entire range of

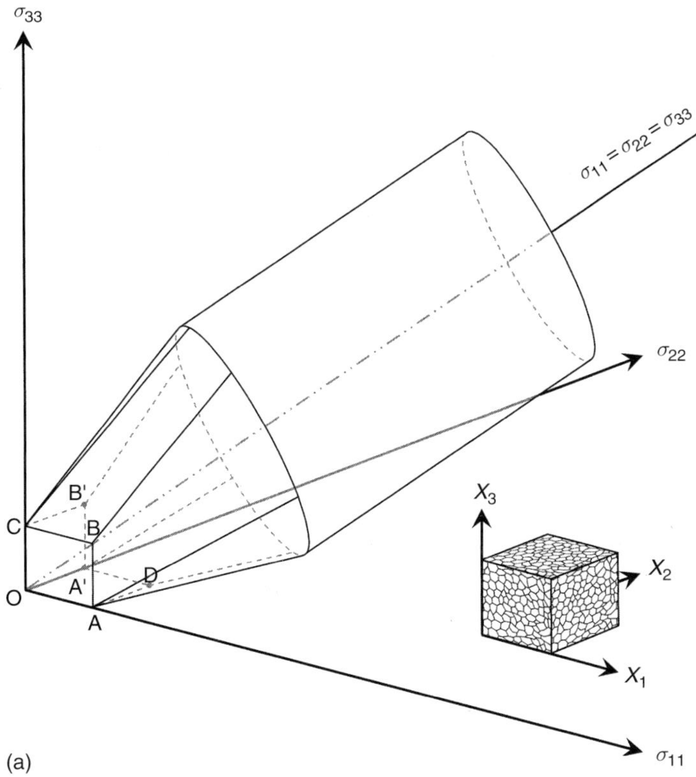

Figure 12.11. Schematic sketch of proposed failure surfaces for (a) granular and (b) S2 columnar ice. Brittle failure is described by a faceted, conical surface that expands upon increasing pressure. Ductile failure is described by a surface whose deviatoric component is independent of pressure.

confinement, in which case brittle behavior is suppressed. Conversely, as suggested by the observations of Rist and Murrell (1994) and Sammonds *et al.* (1998), the ductile surface is expected to expand with increasing strain rate, even though flow tends to be localized, resulting in a greater range of confinement over which brittle behavior dominates. The inelastic *strain vectors*[3] are expected to be normal to the surface for ductile failure, based upon measurements under biaxial loading (Schulson and Nickolayev, 1995). For brittle failure, the strain vectors are expected to be non-normal to the surface, based upon similar measurements (Weiss *et al.*, 2007). This deviation from a normal flow rule reflects the fact that volume is not conserved within the regime of brittle behavior.

[3] Although strain is a quantity defined by a second-order tensor, for the purpose of defining the failure surface it is useful to consider strain as a vector (Backofen, 1972). Accordingly, if the principal directions are defined by unit vectors **i**, **j**, **k**, then the "strain vector" is written as $d\varepsilon = d\varepsilon_1 i + d\varepsilon_2 j + d\varepsilon_3 k$.

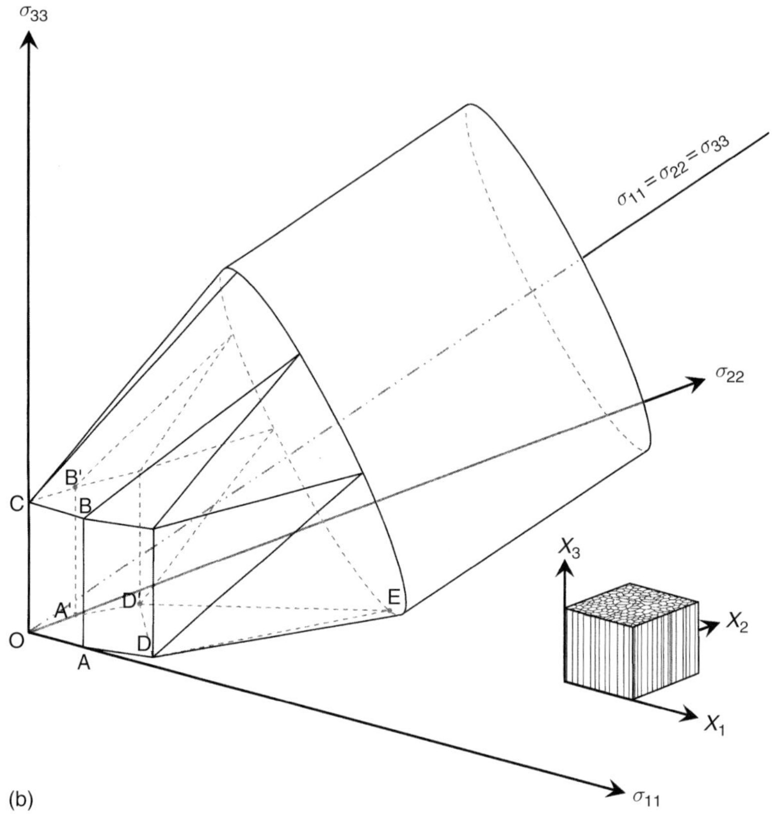

Figure 12.11. (cont.)

The boundaries and facets of the brittle-failure surface define the failure modes. In granular ice two brittle modes operate (Figure 12.11a): splitting along the direction of greatest principal stress (i.e., the most compressive applied stress) and Coulombic faulting. Splitting limits strength when the loading path terminates on the boundaries AB (A′B′), BC (B′C), AD (A′D) and when the path is characterized by a very small degree of triaxiality (not shown in the figure). Under higher degrees of triaxial confinement, faulting operates – Coulombic within the brittle regime and plastic within the ductile regime. In S2 ice (Figure 12.11b) three brittle modes operate: splitting, faulting and across-column spalling. Splitting operates along AB(A′B′) and CB(CB′). We expect it to operate as well under a very small degree of triaxiality (again not shown), as in granular ice. Spalling operates along the boundary DD′ and, we propose, when the loading path terminates on the facet DED′. When the path terminates on any other facet, Coulombic faulting operates, its orientation depending upon which of the applied stresses is the most compressive.

We consider the nature of the multi-axial failure modes below – plastic faulting in Section 12.7, Coulombic faulting in Section 12.8 and spalling in Section 12.9.

12.6 Relationship between compressive fracture and friction

Consider next the relationship between brittle compressive strength under moderate confinement and internal friction. Specifically, consider the two-dimensional deformation of S2 ice and the relationship between the parameters q and k'' of Equations (12.2) and (12.4) and the coefficient of internal friction μ_i. The treatment is similar for the three-dimensional deformation of granular ice under triaxial loading: the only difference is the use of the third principal stress σ_3 instead of the second principal stress $\sigma_2 = \sigma_{22}$.

From the phenomenological theory of brittle compressive failure (e.g., see Jaeger and Cook, 1979) the resistance to frictional sliding across a macroscopic plane – in the present instance, a plane that marks an incipient fault – is controlled by the cohesion of the material τ_o and by the product of the normal stress across that plane σ_n and the internal friction coefficient. Like the case for crack sliding (Section 11.7), the failure criterion is given by Coulomb's relationship:

$$| \tau | - \mu_i \sigma_n = \tau_o \tag{12.5}$$

where τ is the applied shear stress across the plane. Again taking compressive normal stresses to be positive and expressing τ and σ_n in terms of the principal stresses in which the failure envelope is expressed, the left-hand side of Equation (12.5) may be written:

$$|\tau| - \mu_i \sigma_n = \frac{1}{2}(\sigma_{11} - \sigma_{22})(\sin 2\theta + \mu \cos 2\theta) - \frac{1}{2}\mu_i(\sigma_{11} + \sigma_{22}) \tag{12.6}$$

where θ denotes the angle between the maximum principal stress σ_{11} and the incipient failure plane. Upon differentiating Equation (12.6) with respect to θ and setting the result to zero, this function attains its maximum value when:

$$\tan 2\theta = 1/\mu_i \tag{12.7}$$

or when $\sin 2\theta = 1/(\mu_i^2 + 1)^{0.5}$ and $\cos 2\theta = (\mu_i/(\mu_i^2 + 1)^{0.5}$. The maximum stress at terminal failure is given by:

$$\sigma_{11} = \frac{2\tau_o + \sigma_{22}\left[\left(\mu_i^2 + 1\right)^{1/2} + \mu_i\right]}{\left[\left(\mu_i^2 + 1\right)^{1/2} - \mu_i\right]}. \tag{12.8}$$

Thus, the slope of the low-confinement segment of the brittle compressive failure envelope is given by the relationship:

Table 12.3. *Internal friction coefficients deduced from the brittle compressive failure envelope for S2 columnar ice*

Material	Slope of failure envelope, q	Internal friction coefficient μ_i from Equation (12.9) and q	Coefficient of friction across Coulombic fault	Reference
S2 fresh-water ice at −3 °C	3.5 ± 0.2	0.66 ± 0.04	0.69 ± 0.07	Schulson *et al.* (2006b)
S2 fresh-water ice at −10 °C	5.7 ± 0.3	0.98 ± 0.04	0.89 ± 0.10	Schulson *et al.* (2006b)
S2 first-year sea ice at −10 °C	5.2 ± 0.7	0.92 ± 0.07	0.98 ± 0.07	Schulson *et al.* (2006a)

$$q = \frac{d\sigma_{11}}{d\sigma_{22}} = \frac{\left(\mu_i^2 + 1\right)^{1/2} + \mu_i}{\left(\mu_i^2 + 1\right)^{1/2} - \mu_i} = \left[\left(\mu_i^2 + 1\right)^{1/2} + \mu_i\right]^2. \tag{12.9}$$

From Equation (12.9) and from the q-values listed in Table 12.1, the internal friction coefficients can be obtained. Values for fresh-water and salt-water S2 ice loaded across the column are listed in Table 12.3.

Interestingly, and reminiscent of an earlier observation by Sammonds *et al.* (1998), the internal friction coefficient deduced from the terminal failure envelope is closely similar to the kinetic coefficient of friction for sliding along a Coulombic fault once it has formed (Schulson *et al.*, 2006a,b; see Section 12.10). This similarity probably reflects the fact that in both the development of the fault and the subsequent sliding across it, shear deformation is concentrated within a narrow zone of material whose structure and properties are similar. The cohesion, of course, is different: barring healing, it has a value close to zero once the fault has formed.

Incidentally, the confinement that marks the transition from Coulombic to plastic faulting can be obtained from the foregoing analysis. By writing $\sigma_{22} = R\sigma_{11}$ and by noting that the value of σ_{11} for frictional sliding tends to infinity as R becomes large enough to suppress frictional sliding, Equation (12.1) is obtained from Equation (12.8). The confinement so obtained must be viewed as an upper limit, for the plastic process becomes activated before σ_{11} reaches infinity.

The relationship given by Equation (12.8) between fracture and friction is not limited to ice. It appears to be characteristic of rock as well (Byerlee, 1967; Scott and Nielsen, 1991a,b; Schulson *et al.*, 2006b; Costamagna *et al.*, 2007), judging from the behavior of granite and other materials.

12.7 Nature of plastic faults

Why does plastic flow at higher strain rates become localized, when at lower rates it is more uniformly distributed on a macroscopic scale?

12.7.1 Adiabatic heating

Localization implies dynamic instability. Transformation to a phase of higher density under confining pressure, as may happen, for instance, as olivine transforms to spinel deep within Earth's crust (Kirby, 1987), is not the answer here: the mean stress under which the Coulombic (C) to plastic (P) transition occurs in ice is much lower than the ice Ih to ice II or the ice Ih to ice III phase-transition pressures. Those pressures are of the order of 200 MPa (Petrenko and Whitworth, 1999), whereas the pressure marking the C–P transition is around a few MPa (Rist and Murrell, 1994; Gagnon and Gammon, 1995; Schulson and Gratz, 1999). Instead, *adiabatic heating* may be the cause (Schulson, 2002). The phenomenon is well known in metals and alloys when deformed at high rates, particularly in poor thermal conductors like titanium (Winter, 1975) and bulk metallic glasses (Liu *et al.*, 1998), and there is a suggestion that it acts to localize plastic flow within rocks and minerals when deformed under high pressure (Griggs and Baker, 1969; Ogawa, 1987; Hobbs and Ord, 1988; Karato *et al.*, 2001; Wiens, 2001; Renshaw and Schulson, 2004). The clue that adiabatic heating may account for plastic faulting in ice is the recrystallized microstructure that accompanies the process (Section 12.3.2, Fig. 12.10).

The idea, following Orowan (1960), Cottrell (1964) and Frost and Ashby (1982), is as follows: once flow initiates, thermal softening exceeds the combined effects of strain hardening and strain-rate hardening. Material within a narrow zone softens and then continues to deform even more, thereby localizing the deformation. To quantify the effect, Schulson (2002) made a conservative estimate of the increase in temperature by assuming that no heat flows out from the plastic zone, that the zone is about one grain diameter in thickness and that the work done equates to the increase in enthalpy. In so doing, he obtained $\Delta T \sim 42°$ for the low-temperature (−196 °C to −160 °C) experiments of Durham *et al.* (1983), $\Delta T \sim 14°$ for the higher-temperature (−40 °C) experiments of Rist and Murrell (1994) and $\Delta T \sim 4°$, 2° and 0.5°, respectively, for the still higher-temperature experiments of Schulson and Gratz (1999), Meglis *et al.* (1999) and Gagnon and Gammon (1995). The lower values at higher temperatures reflect both higher heat capacities (Petrenko and Whitworth, 1999) and lower strength. Should the plastic zone be smaller than one grain diameter, ΔT would be greater than noted.

For adiabatic softening to operate, two criteria must be satisfied (Frost and Ashby, 1982): the plastic strain must exceed a critical value; and the plastic strain rate must excel a critical value. The first criterion ensures that sufficient strain energy in the form of dislocation substructure has been stored to trigger the transformation; and the second, that time is insufficient for heat to flow. Schulson (2002) found that the criteria were well met in one case (Rist and Murell, 1994), but only poorly met in another (Durham *et al.*, 1983). The other cases cited above could not be evaluated because failure strains were not reported.

Further details may be obtained from the references cited. At this juncture, adiabatic heating cum softening should be viewed as a plausible hypothesis for the localization of plastic flow within ice loaded rapidly under a high degree of triaxial confinement. To test the hypothesis further requires additional information. A thermal print of the fault would be informative.

12.7.2 Implications

One scenario concerns rapid indentation, of the kind that could occur during the interaction between an ice sheet and an engineered structure. In that situation the stress state immediately behind the indenter varies from one of high confinement near the center of the contact zone to one of lower confinement near the edges (Johnson, 1985). It seems possible that the more highly confined ice may fail by plastic faulting. Indeed, in experiments designed to simulate such interactions, Jordaan *et al.* (1999, 2001) found regions of intense shear accompanied by recrystallization within high-pressure zones and regions of brittle failure surrounding those zones. We return to this scenario in Chapter 14.

Another scenario is the formation of thrust faults within glaciers. Lliboutry (2002) reviews the evidence for such faults and notes that their origin remains a mystery. Perhaps adiabatic heating might again be at play.

One other point is worth mentioning. Prior to the realization that ice exhibits two kinds of faults when rapidly loaded under triaxial compressive stress states, a question arose concerning the strength of ice and whether it reaches a maximum at the ductile-to-brittle transition, as asserted in Chapter 11. The question was raised by Sammonds *et al.* (1998) who observed that, whereas the compressive strength of unconfined multi-year sea ice reaches a maximum (or at least a constant level) at the transition point, the strength of the ice under the confinement they imposed continued to increase with increasing strain rate, Figure 12.12a, past the point at which faulting set in. They concluded, as did Sammonds and Rist (2001), that the compressive strength does not always reach a maximum at the ductile-to-brittle transition. The problem with that conclusion is that in the experiments cited the ice did not actually fail via a brittle mode, even at the highest strain rate. Under the confinement imposed the material failed via plastic faulting – i.e., via a fault inclined by $\sim 45°$ to the direction of shortening, Figure 12.12b – and that mode

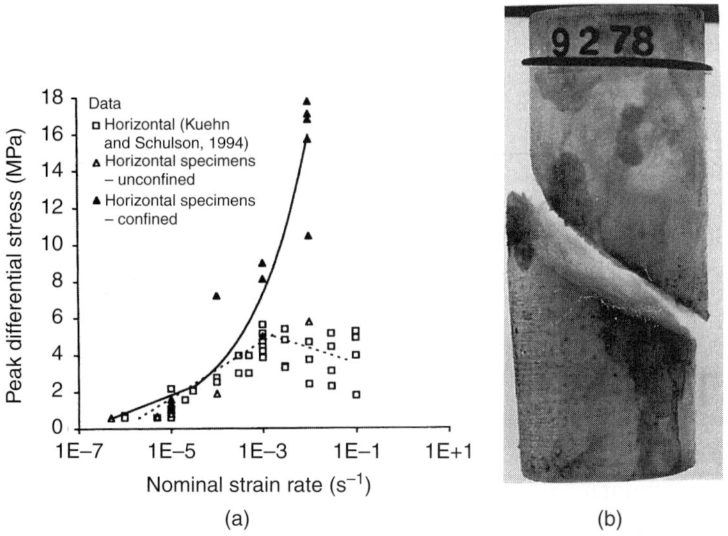

Figure 12.12. (a) Graph of the maximum principal stress difference (or the differential stress) vs. strain rate for sea ice of horizontal orientation loaded at −10 °C without confinement and under triaxial confinement. (b) Photograph showing plastic fault in specimen of multi-year sea ice loaded to terminal failure at −10 °C at $10^{-2}\,s^{-1}$ under triaxial confinement. (From Sammonds *et al.*, 1998.)

of failure, albeit brittle-like in its mechanical expression, is characterized by strain-rate hardening. We hold, therefore, that the compressive strength of ice does indeed reach a maximum at the ductile-to-brittle transition.

12.8 Nature of Coulombic shear faults

To elucidate the development of Coulombic faults, Schulson *et al.* (1999) studied the evolution of deformation damage within S2 ice proportionally loaded biaxially. Columnar ice is conducive to this kind of study because, as mentioned in Chapter 11, the columns restrict the orientation of cracks to planes that are parallel to the long, columnar axes. As a result, cracks do not shield each other when the ice is viewed along the columns. For this work, the specimens were made from fresh-water ice of 10–12 mm column diameter and then pulse-loaded along the path $(1{:}0.1{:}0)\sigma_{11}$ to various fractions of the terminal failure stress at −10 °C at $\dot{\varepsilon}_{11} = 4.8 \times 10^{-3}\,s^{-1}$. To avoid damaging the material upon unloading (Couture and Schulson, 1994) the ice was unloaded between pulses at the rate it was loaded. To minimize specimen-to-specimen variations, sets of microsimilar samples that contained essentially the same set of columnar grains were prepared, by first harvesting thick (~150 mm) blocks from a unidirectionally grown parent ice sheet and then slicing the blocks into sets of thin (25 mm) plates. Schulson *et al.* (1999) made the following observations.

12.8.1 Evolution of the fault

Early in the deformation at $\sigma_{11}/\sigma_{11,f} = 0.17$ (where the subscript f denotes terminal failure) a few inclined primary cracks of grain-size dimensions form either on grain boundaries or in very close proximity to them and are distributed more or less uniformly throughout the matrix, Figure 12.13a. Shortly thereafter ($\sigma_{11}/\sigma_{11,f} = 0.25$) a region of intense damage of width approximately equal to the grain diameter sprouts from one side of one or more primary cracks, such as crack A in Figure 12.13b. Upon further deformation ($\sigma_{11}/\sigma_{11,f} = 0.74$) more damage develops and additional regions of intense damage form, again on one side of parent cracks that had nucleated earlier, such as crack B in Figure 12.13c. Damage continues to develop as the load increases and, in a macroscopic sense, remains more or less uniformly distributed throughout the matrix up to stresses just below the terminal failure stress ($\sigma_{11}/\sigma_{11,f} = 0.96$), Figure 12.13d. At that point intense damage has developed on one side of many of the parent cracks. Finally, upon raising the load a little higher more cracks form and the damage appears to link up in a somewhat atactic manner to form a macroscopic shear fault, Figure 12.13e. The fault is two to three column diameters in thickness (Iliescu and Schulson, 2004) and is inclined by $\sim 28°$ to the direction of maximum principal stress; its surface is decorated by highly fragmented gouge-like material that is evident upon separating the two parts of the faulted specimen.

A remarkable characteristic of this evolution is its reproducibility. From an examination of microstructurally similar plates harvested from the same block of columnar-grained ice and then loaded in the same manner, Iliescu (2000) found that the spatial and temporal evolution of deformation damage is almost identical from one specimen to another within a set, from the first grain boundary to crack to the linking up of damage that constitutes the macroscopic fault. Which particular aspect of microstructure governs the evolution – the structure of the grain boundaries, the orientation of grains and their orientation with respect to their neighbors, the orientation of the boundaries themselves – or even whether there is a governing element as opposed to a collective interaction amongst several elements, is not clear. The important point is that the evolution of damage is not a random process, but one that appears to be prescribed in a manner that is not yet understood.

12.8.2 Wing cracks and comb cracks

Of particular significance are finer features within the fault, Figure 12.14. When observed within the specimen as a whole, the fault can be seen to be bordered by a zigzag edge, Figure 12.14a, and to be composed of highly damaged (whitish) zones mentioned above that had sprouted from one side of a parent crack, such as regions A and B of Figure 12.14a. The zigs and the zags denote wing cracks. The whitish

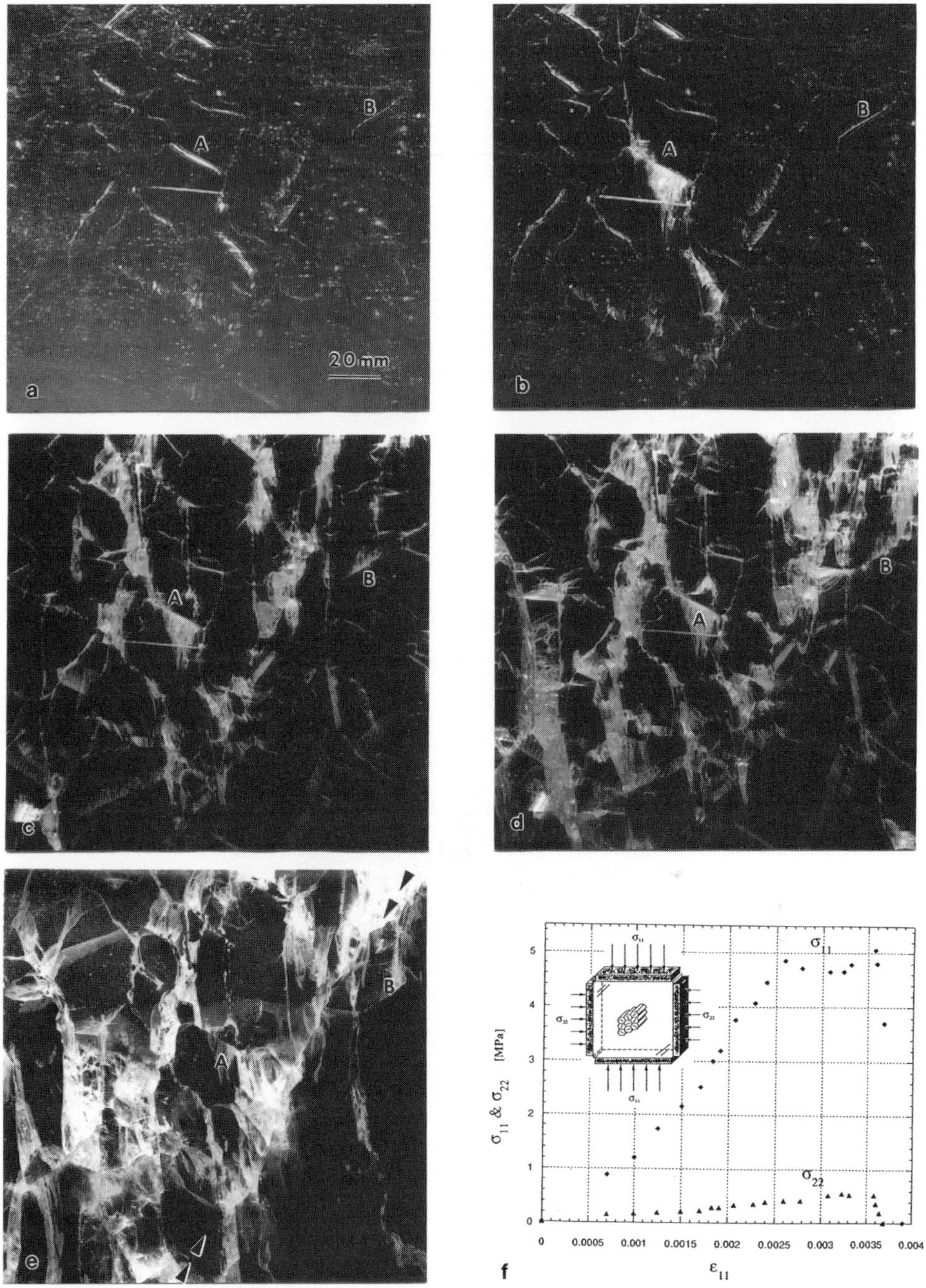

Figure 12.13. Photographs showing the sequential development (a–e) of a Coulombic shear fault in columnar-grained S2 fresh-water ice proportionally pulse-loaded biaxially across the columns under moderate confinement ($\sigma_2/\sigma_1 = 0.1$) at $-10\,°C$ at $\dot{\varepsilon}_{11} = 4.8 \times 10^{-3}\ s^{-1}$, as viewed along the columns. The maximum principal stress was applied in the vertical direction. (f) shows the corresponding stress–strain curve. (From Schulson *et al.*, 1999.)

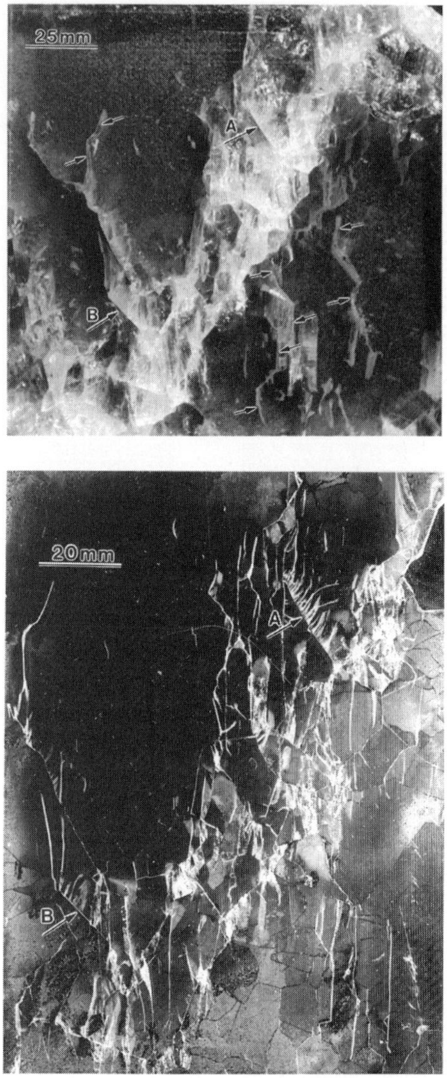

(a)

(b)

Figure 12.14. Photographs of a Coulombic shear fault in the ice described in the caption to Figure 12.13, as viewed along the columns. (a) As seen within the specimen of full thickness (25 mm). Note the zigzag edge. The arrows point to wing cracks. (b) The whitish regions in (a) as seen in a thin (~1 mm) section. Note the set of comb-like, secondary cracks that formed on one side of an inclined parent crack at A. (From Schulson *et al.*, 1999.)

zones, when examined more closely within thin (~1 mm) sections, are seen to be sets of another kind of secondary crack, Figure 12.14b. This other kind is closely spaced and, like wing cracks, tends to align with the direction of maximum principal stress. Schulson (2001) and Renshaw and Schulson (2001) termed them *comb cracks*. In effect, this kind creates sets of slender micro-plates, fixed on one end and free on the other. When loaded by frictional drag across their free ends, the micro-plates bend and then break, rather like teeth in a comb under a sliding thumb. The broken "teeth" can account for the gouge-like material that decorates Coulombic faults.

Comb cracks are not limited to S2 fresh-water ice. They have also been observed in first-year S2 arctic sea ice when loaded biaxially under moderate confinement (Schulson *et al.*, 2006a) and in granular ice when loaded triaxially under moderate confinement (Schulson *et al.*, 1991). However, they have not been seen within ice loaded uniaxially (Iliescu, 2000). The implication is that they constitute an intrinsic deformation feature of all variants of ice when lightly confined and loaded within the regime of brittle behavior.

Comb cracks are significant because they provide a mechanism for triggering destabilization (Schulson *et al.*, 1999; Renshaw and Schulson, 2001). We imagine that the first micro-plate to break sheds load to adjacent ones, which then break, and so on, leading to the creation of additional damage and to a domino-like cascade of events that culminates in a recognized fault, Figure 12.15. While the exact sequence remains to be elucidated, micro-plates located near free surfaces – either external surfaces or internal ones such as cavities or even a wing crack – have a greater likelihood of failing, owing to lesser constraint there. This would generate a kind of "blowout" toward the free surface and could lead to a focusing of damage in the manner of a *process zone*.

The triggering of a fault by micro-plate failure is not a new idea. Peng and Johnson (1972), Ashby and Hallam (1986), Sammis and Ashby (1986) and Bazant and Xiang (1997), for instance, invoked the concept in their treatment of brittle compressive failure. In those mechanisms, however, failure occurs by elastic buckling of columns fixed on both ends, formed, for instance, between adjacent wing cracks. The problem is that when realistic slenderness ratios are employed (measured to be $h/w \sim 3$–7 for ice) the applied stress at failure is estimated to be two to three orders of magnitude greater than the measured strength. Buckling loads scale as $1/(h/w)^2$ and so a ratio closer to $h/w \sim 100{:}1$ would be needed for those mechanisms to comply with observation. Ratios this high seem unlikely, given that in the study cited above (Schulson *et al.*, 1999) the average dimensions of the micro-plates were $h \sim 6.2 \pm 4.0$ mm and $w \sim 1.2 \pm 0.2$ mm. The comb-crack mechanism does not suffer from this limitation. Although the micro-plates are loaded axially as in the earlier models, they are also loaded tangentially through frictional drag across their free ends. Tangential loading is key, for it has the greater effect on the load-bearing ability of the micro-plates.

Comb-like cracks, incidentally, have been seen also in rocks and minerals, in the laboratory and in the field, again emanating from one side of a parent inclined crack, often near its tips. In the laboratory they form on the scale of the grains (Conrad

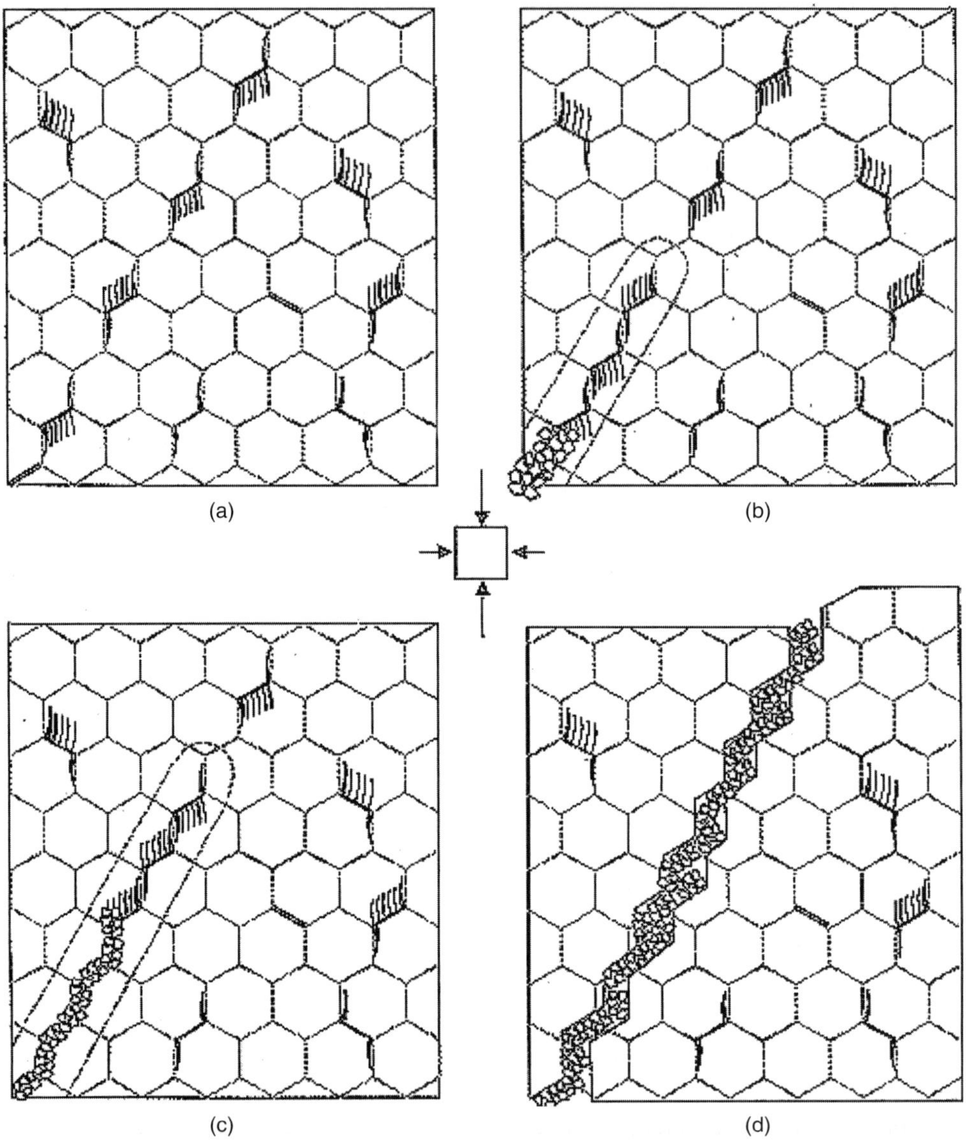

Figure 12.15. Schematic sketch of the initiation and growth of a macroscopic compressive shear fault, from the onset of terminal failure (i.e., from very near the peak of the stress–strain curve). The hexagonal cells denote grain boundaries.

(a) Before initiation, damage is uniformly distributed throughout the body, and consists of a mixture of parent grain boundary cracks inclined by about 45° to the highest principal stress plus parent cracks with wing-crack extensions, and parent cracks with sets of secondary comb cracks stemming from one side.

(b) Near-surface comb crack fails, owing to less constraint there, shedding load and redistributing stress to adjacent material within which new cracks initiate. The nucleus of the eventual shear fault is formed.

and Friedman, 1976; Ingraffea, 1981; Wong, 1982; Gottschalk *et al.*, 1990), while in the field they form on spatial scales up to many times the grain size (Rispoli, 1981; Segall and Pollard, 1983; Granier, 1985; Davies and Pollard, 1986; Martel *et al.*, 1988; Cruikshank *et al.*, 1991). Within the rock mechanics literature the features have been termed either *feather cracks, horsetail cracks* or *splay cracks*, but have not been related directly to the development of Coulombic faults.

How do comb cracks form? Opposing surfaces, it appears, must be in contact. This view is based upon three observations: that this kind of secondary crack tends not to form on pre-engineered parent cracks in ice where Teflon tape is used to simulate the primary, sliding crack (Schulson *et al.*, 1999); that the features are absent from inclined slots in deformed glass (Brace and Bombolakis, 1963); and that they are absent as well from metal- and Teflon-lined slots in deformed Columbia resin (Horii and Nemat-Nasser, 1985, 1986). One explanation is that the comb results from Hertzian stresses that arise from the contact of asperities on cracked grain boundaries. The one-sided character could then be a manifestation of different tensile strengths on either side of the boundary. However, this kind of mechanism seems improbable, for comb cracks emanate not only from cracked grain boundaries, but also from one side of transgranular cracks (Schulson and Gratz, 1999). Moreover, should the origin be similar to that of feather/splay/horsetail cracks in rock, where the features often span many grains, the one-sided character could not be explained in terms of grain-scale crystallography.

A more likely explanation (Schulson, 2001; Renshaw and Schulson, 2001) involves non-uniform sliding, Figure 12.16. The idea, following Martel and Pollard (1989) and Cooke (1997), is that non-uniform displacements create normal stresses that act across planes perpendicular to both the sliding plane and the sliding direction, compressive on one side of the parent and tensile on the other. Provided that the rate of sliding is high enough to prevent significant stress relaxation – a point that is fundamental to the ductile-to-brittle transition (Chapter 13) – the tensile stress is relieved through the formation of secondary cracks. Non-monotonic displacement could lead to tensile fields that alternate from one side to the other of the primary crack and thus to combs that sprout from opposite sides of the sliding interface at different points along it.

Caption for Figure 12.15. (cont.)

(c) More comb cracks fail and localized damage spreads as a band across the section, creating granulated material or gouge in its wake and forming new damage in a "process zone" ahead of the advancing fault front.

(d) A recognizable macroscopic, Coulombic shear fault.

(From Schulson *et al.*, 1999.)

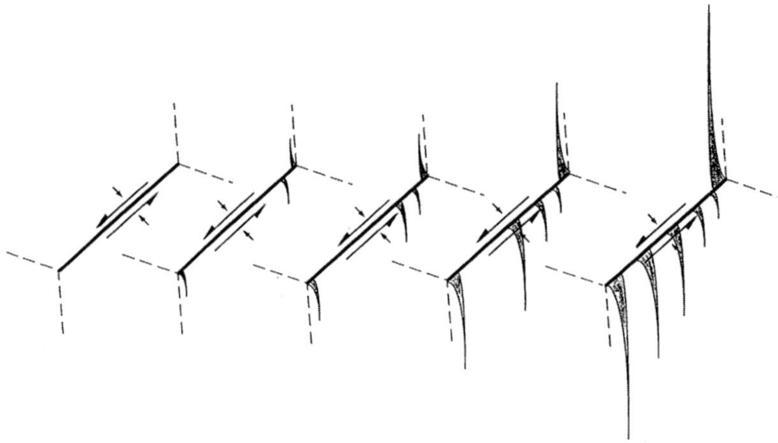

Figure 12.16. Schematic sketch of the proposed evolution (a–e) under compression (applied in the vertical and horizontal directions) of a set of secondary or comb-like cracks. (From Schulson, 2001.)

12.8.3 Comb-crack mechanics

The stress to nucleate a Coulombic fault via the comb-crack mechanism may be estimated as follows (Renshaw and Schulson, 2001): Consider a parent crack of length $2c$ oriented at 45° with respect to the direction of the maximum principal stress, Figure 12.17. From one side extend secondary cracks of length h and spacing w. The cracks create micro-plates of the same dimensions. Each micro-plate is loaded by an axial load P and by a net moment M. The moment is created by frictional drag across the free ends of the plates minus frictional drag against the sides and (per unit thickness) is given by the relationship:

$$M = \mu_i w h \frac{\sigma_1}{2}(1 - R) \tag{12.10}$$

where R denotes the degree of confinement and is lower than that required to suppress frictional sliding (given by Equation (12.1)). For granular ice loaded triaxially $R = \sigma_3/\sigma_1$, while for S2 columnar ice loaded either biaxially across the columns or triaxially across and along the columns $R = \sigma_{22}/\sigma_{11}$, as noted above. As a result of both P and M, material at the tips of the secondary cracks experiences mixed mode-I and mode-II loading, analogous to the loading of an edge crack within a brittle plate. The energy release rate per unit advance per unit thickness (in the long-crack limit) is given by (Thouless *et al.*, 1987):

$$G = \frac{\sigma_1^2 w^2 + 12(M/w)^2}{2wE'} \tag{12.11}$$

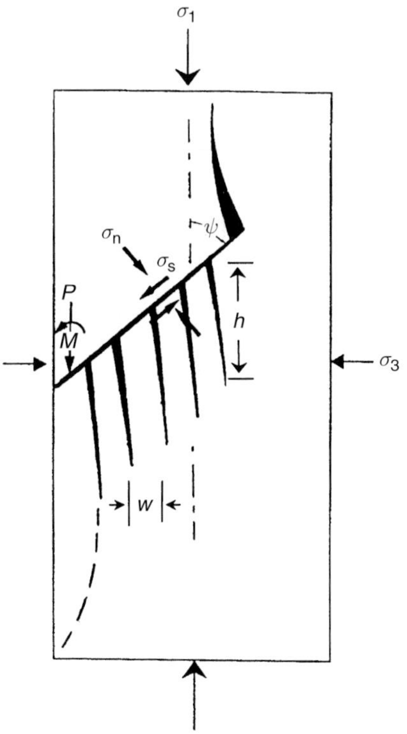

Figure 12.17. Schematic sketch of a comb crack. Secondary cracks initiate from one side of a parent crack that is inclined by an acute angle ψ to the maximum (most compressive) principal stress. The secondary cracks create sets of slender micro-plates of length h and width w, fixed on one end and free on the other. Axial loading produces a load/unit depth, P, on the plates, and frictional drag across their free ends produces a moment/unit depth, M. The moment is of greater importance than the axial load and eventually becomes large enough to "break" the micro-plates via mixed mode-I and mode-II crack propagation. (From Schulson, 2001.)

where $E' = E/(1 - \nu^2)$ and E and ν are Young's modulus and Poisson's ratio, respectively. The micro-plates fail when the energy release rate reaches a critical value, G_c, given by:

$$G_c = \frac{K_{Ic}^2 + K_{IIc}^2}{E'} \tag{12.12}$$

where K_{Ic} and K_{IIc} are the critical stress intensity factors for mode-I and mode-II loading, respectively. We assume that $K_{IIc} \approx K_{Ic}$, which appears to be true for sea ice at least (Shen and Lin, 1986). Upon rearranging, Equations (12.10), (12.11) and (12.12) lead to the non-dimensionalized stress Σ to cause the first micro-plate of the comb crack to break (Renshaw and Schulson, 2001):

$$\Sigma = \frac{\sigma_{\mathrm{f}}\sqrt{c}}{K_{\mathrm{Ic}}} = \frac{2}{\left[\left(1+\left(1-\mu_{\mathrm{i}}\dfrac{1+R}{1-R}\right)^{2/3}\right)^{1/2}-1\right]^{1/2}\left(1+3\mu_{\mathrm{i}}^{2}\alpha^{2}(1-R)^{2}\right)^{1/2}} \tag{12.13}$$

where $\alpha = h/w$ is the slenderness ratio. The more slender the micro-plate, the lower is the failure stress.

To determine α, Renshaw and Schulson (2001) invoked two earlier analyses. They obtained the length from the frictional crack-growth mechanics of Ashby and Hallam (1986) and they obtained the width (i.e., the spacing of the secondary cracks) from Pollard and Segall's (1987) analysis of the shear stress field ahead of a mode-II crack. In calculating length, they noted that crack length approaches an asymptotic limit and used that value in their analysis. In calculating micro-plate width, they imagined a multiple-step process: that the first secondary crack forms soon after a short segment of primary crack forms, once frictional sliding on that segment stops; that the decohesed segment subsequently extends a short distance along the grain boundary (governed by the extent of the zone of shear stress concentration) via viscous-based decohesion of the kind observed by Weiss and Schulson (2000); that the next secondary crack forms upon further sliding and stopping; and so on. In that way Renshaw and Schulson (2001) obtained the relationships (for $R > 0$):

$$\frac{h}{c} = \frac{1 - R - \mu(1+R)}{4.3R} \tag{12.14}$$

and

$$\frac{w}{c} = \left[1 + \left(1 - \mu\frac{1+R}{1-R}\right)^{2/3}\right]^{1/2} - 1. \tag{12.15}$$

The slenderness ratio $\alpha = h/w$ is then obtained from Equations (12.14) and (12.15).

Figure 12.18 compares the model with the biaxial compressive strength (in non-dimensional units) of both S2 fresh-water ice and first-year arctic sea ice loaded across the columns. The strength of the two materials is practically indistinguishable because their fracture toughness (Chapter 9) and friction coefficient (Section 12.10) are almost indistinguishable. Barring the sharp upturn at higher confinement, which reflects the suppression of frictional sliding, the model correctly captures the effect of confinement. However, it underestimates the strength somewhat in two cases. This discrepancy, we believe, is less a problem with the physics of failure

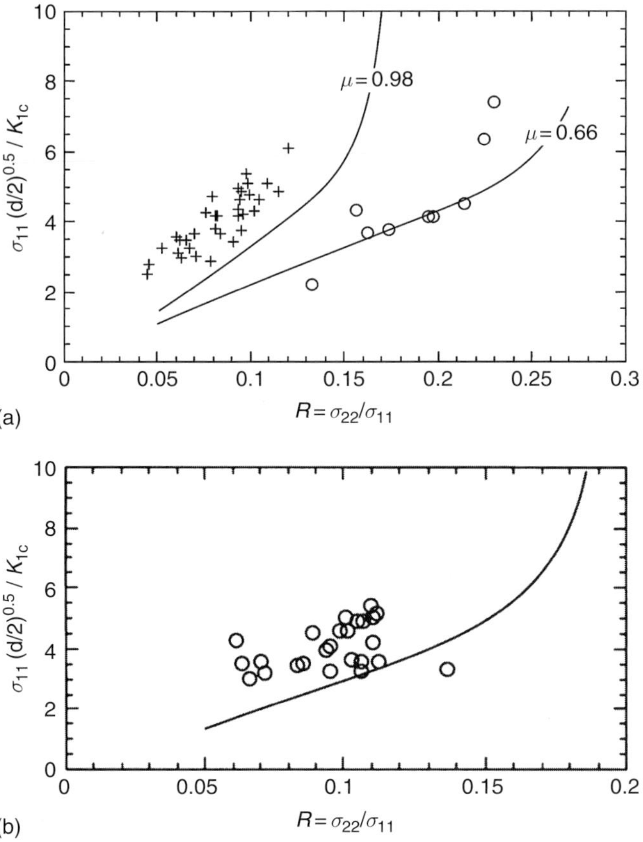

Figure 12.18. Brittle compressive strength (in non-dimensionalized units) (a) of S2 fresh-water ice at −3 °C (lower curve) and at −10 °C and (b) of S2 first-year arctic sea ice at −10 °C vs. confinement ratio $R=\sigma_{22}/\sigma_{11}$. The data were obtained from Figures 12.5 and 12.7 and from Schulson *et al.* (2006a, b). The curves were computed from Equation (12.13) using parametric values listed in Table 12.4. The length of the parent crack grain was equated to the grain size.

and more a problem with the assumptions in the model. A major point in its favor is that unlike earlier analyses of brittle compressive failure under confinement (Horii and Nemat-Nasser, 1985; Peng and Johnson, 1972; Bazant and Xiang, 1997; Gupta and Bergström, 1998) the comb-crack model contains no "free parameters". Each can be and has been measured independently.

The model, incidentally, is not limited to ice. Figure 12.19 shows that it also captures the triaxial strength of a variety of rocks and minerals loaded under moderate confinement (Renshaw and Schulson, 2001). In such cases, comb-like cracks have also been seen, as already noted, and terminal failure is governed by the development of Coulombic shear faults. In other words, the comb-crack model appears to be applicable to the entire class of brittle crystalline materials.

Table 12.4. *Parametric values used to calculate the dimensionless failure stress of ice given by Equation (12.13)*

Material	Internal friction coefficient from Table 12.3	Grain size (mm)	Fracture toughness (MPa m$^{1/2}$) from Chapter 9
S2 fresh-water ice at −3 °C	0.66	5	0.08
S2 fresh-water ice at −10 °C	0.98	5	0.1
S2 first-year arctic sea ice at −10 °C	0.92	4	0.1

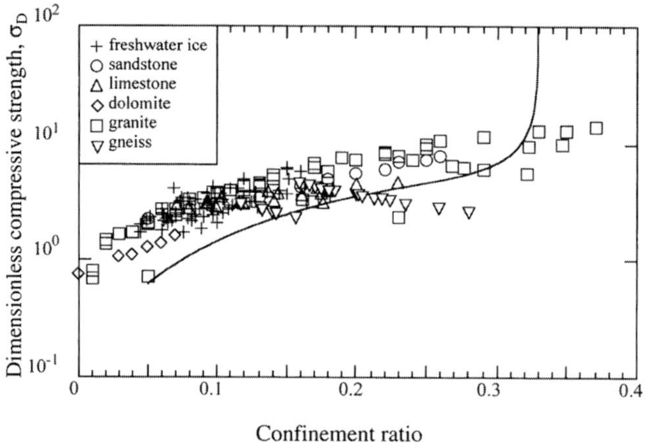

Figure 12.19. Brittle compressive strength of a variety of rocks and minerals vs. confinement ratio $R = \sigma_3/\sigma_1$. The curve was computed from Equation (12.13) using parametric values listed by Renshaw and Schulson (2001). $\sigma_D = \Sigma$. (From Renshaw and Schulson, 2001.)

12.9 Nature of spalling

Spalling of S2 columnar ice loaded across the columns under higher degrees of biaxiality appears to be similar to splitting under uniaxial loading. Wing cracks may again be involved, as suggested by microstructural evidence, Figure 12.20. Their growth could account for the formation of the axial splits. Terminal failure via spalling may correspond to the destabilization of thin plates created by the splits.

The first step appears to be grain boundary sliding. This is suggested by the development of whitish-looking grain boundary features, Figure 12.20a, that resemble the decohesion zones reported by Nickolayev and Schulson (1995), Picu and Gupta (1995) and by Weiss and Schulson (2000) from experiments on the same kind of material. Of interest is the displacement out of the loading plane that is facilitated by the natural taper of the columnar-shaped grains. As sliding

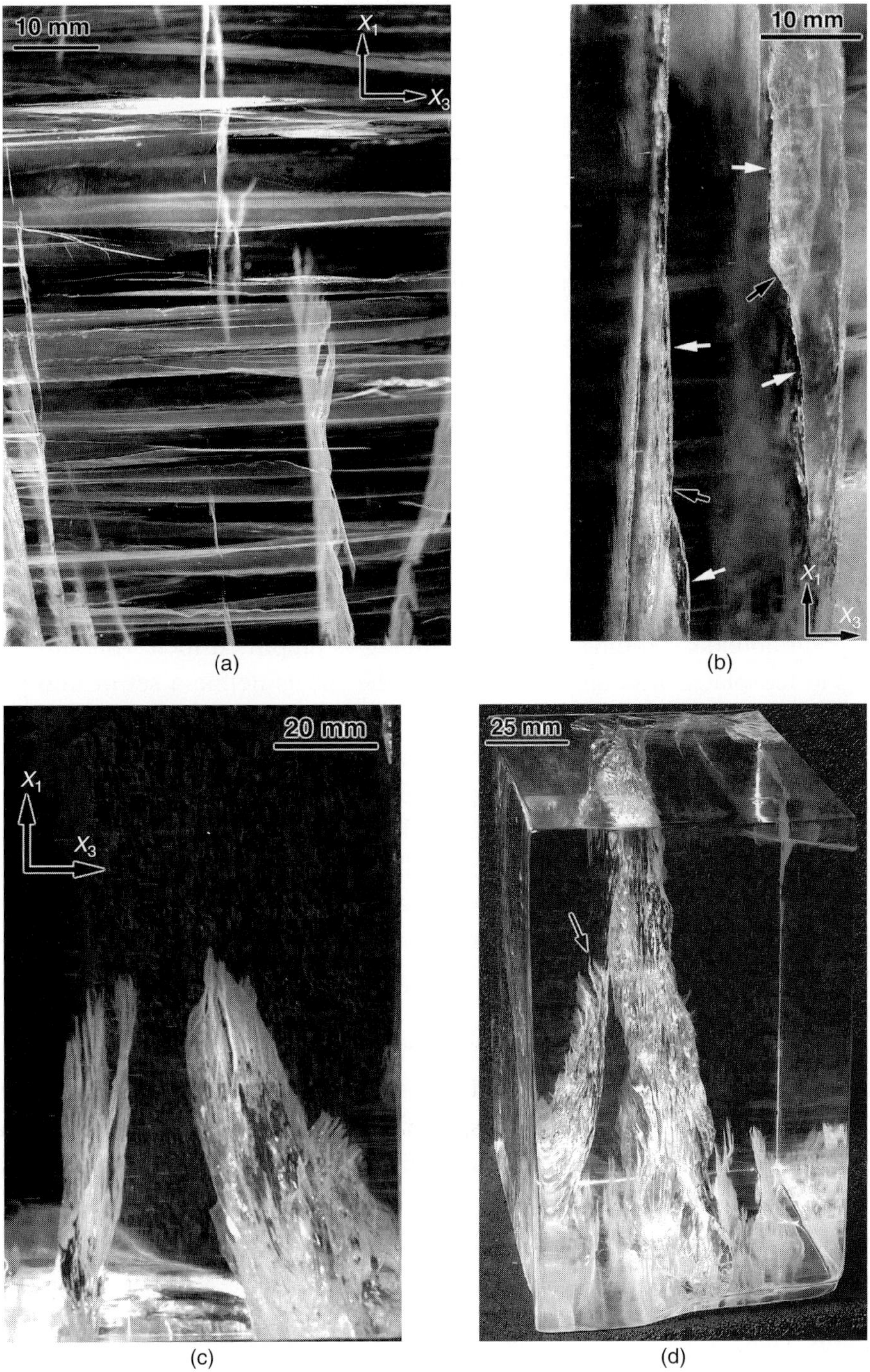

Figure 12.20. Photographs of across-column cracks within S2 columnar-grained fresh-water ice proportionally loaded biaxially across the columns under a higher degree of confinement ($\sigma_{22}/\sigma_{11} \sim 0.5$) at $-10\,°C$ at $\dot{\varepsilon}_{11} = 4 \times 10^{-3}\ s^{-1}$ showing:

occurs, stress intensifies near boundary defects, such as ledges and steps (Liu *et al.*, 1995). The stress is eventually relaxed through the nucleation of across-column cracks that are inclined slightly to the direction of shortening. Within ice loaded at −10 °C, the cracks first nucleate under an applied stress of $\sigma_{11} < 5$ MPa, which is consistent with the upper limit one expects for a taper of $\sim 5°$ and for a crack initiation shear stress of about 0.3 MPa (Picu and Gupta, 1995; Weiss and Schulson, 2000).

The next steps involve wing-crack initiation and growth. Across-column cracks serve as sliding surfaces, because they are generally inclined to both directions of loading. Sliding along them creates rather complicated tensile extensions. Because the parent does not penetrate all the way through the ice, but is generally buried, material surrounding the extensions experiences combined mode-II and mode-III loading. As a result, wings sprout and then wrap around the parent, in the manner described in Chapter 11 (see Figure 11.13c). The wings develop into blade-like cracks as sliding continues under increasing load. The blades then interact to create sheets of cracks, Figure 12.20b. The sheets generally initiate near the ice–platen interface, Figures 12.20c and 12.20d, because the shear stress that drives both grain boundary sliding and crack sliding is amplified there (Iliescu, 2000), owing to ice–platen friction. Once formed, the sheets define a set of thin plates, Figure 12.6b, fixed on one end and free on the other. Finally, under a sufficiently high load the plates become unstable, apparently through Euler buckling. The outermost ones fail first, leading to rapid collapse of the material.

To quantify the model, Iliescu and Schulson (2004) invoked a competition between wing-crack growth and Euler buckling. The driving stress for crack growth increases with crack length (Ashby and Hallam, 1986), while that for Euler buckling decreases with crack length (Roark and Young, 1975). Terminal failure is defined as the point at which the growth stress and buckling stress are equal, which occurs at the crack length $2L = 2L^*$. Figure 12.21 shows a schematic sketch: the height and width of the plates are proportional to the length, $L_{1,2}$, of the wing/blade cracks along directions X_1 and X_2; their thickness is set by the sheet-crack spacing, t_0, which is assumed to be set by the spacing of ledges along the grain boundaries; and their orientation is defined by the angle ψ. (In this depiction the length of the parent,

Caption for Figure 12.20. (cont.)
(a) grain boundary decohesion (whitish zones) and short across-column cracks; (b) cross section through two sheet-like cracks where black arrows point to parent, across-column cracks and white arrows point to wing-crack extensions that appear to have nucleated at the tips of the across-column parent cracks; (c) two sheet-like cracks that initiated near the ice–platen interface before terminal failure; and (d) the same two sheet-like cracks of (c) near terminal failure. Direction X_3 defines the long axis of the grains. (From Iliescu and Schulson, 2004.)

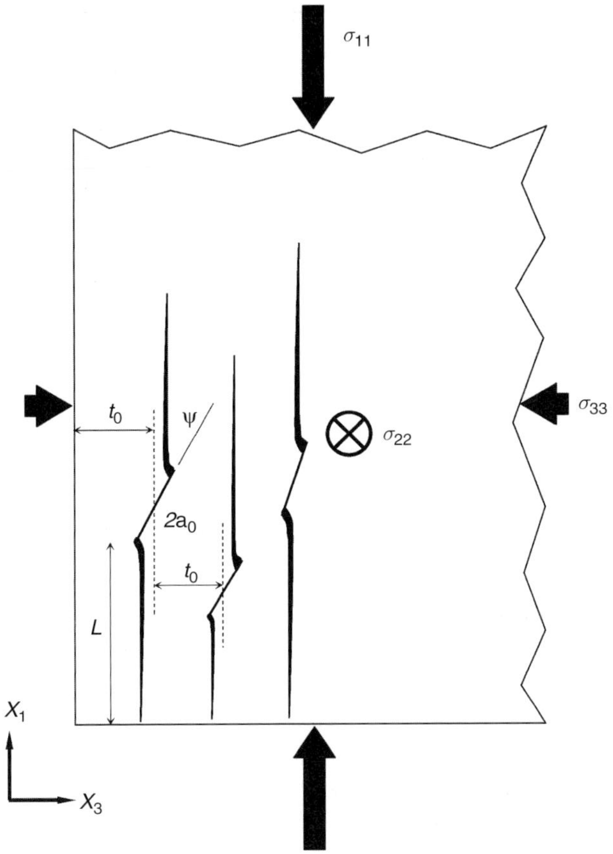

Figure 12.21. Schematic sketch of proposed spalling model. Stresses were applied along directions X_1 and X_2. The component of stress, σ_{33}, was induced through ice–platen interaction.

across-column sliding crack is denoted $2a_0$ instead of $2c$ as above, in deference to easier reference when consulting the original paper.) End constraint impedes wing-crack growth and is denoted by an along-column normal stress, σ_{33}, which is a fraction R_{31} of the major stress σ_{11}. L_2 is linked to L_1 because σ_{22} is linked to σ_{11} through the stress ratio $R_{21} = \sigma_{22}/\sigma_{11}$. Further details are given by Iliescu and Schulson (2004). In short, the model dictates that under biaxial loading along the path $R_{21} = 1$ the two driving stresses are equal when $2L^* = 54$ mm. Correspondingly, the failure stress is $\sigma_{11,\mathrm{f}} = 15$ MPa, Figure 12.22. These results are in reasonably good agreement with experiment, which shows that the plates reach a length of 30 mm to 50 mm (Figure 12.6b) under a balanced biaxial stress of 15 MPa to 20 MPa (Figure 12.5).

It should be noted that this apparent agreement rests to a great extent on the choice of the confining stress σ_{33}. Doubling its magnitude roughly doubles the failure

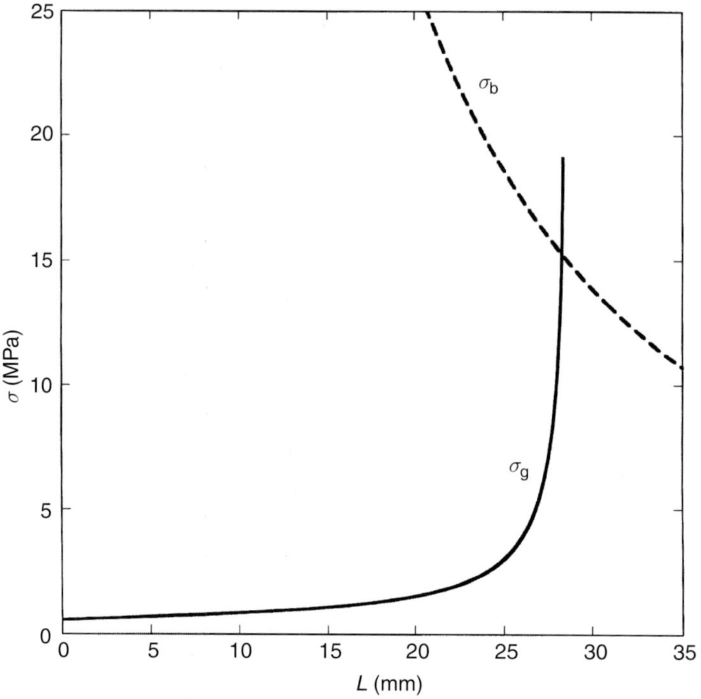

Figure 12.22. Graphs showing calculated stresses for wing-crack growth, σ_g, and buckling stress, σ_b, of a corresponding thin plate formed between two adjacent crack sheets vs. wing-crack length. The intersection is defined as the terminal failure stress. (From Iliescu and Schulson, 2004.)

stress. About all that is known of this parameter is that it must be very small; otherwise, plastic faulting would have set in. As a result of this and of other uncertainties, and of the application of two-dimensional wing-crack mechanics to a three-dimensional situation, caution is appropriate when applying the model.

12.10 Post-terminal failure of faulted ice

Upon reaching terminal failure, ice still possesses strength, as long as it remains confined. Our objective for the remainder of this chapter is to consider post-terminal failure strength, specifically of ice that has reached terminal failure through the development of one or more Coulombic faults. Again, frictional sliding is a dominant element. The results we describe are based primarily upon a systematic investigation by Fortt and Schulson (2007b). In that work, plate-shaped specimens of columnar-grained S2 fresh-water ice were proportionally loaded biaxially across the columns to terminal failure at −10 °C, and then, after taking care to avoid ice–platen interaction, slid across the fault under a variety of conditions (temperature: −3, −10 and −40 °C; sliding speed: $10^{-6} - 4 \times 10^{-3}$ m s^{-1}; normal stress across the fault: < 1 MPa).

12.10.1 Ductile-like and brittle-like sliding

Faulted ice exhibits two kinds of behavior, reminiscent of virgin material. When shortened slowly, it deforms in a ductile-like manner: the load increases pseudo-linearly, and then either levels off or, at slightly higher speeds, reaches a broad maximum before leveling off; correspondingly, the ice remains quiet to the unaided ear. On the other hand, when shortened rapidly, the material deforms in a brittle-like manner: the load reaches a sharp maximum and then quickly drops to almost zero, after which deformation occurs in a stick-slip manner; correspondingly, the ice is noisy. Within both regimes, inelastic deformation appears to be limited to sliding within the narrow zone of intense damage that constitutes the fault, evident from the fact that additional damage appears not to be generated within the ice on either side.

Interestingly, the equivalent strain rate that marks the transition from ductile-like to brittle-like sliding is almost the same as the strain rate that marks the ductile-to-brittle transition within virgin material (Chapter 13). Sliding occurs not across a sharply defined plane, but through shear deformation within a narrow zone of thickness w. The shear strain rate within that zone is estimated from the relationship $\dot{\gamma} = V_s/w$, where V_s is the sliding velocity and is given by the relationship $V_s = \dot{\varepsilon}_{11} l_o / \cos\,\theta$, where $\dot{\varepsilon}_{11}$ is the applied strain rate along the direction of shortening of the specimen, l_o is the height of the specimen and θ is the angle between the fault and the direction of shortening. The normal strain rate within the sliding fault is thus estimated from the relationship $\dot{\varepsilon}_f \approx \dot{\gamma}/2 = \dot{\varepsilon}_{11} l_o / 2w \cos\,\theta$. For fresh-water ice $w \sim 3d$ (Iliescu and Schulson, 2004) where d is the diameter of the columnar-shaped grains. For the specimen and fault geometry employed by Fortt and Schulson (2007b), where $l_o = 150$ mm, $d \sim 6$ mm and $\theta = 27°$, the transition at −10 °C occurred at a sliding velocity between $\sim 8 \times 10^{-5}$ m s^{-1} and $\sim 8 \times 10^{-4}$ m s^{-1}. That velocity corresponds to a normal strain rate within the fault of $\dot{\varepsilon}_{f,t} \sim 10^{-3}$ s^{-1} to 10^{-2} s^{-1}, close to the strain rate that marks the ductile-to-brittle transition within virgin ice when deformed under confinement at the same temperature (Chapter 13). A similar result was obtained from experiments on first-year arctic sea ice (Fortt, 2006). For that material, the transition strain rate is greater by about an order of magnitude, owing to its lower creep strength.

12.10.2 Failure envelope of faulted ice

Figure 12.23 shows the failure envelope for faulted ice of the kind described above, at −10 °C at a sliding velocity of 8×10^{-4} m s^{-1}. It is typical of all envelopes obtained

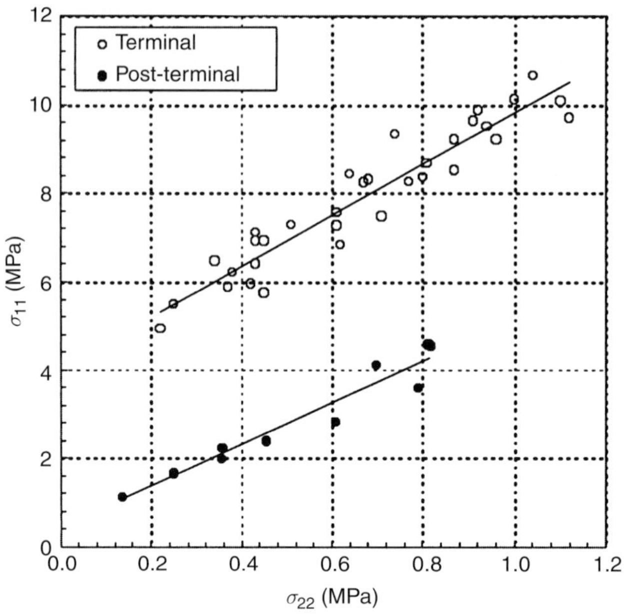

Figure 12.23. Post-terminal and terminal failure envelopes for columnar-grained, S2 fresh-water ice loaded proportionally across the columns at a strain rate of $\dot{\varepsilon}_{11} = 4 \times 10^{-3}\,\mathrm{s}^{-1}$ at $-10\,°\mathrm{C}$. Note the different scales on the abscissa and ordinate. (From Fortt and Schulson, 2007b.)

over the range of velocities from $8 \times 10^{-7}\,\mathrm{m\,s^{-1}}$ to $4 \times 10^{-3}\,\mathrm{m\,s^{-1}}$ at $-3\,°\mathrm{C}$ to $-40\,°\mathrm{C}$ from both fresh-water ice and first-year arctic sea ice. Note that when unconfined the strength of the ice, taken as the value of σ_{11}, is close to but not exactly zero. The non-zero value reflects moderate healing. Upon increasing confinement, the strength increases, scaling linearly with confining stress. A linear relationship was evident for all 15 combinations of temperature and sliding velocity that were examined, with a correlation coefficient that generally exceeded $r^2 = 0.92$ (Fortt and Schulson, 2007b). Such correlations were derived from envelopes that contained at least ten measurements, each corresponding to a different loading path. The linear dependence implies that faulted ice, like its virgin predecessor, obeys Coulomb's failure criterion (Equation 12.5). Note also that the slope of the post-terminal envelope is almost the same as the slope of the Coulombic segment of the failure envelope for virgin material (Figure 12.5), shown again in Figure 12.23. This implies that the coefficient of friction for sliding across the fault (more below) is almost identical to the internal coefficient of friction that resisted the formation of the fault in the first place.

Sliding distance has little effect, if any, on the slope of the post-failure envelope, at least over small distances. For instance, envelopes for sliding up to 8 mm are almost identical to the one shown above, which was obtained at the onset of sliding. This implies that the friction coefficient is essentially independent of displacement, at least of the small displacements examined by Fortt and Schulson (2007b). The only detectable difference is that cohesion tends to zero, particularly at higher sliding speeds and lower temperatures where *dynamic healing* becomes negligible.

On healing per se, preliminary experiments (Fortt, 2006) show that a hold time of up to 100 days at −10 °C does not significantly affect the slope of the failure envelope. However, it does raise the cohesive strength somewhat (i.e., the strength under zero confinement). Although the effect remains to be quantified, it implies that in practical situations involving damaged and partially healed ice, there exists a set of failure envelopes nested within each other. We return to this point in Chapter 15 where we consider the deformation of the faulted, arctic sea ice cover.

12.10.3 Coefficient of friction on Coulombic faults

To obtain the coefficient of friction for sliding across the fault μ_f, Fortt and Schulson (2007b) transformed the applied normal stresses at post-terminal failure to the shear stress τ and the normal stress σ_n acting on the plane of the fault and then obtained the coefficient through the derivative $\mu_f = d\tau/d\sigma_n$. Figure 12.24 shows the deduced values vs. sliding velocity and temperature. Included for completeness is the cohesive strength. Note that the friction coefficient increases with increasing velocity at lower speeds where sliding exhibits quiet, ductile-like behavior, but decreases with increasing velocity at higher speeds where sliding exhibits noisy, brittle-like behavior. Note also that the coefficient generally increases with decreasing temperature.

12.10.4 Interpretation

An increasing coefficient of friction with increasing sliding velocity at low velocities has been observed by other researchers, albeit for polished ice-on-ice (Kennedy *et al.*, 2000; Montagnat and Schulson, 2003), for ice-on-granite and ice-on-glass (Barnes *et al.*, 1971). To account for this behavior Kennedy *et al.* (2000) proposed that friction at low velocities is controlled by creep. This explanation is applicable to sliding along a Coulombic fault as well. Although all three stages of creep are probably involved, only secondary creep is invoked, where the creep rate, $\dot{\varepsilon}$, is related to an effective stress $\bar{\sigma}$ and to temperature T through the power-law relationship (Chapter 6) $\dot{\varepsilon} \approx B\bar{\sigma}^n e^{-Q/RT}$ where B is a material-dependent parameter, n is the stress exponent, Q is the apparent activation energy, and R is the universal gas constant. If the friction coefficient

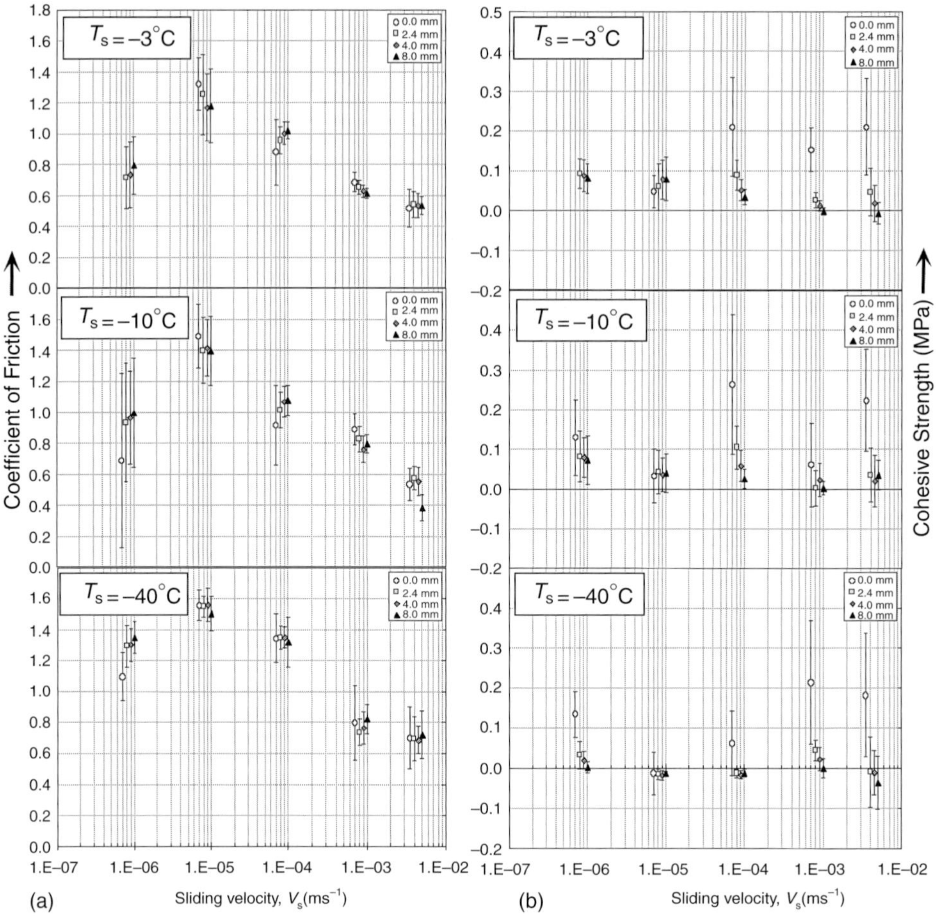

Figure 12.24. Graphs showing (a) the kinetic coefficient of friction across Coulombic shear faults and (b) the internal cohesion vs. sliding velocity at −3, −10 and −40 °C for displacements across the fault from 0 to 8 mm. Note: points are displaced on velocity. (From Fortt and Schulson, 2007b.)

is considered to be proportional to the effective stress, then it follows that $\mu_f \propto \bar{\sigma} \propto (V_s/wB)^{1/n} e^{Q/nRT}$. At low stress rates $n \sim 3$ (Chapter 6), but at the higher stress that are expected to exist within the damaged material of the loaded fault the stress dependence is expected to be greater, possibly as high as $n \sim 5$ to 10 when cognizance is of the high-stress creep measurements of Barnes *et al.* (1971). Thus, over the range of velocities that define the rising branch of the μ_f:V_s curve (i.e., $8 \times 10^{-7}\,\mathrm{m\,s^{-1}}$ to $8 \times 10^{-6}\,\mathrm{m\,s^{-1}}$) this interpretation leads to the expectation that the friction coefficient should increase by a factor of $10^{1/10}$ to $10^{1/5} = 1.3$ to 1.6, in reasonable agreement with observation (Figure 12.24). This interpretation is also in rough agreement with the observed increase in friction with decreasing temperature: taking the values $Q = 78\,\text{kJ mol}^{-1}$ (Chapter 6) and $n \sim 5$ to 10, the coefficient of friction is expected to increase by a factor between 1.6 and 2.5 over the temperature range −10 °C to −40 °C, similar to but greater than the observed increase.

The falling branch of the μ_f:V_s curve has a different origin. Again, an effect of this kind has been observed by a number of investigators from experiments on prepared surfaces (Bowden and Hughes, 1939; Barnes *et al.*, 1971; Tusima, 1977; Colbeck, 1995; Kennedy *et al.*, 2000). The interpretation that is generally advanced relates the reduction in friction to a thin layer of water that forms and then lubricates the sliding interface. Calculations that take into account frictional heating, flash temperature rise, and pressure melting indicate that surface melting should occur at −10 °C at sliding speeds $\geq 10^{-2}\,\text{m s}^{-1}$ (Fortt and Schulson, 2007b). This is higher than the sliding velocities where the coefficient of friction begins to decrease. However, this calculated velocity is for complete melting of the sliding surface. There could still be small areas where melting takes place locally. That localized melting (followed by freezing) may actually occur is evident from globular and somewhat porous features that form during sliding at a higher velocity (Fortt and Schulson, 2007b). In addition to melt-water lubrication – or even in its absence – the other factor that may contribute to the reduction in friction is the reduction upon frictional heating of the compressive strength of contacting asperities.

On the magnitude of the friction coefficient itself, it is interesting to compare values from faulted surfaces to those obtained from polished surfaces, which are summarized in Chapter 4. In those cases, the surface roughness was ~2 μm (Kennedy *et al.*, 2000). In the present cases, the surface roughness[4] for the fault was closer to 2 mm (Fortt and Schulson, 2007). Under every condition examined, the friction coefficient across the fault is greater by a factor of two or more. We attribute this difference to a greater interlocking of asperities.

12.10.5 Non-linear friction?

Before closing, we return to the dependence of post-terminal strength on confining stress. Fortt and Schulson (2007b) observed a linear relationship, as noted above. In comparison, Rist *et al.* (1994), Rist (1997) and Sammonds *et al.* (1998)[5] reported a non-linear relationship. These investigators performed post-terminal triaxial

[4] Roughness of a natural surface is a fractal parameter, and so depends upon the method of measurement. Fortt and Schulson (2007b) discuss this point.

[5] There appears to be an error in the friction coefficient reported in Table 4 of Sammonds *et al.* (1998). In obtaining the value listed, instead of dividing the shear stress on the plane of the fault by the total normal stress acting across the fault, the shear stress appears to have been divided by only part of the normal stress, thus giving values that are too high. That is, the coefficient appears to have been calculated from the ratio $(\sigma_1 - \sigma_3)/(\sigma_1 - \sigma_3 + \sigma_3)$ instead of from the ratio $(\sigma_1 - \sigma_3)/(\sigma_1 + \sigma_3)$. Thus, for instance, the coefficient 0.55 changes to 0.38, and so on.

compression tests on both fresh-water granular ice and multi-year sea ice deformed at a strain rate of $\dot{\varepsilon}_1 = 10^{-2}\,s^{-1}$ at temperatures from −10 °C to −40 °C. They found that when the shear stress acting on the fault is plotted against the normal stress acting on the fault the relationship is of the form $\tau \propto (\bar{\sigma})^{2/3}$. Moreover, when they combined their results with those of Beeman *et al.* (1988), who performed sliding experiments across saw-cuts in fresh-water granular ice at temperatures from −196 °C to −158 °C, the data fit upon a single curve (Rist *et al.*, 1994) that is characterized by the non-linear functionality noted. Beeman's experiments were performed not only on colder ice, but also at lower sliding speeds. That his results fit upon Rist's curve implies that the friction coefficient so determined under the conditions of those experiments depends upon the normal stress, but not upon either temperature or sliding speed. This is quite different from the behavior described above where temperature and sliding speed have major effects.

Several factors may account for this difference. One is the character of the fault. Fortt and Schulson examined Coulombic faults that were inclined by 27 ± 4° to the direction of shortening. Rist and Sammonds appear to have examined plastic faults, judging from their ~45° inclination. Another factor is the magnitude of the applied stress. The across-fault normal stress in the experiments of Fortt and Schulson was $\sigma_n < 1$ MPa. In the Rist and the Sammonds experiments the normal stress was more than an order of magnitude greater, $\sigma_n \sim 10$–100 MPa. So, too, was the normal stress in Beeman's experiments, where $\sigma_n \sim 5$–200 MPa. The other factor is that the non-linear relationship was based upon relatively few measurements.

Given the important role of friction in brittle compressive failure, more work is needed to better understand this property. We return to the subject in Chapter 15, where we consider the sea ice cover on the Arctic Ocean.

References

Ashby, M. F. and S. D. Hallam (1986). The failure of brittle solids containing small cracks under compressive stress states. *Acta Metall.*, **34**, 497–510.

Backofen, W. A. (1972). *Deformation Processing*. Reading, Mass.: Addison-Wesley Publishing Co.

Barnes, P., D. Tabor, F. R. S. Walker and J. F. C. Walker (1971). The friction and creep of polycrystalline ice. *Proc. R. Soc. Lond. A*, **324**, 127–155.

Bazant, Z. P. and Y. Xiang (1997). Size effect in compression fracture: Splitting crack band propagation. *J. Eng. Mech.*, February, 162–172.

Beeman, M., W. B. Durham and S. H. Kirby (1988). Friction of ice. *J. Geophys. Res.*, **93**, 7625–7633.

Bowden, F. P. and T. P. Hughes (1939). The mechanism of sliding on ice and snow. *Proc. R. Soc. A*, **172**, 280–298.

Brace, W. F. and E. G. Bombolakis (1963). A note of brittle crack growth in compression. *J. Geophys. Res.*, **68**, 3709–3713.

Byerlee, J. D. (1967). Frictional characteristics of granite under high confining pressure. *J. Geophys. Res.*, **72**, 3639–3648.

Colbeck, S. C. (1995). Pressure melting and ice skating. *Amer. J. Phys.*, **63** (10), 888–890.

Conrad, R. E., II and M. Friedman (1976). Microscopic feather fractures in the faulting process. *Tectonophysics*, **33**, 187–198.

Cooke, M. L. (1997). Fracture localization along faults with spatially varying friction. *J. Geophys. Res.*, **102**, 24,425–24,434.

Costamagna, R., J. Renner and O. T. Bruhns (2007). Relationship between fracture and friction for brittle rocks. *Mech. Mater.*, **39**, 291–301.

Cottrell, A. H. (1964). *The Mechanical Properties of Matter.* Wiley Series on the Science and Technology of Materials. New York: John Wiley & Sons, Inc.

Couture, M. L. and E. M. Schulson (1994). The cracking of ice under rapid unloading. *Phil. Mag. Lett.*, **69**, 865–886.

Cruikshank, K. M., G. Zhao and A. Johnson (1991). Analysis of minor fractures associated with joints and faulted joints. *J. Struct. Geol.*, **13**, 865–886.

Davies, R. K. and D. D. Pollard (1986). Relations between left-lateral strike-slip faults and right-lateral kink bands in granodiorite, Mt. Abbot Quadrangle, Sierra Nevada, California. *Pure Appl. Geophys.*, **124**, 177–201.

Durham, W. B., H. C. Heard and S. H. Kirby (1983). Experimental deformation of polycrystalline H_2O ice at high pressure and low temperature: Preliminary results. *J. Geophys. Res.*, **88**, B377–B392.

Fortt, A. (2006). The resistance to sliding along coulombic shear faults in columnar S2 ice. Ph.D. thesis, Thayer School of Engineering, Dartmouth College.

Fortt, A. and E. M. Schulson (2007a). Do loading path and specimens thickness affect the brittle compressive strength of ice? *J. Glaciol.*, **53**, 305–309

Fortt, A. L. and E. M. Schulson (2007b). The resistance to sliding along coulombic shear faults in ice. *Acta Mater.*, **55**, 2253–2264.

Frederking, R. (1977). Plane-strain compressive strength of columnar-grained and granular-snow ice. *J. Glaciol.*, **18**, 505–516.

Frost, H. J. and M. F. Ashby (1982). *Deformation Mechanisms Maps*. Oxford: Permagon Press.

Gagnon, R. E. and P. H. Gammon (1995). Triaxial experiments on iceberg and glacier ice. *J. Glaciol.*, **41**, 528–540.

Golding, N. (2009). M.Sc. thesis. Hanover, Dartmouth College (in preparation).

Gottschalk, R. R., A. K. Kronenberg, J. E. Russel and J. Handin (1990). Mechanical anisotropy of gneiss: Failure criterion and textural sources of directional behavior. *J. Geophys. Res.*, **95**, 613–621.

Granier, T. (1985). Origin, damping and pattern development of faults in granite. *Tectonics*, **4**, 721–737.

Gratz, E. T. and E. M. Schulson (1997). Brittle failure of columnar saline ice under triaxial compression. *J. Geophys. Res.*, **102**, 5091–5107.

Griggs, D. T. and D. W. Baker (1969). The origin of deep focus earthquakes. In *Properties of Matter Under Unusual Conditions*, eds. H. Mark and S. Fernback. Hoboken, N.J.: Wiley Interscience, pp. 23–42.

Gupta, V. and J. S. Bergström (1998). Compressive failure of rocks by shear faulting. *J. Geophys. Res.*, **103**, 23,875–23,895.

Haimson, B. (2006). True triaxial stresses and the brittle fracture of rock. *Pure Appl. Geophys*. **163**, 1101–1130.

Häusler, F. U. (1981). Multiaxial compressive strength tests on saline ice using brush-type loading platens. *IAHR Ice Symposium*, Quebec, Canada.

Häusler, F. U. (1989). Beitrag zur ermittlung der kräfte beim eisbrechen under besonderere berücksichtigung der anisotropie des eises under seinter versagenseigenschaften under mehrachsiger beanspruchung. *Bericht Nr. 494*. Hamburg: Institut für Schiffbau der Universität Hamburg, 142.
Hill, R. (1950). *The Mathematical Theory of Plasticity*. New York: Oxford University Press.
Hobbs, B. E. and A. Ord (1988). Plastic instabilities: Implications for the origin of intermediate and deep focus earthquakes. *J. Geophys. Res.*, **93**, 10,521–10,540.
Horii, H. and S. Nemat-Nasser (1985). Compression-induced microcrack growth in brittle solids: Axial splitting and shear failure. *J. Geophys. Res.*, **90**, 3105–3125.
Horii, H. and S. Nemat-Nasser (1986). Brittle failure in compression: Splitting, faulting and brittle-ductile transition. *Phil. Trans. R. Soc. A*, **319**, 337–374.
Iliescu, D. (2000). Contributions to brittle compressive failure of ice. Ph.D. thesis, Thayer School of Engineering, Dartmouth College.
Iliescu, D. and E. M. Schulson (2004). The brittle compressive failure of fresh-water columnar ice loaded biaxially. *Acta Mater.*, **52**, 5723–5735.
Ingraffea, A. R. (1981). Mixed-mode fracture initiation in Indiana limestone and westerly granite. *Proceedings 22nd U.S. Symposium on Rock Mechanics*, 199–204.
Jaeger, J. C. and N. G. W. Cook (1979). *Fundamentals of Rock Mechanics*, 3rd edn. London: Chapman and Hall.
Johnson, K. L. (1985). *Contact Mechanics*. Cambridge: Cambridge University Press.
Jones, S. J. (1982). The confined compressive strength of polycrystalline ice. *J. Glaciol.*, **28**, 171–177.
Jordaan, I. J. (2001). Mechanics of ice-structure interaction. *Eng. Fract. Mech.*, **68**, 1923–1960.
Jordaan, I. J., D. G. Matskevitch and I. L. Meglis (1999). Disintegration of ice under fast compressive loading. *Int. J. Fract.*, **97**, 279–300.
Karato, S., M. R. Riedel and D. A. Yuen (2001). Rheological structure and deformation of subducted slabs in the mantle transition zone: Implications for mantle circulation and deep earthquakes. *Phys. Earth Planet. Inter.*, **127**, 83–108.
Kennedy, F. E., E. M. Schulson and D. Jones (2000). Friction of ice on ice at low sliding velocities. *Phil. Mag. A*, **80**, 1093–1110.
Kirby, S. H. (1987). Localized polymorphic phase transformations in high-pressure faults and applications to the physical mechanism of deep earthquakes. *J. Geophys. Res.*, **92**, 13,789–13,800.
Kuehn, G. A. and E. M. Schulson (1994). The mechanical properties of saline ice under uniaxial compression. *Ann. Glaciol.*, **19**, 39–48.
Liu, F., I. Baker and M. Dudley (1995). Dislocation-grain boundary interactions in ice crystals. *Phil. Mag. A*, **71**, 15–42.
Liu, C. T., L. Heatherly, D. S. Easton *et al.* (1998). Test environments and mechanical properties of Zr-base bulk amorphous alloys. *Metall. Mater. Trans. A*, **29**, 1811–1820.
Lliboutry, L. (2002). Overthrusts due to easy-slip/poor-slip transitions at the bed: The mathematical singularity with non-linear isotropic viscosity. *J. Glaciol.*, **48**, 109–119.
Martel, S. M. and D. D. Pollard (1989). Mechanics of slip and fracture along small faults and simple strike-slip fault zones in granitic rock. *J. Geophys. Res.*, **94**, 9417–9428.
Martel, S. J., D. D. Pollard and P. Segall (1988). Development of simple strike-slip fault zones in granitic rock, Mount Abbott Quadrangle, Sierra Nevada, California. *Geol. Soc. Am. Bull.*, **100**, 1451–1465.
Meglis, I. L., P. M. Melanson and I. J. Jordaan (1999). Microstructural change in ice: II. Creep behavior under triaxial stress conditions. *J. Glaciol.*, **45**, 438–448.

Melanson, P. M., I. L. Meglis, I. J. Jordaan and B. M. Stone (1999). Microstructural change in ice: I. Constant-deformation-rate tests under triaxial stress conditions. *J. Glaciol.*, **45**, 417–455.

Melton, J. S. and E. M. Schulson (1998). Ductile compressive failure of columnar saline ice under triaxial loading. *J. Geophys. Res.*, **103**, 21,759–21,766.

Montagnat, M. and E. M. Schulson (2003). On friction and surface cracking during sliding. *J. Glaciol.*, **49**, 391–396.

Nadreau, J.-P. and B. Michel (1986). Yield and failure envelope for ice under multiaxial compressive stresses. *Cold Reg. Sci. Technol.*, **13**, 75–82.

Nickolayev, O. Y. and E. M. Schulson (1995). Grain-boundary sliding and across-column cracking in columnar ice. *Phil. Mag. Lett.* **72**, 93–97.

Ogawa, M. (1987). Shear instability in a viscoelastic material as the cause for deep focus earthquakes. *J. Geophy. Res.*, **92**, 13,801–13,810.

Orowan, E. (1960). Mechanism of seismic faulting. In *Rock Deformation*, eds. D. T. Griggs and J. Handin. New York: Memoirs of the Geological Society of America, pp. 323–345.

Paterson, M. S. and T.-F. Wong (2005). *Experimental Rock Deformation: The Brittle Field*, 2nd edn. New York: Springer-Verlag.

Peng, S. and A. M. Johnson (1972). Crack growth and faulting in cylindrical specimens of chelmsford granite. *Int. J. Rock. Mech. Min. Sci.*, **9**, 37–86.

Petrenko, V. F. and R. W. Whitworth (1999). *Physics of Ice*. New York: Oxford University Press.

Picu, R. C. and V. Gupta (1995). Crack nucleation in columnar ice due to elastic anistropy and grain boundary sliding. *Acta Metall. Mater.*, **43**, 3783–3789.

Pollard, D. D. and P. Segall (1987). Theoretical displacements and stresses near fractures in rocks: With applications to faults, joints, veins, dikes, and solution surfaces. In *Fracture Mechanics of Rock*, ed. B. K. Atkinson. San Diego: Academic Press, pp. 277–349.

Renshaw, C. E. and E. M. Schulson (2001). Universal behavior in compressive failure of brittle materials. *Nature*, **412**, 897–900.

Renshaw, C. E. and E. M. Schulson (2004). Plastic faulting: Brittle-like failure under high confinement. *J. Geophys. Res.*, **109**, 1–10.

Richter-Menge, J. (1991). Confined compressive strength of horizontal first-year sea ice samples. *J. Offshore Mech. Arctic Eng.*, **113**, 344–351.

Richter-Menge, J. A. and K. F. Jones (1993). The tensile strength of first-year sea ice. *J. Glaciol.*, **39**, 609–618.

Rispoli, R. (1981). Stress fields about strike-slip faults from stylolites and tension gashes. *Tectonophysics*, **75**, T29–T36.

Rist, M. A. (1997). High stress ice fracture and friction. *J. Phys. Chem. B*, **101**, 6263–6266.

Rist, M. A. and S. A. F. Murrell (1994). Ice triaxial deformation and fracture. *J. Glaciol.*, **40**, 305–318.

Rist, M. A., S. J. Jones and T. D. Slade (1994). Microcracking and shear fracture in ice. *Ann. Glaciol.*, **19**, 131–137.

Roark, R. J. and W. C. Young (1975). *Formulas for Stress and Strain*, 5th edn. New York: McGraw-Hill Book Co., p. 550.

Rudnicki, J. W. and J. R. Rice (1975). Conditions for the localization of deformation in pressure-sensitive dilatant materials. *J. Mech. Phys. Sol.*, **23**, 371–394.

Sammis, C. G. and M. F. Ashby (1986). The failure of brittle porous solids under compressive stress states. *Acta Metall.*, **34**, 511–526.

Sammonds, P. R. and M. A. Rist (2001). Sea ice fracture and friction. In *Scaling Laws in Ice Mechanics*, eds. J. P. Dempsey and H. H. Shen. Dordrecht: Kluwer Academic Publishing, pp. 183–194.

Sammonds, P. R., S. A. F. Murrell and M. A. Rist (1989). Fracture of multi-year sea ice under triaxial stresses: Apparatus description and preliminary results. *J. Offshore Mech. Arctic Eng.*, **111**, 258–263.

Sammonds, P. R., S. A. F. Murrell and M. A. Rist (1998). Fracture of multi-year sea ice. *J. Geophys. Res.*, **103**, 21,795–21,815.

Schulson, E. M. (2001). Brittle failure of ice. *Eng. Fract. Mech.*, **68**, 1839–1887.

Schulson, E. M. (2002). Compressive shear faulting in ice: Plastic vs. Coulombic faults. *Acta Mater.*, **50**, 3415–3424.

Schulson, E. M. and S. E. Buck (1995). The brittle-to-ductile transition and ductile failure envelopes of orthotropic ice under biaxial compression. *Acta Metall. Mater.*, **43**, 3661–3668.

Schulson, E. M. and E. T. Gratz (1999). The brittle compressive failure of orthotropic ice under triaxial loading. *Acta Mater.*, **47**, 745–755.

Schulson, E. M. and D. Iliescu (2006). Brittle compressive failure of ice: Proportional straining vs. proportional loading. *J. Glaciol.*, **52**, 248–250.

Schulson, E. M. and O. Y. Nickolayev (1995). Failure of columnar saline ice under biaxial compression: failure envelopes and the brittle-to-ductile transition. *J. Geophys. Res.*, **100**, 22,383–22,400.

Schulson, E. M., D. E. Jones and G. A. Kuehn (1991). The effect of confinement on the brittle compressive fracture of ice. *Ann. Glaciol.*, **15**, 216–221.

Schulson, E. M., D. Iliescu and C. E. Renshaw (1999). On the initiation of shear faults during brittle compressive failure: A new mechanism. *J. Geophys. Res.*, **104**, 695–705.

Schulson, E. M., A. Fortt, D. Iliescu and C. E. Renshaw (2006a). Failure envelope of first-year arctic sea ice: The role of friction in compressive fracture. *J. Geophys. Res.*, **111**, doi: 10.1029/2005JC003234186.

Schulson, E. M., A. Fortt, D. Iliescu and C. E. Renshaw (2006b). On the role of frictional sliding in the compressive fracture of ice and granite: Terminal vs. post-terminal failure. *Acta Mater.*, **54**, 3923–3932.

Scott, T. E. and K. C. Nielsen (1991a). The effects of porosity on the brittle-ductile transition in sandstones. *J. Geophys. Res.*, **96**, 405–414.

Scott, T. E. and K. C. Nielsen (1991b). The effects of porosity on fault reactivation in sandstones. *J. Geophy. Res.*, **96**, 2352–2362.

Segall, P. and D. D. Pollard (1983). Nucleation and growth of strike-slip faults in granite. *J. Geophys. Res.*, **88**, 555–568.

Shen, W. and S. Z. Lin (1986). Fracture toughness of Bohai Bay sea ice. *5th International Offshore Mechanics and Arctic Engineering Symposium*, OMAE-AIME.

Smith, T. R. and E. M. Schulson (1993). The brittle compressive failure of fresh-water columnar ice under biaxial loading. *Acta Metall. Mater.*, **41**, 153–163.

Smith, T. R. and E. M. Schulson (1994). The brittle compressive failure of columnar salt-water ice under biaxial loading. *J. Glaciol.*, **40**, 265–276.

Taylor, K. (2005). Faulting under high-confinement conditions: An experimental study of compressive failure in granular ice. M.Sc. thesis, Earth Sciences, Dartmouth College, 87.

Thouless, M. D., A. G. Evans, M. F. Ashby and J. W. Hutchinson (1987). The edge cracking and spalling of brittle plates. *Acta. Metall.*, **35**, 1333–1341.

Timco, G. W. and R. M. W. Frederking (1986). Confined compression tests: Outlining the failure envelope of columnar sea ice. *Cold Reg. Sci. Technol.*, **12**, 13–28.

Tusima, K. (1977). Friction of a steel ball on a single crystal of ice. *J. Glaciol.*, **19**, 225–235.

von Mises, R. (1928). Mechanics of the ductile form changes of crystals. *Z. Angew. Math Mech.*, **8**, 161–185.

Wachter, L., C. E. Renshaw and E. M. Schulson (2008). Transition in brittle failure mode in ice under low confinement. *Acta Mater.*, in press.

Weiss, J. and E. M. Schulson (1995). The failure of fresh-water granular ice under multiaxial compressive loading. *Acta Metall. Mater.*, **43**, 2303–2315.

Weiss, J. and E. M. Schulson (2000). Grain boundary sliding and crack nucleation in ice. *Phil. Mag.*, **80**, 279–300.

Weiss, J., E. M. Schulson and H. L. Stern (2007). Sea ice rheology in-situ, satellite and laboratory observations: Fracture and friction. *Earth Planet. Sci. Lett.*, doi: 10.1016/j.epsl.2006.11.033.

Wiens, D. A. (2001). Seismological constraints on the mechanism of deep earthquakes: Temperature dependence of deep earthquake source properties. *Phys. Earth Planet. Inter.*, **127**, 145–163.

Winter, R. E. (1975). Adiabatic shear of titanium and polymethylmethacrylate. *Phil. Mag.*, **31**, 765–773.

Wong, T.-F. (1982). Micromechanics of faulting in Westerly granite. *Int. J. Rock Mech. Min. Sci.*, **19**, 49–64.

13

Ductile-to-brittle transition under compression

13.1 Introduction

Our objective in this chapter is to consider the ductile-to-brittle transition within polycrystalline ice loaded under compression. We describe a physical model that accounts quantitatively for the effects (described below) of strain rate, temperature, grain size, confinement, salinity and damage and which accounts as well for the transition within a number of rocks and minerals. We express the transition in terms of a critical strain rate, although it could just as well be expressed in terms of other factors. However, strain rate seems to be the most useful, because it spans the widest range of environmental and microstructural conditions encountered in practice.

The critical strain rate marks the point where compressive strength (Chapters 11 and 12) and indentation pressure (Chapter 14) reach a maximum. It is thus a point of considerable importance in relation to ice loads on offshore structures. It is of interest as well in relation to the rheological behavior of the sea ice cover on the Arctic Ocean (Chapter 15) and to the deformation of icy material of the outer Solar System.

13.2 Competition between creep and fracture

Three explanations have been offered. An early view, particularly in relation to the transition during indentation failure in the field (Sanderson, 1988), was that the transitional strain rate marks the conditions under which the first cracks nucleate, somewhat analogous to the transition under tension. Under compression, however, cracks form early in the deformation history at strain rates two to three orders of magnitude lower than the transition rate. Crack nucleation itself, therefore, cannot be the explanation. Another view (Gold, 1997), based upon the observation that the relative number of intergranular to transgranular cracks increases with increasing strain rate, was that the transition coincides in some way with a certain proportion of intergranular cracks. The problem with that interpretation is that the strain rate marking any given ratio of grain boundary cracks to total cracks increases with increasing grain size (Gold, 1997), implying that the transition strain rate should increase with increasing grain size. Ice exhibits the opposite effect (Batto and Schulson, 1993), as will become apparent.

The other explanation, and the one we develop below, is based upon a competition between crack blunting and crack propagation (Schulson, 1990; Renshaw and Schulson, 2001). Accordingly, the ductile-to-brittle transition marks the point where stress buildup dominates stress relaxation to the point that the mode-I stress intensity factor reaches the critical level. The propagation that ensues occurs in a macroscopically stable or quasi-stable manner – i.e., the applied load must increase – because under compressive stress states the strain energy release rate is lower than the increase in the resistance to crack growth; i.e., $dG/dc < dR/dc$ (Chapter 9). Direct support for this view is the observation (Batto and Schulson, 1993) that wing cracks that are initiated from the tips of artificial "Teflon cracks" – i.e., from strips of Teflon inclined to the maximum compressive stress – do not propagate as the load increases at strain rates within the ductile regime, but do propagate when the "pre-cracked" ice is loaded at strain rates within the brittle regime. Similarly, comb cracks do not develop on the ductile side of the transition (Iliescu, 2000), presumably because stress relaxation prevents the development of sufficient tensile stresses necessary for their formation. As we show below, a model based upon the competition between creep and fracture captures the transition rather well.

13.3 Micromechanical model

Imagine, as we did in Chapter 11, an infinite plate containing a primary crack of length $2c$ subjected to principal stresses σ_1 and σ_3 where, again, σ_1 is taken to be the highest compressive stress and σ_3 to be the least compressive stress and both stresses are taken to be positive. (For S2 columnar ice loaded biaxially across the columns under a moderate confining stress, σ_3 is replaced by σ_{22} and σ_1 by σ_{11}.) The plane of the crack is inclined at an angle ψ to the X_1 axis, Figure 13.1. The applied stress field generates a shear stress τ on the plane of the crack and a normal stress σ_n perpendicular to the crack plane, given by:

$$\tau = \frac{\sigma_1 - \sigma_3}{2}\sin(2\psi) \tag{13.1}$$

and

$$\sigma_n = \frac{\sigma_1 + \sigma_3}{2} + \frac{\sigma_3 - \sigma_1}{2}\cos(2\psi). \tag{13.2}$$

The shear stress tends to cause the opposing faces of the crack to slide, while the normal stress helps the opposing faces to be in contact. The product $\sigma_n\mu$ resists

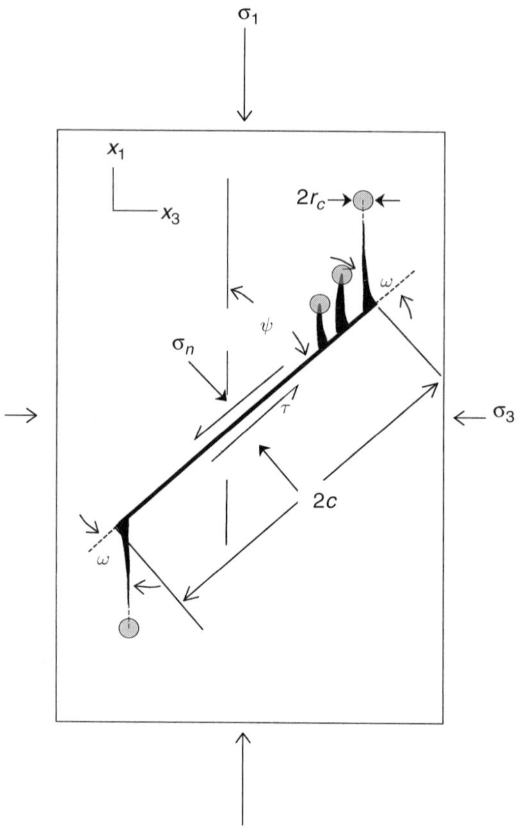

Figure 13.1. Schematic sketch of a primary crack of length $2c$ inclined by an acute angle ψ to the direction X_1 along which the maximum principal stress, σ_1, acts. Out-of-plane secondary cracks stem from the primary crack at an acute angle ω. A creep zone, shaded, of radius r_c is located at the tips of the secondary cracks.

sliding where μ denotes the friction coefficient. The appropriate friction coefficient, we believe, is not the internal coefficient that relates to terminal failure, described in Chapter 12, but the kinetic coefficient that relates to an earlier state in the deformation history, as described in Chapter 4. The excess or effective shear stress τ^* that effects sliding is given by:

$$\tau^* = \tau - \mu\sigma_n. \tag{13.3}$$

The crack intensifies τ^* and this leads to a singular stress field characterized by the quantity $\tau^*\sqrt{\pi c}$ (Ashby and Hallam, 1986; Horii and Nemat-Nasser, 1986). The field is mainly shear in the plane of the primary crack, but contains a normal component on planes that are inclined by the acute angle ω to the crack plane. This normal stress tends to cause mode-I secondary cracks to grow from the primary crack tips, whose stress intensity factor is given by the relationship (Horii and Nemat-Nasser, 1986):

$$K_{\mathrm{I}} = \frac{3}{4}\tau^{*}\sqrt{\pi c}(\sin\omega/2 + \sin 3\omega/2). \tag{13.4}$$

Primary cracks tend to form on grain boundaries inclined by $\psi \sim 45°$ (Cannon *et al.* 1990). Taking that value for ω as well and setting σ_3 to be proportional to σ_1 through the loading path $R = \sigma_3/\sigma_1$, the stress intensity factor becomes:

$$K_{\mathrm{I}} \approx \frac{\sigma_1}{2}\sqrt{\pi c}[(1 - R) - \mu(1 + R)]. \tag{13.5}$$

Two details should be noted. First, the Ashby–Hallam (1986) analysis leads to a factor of $\sqrt{3}$ instead of the factor 2 in Equation (13.5), a difference we consider to be insignificant given the approximations in the analysis. Second, if instead of setting $\psi = \omega = \pi/4$ we first find the plane on which the tensile stress is a maximum and then find the orientation of the plane on which the excess shear stress is a maximum, then:

$$K_{\mathrm{I}} \approx \frac{\sigma_1}{2}\sqrt{\pi c}\left[(1 - R)\left(1 + \mu^2\right)^{1/2} - \mu(1 + R)\right]. \tag{13.6}$$

Under the moderate confinement ($R \leq 0.3$) of interest here – i.e., where frictional sliding is not suppressed (Chapter 12) – Equations (13.5) and (13.6) generate similar values. We adopt the simpler expression in the analysis below.

As sliding proceeds, K_{I} increases. Imagine that this occurs in the absence of creep. When the local stress reaches the elastic limit σ_{e}, the material deforms inelastically. The radius of the inelastic region under plane strain conditions may be estimated from the relationship (Anderson, 1995):

$$r_{\mathrm{e}} \approx \frac{1}{6\pi}\left(\frac{K_{\mathrm{I}}}{\sigma_{\mathrm{e}}}\right)^2. \tag{13.7}$$

Upon equating the elastic limit to tensile strength, which we take from Chapter 10 to be given by $\sigma_{\mathrm{t}} = K_{\mathrm{Ic}}/Y\sqrt{c}$, where again K_{Ic} is the critical stress intensity factor and Y denotes a geometrical factor that we take to be of order unity, the size of the inelastic zone may be estimated from the relationship:

$$r_{\mathrm{e}} \approx \frac{c}{6\pi}\left(\frac{K_{\mathrm{I}}}{K_{\mathrm{Ic}}}\right)^2. \tag{13.8}$$

At the onset of out-of-plane crack propagation $K_{\mathrm{I}} = K_{\mathrm{Ic}}$ and so from Equation (13.8):

$$\mathrm{r}_{\mathrm{e}} \approx c/(6\pi). \tag{13.9}$$

The size of the inelastic zone is thus set by the size of the primary crack.

Whether the local stress actually becomes large enough to cause primary cracks to propagate in an out-of-plane direction, and thus to initiate secondary cracks, depends upon the rate of stress relaxation which occurs via creep. When the ice is loaded slowly, creep occurs rapidly enough to prevent K_I from reaching K_{Ic}, thereby promoting macroscopically ductile behavior. When loaded more rapidly, on the other hand, the creep rate is too low and brittle behavior ensues.

A key point in the model (Schulson, 1990; Renshaw and Schulson, 2001) is the assumption that cracks will propagate and link up only when the size of the creep zone r_c is smaller than the size of the elastic-limit zone; i.e., $r_c < r_e$. The criterion for the ductile-to-brittle transition may then expressed as:

$$r_c = r_e. \tag{13.10}$$

The creep zone is defined as the region within which creep strain exceeds elastic strain. To estimate its size Schulson (1990) and Renshaw and Schulson (2001) invoked the model of Riedel and Rice (1980). Accordingly:

$$r_c = \frac{K_I^2}{2\pi E^2}\left(\frac{(n+1)^2 E^n B t}{2n\varsigma^{n-1}}\right)^{2/(n-1)} F \tag{13.11}$$

where F denotes an angular function of order unity, ς is a dimensionless factor also of order unity, E is Young's modulus, t is time and B and n are the materials parameters in the power-law creep expression $\dot{\varepsilon} = B\sigma^n$ (see Chapter 6).The loading time t may be approximated as the ratio of the critical stress intensity factor to the loading rate, $t = K_{Ic}/\dot{K}_I$, where the loading rate is determined from the product of the partial derivatives $\dot{K}_I = \left(\frac{\partial K_I}{\partial \sigma}\right)\left(\frac{\partial \sigma}{\partial \varepsilon}\right)\left(\frac{\partial \varepsilon}{\partial t}\right)$. The derivative $\partial\sigma/\partial\varepsilon = E^*$ is a modified Young's modulus (i.e., reduced by damage) and $\partial\varepsilon/\partial t = \dot{\varepsilon}$ is the applied strain rate; the other derivative is obtained from Equation (13.5) and is given by $\frac{\partial K_I}{\partial \sigma} = \frac{\sqrt{\pi c}}{2}[(1-R) - \mu(1+R)]$. At the ductile-to-brittle transition $K_I \approx K_{Ic}$. Upon inserting these relationships into Equation (13.11) and writing $E^* \approx E$, and upon setting Equation (13.11) equal to Equation (13.9) and rearranging, the transition strain rate under compression $\dot{\varepsilon}_{tc}$ is given by the relationship:

$$\dot{\varepsilon}_{tc} = \frac{(n+1)^2 (3)^{\frac{n-1}{2}} B K_{Ic}^n}{n\sqrt{\pi}[(1-R) - \mu(1+R)]c^{n/2}} \tag{13.12}$$

which is subject to the limitation $R < (1-\mu)/(1+\mu)$. Note that the functionality $\dot{\varepsilon}_{tc} \propto BK_{Ic}^n/c^{n/2}$ is essentially the same as that for the transition strain rate under tension (Equation 10.5). More often than not, under ambient conditions $n=3$ (Chapter 6), in which case:[1]

$$\dot{\varepsilon}_{tc} = \frac{9BK_{Ic}^3}{c^{3/2}[(1-R)-\mu(1+R)]}. \tag{13.13}$$

It is generally found that crack size is set by grain size, i.e., $2c=d$, and so:

$$\dot{\varepsilon}_{tc} = \frac{25BK_{Ic}^3}{d^{3/2}[(1-R)-\mu(1+R)]}. \tag{13.14}$$

Equation (13.14) thus describes the ductile-to-brittle transition strain rate in terms of three separate processes – crack propagation, creep, and frictional sliding – as well as the microstructure of the ice and the degree of confinement. The resistance to each process can be measured independently, as can the microstructure of the material and the stress ratio at terminal failure. The model, in other words, incorporates no adjustable parameters.

That said, we recognize that the friction coefficient of ice on ice depends upon sliding speed (Chapter 4). If we follow the practice in Chapter 11 and estimate sliding speed from the product of crack size and applied strain rate, then in a situation where ice deforms at a given rate under known conditions, one must first choose the appropriate value for the friction coefficient and then, in conjunction with the other parametric values, determine from Equation (13.14) whether the transition strain rate is greater or less than the applied rate. In practice, however, the range of choices is generally not so large as to change the order of magnitude of the calculation. At −10 °C, for instance, the friction coefficient ranges between about $\mu=0.1$ and $\mu=0.6$ (Chapter 4) over which range the calculated transition strain rate, say under a moderate degree of confinement of $R=0.1$, varies by about a factor of three.

13.4 Comparison with experiment

Fresh-water and salt-water ice Table 13.1 lists experimentally measured values of the three materials parameters at −10 °C for columnar-grained fresh-water ice and salt-water ice each of ~6 mm grain size (i.e., column diameter, as measured using the method of linear intercepts). Both materials were prepared in the laboratory and both possessed the S2 growth texture. The values of the friction coefficients listed there were obtained from double-shear experiments by Kennedy *et al.* (2000),

[1] An error of a factor of 2 appeared in the original paper (Renshaw and Schulson, 2001). Equation (A.7) of that paper was equated to Equation (A.9) to obtain the transition strain rate. As a result, the factor 8 in original Equation (A.10.b) (and in Equation (1) in the original text) should be 16. Thus $16/\sqrt{\pi} \sim 9$.

Table 13.1. *Parametric values at −10 °C for Equation (13.14)*

Parameter	Units	S2 fresh-water ice[a]	Reference	S2 salt-water ice[a,b]	Reference
B	$\mathrm{MPa^{-3}\,s^{-1}}$	4.3×10^{-7}	Schulson and Buck (1995)	5.1×10^{-6}	Sanderson (1988)
K_{Ic}	$\mathrm{MPa\,m^{1/2}}$	0.1	Chapter 9	0.1	Chapter 9
μ	–	0.5	Kennedy *et al.* (2000)	0.5	Kennedy *et al.* (2000)

[a] Loaded across the columns.
[b] Melt-water salinity = 4–5 ppt; density = $907 \pm 3\ \mathrm{kg\,m^{-3}}$.

Table 13.2. *Ductile-to-brittle transition strain rate ($T = -10$ °C, $d = 6$ mm, $R = 0$)*

Material	Calculated	Experimental measurement	Reference
Fresh-water ice[a]	$0.5 \times 10^{-4}\ \mathrm{s^{-1}}$	$0.7 \pm 0.3 \times 10^{-4}\ \mathrm{s^{-1}}$	Qi and Schulson (1998)
Salt-water ice[a]	$0.6 \times 10^{-3}\ \mathrm{s^{-1}}$	$0.8 \pm 0.2 \times 10^{-3}\ \mathrm{s^{-1}}$	Qi and Schulson (1998)

[a] Loaded across the columns.

described in Chapter 4, and correspond to a sliding speed of 0.4 to $4 \times 10^{-6}\ \mathrm{m\,s^{-1}}$. As mentioned in Chapter 11, this speed was obtained from the product of grain size and applied strain rate and is assumed to approximate the speed with which cracks slide near the transition. When the values are inserted into Equation (13.14), the predicted transition strain rate for unconfined ice ($R = 0$) of both types is found to be in good agreement with values measured through systematic experiment, listed in Table 13.2. The transition strain rate for the salt-water ice is greater by about a factor of 10, because salt-water ice of the salinity noted creeps more rapidly than fresh-water ice by about the same factor (de La Chapelle *et al.*, 1995; Cole *et al.*, 1998).

Grain size Figure 13.2 shows the effect of grain size, at −10 °C. The curve was calculated from Equation (13.14) using the parameters from Table 13.1. With one exception, the data were obtained by Batto and Schulson (1993), who performed compression tests on unconfined S2 fresh-water ice loaded across the columns at a variety of strain rates that straddled the transition. The bars mean that specimens that were compressed at rates equal to or lower than the bottom of the bar exhibited macroscopically ductile behavior; likewise, samples compressed at rates equal to or greater than the top of the bar exhibited brittle behavior. The exception is the single point: it was obtained from experiments on granular ice of 1.1 mm grain size (Schulson and Cannon, 1984) and appears to be in line with the results from the

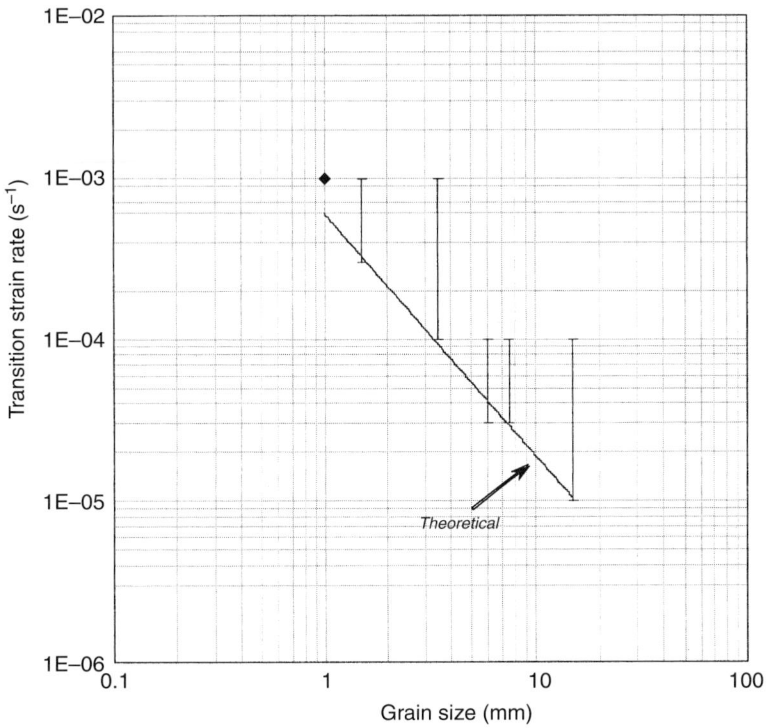

Figure 13.2. Ductile-to-brittle transition strain rate vs. grain size for unconfined, laboratory-grown, columnar-grained S2 fresh-water ice ($d \sim 6$ mm) compressed across the columns at −10 °C. The theoretical line was computed using Equation (13.14). (Data from Batto and Schulson, 1993.)

columnar ice. The model dictates that $\dot{\varepsilon}_{tc}$ scales as (grain size)$^{-1.5}$. Linear regression analysis of the data in Figure 13.2, using the midpoints of the ranges noted, indicates that $\dot{\varepsilon}_{tc}$ scales as (grain size)$^{-1.3}$ with a regression coefficient of $r^2 = 0.79$. The agreement between theory and experiment thus appears to be reasonably good.

The model also accounts for a related observation. In early work on compressive failure of unconfined fresh-water granular ice, Hawkes and Mellor (1972) did not report a sudden ductile-to-brittle transition, at strain rates up to $10^{-2}\,s^{-1}$ at −7 °C. In light of the above discussion, the most likely explanation for its absence is that the grain size of the material (~0.7 mm) was so fine that it probably raised the transition strain rate to a level very near to the highest rate explored. Given the limited data and given the fact that the transition occurs over a range of strain rates, it seems probable that their experiments approached, but did not fully enter, the brittle regime. Indeed, Hawkes and Mellor (1972) noted that the compressive strength appeared to reach a "limiting value" at $10^{-2}\,s^{-1}$ where the ice failed by splitting; i.e., via the expected mode of brittle failure within unconfined material (Chapter 11).

Temperature Of the materials parameters in the model, the creep parameter B exhibits the greatest sensitivity to temperature. It varies exponentially, scaling as $B \sim \exp(-Q/RT)$ where Q is the apparent activation energy for secondary creep, R is the universal gas constant and T is absolute temperature. If we take $Q = 78\,\mathrm{kJ\,mol^{-1}}$ (Chapter 6), a value within the range $Q = 60$–$84\,\mathrm{kJ\,mol^{-1}}$ measured for granular ice at temperatures below −10 °C (Barnes *et al.*, 1971; Weertman, 1983), then B is expected to decrease by a factor of ~90 upon lowering the temperature from −10 °C to −40 °C. Over this same range of temperature friction also increases (Kennedy *et al.*, 2000), but by a much lower factor of ~1.6, from $\mu \sim 0.5$ at −10 °C to $\mu \sim 0.8$ at −40 °C. Similarly, the resistance to crack propagation also appears to increase slightly, from $K_{\mathrm{Ic}} \sim 0.1\,\mathrm{MPa\,m^{1/2}}$ at −10 °C to $K_{\mathrm{Ic}} \sim 0.12\,\mathrm{MPa\,m^{1/2}}$ at −40 °C (Chapter 9). Thus, the model predicts that the transition strain rate is expected to decrease by about a factor of 40 upon lowering the temperature from −10 °C to −40 °C. In comparison, experiment indicates a reduction by about a factor of 10 for both fresh-water and salt-water ice, Figure 13.3.

Confinement The model dictates that the transition strain rate increases with increasing confinement. For example, for $\mu = 0.5$, $\dot{\varepsilon}_{\mathrm{tc}}$ scales as $1/(1-3R)$ for $R < 0.33$. (Given that frictional sliding is suppressed at confinements around this value or lower, as discussed in Chapter 11, this limitation is of little practical concern.) Thus, an increase from $R = 0$ to $R = 0.2$, for instance, implies an increase in the transition strain rate by about a factor of 2.5. This prediction is in good agreement with measurements on both fresh-water and salt-water columnar grained S2 ice of the kind described in the first paragraph of this section (Schulson and Buck, 1995; Schulson and Nickolayev, 1995).

Damage Deformation of the kind that induces grain-sized microcracks increases creep rates under both uniaxial (Sinha, 1988) and triaxial (Meglis *et al.*, 1999) loading, by as much as a factor of 100 (Jordaan, 2001). Specifically, damage increases the creep constant B without changing the stress exponent n in the power-law creep relationship (Meyssonnier and Duval, 1989; Weiss, 1999). Deformation damage also appears to reduce the fracture toughness of ice, albeit only slightly (Chapter 9). The friction coefficient, we expect, is essentially insensitive to damage, given that this property appears to be more dependent upon the structure of the surface of ice than upon its microstructure (Kennedy *et al.*, 2000). Qualitatively, therefore, if we assume that deformation damage raises B by a greater factor than it lowers K_{Ic}, as appears to be the case, then even though fracture toughness is raised to the power 3 in Equation (13.14), we expect damage to increase the transition strain rate. Observations support that expectation. For instance, the transition strain rate at −10 °C of S2 salt-water ice (4.3 ± 0.2 ppt melt-water salinity, $914 \pm 3\,\mathrm{kg\,m^{-3}}$

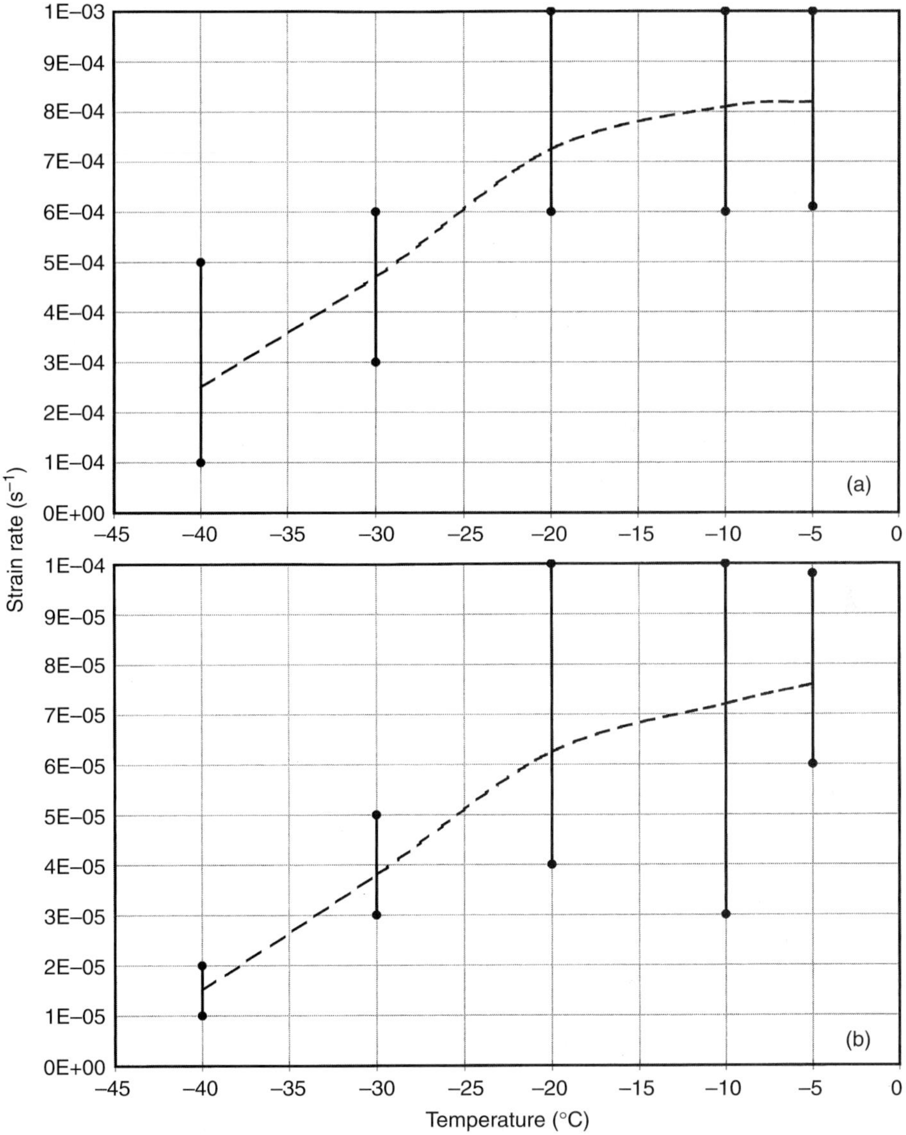

Figure 13.3. Ductile-to-brittle transition strain rate vs. temperature for unconfined, laboratory-grown, columnar-grained S2 ice ($d \sim 6$ mm) compressed across the columns: (a) salt-water ice; (b) fresh-water ice. (From Qi and Schulson, 1998.)

density) loaded across the columns is increased by about an order of magnitude when slowly pre-strained by 3.5% (Kuehn *et al.*, 1988).

Size If we assume that within natural bodies the size of the governing defect scales with the size of the body, then we would expect the transition strain rate to decrease with increasing size, scaling as $(\text{size})^{-3/2}$. We are not aware of any experiments to test this point. However, from the literature on ice–structure interactions (Chapter 14) there is some evidence of an effect. For instance, Sanderson (1988) noted that during indentation at the small scale (~1 m) the transition occurs at a strain rate of ~10^{-4}–10^{-2} s^{-1}; that at the intermediate scale (~10 m) it occurs at 10^{-5}–10^{-4} s^{-1}; and that at the full scale (~100 m) of interactions of ice sheets with artificial islands the transition occurs at a strain rate of about 10^{-7}–10^{-6} s^{-1}. Such behavior is roughly consistent with (size)$^{-3/2}$ functionality.

Comment Theory and experiment thus appear to be in reasonable agreement. We note, however, that the agreement is probably better than it should be, given the assumptions in the model. Specifically, the Riedel–Rice analysis was developed for creep at the tips of non-interacting cracks within an isotropic, homogeneous material loaded under far-field tension. Consequently, their analysis may not be strictly applicable to crack-tip creep within tensile zones induced locally within a damaged, plastically anisotropic aggregate loaded under far-field compression. Also, stress relaxation within the crack-tip region probably occurs through a combination of primary and secondary creep, and not just secondary creep as assumed. There is also uncertainty in the size of the elastic region, and there is the question of friction and whether coefficients obtained from ice sliding upon ice across a smooth (~2 μm) interface underestimate the coefficient of friction across the faces of cracks. Nevertheless, the model captures rather well the essential character of the ductile-to-brittle transition.

13.5 Dirty ice

The discussion so far, without being explicit, has addressed the behavior of "clean ice"; i.e., material free from a significant volume fraction of hard particles. Yet "dirty ice" also exists, at the base of glaciers (Chapter 6) and within silicate-rich extra-terrestrial bodies such as ice-rich permafrost on Mars (Squyres, 1989). However, there is little literature on the fracture of such material. About the only related discussion pertains to ice-lean mixtures of ice and rock. For instance, based upon triaxial compression experiments on sand-rich mixtures (52–75% sand by volume) confined under 12 MPa and deformed at 4.6×10^{-6} s^{-1} at −10 °C, Mangold *et al.* (2002) claim that a ductile-to-brittle transition occurs when the ice content is lower than ~28% by volume. Given the likelihood of particle contact at low volume fractions of ice, the nature of that transition is probably quite different from the transition that occurs within ice itself.

What does the above micromechanical model predict about ice-rich, dirty material? On the creep component, a considerable body of literature suggests that, barring low concentrations of ultrafine (~15 nm diameter) silica particles which appear to strengthen ice (Nayar *et al.*, 1971; Lange and Ahrens, 1983), hard particles of the order of a small volume percentage and of larger size (a few hundred micrometers) actually weaken ice (Baker and Gerberich, 1979; Shoji and Langway, 1985; Song *et al.*, 2005). On the fracture component, other than preliminary work by Smith *et al.* (1990) that revealed little effect of kaolinite (up to 3% by volume), systematic study has not been reported. Similarly, the role of hard particles on the friction coefficient is not known, although one might expect a positive effect. There is also the question of grain size and whether particles might actually lead to some refinement. In other words, it is difficult to predict with much confidence the role hard particles might play on the ductile-to-brittle transition.

13.6 Application to rocks and minerals

Finally, we consider the application of the above model to rocks and minerals. In dimensionless form the transition strain rate from Equation (13.13) may be expressed by the relationship (Renshaw and Schulson, 2001):

$$\dot{\varepsilon}_{\mathrm{D}} = \frac{\dot{\varepsilon}_{\mathrm{tc}} c^{3/2}}{B K_{\mathrm{Ic}}^{3}} = \frac{9}{[(1-R)-\mu(1+R)]}. \tag{13.15}$$

Again, the crack size is equated to the grain size; i.e., $2c = d$. Upon inserting appropriate values for the material constants B, K_{Ic}, and μ, listed in Table 13.3, we obtain the curve shown in Figure 13.4. The data were obtained by the investigators cited in Table 13.3. The calculated curve clearly marks the boundary

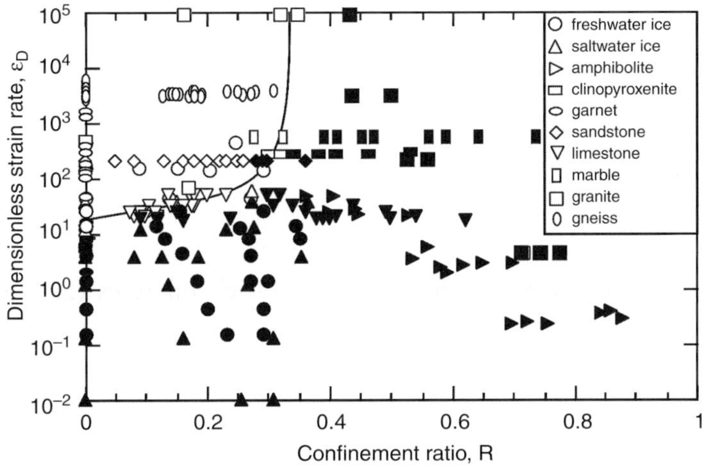

Figure 13.4. Dimensionless ductile-to-brittle transition strain rate vs. compressive stress ratio $R = \sigma_3/\sigma_1$, for a variety of polycrystalline rocks and minerals. Open symbols indicate brittle behavior and closed symbols indicate ductile behavior. (From Renshaw and Schulson, 2001.)

Table 13.3. *Material properties for rocks and minerals*

Material	References	$2c$ (mm)	K_{Ic} (MPa m$^{1/2}$)	Creep constant (MPa^{-3} s^{-1})[a]	Temperature (K)
Fresh-water ice	Schulson and Buck (1995)	1–4	0.1	4.3×10^{-7}	263
Salt-water ice	Schulson and Nickolayev (1995)	1–4	0.1	5.1×10^{-6}	263
Clinopyroxenite	Kirby (1980)	0.1	2.0	$\log(A) = 8.04$, $Q = 380$	873
Amphibolite	Hacker and Christie (1990)	0.1	2.0	$\log(A) = -1.60$, $Q = 234$	923–1223
Garnet	Wang and Ji (1999)	0.1	1.0	$\log(D) = 17.42$, $G = 8.5$–11, $g = 32$, $T_m = 1510$–1670	1223–1528
Sandstone	Gowd and Rummel (1980)	1.0	1.0	2.63×10^{-12}	298
Limestone and dolomite	Heard (1960); Mogi (1966)	0.1	1.0	$\log(A) = -11.15$, $Q = 4$	298–873
Marble	Kirby (1980)	0.1	2.0	$\log(A) = 31.77$, $Q = 259$	298
Granite	Mogi (1966); Shock and Heard (1974); Tullis and Yund (1977); Brace and Byerlee (1967)	0.1	2.0	$\log(A) = -5.7$, $Q = 187$	298–1173
Gneiss	Gottschalk *et al.* (1990)	0.1	2.0	$\log(A) = -5.7$, $Q = 187$	298

[a] When data were collected at a single temperature, only the creep constant B is given. When data were collected at multiple temperatures, creep coefficients are given and the creep constant may be calculated using either $B = A \exp(-Q/(RT)$, where A is an experimental constant (MPa^{-3} s^{-1}), Q is the activation energy (kJ mol^{-1}), R is the universal gas constant and T the absolute temperature, or $B = DG^{-3} \exp(gT_m/T)$, where D (s^{-1}) and g are experimental constants, G is the shear modulus (GPa), and T_m is the melting temperature (in K).

between the two kinds of observed behavior. We conclude, therefore, that the model accounts reasonably well for the ductile-to-brittle transition in a variety of crystalline materials.

References

Anderson, T. L. (1995). *Fracture Mechanics. Fundamentals and Applications*, 2nd edn. Boca Raton: CRC Press.

Ashby, M. F. and S. D. Hallam (1986). The failure of brittle solids containing small cracks under compressive stress states. *Acta Metall.*, **34**, 497–510.

Baker, R. W. and W. W. Gerberich (1979). The effect of crystal size and dispersed-solid inclusions on the activation energy for creep of ice. *J. Glaciol.*, **24**, 179–194.

Barnes, P., D. Tabor, F. R. S. Walker and J. F. C. Walker (1971). The friction and creep of polycrystalline ice. *Proc. R. Soc. Lond. A*, **324**, 127–155.

Batto, R. A. and E. M. Schulson (1993). On the ductile-to-brittle transition in ice under compression. *Acta Metall. Mater.*, **41**, 2219–2225.

Brace, W. F. and J. D. Byerlee (1967). Recent experimental studies of brittle fracture of rocks. In *Failure and Breakage of Rock*, ed. C. Fairhurst. New York: American Institute of Mining, Metallurgical and Petroleum Engineers Inc.

Cannon, N. P., E. M. Schulson, T. R. Smith and H. J. Frost (1990). Wing cracks and brittle compressive fracture. *Acta Metall. Mater.*, **38**, 1955–1962.

Cole, D. M., R. A. Johnson and G. D. Durell (1998). Cyclic loading and creep response of aligned first-year sea ice. *J. Geophys. Res.*, **103**, 21,751–21,758.

de La Chapelle, S., P. Duval and B. Baudelet (1995). Compressive creep of polycrystalline ice containing a liquid phase. *Scr. Metall. Mater.*, **33**, 447–450.

Gold, L. W. (1997). Statistical characteristics for the type and length of deformation-induced cracks in columnar-grain ice. *J. Glaciol.*, **43**, 311–320.

Gottschalk, R. R., A. K. Kronenberg, J. E. Russel and J. Handin (1990). Mechanical anisotropy of gneiss: Failure criterion and textural sources of directional behavior. *J. Geophys. Res.*, **95**, 613–621.

Gowd, T. N. and F. Rummel (1980). Effect of confining pressure on the fracture-behavior of a porous rock. *Int. J. Rock Mech. Mining Sci.*, **17**, 225–229.

Hacker, B. R. and J. M. Christie (1990). Brittle/ductile and plastic/cataclastic transitions in experimentally deformed and metamorphosed amphibolite. In *The Brittle-Ductile Transition in Rocks*, eds. A. G. Duba, W. B. Durham, J. W. Handin and H. F. Wang. Washington, D.C.: American Geophysical Union, 243.

Hawkes, I. and M. Mellor (1972). Deformation and fracture of ice under uniaxial stress. *J. Glaciol.*, **11**, 103–131.

Heard, H. C. (1960). Transition from brittle fracture to ductile flow in solnhofen limestone as a function of temperature, confining pressure and interstitial fluid pressure. In *Rock Deformation*, eds. D. T. Griggs and J. Handin. New York: Geological Society of America, 193–226.

Horii, H. and S. Nemat-Nasser (1986). Brittle failure in compression: Splitting, faulting and brittle-ductile transition. *Phil. Trans. R. Soc. A*, **319**, 337–374.

Iliescu, D. (2000). Contributions to brittle compressive failure of ice. Ph.D. thesis, Thayer School of Engineering, Dartmouth College.

Jordaan, I. J. (2001). Mechanics of ice-structure interaction. *Eng. Fract. Mech.*, **68**, 1923–1960.

Kennedy, F. E., E. M. Schulson and D. Jones (2000). Friction of ice on ice at low sliding velocities. *Phil. Mag. A*, **80**, 1693–1110.

Kirby, S. H. (1980). Tectonic stresses in the lithosphere – constraints provided by the experimental deformation of rocks. *J. Geophy. Res.*, **85**, 6353–6363.

Kuehn, G. A., R. W. Lee, W. A. Nixon and E. M. Schulson (1988). The structure and tensile behavior of first year sea ice and laboratory grown saline ice. In *Proceedings of Seventh International Conference on Offshore Mechanics and Arctic Engineering*, IV, pp. 11–17.

Lange, M. A. and T. J. Ahrens (1983). The dynamic tensile-strength of ice and ice-silicate mixtures. *J. Geophys. Res.*, **88**, 1197–1208.

Mangold, N., P. Allemand, P. Duval, Y. Geraud and P. Thomas (2002). Experimental and theoretical deformation of ice-rock mixtures: Implications on rheology and ice content of Martian permafrost. *Planet. Space Sci.*, **50**, 385–401.

Meglis, I. L., P. M. Melanson and I. J. Jordaan (1999). Microstructural change in ice: II. Creep behavior under triaxial stress conditions. *J. Glaciol.*, **45**, 438–448.

Meyssonnier, J. and P. Duval (1989). Creep behaviour of damaged ice under uniaxial compression: A preliminary study. *POAC 89; Port and Ocean Engineering Under Arctic Conditions*, Luleaa, Sweden, Luleaa University of Technology.

Mogi, K. (1966). Some precise measurements of fracture strength of rocks under uniform compressive stress. *Rock Mech. Eng. Geol.*, **4**, 41–55.

Nayar, H. S., F. V. Lenel and G. S. Ansell (1971). Creep of dispersions of ultrafine amorphous silica in ice. *J. Appl. Phys.*, **42**, 3786–3789.

Qi, S. and E. M. Schulson (1998). The effect of temperature on the ductile-to-brittle transition in columnar ice. *14th International Symposium on Ice*, Clarkson University, Potsdam, New York; A. A. Balkema Publishers.

Renshaw, C. E. and E. M. Schulson (2001). Universal behavior in compressive failure of brittle materials. *Nature*, **412**, 897–900.

Riedel, H. and J. R. Rice (1980). Tensile cracks in creeping solids. *ASTM-STP*, **7700**, 112–130.

Sanderson, T. J. O. (1988). *Ice Mechanics: Risks to Offshore Structures*. London: Graham & Trotman.

Schulson, E. M. (1990). The brittle compressive fracture of ice. *Acta Metall. Mater.*, **38**, 1963–1976.

Schulson, E. M. and S. E. Buck (1995). The brittle-to-ductile transition and ductile failure envelopes of orthotropic ice under biaxial compression. *Acta Metall. Mater.*, **43**, 3661–3668.

Schulson, E. M. and N. P. Cannon (1984). The effect of grain size on the compressive strength of ice. *IAHR Ice Symposium*, Hamburg.

Schulson, E. M. and O. Y. Nickolayev (1995). Failure of columnar saline ice under biaxial compression: Failure envelopes and the brittle-to-ductile transition. *J. Geophys. Res.*, **100**, 22,383–22,400.

Shock, R. N. and H. C. Heard (1974). Static mechanical properties and shock loading response of granite. *J. Geophy. Res.*, **79**, 1662–1666.

Shoji, H. and C. C. J. Langway (1985). Comparison of mechanical test on the Dye-3, Greenland ice core and artificial laboratory ice. *Ann. Glaciol.*, **6**, 306–308.

Sinha, N. K. (1988). Crack-enhanced creep in polycrystalline material: strain-rate sensitive strength and deformation of ice. *J. Mater. Sci.*, **23**, 4415–4428.

Smith, T. R., M. E. Schulson and E. M. Schulson (1990). The fracture toughness of porous ice with and without particles. *9th International Conference on Offshore Mechanics and Arctic Engineering*.

Song, M., I. Baker and D. M. Cole (2005). The effect of particles on dynamic recrystallization and fabric development of granular ice during creep. *J. Glaciol.*, **51**, 377–382.
Squyres, S. W. (1989). Urey Price Lecture: Water of Mars. *Icarus*, **79**, 229–288.
Tullis, J. and R. A. Yund (1977). Experimental deformation of dry westerly granite. *J. Geophy. Res.*, **82**, 5705–5718.
Wang, Z. and S. Ji (1999). Deformation of silicate garnets: Brittle-ductile transition and its geological implications. *Can. Mineral.*, **37**, 525–541.
Weertman, J. (1983). Creep deformation of ice. *Ann. Rev. Earth Planet. Sci.*, **11**, 215–240.
Weiss, J. (1999). The ductile behaviour of damaged ice under compression. *POAC*, Helsinki, Finland.

14

Indentation fracture and ice forces on structures

14.1 Introduction

We turn now to the indentation of ice. This kind of deformation can occur, for instance, during the interaction between a floating ice feature and a bridge pier or an offshore platform. In those situations and others like them, the primary issue is the interaction force. If too high, the structure will fail, as exemplified by the lighthouse shown in Figure 14.1. The limiting load is set by the compressive strength of the ice and, as noted in Chapter 11, can exceed the 100-year wave force (API, 1995).

Ice loading per se is beyond the scope of this book. Loading has been discussed by Sanderson (1988), who provides an account of a variety of field studies, and more recently by Bazant (2001), Blanchet and Defranco (2001), Jordaan (2001), Jordaan and Pond (2001), Masterson and Spencer (2001), Palmer and Johnston (2001), Sodhi (2001), Takeuchi *et al.* (2001), Tuhkuri (2001), Timco and Johnston (2003, 2004) and Frederking and Sudom (2006). Ice loading is discussed also in a report by the American Petroleum Institute (API, 1995) and more recently by the Canadian Standards Association Code (see Masterson and Frederking, 2006), by the International Standards Organization (Blanchet *et al.*, 2007) and by Masterson *et al.* (2007).

Our objective is to consider the physical processes that underlie indentation failure. The process is a complicated one. Stress and strain states within the contact zone vary spatially and they vary temporally as well once indentation has been initiated; and the contact zone within floating ice sheets experiences a thermal gradient which, within sea ice, lowers the stiffness near the bottom owing to the increase in porosity. As a result, multiple modes of failure are generally activated. Our premise, however, is that the fundamental mechanisms are essentially the same as the ones that operate under homogeneous loading, which we discussed in Chapter 12.

We focus on brittle failure – i.e., on the kind of failure that accompanies rapid indentation (defined below). Pressures within that regime reach higher levels than

Figure 14.1. Photograph showing the destructive effect of a moving sea ice cover on a lighthouse in the Baltic Sea.

within the ductile regime and, hence, limit design loads. We focus also on the initiation stage of indentation where the load reaches its highest level. During the subsequent or continuous stage, the loads are lower. This is not to say that the continuous stage is of little interest. On the contrary, owing to the cyclic building-up and sudden dropping of ice force as the ice advances, continuous indentation within the brittle regime can lead to significant vibration of a compliant structure (Kärnä and Turunen, 1989; Jordaan, 2001; Sodhi, 2001), as it did in the Molikpaq (Wright and Timco, 1994). Live loading, however, is also beyond the scope of this book.

A secondary ice–structure issue is abrasion. Concrete is the structural material of choice and typically abrades at the rate of about 0.05 mm/km of ice movement (Huovinen, 1990; Ito *et al.*, 1995; Itoh *et al.*, 1996; Moen and Jacobsen, 2008). Over a service life of tens of years, the process can lead to significant deterioration. Whether the attendant roughening of the surface also leads to an increase in ice loading is not clear. For the present discussion we assume that it does not.

14.2 Ductile-to-brittle transition

14.2.1 Characteristics

Over the ranges of temperature and types of ice encountered, the principal factor that governs the behavior of the ice is the indentation velocity (Sanderson, 1988; Takeuchi *et al.*, 1997; Sodhi, 1998; Sodhi *et al.*, 1998). At lower velocities indentation occurs in a ductile manner: the load rises monotonically and tends towards a constant level as indentation continues, Figure 14.2a; correspondingly,

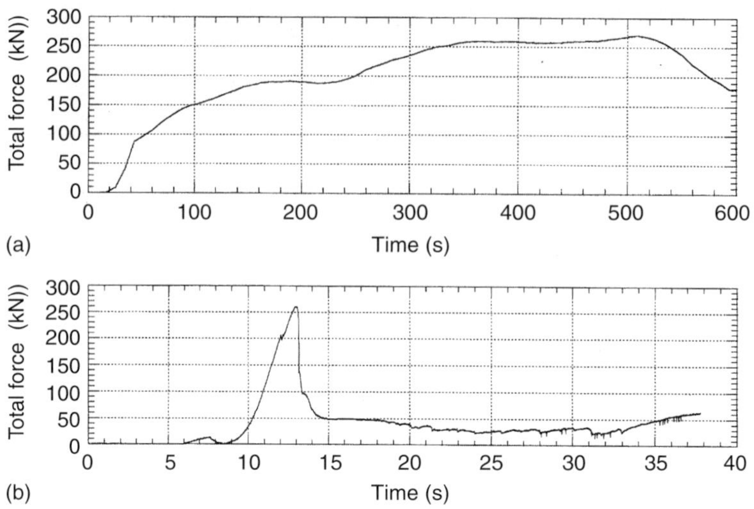

Figure 14.2. Graphs of indentation force vs. time during interactions between a medium-scale indenter (1.5 m wide) and a floating sheet of first-year sea ice of average thickness about 300 mm and of average temperature about −3 °C, at indentation speeds of (a) 0.3 mm s^{-1} and (b) 30 mm s^{-1}. The lower speed leads to ductile behavior and the higher speed, to brittle behavior. (From Sodhi *et al.*, 1998.)

the ice flows plastically and failure occurs via creep buckling. At higher velocities indentation occurs in a brittle manner: the load rises in a pseudo-linear manner until it reaches a sharp maximum after which it drops suddenly and becomes jerky as indentation continues, Figure 14.2b; correspondingly, the ice fractures and fragments are ejected from the contact zone in the wake of the indenter, as evident from Figure 14.1.

The relationship between failure pressure and velocity or strain rate is different in the two regimes. Within the ductile regime the failure pressure P increases with increasing strain rate, Figure 14.3. It scales as $P \sim (U/W)^{1/n}$ where U denotes velocity and W denotes the width or diameter of the indenter; the exponent has the value $n \sim 3$ owing to the governing role of dislocation creep. Within the brittle regime the failure pressure decreases slightly (Timco, 1986), Figure 14.4, reminiscent of the strain-rate softening that is observed under both uniaxial loading (Chapter 11) and biaxial loading (Gratz and Schulson, 1997) of columnar-grained ice when rapidly compressed under a relatively homogeneous stress state. The brittle failure pressure is rather scattered and varies between ~10 MPa and ~20 MPa for small-scale structures (~1 m) loaded by ice at temperatures around −10 °C. The ductile-to-brittle transition marks the point where the resistance to deformation reaches its highest level, again in keeping with behavior under homogeneous loading.

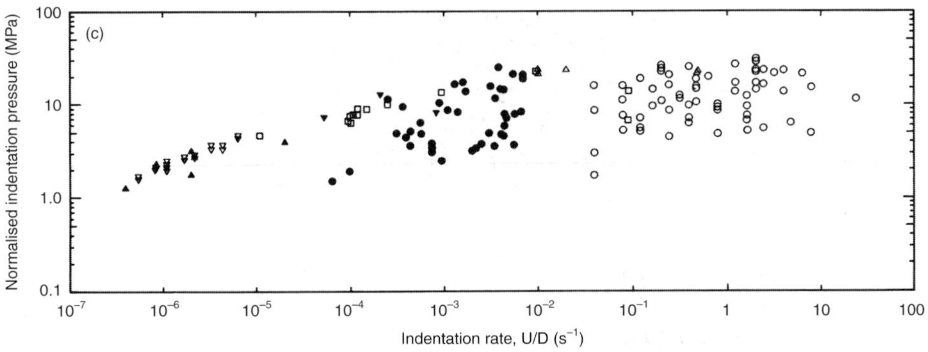

Figure 14.3. Graph of indentation pressure of edge-loaded sheets of columnar-grained ice vs. indentation rate. The data were obtained from laboratory experiments on S2 fresh-water ice and from field tests on sea ice, and were normalized to zero salinity and to −10 °C. The indentation rate is defined as indentation velocity U divided by the width W or the diameter D of the indenter. The indentation pressure is defined as the load on the indenter divided by the nominal or gross contact area. Each point corresponds to the maximum load registered during each test. (From Sanderson, 1988.)

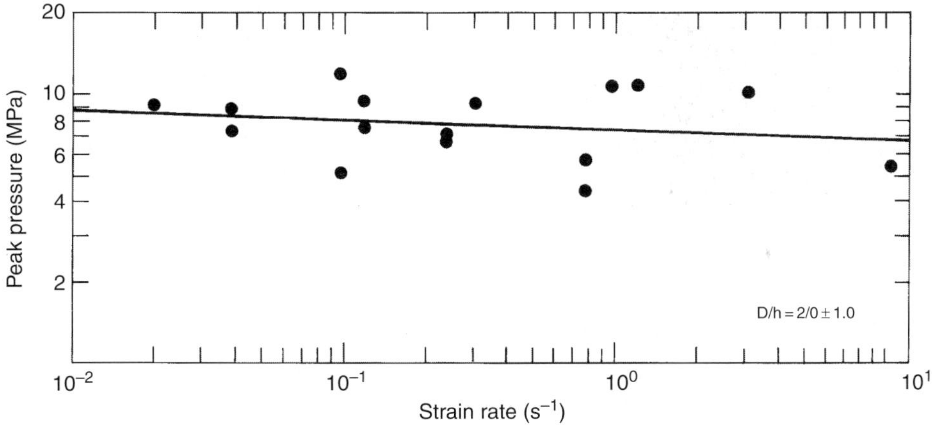

Figure 14.4. Graph of brittle indentation pressure vs. strain rate for a floating sheet of columnar-grained S2 fresh-water ice loaded on edge at an average temperature of ~−3 °C, aspect ratio $W/h = 2.0 \pm 1.0$. (From Timco, 1986.)

The transition strain rate appears to depend upon size. In the small-scale (~1 m) example depicted in Figure 14.3, where we equate the ratio U/W to a sort of strain rate $\dot{\varepsilon}_{ind}$, the transition occurs over the range $\dot{\varepsilon}_{ind} \sim 10^{-4}$–$10^{-2}\ s^{-1}$. In comparison, Sanderson (1988) noted that during indentation at the intermediate scale (~10 m) the transition occurs at $\dot{\varepsilon}_{ind} \sim 10^{-5}$–$10^{-4}\ s^{-1}$; and at the full scale (~100 m) of interactions of ice sheets with artificial islands in the Canadian Beaufort Sea the transition occurs at a strain rate of $\dot{\varepsilon}_{ind} \sim 10^{-7}$–$10^{-6}\ s^{-1}$.

Thus, the transition from ductile to brittle behavior during indentation exhibits characteristics similar to those exhibited during homogeneous loading. The transition strain rate is similar (at least on the smaller scale), although a little on the higher side owing to the effect of confinement, and it is spread over a broader range, owing perhaps to the gradients of stress and strain within the contact zone. The implication is that the underlying physical mechanisms are also similar.

Indentation within a natural setting, incidentally, can lead to an interesting response. The velocity of a floating ice feature is governed mainly by wind that builds up as a weather system develops and then drops as the system passes. As a result, an interaction generally begins slowly. Correspondingly, the ice begins to deform within the ductile regime. As the velocity increases, so does the indentation pressure, but only up to a point: once the velocity/strain rate reaches and then exceeds the transition level, the indentation pressure decreases slightly, even though the velocity may continue to increase for a time. In other words, as Sanderson (1988) notes, the peak in ice stress may not coincide with a peak in ice velocity.

14.2.2 Origin

Two models have been proposed, both based upon the concept that the ductile-to-brittle transition may be viewed as a competition between creep and fracture. The micromechanical model, described in Chapter 13, holds that the transition marks the point at which cracks begin to propagate; i.e., when creep blunting is less favorable than stable crack growth (Schulson, 1990; Renshaw and Schulson, 2001). There is no direct support for this interpretation from indentation studies in the way there is from studies under homogeneous loading, but there is little reason to question it, given the similarity in both the transition strain rate and the failure modes (described below). Thus, again the important materials parameters are considered to be the resistance to crack propagation as measured by fracture toughness K_{Ic} the resistance to creep as measured by the reciprocal of the temperature-dependent materials constant B in the power-law creep equation $\dot{\varepsilon} = B\sigma^3$ (where $\dot{\varepsilon}$ denotes the equivalent secondary creep rate and σ denotes the equivalent or effective stress) plus the resistance to frictional sliding as measured by the kinetic coefficient of friction μ. In short, from the discussion and derivation in Chapter 13 we expect the indentation transition strain rate to scale as $\dot{\varepsilon}_{tind} \sim BK_{Ic}^3/f(\mu, R)c^{3/2}$ where $f(\mu, R)$ is defined in Equation (13.14) and c is a measure of crack length. Should the length of the governing crack scale with the width or diameter of the contact zone, the model predicts that the transition strain rate should scale as $(\text{size})^{-3/2}$, roughly in agreement with Sanderson's (1988) observation; i.e., $\dot{\varepsilon}_{tind}(100\,\text{m}) \sim 10^{-3}\dot{\varepsilon}_{tind}(1\,\text{m})$.

The other model is more phenomenological in nature. Following a series of indentation tests on naturally grown sheets of columnar-grained sea ice at an average temperature of −2 °C, Sodhi *et al.* (1998) proposed that the transition occurs when the pressure to maintain creep indentation equals the pressure to activate spalling through fast-crack propagation. Accordingly, they invoked both the power-law creep relationship and Tuhkuri's (1996) model of crack propagation and obtained for the transition velocity the relationship:[1]

$$v_t = B\lambda I_n \left(\frac{55 K_Q}{\sqrt{a}} \right)^n \tag{14.1}$$

where λ denotes the length of an observed wedge-shaped front to the sheet, I_n is a dimensionless integration parameter[2] of order unity that incorporates the thicknesses of the ice at the front of the wedge a and at the back h, K_Q is an apparent fracture toughness and n is the power-law exponent in the secondary creep equation. Upon equating the term in parentheses to the failure pressure (6–8 MPa) and inserting for the other parameters the values $n = 3$, $B = 1.1 \times 10^{-4}\,\text{MPa}^{-3}\,\text{s}^{-1}$, $\lambda = 0.06$ m, $a = 0.01$ m, $h = 0.2$ m and $I_n = 0.263$, Sodhi *et al.* (1998) obtained the value $v_t = 0.4$–$0.9\,\text{mm s}^{-1}$. Barring a question about the value of B, which is discussed below, the calculated transition velocity appears to bracket the experimental observations (Sodhi *et al.*, 1998) – namely, of ductile indentation at a velocity of $0.3\,\text{mm s}^{-1}$ and brittle indentation at $3\,\text{mm s}^{-1}$.

Although the two models are similar in that they both incorporate creep and fracture, they differ in significant ways. The micromechanical model contains only materials parameters that can and have been measured independently. Also, it incorporates friction, in keeping with the fundamental role of frictional sliding in brittle compressive failure (Chapter 12). In addition, through the crack size, the model includes the possibility that the transition strain rate is scale dependent should the size of the governing crack scale with the size of the ice sheet. The model is also independent of the geometry of indentation. In comparison, the phenomenological model contains parameters (a, λ) that are difficult to specify a priori, and it requires prior knowledge of the indentation pressure. The model is also limited to a specific geometry; namely, edge indentation of a floating ice sheet. In addition, unlike the micromechanical model which dictates a relatively small effect of temperature – B and μ are the only parameters that vary significantly with temperature and they vary in the opposite sense (Chapter 13) – the

[1] The exponent n of Equation (14.1) was inadvertently dropped from the original equation (see Equation (8) by Sodhi *et al.* (1998)), although it follows directly from the derivation presented. It follows also from dimensional analysis.

[2] The relationship from which is derived the integration parameter I_n (see Equation (3) of Sodhi *et al.* (1998)) describes the stress in the direction of loading as increasing instead of decreasing along the wedge. The error lies in the associated figure (see Figure 13 of Sodhi *et al.* (1998)) in which the variable x is shown to increase from the front of the wedge instead of from the back.

phenomenological model dictates a relatively large effect. That follows from the fact that the creep constant scales as $B \sim \exp(-Q/RT)$, where $Q \sim 78\,\mathrm{kJ\,mol^{-1}}$ is the apparent activation energy (Chapter 6) and R is the gas constant. A reduction in temperature from −10 °C to −40 °C, for instance, is expected to lower the transition velocity by a factor of ~100. We are not aware that such a reduction actually happens. There is also an issue with the value of B. The value used by Sodhi *et al.* (1998) is about an order of magnitude greater than one would expect for warm sea ice (−2 °C), based upon measurements on columnar-grained sea ice at −10 °C (where $B = 5.1 \times 10^{-6}\,\mathrm{MPa^{-3}\,s^{-1}}$, Sanderson, 1988). A lower and more realistic value leads to a transition strain rate an order of magnitude lower than observed.

Despite these differences, both the physical model and the phenomenological model dictate that the transition strain rate or transition velocity scales directly with the product of the creep constant and the cube of the fracture toughness. This functionality follows from the fact that both models are based upon a competition between creep and fracture.

14.3 Brittle failure modes

We consider two scenarios, the floating plate of columnar-grained ice loaded along its edge throughout its thickness by a vertical indenter and the thick wall loaded by a blunt indenter. The first scenario is relevant to offshore structures situated within ice-covered waters such as the Sea of Okhotsk; the latter, to structures vulnerable to attack by icebergs such as the Hibernia platform. As will become apparent, brittle failure during indentation is accompanied by modes similar to those that operate during more homogeneous loading. The difference is that multiple as opposed to single modes are generally observed.

14.3.1 Edge-loaded plates

In a series of systematic experiments in an ice tank, Timco (1986) edge-loaded sheets of finely grained (1–3 mm) fresh-water columnar S2 ice using a flat-faced indenter. He varied the thickness of the sheet ($h = 6.4$–35 mm) and the width of the indenter ($W = 25$–127 mm) and hence the aspect ratio ($W/h = 0.6$–21). He also varied the indentation strain rate ($\sim 10^{-2}\,\mathrm{s^{-1}}$ to $\sim 10\,\mathrm{s^{-1}}$), defined in this case as the velocity of the indenter divided by twice the indenter width. The temperature across the sheet varied from −10 °C at the top to 0 °C at the bottom.

Timco (1986) identified four brittle modes: crushing, spalling, radial cracking and circumferential cracking. He also observed creep buckling, at the lowest deformation rate. Although identified from smaller-scale observations, the modes appear to operate on the larger scale as well. They may be described as follows:

- Crushing develops immediately ahead of the indenter, as apparent from Figure 14.1. Fragments are produced of many sizes. The size distribution appears to obey a power law (Timco and Jordaan, 1987) and, thus, to exhibit the character of fractal geometry (Palmer and Sanderson, 1991), a point to which we return in Section 14.5. Crushing occurs over all combinations of indentation rate and aspect ratio. It leads to cyclic loading, owing to the expulsion of fragments from within the contact zone (Jordaan, 2001). This mode of failure is analogous to the post-terminal mode that operates under homogeneous loading (Chapter 12) and, like that mode, is governed by friction as fragments slide past each other (Singh *et al.*, 1995).
- Spalling, like crushing, develops immediately ahead of the indenter. It is manifested by disc-shaped plates of diameter similar to the width of the indenter (Croasdale *et al.*, 1977; Timco, 1986). The discs form through across-column cleavage of the columnar-shaped grains that constitute the cover and through the subsequent linking up of the cleavage cracks into thin sheets that fail, we imagine, via the combination of crack growth and Euler buckling described in Chapter 12 (Section 12.9). Presumably, ice is expelled not only from the upper surface, but also from the lower surface where the hydrostatic pressure is too low to suppress the process. Spalling operates in concert with crushing at all rates within the brittle regime, but not under all aspect ratios. It appears to be limited to lower ratios of around $W/h \leq 3$; i.e., to brittle indentation by a relatively narrow structure. The reason is that material within a narrow contact zone experiences a higher degree of confinement (Johnson, 1985); and spalling, as evident from experiments (Iliescu and Schulson, 2004) on the same kind of ice biaxially loaded across the columns under a homogeneous stress state, is a mode of failure limited to higher degrees of biaxial confinement; e.g., to $\sigma_{22}/\sigma_{11} > 0.2$ at $-10\,°\mathrm{C}$ (Section 12.4.1).
- Radial cracking occurs in concert with crushing as well. The cracks extend ahead of the indenter by several indenter widths, generally in a start-stop manner, and are inclined to the direction of loading. The cracks do not generally emanate from the corners of the indenter, but from a variety of sites along the interface. They form during indentations characterized by higher aspect ratios (i.e., ahead of a relatively wide structure) and, hence, by lower degrees of confinement. Radial cracks do not reduce ice loads, unless a free boundary is nearby: in that case the cracks tend toward the boundary, thereby reducing confinement and lowering the load (Timco, 1987). These features form to relax tensile stresses that develop ahead of the indenter as a result of its wedging action.
- Circumferential cracking occurs in concert with crushing and radial cracking. This is not a compressive failure mode per se, but a tensile mode that is induced by bending of sectors created by the radial cracks. There is no analogue under homogeneous loading.

Figure 14.5 summarizes Timco's (1986) observations. Although specific for the conditions investigated, based upon earlier work by Michel and Blanchet (1983) and upon more recent laboratory and medium-scale field studies by Sodhi (1998)

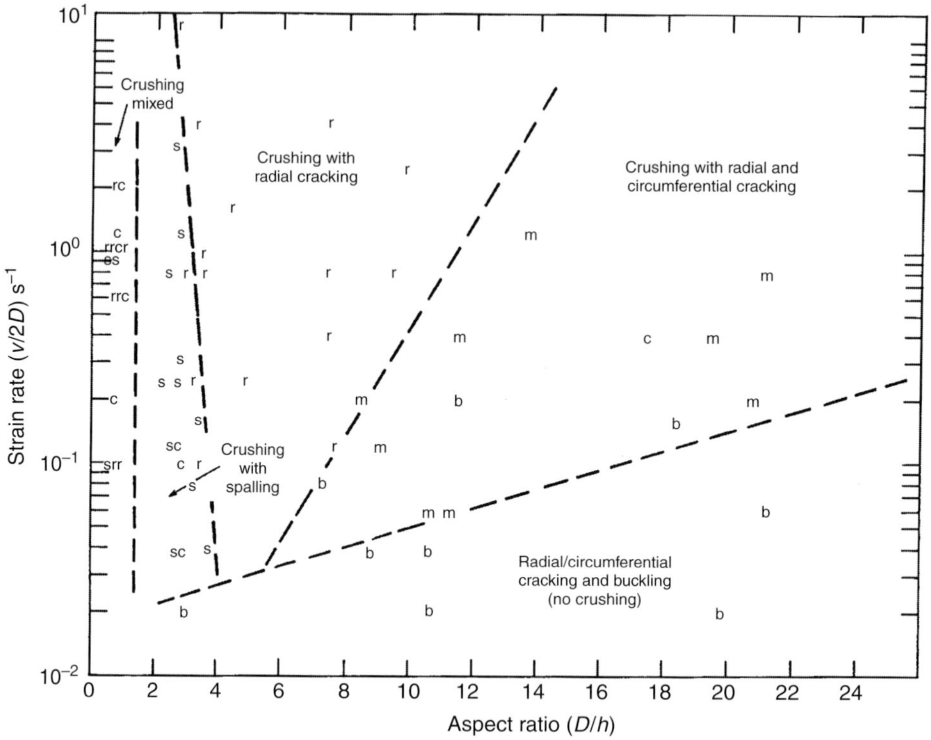

Figure 14.5. Indentation failure mode map for a floating sheet of columnar-grained S2 fresh-water ice loaded on edge at an average temperature of ~−3 °C. The letters denote the mode: c, crushing; s, spalling; r, radial cracking; m, circumferential cracking; b, creep buckling. *D* denotes width. (From Timco, 1986.)

and by Sodhi *et al.* (1998), our sense is that Timco's map applies qualitatively to larger scales as well. Aspect ratio is a key factor. As already noted, the higher it is the lower is the degree of confinement. Correspondingly, it affects the failure pressure: as W/h increases over the range $0 < W/h < 5$ the pressure falls dramatically and then tends to level off (Timco, 1986; Masterson and Spencer, 2001), Figure 14.6. Interestingly, the indentation pressure falls by about the same factor of about five as does the biaxial compressive strength over the range of confinement where the failure mode changes from spalling to splitting (Figures 12.5 and 12.7).

First cracks initiate behind the indenter, at least when the edges are rounded. This point was noted by Grape and Schulson (1992) who pressed a narrow indenter with rounded edges ($W = 38$ mm) into small ($203 \times 142 \times 38$ mm^3) non-floating plates ($W/h = 1$) of S2 fresh-water ice, at −10 °C at 10^{-3} s^{-1}. The plates were confined across the columns to simulate larger bodies and the indenter acted along their full thickness. From pressure-synchronized photography, Grape and Schulson found that the first cracks initiated at a distance of about one indenter half-width behind the

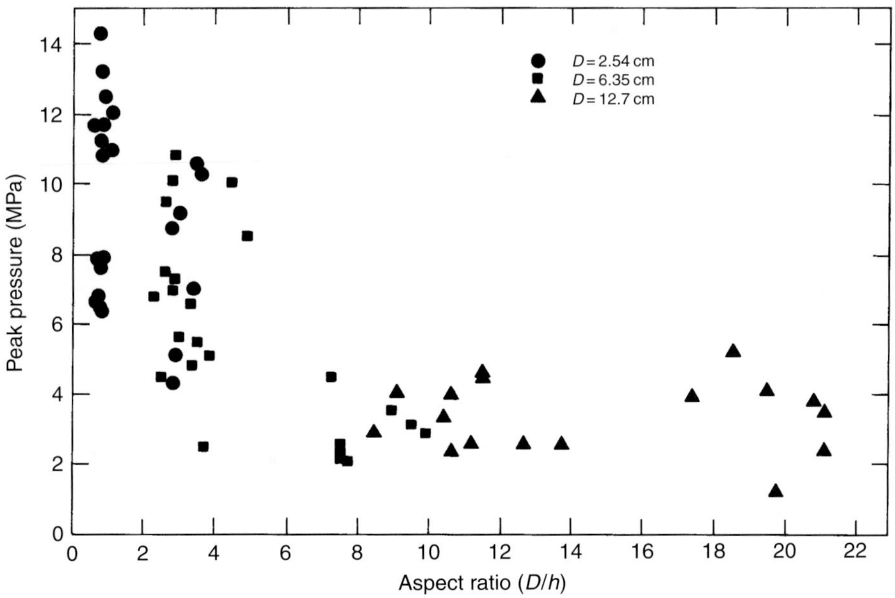

Figure 14.6. Graph of brittle indentation pressure vs. aspect ratio (where *D* denotes width) for a floating sheet of columnar-grained S2 fresh-water ice loaded on edge at an average temperature of ~−3 °C. (From Timco, 1986.)

interface, Figure 14.7. Analysis of the principal stresses within the plane of the plate showed that the in-plane shear stress, taken as one-half the difference between the two in-plane principal stresses, reached a maximum at about the same point. (This observation should not be taken as a contradiction of Gold's (1990) prediction that cracks initiate at the edge of the indenter, for his calculations are based upon the assumption that the indenter has sharp edges.) The first crack formed under ~20% of the indentation failure pressure and new ones nucleated as the pressure increased. The applied stress reached a peak when a macroscopic fracture developed, via a combination of spalling and radial cracking; i.e., via modes consistent with Timco's (1986) map for the aspect ratio of unity that was employed. Grape and Schulson (1992) also noted that terminal failure corresponds to a sudden drop in applied pressure and to the onset of crushing. In other words, the development of damage during the early stage of indentation appears to be similar to its development under homogeneous loading (Sections 11.6 and 12.8).

14.3.2 Indented walls

Crushing is the dominant failure mode of indented walls. From field experiments in Canada's High Arctic, Frederking *et al.* (1990) and Masterson and Frederking

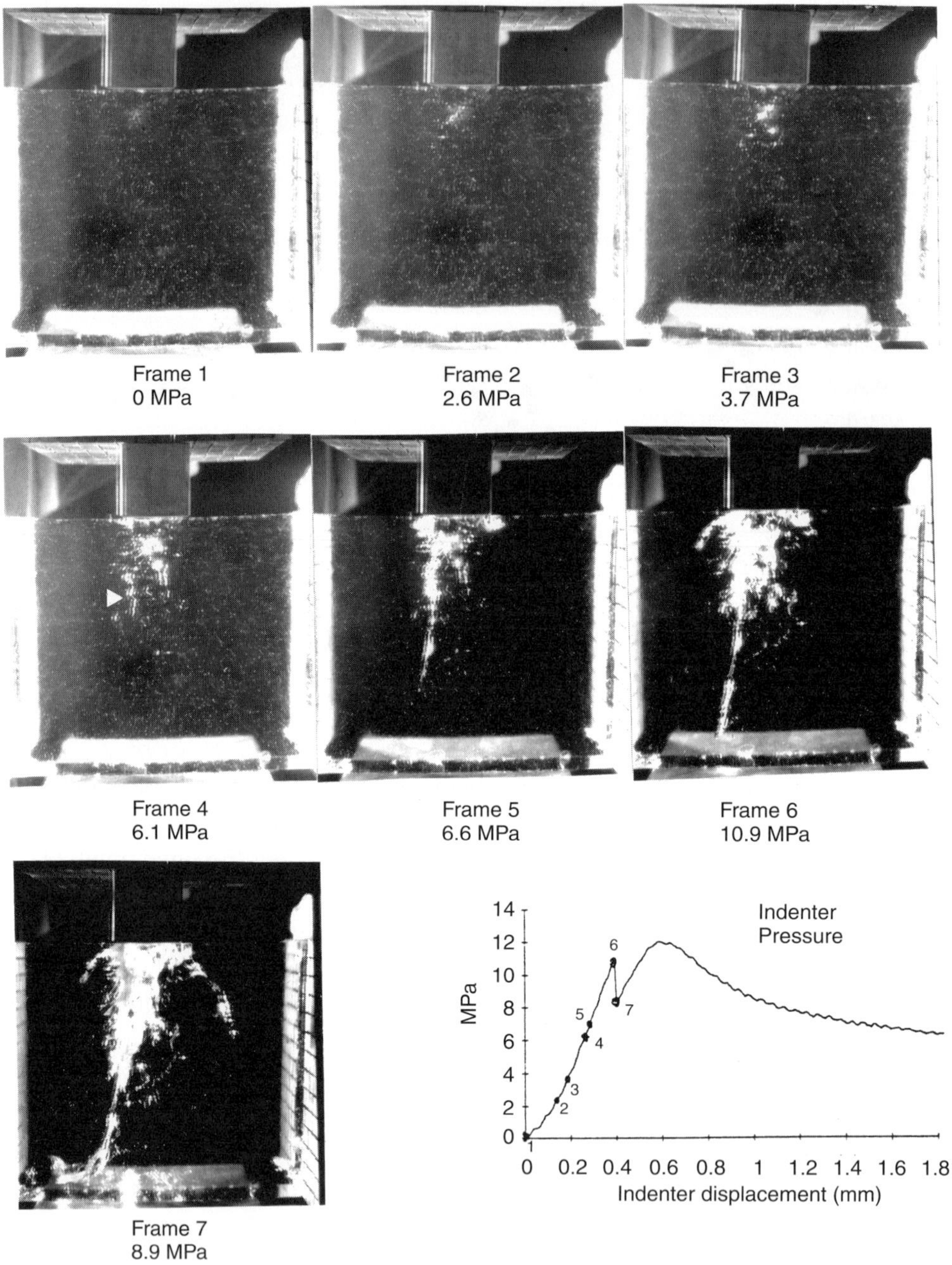

Figure 14.7. Photographs showing the development of damage within columnar-grained S2 fresh-water ice indented on edge at −10 °C at 10^{-3} s^{-1}. The long axes of the grains are perpendicular to the plane of the paper. The ice was confined under 0.6 MPa applied horizontally to simulate the behavior of a large plate. (From Grape and Schulson, 1992.)

(1993) noted that when multi-year sea ice (of 1 ppt salinity, 890 kg m^{-3} density and remnant columnar-shaped grains of 2–7 mm column diameter) at a temperature between −20 °C and −14 °C is rapidly (> ~3 mm s^{-1}) indented across the columns using a medium-sized (0.8 m^2) spherically shaped or flat indenter, post-terminal failure (our words) is again accompanied by the stream-like expulsion of fragmented material from within the contact zone. The fragments originate from fracture that is concentrated within a whitish-looking layer that develops immediately in front of the indenter. Beyond this layer, but more centrally located, is a relatively intact "blue zone" (more below). Crushing occurs on the larger scale as well, as evident from the Molikpaq event (Wright and Timco, 1994) mentioned in Chapter 11.

The whitish layer is an interesting feature. Its snow-like appearance results from the scattering of light from a multitude of interfaces (cracks and surface of fragments) created during indentation. More importantly, its mechanics appear to account for the cyclic character of loading during the continuous stage of indentation (Jordaan, 2001), at least under higher degrees of triaxial confinement. The idea is that once deformation damage that is created during uploading becomes high enough to pulverize the ice, fragments suddenly extrude when the central zone fails, thereby allowing the indenter to quickly advance a short distance under the lower compliance of the material and, hence, under lower load. The process repeats itself in a regular manner (more below). Again, friction is at play (Singh *et al.*, 1995) and the load does not fall to zero under the action of the confining stresses within the contact zone. Melting could be a factor as well (Kheisin and Likhomanov, 1973; Gagnon, 1994; Gagnon and Daley, 2005), induced more by friction than by pressure-related reduction in the equilibrium melting point.

Another interesting feature, and an important one from the perspective of local loads (more below), is the so-called "blue[3] zone" (Frederking *et al.*, 1990). The zones are relatively transparent and form near the center of impact. Both their size O(10 mm) and number density vary, as does their shape. They are composed of small (< 0.5 mm) recrystallized grains and are reminiscent of features reported by Kheisin and Likhomanov (1973) upon impacting ice in a drop-ball experiment. Muggeridge and Jordaan (1999) observed similar finely grained recrystallized regions within the contact zone between a small iceberg (500–1400 tonnes) moving at ~1.5 m s^{-1} and an instrumented steel panel (6 × 6 m^2) mounted on a near-vertical cliff on the coast of Labrador, suggesting that "blue zones" are a general characteristic of thick-wall indentation. They have been observed (Melanson *et al.*, 1999) as well within fresh-water granular ice that had been compressed under homogeneous stress states characterized by a high degree of triaxial confinement, at −10 °C at

[3] The ice appears to be blue for the same reason that water appears to be blue. Braun and Smirnov (1993) showed that the color is the result of selective absorption in the red part of the visible spectrum.

strain rates of $\sim 10^{-3}\ s^{-1}$, implying that the features have little to do with the stress–strain gradients that exist within contact zones. Rather, "blue zones" appear to be a fundamental aspect of the rapid deformation of highly confined ice at temperatures close to the melting point (more below). They have also been termed "hot spots" and "high-pressure zones" where pressures up to 30–50 MPa were recorded (Frederking *et al.*, 1990; Jordaan, 2001).

To account for this behavior, we imagine a sequence of events that includes both Coulombic (C) and plastic (P) faulting. We suggest that C-faulting initiates near the free edges of the contact zone where the degree of triaxial confinement is the lowest. Faulting of that kind is accompanied by fragmentation, for the reasons discussed in Chapter 12. The process propagates toward the more highly confined regions within the contact zone, creating more fragments but not yet ejecting many of them. At some position within the zone, the degree of confinement becomes high enough to suppress frictional sliding and Coulombic faulting is suppressed. That position is defined as the point where the principal stress ratio is given by Equation (12.1). At a sliding velocity around 0.01 m s^{-1} at ice temperatures around -3 °C, the friction coefficient in that relationship has the value $\mu \sim 0.6$ (Fortt and Schulson, 2007), leading to the stress ratio $\sigma_3/\sigma_1 \sim 0.3$. From that point inward, we imagine that P-faults develop. Plastic faulting is essentially a time/temperature-dependent, volume-conserving process that could be triggered by adiabatic heating (Section 12.7). The plasticity and attendant recrystallization account for the relatively intact central blue zones. Once the "core" becomes destabilized the entire contact zone becomes unstable: fragments from the C-faults flow outward, the load drops and the indenter advances rapidly as the fragments are cleared. Subsequently, the ice becomes loaded once again and the sequence repeats itself. This picture is conceptually similar to one created by Jordaan (2001), except for the incorporation of more specific failure modes and the inward progression of the failure process.

An implication of this analysis is that owing to the change in stress state within the contact zone the indentation pressure results not from the operation of a single mechanism, but from the operation of two or more. The Coulombic process is pressure sensitive (Chapter 12) and thus depends upon both the hydrostatic part of the stress tensor, $\sigma_h = \sigma_{kk}/3$, and the deviatoric part, $\sigma^*_{ij} = \sigma_{ij} - \sigma_{kk}\delta_{ij}/3$ where δ_{ij} is the Kronecker delta. The plastic process, on the other hand, is driven by shear stresses and thus depends primarily upon the deviatoric part of the tensor. Melting may also play a role, as already noted, suggested by slushy looking material that is ejected from single crystals of relatively warm ice (at -5 and -10 °C) when indented at velocities of 1.5–20 mm s^{-1} (Gagnon 1994; Gagnon and Daley, 2005). As a result of this complexity, the failure pressure is a property that cannot be modeled using a single failure criterion. Instead, a model that incorporates multiple modes is required.

14.4 Non-simultaneous failure and local vs. global loads

From the discussion so far, the reader may have gained the impression that pressure is constant across the ice–indenter interface. While that is more or less true for indentation within the ductile regime (Sodhi, 1998), it is not true for brittle indentation once the continuous stage has begun. In the latter case, as pointed out by Kry (1978) and developed by others (Ashby *et al.*, 1986; Joensuu and Riska, 1989; Frederking *et al.*, 1990; Sodhi, 1998; Sodhi *et al.*, 1998) pressure varies both spatially and temporally. For instance, Sodhi (1998, 2001) observed spatial variations upon rapidly deforming on edge, with a flat segmented indenter, sheets of both columnar-grained fresh-water ice and columnar-grained sea ice, where the aspect ratio was $W/h \sim 5$–10: during indentation at speeds sufficiently high to induce brittle failure, high-pressure zones developed and then shifted continuously. The zones were generally located near the centerline of the sheet, owing to greater degree of triaxial constraint there, reminiscent of the fluctuating, line-like high-pressure zones observed by Joensuu and Riska (1989) who pushed a transparent indenter into a block of ice. High-pressure zones that shift around were observed also by Frederking *et al.* (1990) who inserted transducers into the face of the spherical and flat indenters mentioned in Section 14.3.2. In other words, brittle indentation occurs non-simultaneously, implying that once terminal failure sets in, the real contact area is smaller than the apparent contact area.

Non-simultaneity within engineered structures induces local loads that can have a different effect from global loads. For instance, a global load toppled the lighthouse shown in Figure 14.1, while local loads ruptured the rivets in the hull of the Titanic (Hooper *et al.*, 2003). Presumably, local loads also govern the structural abrasion noted in Section 14.1. Design now incorporates both kinds, as exemplified by the Hibernia structure (Section 11.1) that was built to withstand a local ($A_{local} \sim 1\ m^2$) pressure of 11 MPa and a global pressure of 6 MPa (Jordaan, 2001). "Local", of course, exists on many different scales and so the practical definition depends upon the problem at hand.

14.5 Pressure–area relationship

14.5.1 Observations

Sanderson (1988) made a rather unexpected observation. Upon compiling a large number of measurements (from a variety of sources, from different types of ice and from different geometries) that were made on scales large and small and then plotting the failure pressure P vs. the apparent or gross contact area A he found that the

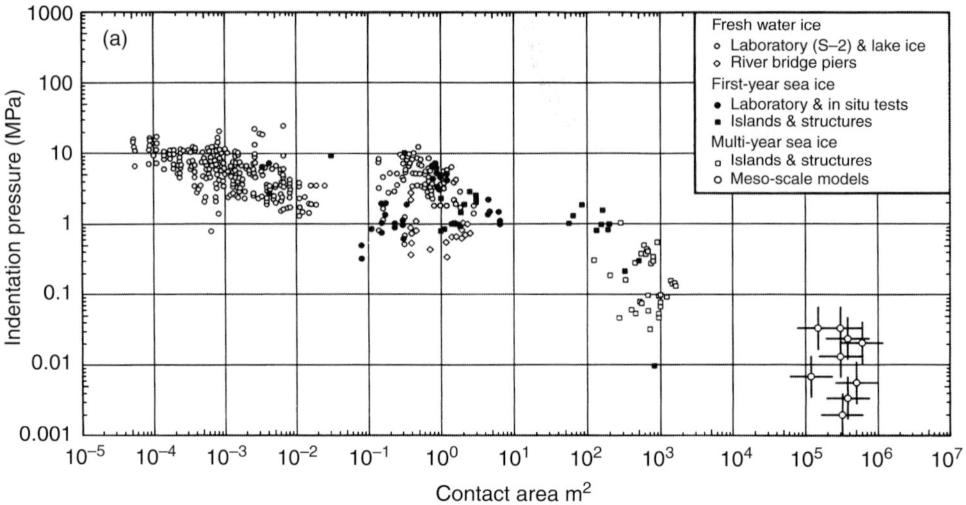

Figure 14.8. Graph of brittle indentation pressure of edge-loaded, floating ice sheets vs. apparent (gross) contact area, for S2 fresh-water ice, first-year sea ice and multi-year sea ice. The data were obtained under a variety of conditions (laboratory experiments; interactions with bridge piers, lighthouses, offshore structures and ice islands). No correction was made for differences in temperature, salinity, microstructure or geometry of the indenter. The entries for the largest area were obtained not from direct measurement, but from computer-based models of ice dynamics on the scale 40–125 km. (From Sanderson, 1988.)

failure pressure, although scattered, decreased as the area increased, Figure 14.8. Over a range of area that varied by nine orders of magnitude ($10^{-4}\,m^2$ to $10^5\,m^2$), the pressure scaled roughly as:

$$P \sim A^q \tag{14.2}$$

where P was defined as the load at failure divided by the apparent contact area, F/A, and where $-0.5 \leq q \leq -0.25$. Although there had been hints of an effect of contact area from earlier work (Kry, 1978; Iyer, 1983; Bercha and Brown, 1985), until Sanderson's compilation the evidence was less compelling. At the time, the prevalent view seemed to have been that ice possessed a scale-independent strength – a view that was based in part upon experimental data (Chen and Lee, 1986; Wang and Poplin, 1986) that showed that the unconfined compressive strength of multi-meter-sized floating blocks of first-year sea ice, when deformed at strain rates up to $8 \times 10^{-5}\,s^{-1}$, is statistically the same as that of sub-meter-sized specimens, which were smaller by a factor of 20 in linear dimensions. Sanderson challenged that position. The effect that he reported has now been confirmed (Masterson and Spencer, 2001; Masterson *et al.*, 2007), albeit from measurements made over a smaller size range (0.1–10 m^2), Figure 14.9,

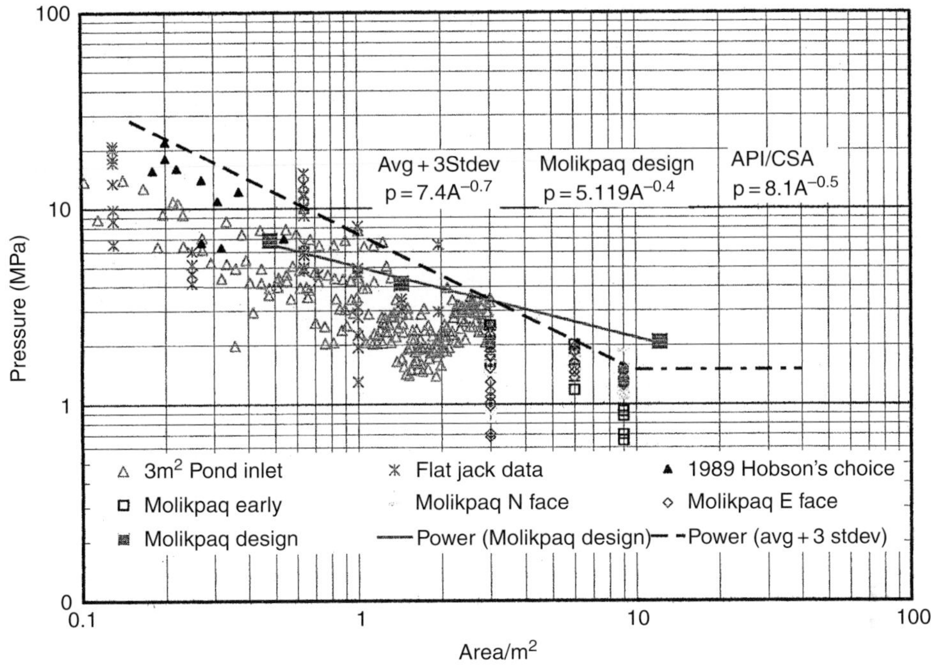

Figure 14.9. Graph of indentation pressure vs. contact area from several field studies. (From Masterson *et al.*, 2007.)

where $-0.7 \leq q \leq -0.4$. The international engineering community now appears to have accepted the effect.

Nevertheless, skeptics remain. The concern appears to hinge on the possibility of an indiscriminate use of Sanderson's curve. As implied above, that compilation contains measurements not just from different contact areas, but from many different indentation conditions – ice type, salinity, temperature, interaction velocity, indenter geometry and aspect ratio. The compilation also includes "data" (at the largest area) that are not measurements at all, but are calculated pressures based upon ice dynamics models. Heightening the concern is the risk that, driven by economic considerations, the pressure–area curve will be used unquestioningly to justify the design and construction of offshore structures and ships that in the end will prove to be incapable of resisting ice forces. Skepticism should not be bolstered, however, with the observations of Chen and Lee (1986) and Wang and Poplin (1986): their compression experiments were performed within the ductile regime where there is little evidence from ice or from any other material that size is a significant factor. Also, differences in ice type, salinity and temperature should not be reasons for serious concern, for such differences do not exhibit order-of-magnitude effects on compressive strength. While there is more at play than gross contact area, that factor appears to be an important one.

14.5.2 Explanations of the pressure–area relationship

Fracture is the key to understanding. Several analyses lead to a relationship of the kind Sanderson proposed, thus providing a theoretical foundation for the effect.

Dimensional analysis Palmer and Sanderson (1991) noted that a simple dimensional argument could account for the size effect, independent of mechanisms or of any other kind of analysis. At root, they argued, is fracture toughness K_{Ic}: they took this property to be scale independent, and there is no evidence to the contrary (Chapter 9). From this and from the parameters of indentation failure, the only dimensionless group that can be formed is $F/A^{3/4}K_{Ic}$. If that group has a fixed value for geometrically similar scenarios, then it follows that $P \sim A^{-1/4}$.

Statistical analysis In an earlier analysis Sanderson (1988) invoked Weibull's (1951) theory of brittle failure. Although developed for tensile failure in which internal cracks do not interact (Chapter 10), he assumed that it might apply as well under compression. Accordingly, he imagined the ice sheet to contain a population of cracks of various sizes, statistically distributed throughout the body. If divided into many imaginary volume elements, each containing a single crack, then the theory holds that the overall strength is governed by the element that contains the largest defect; i.e., not by an average value of the distribution, but by the most dangerous flaw. This means that larger bodies have a greater probability of containing flaws from the tail of the distribution, and are thus weaker. The probability of failure, P_f, is given by the relationship:

$$P_f = 1 - \exp\left\{\left(-\frac{V}{V_o}\right)\left(\frac{\sigma}{\sigma_o}\right)^b\right\} \tag{14.3}$$

where V is the volume of the body, V_o is a reference volume, σ and σ_o, respectively, denote the applied stress and a reference stress and b is termed the Weibull modulus. It follows that the strength scales as $\sigma \sim V^{-1/b}$. Since $V \sim A^{3/2}$, the analysis dictates that indentation pressure scales as $P \sim A^{-3/2b}$. Sanderson chose the value $b = 3$, based upon some rather questionable references, and this led to the exponential value $q = -0.5$. If instead we apply the Weibull modulus obtained directly by Kuehn *et al.* (1993) from measurements of the brittle compressive strength of unconfined ice cubes of different sizes (10^3 mm^3 to 3.4×10^6 mm^3), namely $b = 5.5 \pm 0.8$, then we obtain the exponential value $q = -0.27 \pm 0.04$. In other words, Weibull's theory, questionable though its application may be in this situation, can account for the observed effect.

Geometrical analysis An even simpler explanation, stemming from Griffith's theory of brittle failure (Chapter 9), holds that the indentation pressure is inversely related to the square root of the size c of the largest flaw within the contact zone. If we assume that $c \propto W$ and that $W = A/h$, then for a given plate thickness h it follows that $P \propto A^{-1/2}$.

Ductile–brittle analysis Another explanation relates to the ductile-to-brittle transition and to the fact that the ductile indentation pressure is scale independent. Let us suppose that the transition strain rate truly scales with indenter width in the manner suggested in Section 14.2.2, namely $\dot{\varepsilon}_{\rm tind} \sim W^{-3/2}$, where $W \propto A$ for constant plate thickness. And suppose that within the brittle regime strain-rate softening is insignificant from a practical perspective. Then, from the power-law relationship between the ductile indentation pressure and strain rate, namely $P \sim \dot{\varepsilon}^{1/3}$, it follows that the failure pressure of an edge-loaded plate at the ductile-to-brittle transition scales as $P \sim A^{-1/2}$.

Fractal analysis The concept of fractals has also been employed to explain the pressure–area relationship. Bhat (1990), for instance, considered the external crushing of a fractal surface, and Palmer and Sanderson (1991) considered the fractal crushing of a collection of fragments and did not address surface behavior. While this kind of analysis has been criticized on the basis that it is a mathematical device that offers little physical insight, it introduces a way of thinking that has considerable value. The treatment by Palmer and Sanderson (1991) is particularly lucid, so we summarize it below.

The treatment is based upon the observation that fragments of broken material exhibit a size distribution characterized by a power law, where the number of fragments that have a diameter (or other linear dimension) greater than d scales as:

$$N(d) \sim d^{-D} \tag{14.4}$$

where D is the fractal dimension. Correspondingly, the ratio of the number of particles larger than diameter d_1 to the number larger than diameter d_2 is:

$$\frac{N(d_1)}{N(d_2)} = \left(\frac{d_1}{d_2}\right)^{-D}. \tag{14.5}$$

There is some evidence from rather limited fragment-size measurements by Timco and Jordaan (1987) and from the analysis of those measurements by Palmer and Sanderson (1991) that ice is characterized by $D = 2.5$, as are many brittle materials. Other evidence offered by Weiss (2001) suggests a value closer to $D \sim 2$. The important point is the idea that during crushing a hierarchy of events progressively breaks the ice into smaller pieces: large fragments of diameter (or any linear dimension) d_r fracture with a probability p into n smaller ones of dimension d_{r-1}, and the members of the new generation of smaller fragments break with the same probability p into n still smaller ones, and so on, such that:

$$\frac{d_r}{d_{r-1}} = n^{1/3}. \tag{14.6}$$

There are many levels of hierarchy, each defined by order r, numbered from $r=0$ for the smallest fragments, such that the ratio of the number of fragments of order $r-1$ to the number of fragments of order r is np. From this ratio and from Equations (14.5) and (14.6) it follows (by taking logs and rearranging) that:

$$D = 3(1 + \ln p/\ln n). \tag{14.7}$$

The parameters p and n define the size distribution of the fragments and are assumed to have the same values for each level. This scale invariance is the essential feature of the model and leads to self-similarity of the crack and fragment size distribution.

In developing their model, Palmer and Sanderson (1991) employed for the purpose of illustration the highly idealized fractal geometry shown in Figure 14.10a. Starting from the level r, the sketch illustrates four levels of hierarchy from a very large number of subdivided cubic elements. For the case shown, n is 8 and p is 0.75. Also, they defined a planar section, Figure 14.10b, that supports the total force F_r that is carried by the fractal hierarchy of elements of dimensions d_r or smaller. The section is composed of different numbers of fragments of different orders (e.g., $(1-p)$ of order r, $(1-p)n^{2/3}p$ of order $r-1$, $(1-p)(n^{2/3}p)^2$ of order $r-2$, etc.) the sum of whose areas (d_r^2, d_{r-1}^2 etc.) makes up the area of the load-bearing section. If each fragment can support a load S_i that is governed by its size d_i and by the fracture toughness of the ice, such that $S_i \sim K_{\text{Ic}} d_i^{3/2}$, then through further analysis in which the relative size of the particles is again related to their number through Equation (14.6), Palmer and Sanderson (1991) showed that the total carrying force for the section of order r as a whole is given by:

$$F_r \sim K_{\text{Ic}} d_r^{3/2} \{[1-p][1 + n^{1/6}p + (n^{1/6}p)^2 + \cdots + (n^{1/6}p)^{r-1}] + (n^{1/6}p)^r\}. \tag{14.8}$$

Thus, the crushing force depends on $n^{1/6}$. If $n^{1/6}p<1$ the crushing force for a cell of order r is determined mainly by the fracture of the larger fragments and (from Equation 14.7) $D>2.5$; if $n^{1/6}p>1$ the crushing force is determined mainly by the fracture of the smaller fragments and $D<2.5$; and if $n^{1/6}p=1$, then fragments of each order contribute equally and $D=2.5$. For the idealized model of Figure 14.9, $n^{1/6}p=1.06$. Whichever case prevails, $F_r \sim d_r^{3/2} \sim A_r^{3/4}$. Therefore:

$$\frac{P_r}{P_{r-1}} = \frac{F_r/A_r}{F_{r-1}/A_{r-1}} = \left(\frac{A_r}{A_{r-1}}\right)^{-1/4}. \tag{14.9}$$

Thus, again, analysis leads to a relationship that is consistent with observation.

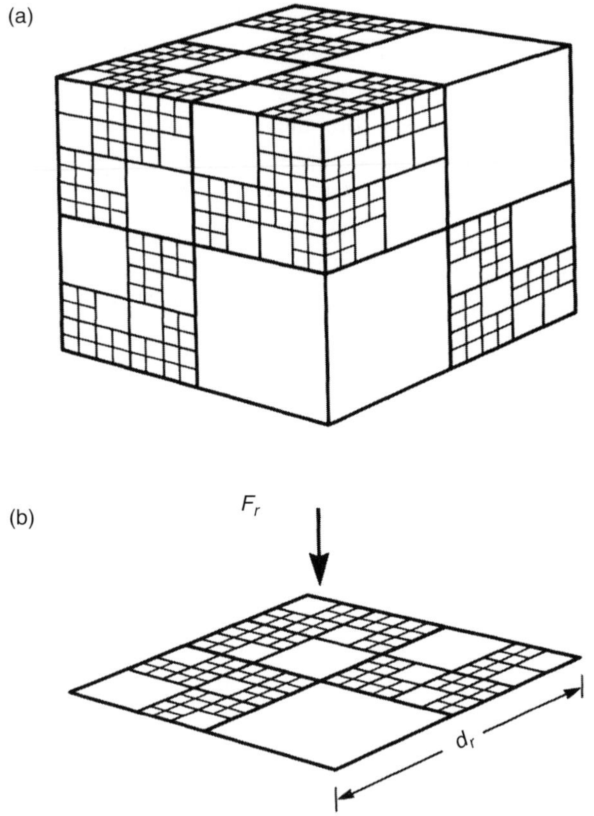

Figure 14.10. Schematic diagram of the idealized fractal crushing of ice. (a) Each cubic element breaks into n smaller elements with a probability p. In this scenario $p=0.75$ and $n=8$. (b) Section through the cubic element in (a) that contains a fractal hierarchy of elements of dimensions d_r or less. The elements support the total force F_r. (From Palmer and Sanderson, 1991.)

14.6 Impact failure

Finally, a few words on impact indentation. Deformation of that kind is relevant, for instance, to the ramming by ships of ice covers and other floating features (Michel, 1978) and to the formation of impact craters within the Martian ice caps (Mangold, 2005) and within other icy bodies of the outer Solar System (Kato *et al.*, 1995). Ice strength is a key parameter. In the case of cratering, a compressive wave develops upon impact and this leads to the creation of pit-like features near the center of the contact zone; when reflected, a tensile wave develops and this leads to the flaking off of thin fragments (through a process termed "spallation" by the planetary science community) and to the formation of a "halo".

Consider some early observations by Kawakami *et al.* (1983). Through a systematic study in which cylindrical projectiles (~10 mm diameter and ~100 mm length) of different materials (aluminum, Teflon, polycarbonate and pyrophyllite) were shot at velocities between 110 m s^{-1} and 680 m s^{-1} against polycrystalline blocks ($280 \times 280 \times 340$ mm^3) of warm (−10 °C) fresh-water ice whose *c*-axes were parallel to the impact direction, they observed bowl-shaped craters. From the pits extended long radial cracks, perpendicular to the impact surface, plus fine concentric cracks whose maximum diameter was about three times the diameter of the crater. The pattern resembles one that forms within the ice cover on a pond when impacted with a walking stick, and is reminiscent as well of cracks that develop within glass upon impact (Lawn, 1995).

Timco and Frederking (1993) performed a different kind of study. They impacted floating sheets of both S1 and S2 fresh-water ice (defined in Chapter 3) by dropping from different heights an instrumented (pressure cells, accelerometer, load cell) cylindrical projectile (177 mm diameter × 445 mm length) of variable mass (35 to 65 kg) whose head was either flat or spherically shaped. The sheet thickness was varied from 178 to 600 mm and the drop height, from 0.14 to 1.6 m; the calculated speed upon impact varied from 1.7 to 5.9 m s^{-1}. The maximum force was found to be a function of loading rate, scaling as (loading rate)$^{0.4}$. The highest pressure measured was 42 MPa. Upon impacting thinner (200–280 mm) sheets of S2 ice, the pressure fell rapidly to a negligible level after reaching a peak, owing to through-thickness displacement, termed punching, that occurred via sliding along the vertically oriented boundaries of the columnar-shaped grains. Thicker sheets of S2 ice as well as S1 sheets also exhibited the peak in pressure, but not the near-zero fall-off. In these cases, crushing was more prevalent than punching. Again, long radial cracks emanated from the impact zone. In addition, Hertzian-like cone-shaped cracks developed within thicker sheets when impacted with the flat-headed indenter.

In an attempt to better understand the relationship between crater dimensions and strength of the ice, Hiraoka *et al.* (2008) measured both the quasi-static tensile strength and the quasi-static compressive strength at −10 °C of ice containing silicates (up to 50% by weight, from powder of serpentine) and then applied dimensional analysis. They found that both the tensile and compressive strength increase with increasing silicate content, the tensile strength being the more sensitive. From dimensional analysis, crater depth d_c when normalized by projectile radius r appears at lower impact velocities ($U < 700$ m s^{-1}) to scale inversely with compressive strength σ_c:

$$\frac{d_c}{r} \sim \left(\frac{\rho U^2}{\sigma_c}\right) \tag{14.10}$$

where ρ is the density of the ice–silicate mixture. At higher velocities, corresponding to kinetic energies up to 10^4 kJ and to depths in silicate-free ice up to ~1 m, the depth appears to scale with (energy)$^{0.4 \pm 0.1}$ (Jordaan and McKenna, 1988). Crater diameter D_c appears to scale inversely with the square root of tensile strength σ_T:

$$\frac{D_c}{r} \sim \left(\frac{\rho U^2}{\sigma_T}\right)^{0.5} \tag{14.11}$$

in keeping with the view that this dimension is governed by "spallation". Similarly, crater volume V scales inversely with tensile strength (Hiraoka *et al.*, 2008). With the caveat that quasi-static strength is probably a poor measure of dynamic strength, these relationships help to account for the observation (Lange and Ahrens, 1982; Koschny and Grün, 2001; Hiraoka *et al.*, 2007) that crater volume decreases with increasing silicate content.

References

API (1995). *Recommended Practice 2N – Planning, Designing and Constructing Structures and Pipelines for Arctic Conditions*, American Petroleum Institute, Dallas, Texas, 114.

Ashby, M. F., A. C. Palmer, M. Thouless *et al.* (1986). Nonsimultaneous failure and ice loads on Arctic structures. *Eighteenth Annual Offshore Technology Conference*, 5–8 May, Houston, Texas.

Bazant, Z. P. (2001). Scaling laws for sea ice fracture. *IUTAM Symposium in Ice Mechanics and Ice Dynamics*, Fairbanks, Alaska. Dordrecht: Kluwer Academic Publishers.

Bercha, F. G. and T. G. Brown (1985). Scale effect in ice/structure interactions. *Offshore Mechanics and Arctic Engineering Conference*, Dallas, Texas.

Bhat, S. U. (1990). Modeling of size effect in ice mechanics using fractal concepts. *Trans. ASME*, **112**, 370–376.

Blanchet, D. and S. Defranco (2001). On the understanding of the local ice pressure area curve. *IUTAM Symposium on Scaling Laws in Ice Mechanics and Ice Dynamics*, Fairbanks, Alaska. Dordrecht: Kluwer Academic Publishers.

Blanchet, D., K. R. Croasdale, W. Spring and G. A. N. Thomas (2007). An international standard for Arctic offshore structures. *POAC 2007*, Dalian, China.

Braun, C. L. and S. N. Smirnov (1993). Why is water blue? *J. Chem. Edu.*, **70**, 612–615.

Chen, A. C. T. and J. Lee (1986). Large-scale ice strength tests at slow strain rates. *Fifth International Offshore Mechanics and Arctic Engineering Symposium*, Tokyo, Japan.

Croasdale, K. R., N. R. Morgenstern and J. B. Nuttall (1977). Indentation tests to investigate ice pressures on vertical piers. *J. Glaciol.*, **81**, 301–312.

Fortt, A. L. and E. M. Schulson (2007). The resistance to sliding along Coulombic shear faults in ice. *Acta Mater.*, **55**, 2253–2264.

Frederking, R. and D. Sudom (2006). Maximum ice force on the Molikpaq during the April 12, 1986 event. *Cold Reg. Sci. Technol.*, **46**, 147–166.

Frederking, R., I. J. Jordaan and J. S. McCallum (1990). Field tests of ice indentation at medium scale, Hobson's Choice Ice Island. *IAHR Symposium on Ice*, Espoo, Finland, Helsinki University of Technology.

Gagnon, R. E. (1994). Generation of melt during crushing experiments on freshwater ice. *Cold Reg. Sci. Technol.*, **22**, 385–398.

Gagnon, R. E. and C. Daley (2005). Dual-axis video observations of ice crushing utilizing high-speed video for one perspective. *18th International Conference on Port and Ocean Engineering Under Arctic Conditions*, Potsdam, NY.

Gold, L. W. (1990). Ductile to brittle transition during indentation of ice. *Can. J. Civ. Eng.*, **18**, 182–190.

Grape, J. A. and E. M. Schulson (1992). Effect of confining stress on brittle indentation failure of columnar ice. *J. Offshore Polar Eng.*, **2**, 212–221.

Gratz, E. T. and E. M. Schulson (1997). Brittle failure of columnar saline ice under triaxial compression. *J. Geophys. Res.*, **102**, 5091–5107.

Hiraoka, K., M. Arakawa, K. Yoshikawa and A. M. Nakamura (2007). Laboratory experiments of crater formation on ice-silicate mixture targets. *Adv. Space Res.*, **39**, 392–399.

Hiraoka, K., M. Arakawa, M. Setoh and A. M. Nakamura (2008). Measurements of target compressive and tensile strength for application to impact cratering on ice-silicate mixtures. *J. Geophys. Res.*, **113**, doi: 10.1029/2007JE002926.

Hooper, J. J., T. Foecke, L. Graham and T. P. Weihs (2003). The metallurgical analysis of wrought iron from the RMS *Titanic*. *Meas. Sci. Technol.*, **14**, 1556–1563.

Huovinen, S. (1990). Abrasion of concrete by ice in Arctic sea structures. *ACI Materials J.*, **87**, 266–270.

Iliescu, D. and E. M. Schulson (2004). The brittle compressive failure of fresh-water columnar ice loaded biaxially. *Acta Mater.*, **52**, 5723–5735.

Ito, Y., F. Hara, H. Saeki and H. Tachibana (1995). The mechanism of the abrasion of concrete structures due to the movement of ice sheets. *International Conference on Concrete under Severe Conditions, CONSEC '95*, Sapporo, Japan. London: E. & F. N. Spon.

Itoh, Y., Y. Tanaka, A. Delgado and H. Saeki (1996). Abrasion depth of a cylindrical concrete structure due to sea ice movement. *Intern. J. Offshore Polar Eng.*, **6**, 2.

Iyer, S. H. (1983). Size effects in ice and their influence on the structural design of offshore structures. *7th International Conference on Port and Ocean Engineering Under Arctic Conditions*, Helsinki, Finland.

Joensuu, A. and K. Riska (1989). Contact between ice and a structure. *Report M-88*. Espoo, Finland, Helsinki University of Technology, Ship Laboratory, 57.

Johnson, K. L. (1985). *Contact Mechanics*. Cambridge: Cambridge University Press.

Jordaan, I. J. (2001). Mechanics of ice-structure interaction. *Eng. Fract. Mech.*, **68**, 1923–1960.

Jordaan, I. J. and R. F. McKenna (1988). *Ice Crushing by Impact and Indentation: A Literature Review*. St. John's, Newfoundland, National Research Council.

Jordaan, I. and J. Pond (2001). Scale effects on and randomness in the estimation of compressive ice loads. *IUTAM Symposium on Scaling Laws in Ice Mechanics and Ice Dynamics*, Fairbanks, Alaska. Dordrecht: Kluwer Academic Publishers.

Kärnä, T. and R. Turunen (1989). Dynamic response of narrow structures to ice crushing. *Cold Reg. Sci. Technol.*, **17**, 173–187.

Kato, M., Y. Iijima, M. Arakawa *et al.* (1995). Ice-on-ice impact experiments. *Icarus*, **133**, 423–441.

Kawakami, S., H. Mizutami, Y. Tagaki, M. Kato and M. Kamazawa (1983). Impact experiments on ice. *J. Geophys. Res.*, **88**, 5806–5814.

Kheisin, D. E. and V. A. Likhomanov (1973). An experimental determination of the specific energy of mechanical crushing of ice by impact. *Probl. Arkt. Antarkt.*, **41**, 69–77.

Koschny, D. and E. Grün (2001). Impact into ice-silicate mixtures: Crater morphologies, volumes, depth-to-diameter ratios, and yield. *Icarus*, **154**, 391–401.

Kry, P. R. (1978). A statistical prediction of effective ice crushing stresses on wide structures. *IAHR Symposium on Ice*, Lulea, Sweden.

Kuehn, G. A., E. M. Schulson, D. E. Jones and J. Zhang (1993). The compressive strength of ice cubes of different sizes. *11th International Symposium and Exhibit on Offshore Mechanics and Arctic Engineering*, Calgary, Alberta, Canada, Transactions of the ASME.

Lange, M. A. and T. J. Ahrens (1982). Impact cratering in ice- and ice-silicate targets: An experimental assessment. *Lunar Planet. Sci.*, **XIII**, 415–416.

Lawn, B. (1995). *Fracture of Brittle Solids*, 2nd edn. Cambridge Solid State Science. Cambridge: Cambridge University Press.

Mangold, N. (2005). High latitude patterned grounds on Mars: Classification, distribution and climatic control. *Icarus*, **174**, 336–359.

Masterson, D. M. and R. Frederking (1993). Local contact pressures in ship ice and structure ice interactions. *Cold Reg. Sci. Technol.*, **21**, 169–185.

Masterson, D. M. and R. Frederking, Eds. (2006). Experience with the Canadian Standards Association Offshore Structures Code. *16th International Offshore and Polar Engineering Conference*, ISOPE'06, San Francisco, CA, ISOPE.

Masterson, D. M. and P. A. Spencer (2001). Ice force versus aspect ratio. *IUTAM Symposium on Scaling Laws in Ice Mechanics and Ice Dynamics: Solid Mechanics and Its Applications*, Fairbanks, Alaska. Dordrecht: Kluwer Academic Publishers.

Masterson, D. M., R. Frederking, B. Wright, T. Karna and W. P. Maddock (2007). A revised ice pressure-area curve. *POAC 2007*, Dalian, China.

Melanson, P. M., I. L. Meglis and I. J. Jordaan (1999). Microstructural change in ice: I. Constant-deformation-rate tests under triaxial stress conditions. *J. Glaciol.*, **45**, 417–422.

Michel, B. (1978). *Ice Mechanics*. Quebec: Laval University Press.

Michel, B. and D. Blanchet (1983). Indentation of an S2 floating ice sheet in the brittle range. *Ann. Glaciol.*, **4**, 180–187.

Moen, E. and S. Jacobsen (2008). Ice abrasion data on concrete structures – an overview. *Proc. Workshop on Ice Abrasion*, Helsingfors, Finland, Oct. 25–26, 2007, The Nordic Concrete Federation, 2008.

Muggeridge, K. J. and I. J. Jordaan (1999). Microstructural change in ice: III: Observations from an iceberg impact zone. *J. Glaciol.*, **45**, 449–455.

Palmer, A. C. and I. Johnston (2001). Ice velocity effects and ice force scaling. *IUTAM Symposium on Scaling Laws in Ice Mechanics and Ice Dynamics*, Fairbanks, Alaska. Dordrecht: Kluwer Acadmic Publishers.

Palmer, A. C. and T. J. O. Sanderson (1991). Fractal crushing of ice and brittle solids. *Proc. R. Soc. Lond. A*, **433**, 469–477.

Renshaw, C. E. and E. M. Schulson (2001). Universal behavior in compressive failure of brittle materials. *Nature*, **412**, 897–900.

Sanderson, T. J. O. (1988). *Ice Mechanics: Risks to Offshore Structures*. London: Graham & Trotman.

Schulson, E. M. (1990). On the ductile to brittle transition in ice under compression: Physical processes. *Ice Technology Conference*, Cambridge, UK.

Schulson, E. M. and E. T. Gratz (1999). The brittle compressive failure of orthotropic ice under triaxial loading. *Acta Mater.*, **47**, 745–755.

Singh, S. K., I. J. Jordaan, J. Xiao and P. A. Spencer (1995). The flow properties of crushed ice. *Trans. ASME*, **117**, 276–282.

Sodhi, D. S. (1998). Nonsimultaneous crushing during edge indentation of freshwater ice sheets. *Cold Reg. Sci. Technol.*, **27**, 179–195.

Sodhi, D. S. (2001). Crushing failure during ice–structure interactions. *Eng. Fract. Mech.*, **68**, 1889–1921.

Sodhi, D. S., T. Takeuchi, N. Nakazawa, S. Akagama and S. Hiroshi (1998). Medium-scale indentation tests on sea ice at various speeds. *Cold Reg. Sci. Technol.*, **28**, 161–182.

Takeuchi, T., T. Masaki, S. Akagama *et al.* (1997). Medium-scale field indentation tests (MSFIT): ice failure characteristics in ice/structure interactions. *7th International Offshore and Polar Engineering Conference*, Honolulu, Hawaii.

Takeuchi, T., M. Sakai, S. Akagama, N. Nakazawa and H. Saeki (2001). On the factors influencing the scaling of ice forces. *IUTAM Symposium on Scaling Laws in Ice Mechanics and Ice Dynamics*, Fairbanks, Alaska. Dordrecht: Kluwer Academic Publishers.

Timco, G. W. (1986). Indentation and penetration of edge-loaded freshwater ice sheets in the brittle range. In *Offshore Mechanics and Arctic Engineering Symposium*, pp. 444–452.

Timco, G. W. (1987). Ice structure interaction tests with ice containing flaws. *J. Glaciol.*, **33**, 186–194.

Timco, G. W. and R. M. W. Frederking (1993). Laboratory impact tests on freshwater ice. *Cold Reg. Sci. Technol.*, **22**, 77–97.

Timco, G. W. and M. Johnston (2003). Ice loads on the Molikpaq in the Canadian Beaufort Sea. *Cold Reg. Sci. Technol.*, **37**, 51–68.

Timco, G. W. and M. Johnston (2004). Ice loads on the caisson structures in the Canadian Beaufort Sea. *Cold Reg. Sci. Technol.*, **38**, 185–209.

Timco, G. W. and I. J. Jordaan (1987). Time-series variation in ice crushing. *9th International Conference on Port and Ocean Engineering under Arctic Conditions*, Fairbanks, Alaska.

Tuhkuri, J. (1996). *Experimental Investigations and Computational Fracture Mechanics Modelling of Brittle Ice Fragmentation*. Mechanical Engineering Series 120. Helsinki, Finland: ACTA Polytechnica Scandinavica.

Tuhkuri, J. (2001). Deformation and failure of ice cover under compression. *IUTAM Symposium on Scaling Laws in Ice Mechanics and Ice Dynamics*, Fairbanks, Alaska. Dordrecht: Kluwer Academic Publishers.

Wang, Y. S. and J. P. Poplin (1986). Laboratory compressive tests of sea ice at low strain rates from a field test program. *Fifth International Offshore Mechanics and Arctic Engineering Conference*, New Orleans, LA, ASME.

Weibull, W. (1951). A statistical distribution function of wide applicability. *J. Appl. Mech.*, **18**, 293–297.

Weiss, J. (2001). Fracture and fragmentation of ice: a fractal analysis of scale variance. *Eng. Fract. Mech.*, **68**, 1975–2012.

Wright, B. D. and G. W. Timco (1994). A review of ice forces and failure modes on the Molikpaq. *12th IAHR Symposium on Ice*, Trondheim, Norway, IAHR.

15

Fracture of the ice cover on the Arctic Ocean

Prologue In the late 1980s, following a presentation by one of us during a meeting of the Advisory Board of Dartmouth's Ice Research Laboratory, R. Weaver of the University of Colorado pointed out that he had seen in Defense Meteorological Satellite images of the winter sea ice cover on the Beaufort Sea features that resembled the wing cracks that one of us had just seen in laboratory specimens and had reported to the Board. The features, Figure 15.1, were dispersed amongst a variety of leads and open cracks and had the distinct markings of wing cracks: extensions, wide at the mouth and sharp at the tip, that had formed out-of-plane from the tips of inclined primary cracks. We estimated from wing-crack mechanics (Chapter 11) that a far-field compressive stress of around 3 kPa would have been needed to create them. W. D. Hibler then calculated the probable stress within the ice cover at the time of the sighting, from the historical record of wind fields, and obtained a value of between 7 and 14 kPa. We published the results shortly thereafter (Schulson and Hibler, 1991). Thus began an enquiry that forms the basis of this chapter and continues as we write: Is the physics of fracture independent of spatial scale?

15.1 Introduction

The sea ice cover on the Arctic Ocean during winter, as revealed through satellite images, generally contains oriented features. They appear as long narrow lineaments within both the seasonal and the perennial covers (Marko and Thomson, 1977; Erlingsson, 1988; Walter and Overland, 1993; Lindsay and Rothrock, 1995; Kwok *et al.*, 1995; Overland *et al.*, 1995, 1998; Stern and Rothrock, 1995; Walter *et al.*, 1995; Kwok, 1998, 2001, 2006; Richter-Menge *et al.*, 2002; Stern and Moritz, 2002; Schulson, 2004; Kwok and Coon, 2006) and can run hundreds and occasionally thousands of kilometers across the Arctic Basin. The lower limit on their length is thought to be below the resolution capability ($\sim$10 m) of current satellite imaging

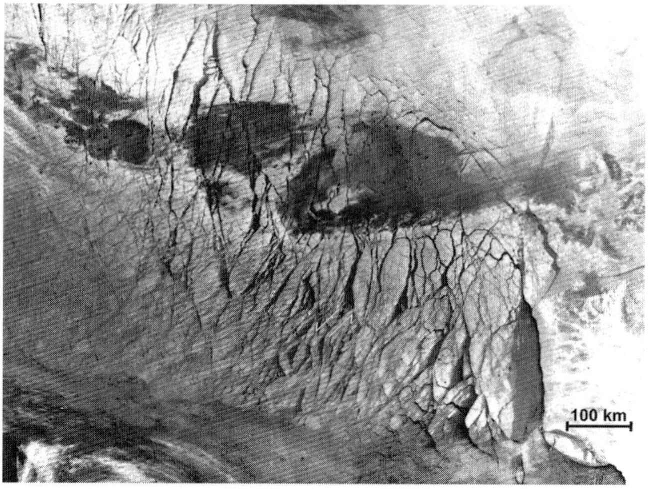

Figure 15.1. Photograph of cracks within the first-year sea ice cover on the Beaufort Sea. Banks Island is shown near the right-hand side under the scale mark. Note the wing-like character of the triangular-shaped features stemming up and down from the opposite tips of an inclined, primary crack between the arrows. The photograph was obtained on 11 February 1983 from a satellite using infrared sensors of 0.6 km resolution. The image was obtained from film transparencies provided through the U.S. Air Force Defense Meteorological Satellite Program. (The transparencies are archived at NOAA and at the University of Colorado, CIRES/National Snow and Ice Data Center. From Schulson and Hibler, 1991.)

systems. Lineaments often intersect in an acute angle (typically $2\theta \sim 20$–$50°$) to form diamond-shaped patterns that appear to be independent of spatial scale. The features are created through the relative movement of the ice cover, mainly under the force of wind during the passage of weather systems (Richter-Menge *et al.*, 2002) – i.e., by opening, closing or shear, or through combinations of these processes (Kwok, 2001). Consequently, they contain either open water, new ice, rafted ice or ridged ice. As a result, they may be viewed as an expression of the way the ice deforms. Kwok (2001) terms them *linear kinematic features* or *LKFs*. Others (Erlingsson, 1988; Overland *et al.*, 1998) term them *slip lines*.[1] We term them *fractures*. In short, lineaments appear to be ubiquitous features within the sea ice cover during winter (from mid September to mid June), marking localized zones of brittle failure where velocity gradients are spatially discontinuous and, hence, where the rates of shear and divergence/convergence and the rate of rotation or vorticity (defined below) are concentrated. For example, the strain rate typically reaches around $3\%\,d^{-1}$ or $3 \times 10^{-7}\,s^{-1}$, when computed from displacements that

[1] The term "slip line" as used within the present context should not be confused with the same term that is used within the context of crystal plasticity.

occur within a 10 km cell over a period of three days (Kwok, 2001; 2006). When computed on smaller spatial and/or temporal scales, the average rate as well as the dispersion about the mean is significantly greater (Marsan *et al.*, 2004). The lineaments are important, because they delineate regions where inelastic deformation affects the ice-thickness distribution (Hibler, 2003; Vavrus and Harrison, 2003; Kwok, 2006) which, in turn, affects the transfer of heat from the ocean to the atmosphere (Maykut, 1982; Heil and Hibler, 2002). Ice thickness also affects the rate of freezing and hence the oceanic salt flux. In addition, localized shearing appears to cause upwelling of the pycnocline[2] (McPhee *et al.*, 2005) which may enhance the transfer of heat from the ocean to the ice and thereby accelerate the rate of thinning (Kwok, 2006).

The questions are: how do lineaments/fractures form; what role do they play in the subsequent deformation of the cover; and, from the materials perspective, why does a body whose temperature is so close to its melting point ($T > 0.9\ T_{mp}$) exhibit brittle behavior when deforming so slowly? We consider each in turn.

15.2 Formation of sliding lineaments/shear faults

Observations are few of the actual processes that lead to the creation of the deformation features. Generally, they are detected some time after they form. Here, we review the observations. They include fortuitously one example of the processes in play, an example that offers significant physical insight.

15.2.1 Observations and interpretation

Figure 15.2 shows a high-resolution scene within the pack ice near the Canadian Archipelago (Schulson, 2004). The image is of 29 m pixel size and was produced on 25 March 2000 by Landsat-7, visible band. The scene encompasses an area 183 km in width (ENE–WSW) by 170 km in length (NNW–SSE). Long lineaments appear dark, because they delineate lines of low albedo associated with either open water or thin ice. Fifteen of the more prominent ones are labeled L1 through L15. Although labeled separately, L2 and L10 may actually be part of the same feature. All appear to be oriented: the majority exhibit a strong N–S component; the minority (L1–L3 and L10), a strong E–W component. The fact that the features are oriented is fundamental to their origin, but the orientation itself is probably not fundamental: it depends, we expect, primarily on the stress state within the cover and not on the cardinal points of the compass, as we discuss below.

[2] The pycnocline is a stable layer across which the density of seawater changes rapidly, owing to changes in temperature and/or salinity.

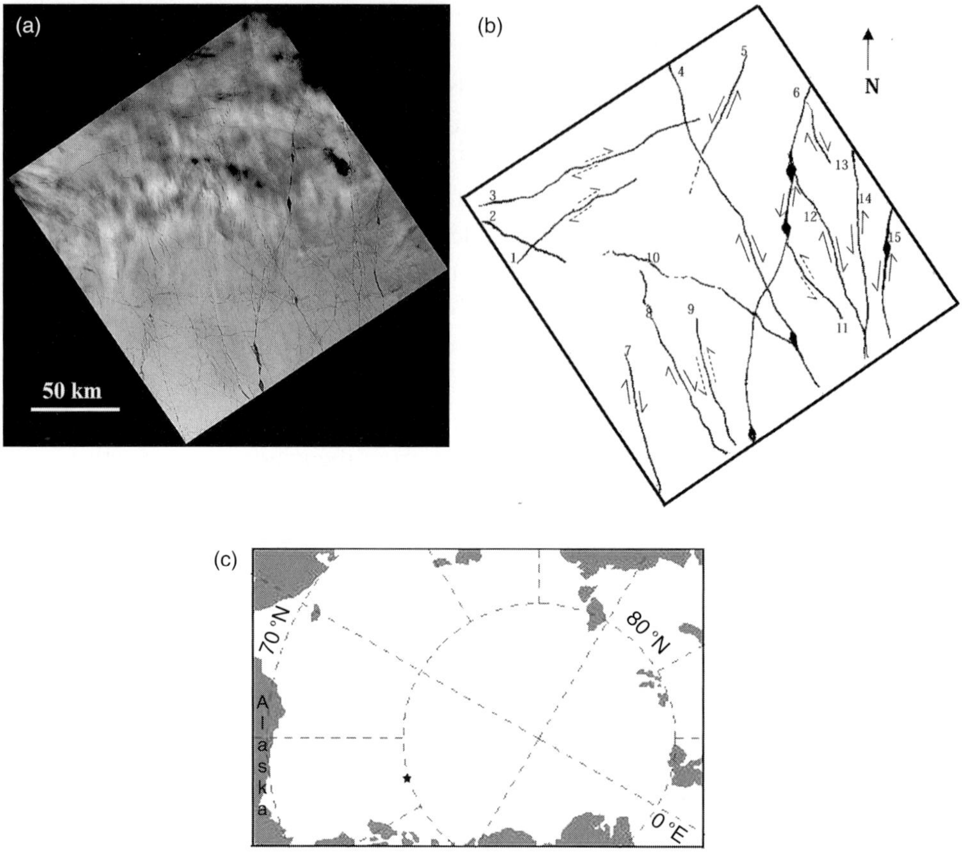

Figure 15.2. (a) Landsat-7 image of the sea ice cover on the Arctic Ocean, 25 March 2000, centered approximately at 80.0° N/135.7° W. North is up. Pixel size is 29 m. The scale is an average of the horizontal and vertical scales. (b) Sketch of (a) where the most prominent lineaments are labeled L1–L15. The half-arrows denote either right-lateral or left-lateral relative movement. (c) Map showing the location (star) of the scene in (a). (All from Schulson, 2004.)

Also evident within Figure 15.2 are rhomboidal-shaped openings: three on L6, one on L15 and a deformed opening on the southern segment of L4. Smaller rhomboids decorate the other lineaments, as exemplified in Figure 15.3 on L7. While the lineaments themselves indicate local divergence (otherwise they could not be seen), the rhomboids indicate relative shear displacement. As evident from their shape, the displacement is of right-lateral (RL) character for the sub-parallel set composed of L4, L7, L8, L12 and L13, and of left-lateral (LL) character for the set composed of L5, L6, L14 and L15. (Handedness is defined by looking down and across the feature.) Left-lateral

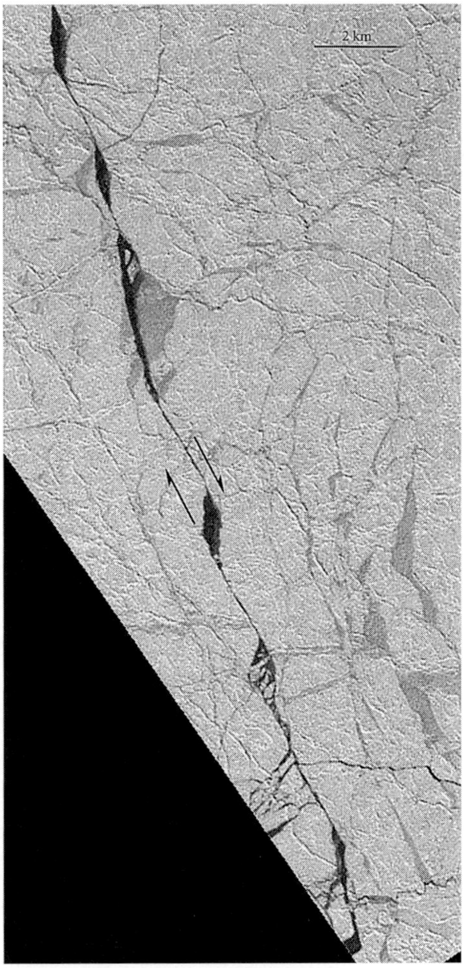

Figure 15.3. Rhomboidal openings on L7 of Figures 15.2(a) and (b), resulting from right-lateral shearing. (From Schulson, 2004.)

displacement plus instances of right-lateral displacement are evident within the scene shown in Figure 15.1 as well. Higher-resolution images of the Landsat-7 scene (see Schulson, 2004) reveal RL displacement along L1 and L3, which are almost orthogonal to the first RL set, plus LL displacement along L9 and L11, which intersect the first LL set at an acute angle of 50° to 60°. The sliding appears to have occurred episodically or intermittently, as evident from the longitudinal variation in contrast within most of the rhomboids (evident more clearly on the original images).

Conjugate fracture is apparent within the Landsat-7 scene. This is evident from the fact that the members of the first set of RL and LL lineaments intersect at an average acute angle of $2\theta = 38 \pm 5°$. Particularly striking is the similarity in appearance of conjugate lineaments L4 and L6 to conjugate Coulombic shear faults that developed within specimens of first-year arctic sea ice harvested from another winter cover and then biaxially compressed under a small degree of confinement to terminal failure within the brittle regime (Schulson, 2006b), Figure 15.4.

The other significant features within the Landsat scene are two kinds of shorter, secondary lineaments. One kind, shown in Figure 15.5a from near the midpoint of L8, resembles the arctic wing crack mentioned in the Prologue. The wings are oriented approximately in the N–S direction and appear to have formed through RL sliding. The other kind forms as a set on one side of a parent lineament, as shown in Figure 15.5b from L4 just above the intersection with L6, and are again oriented mainly in the N–S direction. In effect, this second kind of secondary crack creates a series of slender columns, fixed on one end and free on the other, reminiscent of the comb crack seen in laboratory specimens loaded under biaxial compression (Chapter 12).

Although the Landsat-7 scene is specific in terms of space and time, the features are not unique. They are reminiscent of those seen in Advanced Very High Resolution Radar-Infrared (AVHRR-IR) satellite imagery of the ice cover on the Beaufort Sea during earlier winters (Walter and Overland, 1993; Walter *et al.*, 1995; Overland *et al.*, 1995) and of those seen in five years (1996–2000) of RADARSAT Geophysical Processor System (RGPS) data on ice sheet deformation within the Arctic Basin (Kwok, 2006). They are reminiscent as well of smaller features within the basin, seen by Schulson (2004) from a Twin Otter aircraft when flying iso-longitudinally (148° W) along the path 70° N to 73° N during the period 6–10 April 2003.

Under what stress state did they form? We do not know with certainty. However, we suspect that compressive states dominated. We base this view on two points: on the recurrence of compressive stress states within the winter sea ice cover (Coon *et al.*, 1998; Richter-Menge and Elder, 1998; Richter-Menge *et al.*, 2002); and on the similarity in appearance of the arctic features to ones that form within laboratory specimens of sea ice when loaded to terminal failure under compression (Schulson, 2004). The divergence that is evident from the image is not inconsistent with compressive loading, because, as determined from an examination of a separate time-series of SAR images obtained during the same period (days 81, 84 and 86 of 2000) of the same place (Schulson, 2004), most of the lineaments had formed and were inactive when the scene was surveyed, implying that the driving meteorological event(s) had passed, thereby allowing the cover to open up.

Thus, following Schulson (2004), we propose that sliding lineaments form during an event or series of events that induces within the ice cover a

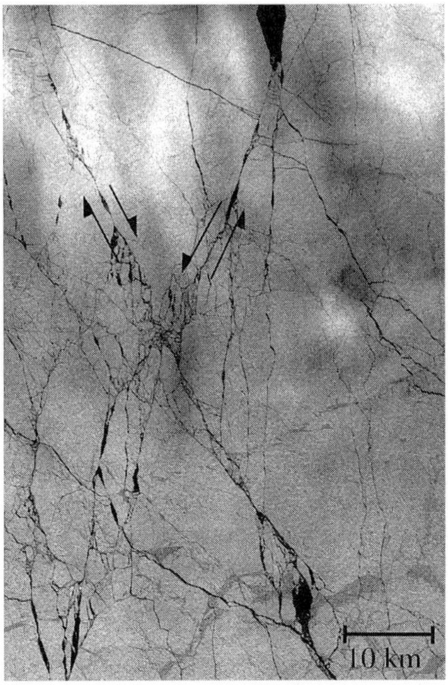

(a)

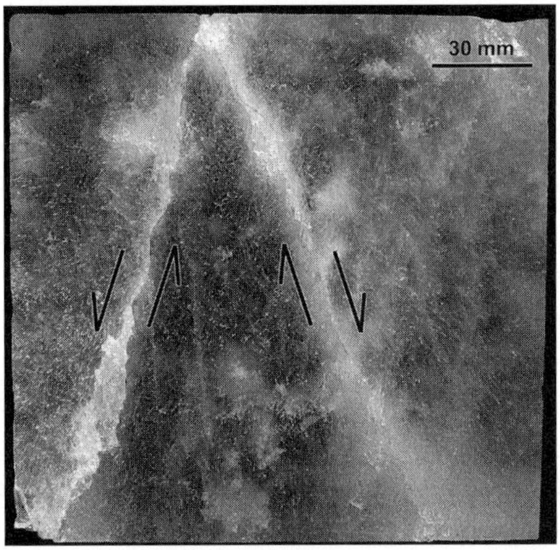

(b)

Figure 15.4. (a) Intersecting Landsat-7 lineaments L4 (right-lateral) and L6 (left-lateral) from Figure 15.2(a). (From Schulson, 2004.) (b) Intersecting pair of right-lateral and left-lateral Coulombic shear faults in a specimen of first-year, columnar-grained S2 sea ice harvested from the winter ice cover on the Arctic Ocean and biaxially compressed across the columns along the loading path $\sigma_{22}=0.1\sigma_{11}$ to terminal failure within the brittle regime. The long axes of the columnar grains are oriented normal to the page. The maximum principal stress was applied in the vertical direction and the confining stress was applied in the horizontal direction. (From Fortt, 2006.)

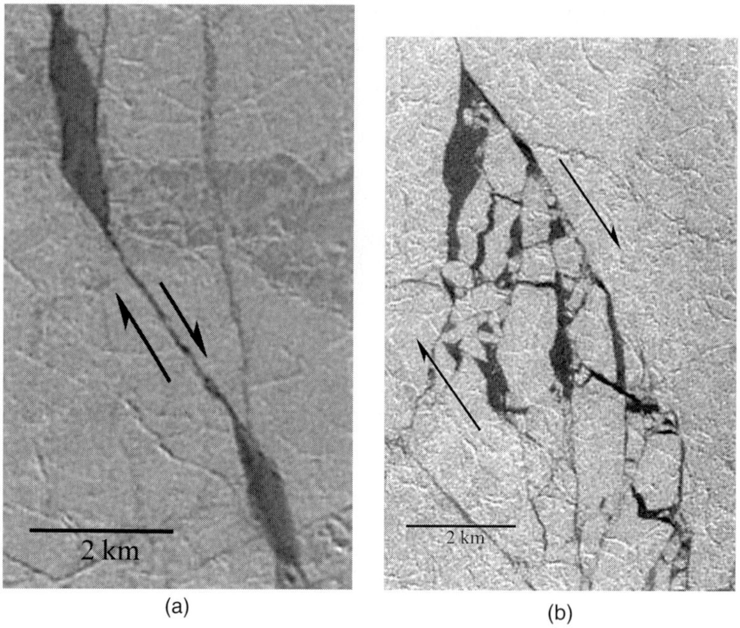

Figure 15.5. (a) Wing-like secondary cracks on lineament L8 of Figure 15.2(a). (b) Comb-like secondary crack on lineament L4 of Figure 15.2(a). (Both from Schulson, 2004.)

far-field compressive stress state in which the maximum principal stress σ_1 (i.e., the most compressive stress) is oriented counter-clockwise by an acute angle θ from left-lateral lineaments and clockwise by the same angle from right-lateral lineaments (more below). On this basis, the N–S oriented features within the Landsat-7 scene probably formed under a state of stress in which the direction of maximum principal stress was approximately parallel to the N–S direction. Similarly, the features bearing a more E–W orientation probably formed during a different event, where the maximum principal stress within the cover was oriented approximately in the E–W direction. In other words, and in keeping with an interpretation by Marko and Thomson (1977) following an examination of an earlier ice cover, sliding lineaments appear to be analogous to strike-slip faults within Earth's crust.

The skeptic might disagree with this interpretation. S/he might argue that the larger-scale fracture features developed not under biaxial compression, but under combined compressive and tensile loading, aided perhaps by thermal contraction, in recognition of an interpretation by Muehlberger (1961) of small-angle conjugate fractures in rock. We cannot rule out this possibility, although it seems unlikely. Under combined tension and compression applied orthogonally to each other, we would expect ice to fracture along planes perpendicular to the tensile direction and parallel to the compressive direction, not along conjugate sets. On the possibility of tensile stresses of thermal origin themselves being the primary driving force, we rule that out because the lineaments exhibit organization: thermal cracks are expected to be randomly oriented. This is not to say that thermal cracks play no role at all, for as we note below, ones that are appropriately oriented with respect to the far-field stress state may act as primary cracks from which secondary cracks initiate, in the way that grain boundary cracks in laboratory specimens act as primary cracks (Chapters 11 and 12).

The skeptic might also argue that elastic buckling is a more likely mode of compressive failure than faulting, given the extreme slenderness of the ice cover: thickness O(1 m) and area O(10^7 km^2). The problem with that argument is that the expected buckling stress is significantly higher than the highest stress measured *in situ*. Specifically, if the floating ice sheet behaves like a plate on an elastic foundation, then under the restorative force arising from the potential energy gradient the buckling stress may be estimated from the relationship (Kerr, 1978; Sanderson, 1988):

$$\sigma_{\mathrm{b}} = 2.6\left\{\frac{Et\rho g}{12(1-\nu^2)}\right\}^{1/2} \tag{15.1}$$

where E denotes Young's modulus, t thickness, ρ the density of seawater, g the acceleration of gravity and ν Poisson's ratio. Appropriate values ($t = 1.8$–2.8 m (Laxon *et al.*, 2003), $\rho = 1028$ kg m^{-3}, $g = 9.8$ m s^{-2}, $\nu = 0.3$ and $E = 8$ GPa (Chapter 4) give $\sigma_{\mathrm{b}} \sim 10$–$12$ MPa. The highest stress measured *in situ* (Richter-Menge *et al.* 1998, 2002) is lower than this by about a factor of 40.

15.2.2 Mechanism of fault initiation

On the mechanism through which sliding lineaments/shear faults initiate, RADARSAT-RGPS images shown in Figure 15.6 offer some insight. They were obtained by Kwok (personal communication, 2003) of the winter cover during the 1997–8 SHEBA experiment (Surface Heat Energy Budget of the Arctic, see Perovich *et al.*, 1999). The SHEBA ice camp is centered in the images and was located on day 318 of 1997 at 76.1 °N/145.8 °W. Figure 15.6a was obtained on day 302 of 1997, Figure 15.6b on day 314, and Figure 15.6c on day 317; Figure 15.6d shows ice motion vectors that Kwok (personal communication, 2003) determined from displacements that occurred between days 314 and 317. Animation (see www-radar.jpl.nasa.gov/rgps/image_files/combine_small.gif) shows that at the time the images were obtained the field was rotating clockwise. Inhomogeneous deformation is evident from the distortion in the superimposed 50 km × 50 km boundary that

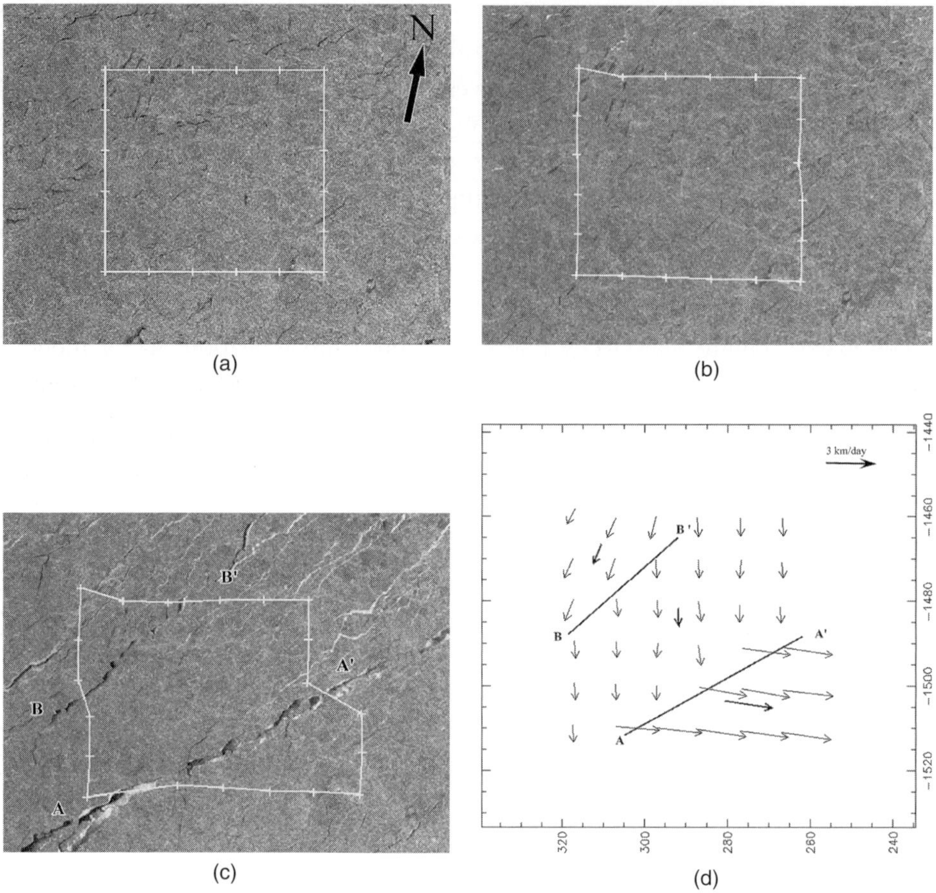

Figure 15.6. RADARSAT images of sea ice taken during the SHEBA experiment. (a) The image was taken on day 302 of 1997.The square is 50 km on side. (b) Day 314 of 1997. Note the distortion in the grid. (c) Day 317 of 1997. Note the two left-lateral faults A–A′ and B–B′ and the two wing-like cracks on fault B–B′ near its intersection with the northern border of the grid. (d) Ice motion vectors from RGPS data of the deformation are localized around faults A–A′ and B–B′ for a time step of 3.0087 days between days 314 and 317 of 1997. The vectors are drawn on a Cartesian coordinate system with the origin at the North Pole; the x-axis is parallel to the 45 east meridian and the y-axis is parallel to the 145 east meridian. The numbers are in units of kilometers. The center of the plot is at approximately (−1480, 290) kilometers. (All courtesy of R. Kwok of Jet Propulsion Laboratory and H. Stern of Applied Physics Laboratory, University of Washington. RADARSAT imagery copyright CSA 2002.)

surrounds the camp. Of interest are the two sub-parallel faults that formed between days 314 and 317, denoted A–A′ and B–B′. Both reflect the discontinuities in the velocity gradients evident in Figure 15.6d and both exhibit left-lateral sliding. Of particular interest in relation to their formation is the en echelon pair of wing-like cracks that appear along fault B–B′, on either side of its intersection with the northern segment of the boundary. Wing-like cracks also define fault A–A′, although in this case they are distorted owing to the greater opening. The parent/primary cracks from which the B–B′ wings sprouted (Fig. 15.6c) are visible in the image obtained three days earlier (Fig. 15.6b), implying that the faults develop through the linking-up of the secondary features, just as they appear to do within laboratory specimens (Chapter 12).

Until more observations are made of the early stages of lineament/shear fault formation, we do not know whether initiation through wing-crack interaction is typical. What does appear to be typical is the deformation field attending the features. This may be seen as follows (Schulson, 2004): Assuming that the ice sheet deforms inelastically in plane strain (as well as under plane stress), the rates of divergence $\dot{D}$ and shear $\dot{S}$ as well as the vorticity V may be defined by the relationships:

$$\dot{D} = (\delta\dot{u}/\delta x + \delta\dot{v}/\delta y), \tag{15.2}$$

$$\dot{S} = [(\delta\dot{u}/\delta x - \delta\dot{v}/\delta y)^2 + (\delta\dot{u}/\delta y + \delta\dot{v}/\delta x)^2]^{1/2}, \tag{15.3}$$

$$V = \delta\dot{v}/\delta x - \delta\dot{u}/\delta y \tag{15.4}$$

where x and y denote coordinates within a rectangular coordinate system centered on a fault (e.g., on fault A–A′, Fig. 15.6c) and u and v denote relative displacements along and perpendicular to the fault, respectively, where the displacement of the cover below each fault is taken to be zero; i.e., $u = v = 0$ for $y < 0$); the partials in x parallel to the faults are zero, by definition. The partial derivatives are evaluated by assuming that displacement occurs continuously during the period between images, which in the SHEBA case was three days, and by taking the cell size to be the size of the reference grid, which is 10 km in the SHEBA scene. Table 15.1 lists results of such analysis for the SHEBA faults of Figure 15.6. They show

Table 15.1. *The deformation rate localized within the vicinity of faults A–A′ and B–B′ of Figure 15.6c*

	Relative motion (km)		Relative velocity (km/day)		Partial derivatives (day^{-1})				Deformation rate (day^{-1})		
Fault	sliding along fault, u	opening normal to fault, v	sliding $\dot{u}$	opening $\dot{v}$	$\frac{\partial\dot{u}}{\partial x}$	$\frac{\partial\dot{v}}{\partial y}$	$\frac{\partial\dot{u}}{\partial y}$	$\frac{\partial\dot{v}}{\partial x}$	divergence	shear	vorticity
A–A′	11.02	3.25	3.66	1.08	0	0.11	0.36	0	0.11	0.38	0.36
B–B′	1.93	0.88	0.64	0.29	0	0.029	0.064	0	0.029	0.069	0.064

From RGPS data provided by R. Kwok (personal communication, 2003).

significant differences across the scene, implying that the deformation is not uniform, in keeping with the RADARSAT images. More importantly in relation to the point in mind, they show that the relative velocities and deformation rates localized within the vicinity of faults A–A′ and B–B′ are of similar magnitude to those reported for earlier lineaments/faults (Kwok, 2001, 2006; Stern and Moritz, 2002). On this basis, therefore, it seems reasonable to assume that the deformation field attending the faults is also similar.

Was the stress state under which the SHEBA features formed consistent with their identification as compressive shear faults? To address that point, Schulson (2004) first determined from the deformation field the directions of principal strain, which he equated to the directions of principal stress by assuming that the ice cover behaved in an isotropic manner in its plane, and then estimated from wing-crack mechanics the magnitude of the maximum principal stress. Accordingly, the components of the strain rate tensor are defined as:

$$\dot{\varepsilon}_{xx} = \delta\dot{u}/\delta x = 0, \tag{15.5}$$

$$\dot{\varepsilon}_{yy} = \delta\dot{v}/\delta y, \tag{15.6}$$

$$\dot{\varepsilon}_{xy} = (\mathrm{d}\dot{u}/\mathrm{d}y + \mathrm{d}\dot{v}/\mathrm{d}x)/2 = (\mathrm{d}\dot{u}/\mathrm{d}y)/2. \tag{15.7}$$

Values are listed in Table 15.2. From these values, the angle θ between the strike of the fault and the direction of maximum principal strain (taken as the most compressive strain) is obtained from the relationship:

$$\tan 2\theta = 2\dot{\varepsilon}_{xy}/(\dot{\varepsilon}_{xx} - \dot{\varepsilon}_{yy}) = -u/v. \tag{15.8}$$

The analysis shows that the direction of maximum principal strain, and hence of maximum principal stress, is oriented counter-clockwise from the strike of both faults, by $\theta = 36.8°$ from A–A′ and by $\theta = 32.7°$ from B–B′ (i.e., in a roughly northerly direction). Also, the B–B′ wing cracks tend to be parallel to that direction. Thus, the orientation of the maximum principal stress is consistent with the

Table 15.2. *The direction of maximum principal stress for the two faults shown in Figure 15.6c from the data in Table 15.1*

Fault	Strain rate (day^{-1}) $\dot{\varepsilon}_{xx}$	$\dot{\varepsilon}_{yy}$	$\dot{\varepsilon}_{xy}$	$\tan 2\theta = \frac{2\dot{\varepsilon}_{xy}}{(\dot{\varepsilon}_{yy} - \dot{\varepsilon}_{xx})}$	θ (degrees)
A–A′	0	0.11	0.18	3.39	36.8
B–B′	0	0.029	0.032	2.17	32.7

Relative to the strike of the fault the principal direction is rotated counter-clockwise by the acute angle θ.

identification of the faults and with their LL character, and consistent as well with the orientation of the wings.

On the magnitude of the principal stress, crack mechanics dictate that the stress to lengthen a wing may be estimated (for the case of uniaxial loading) from the relationship (Chapter 11):

$$\sigma_c = \frac{\sqrt{3}K_{Ic}(1+L)^{3/2}}{\sqrt{\pi c}(1-\mu_i)(\beta L + 1/(1+L)^{1/2})} \tag{15.9}$$

where again K_{Ic} denotes fracture toughness, $2c$ the length of the parent sliding crack and μ_i the internal friction coefficient and where L denotes the ratio of the wing-crack length l to the half-length of the parent crack $L = l/c$; β is a geometrical factor of value ~0.4 (Ashby and Hallam, 1986). From the geometry of the pair of wing cracks noted along fault B–B′ (i.e., $L \sim 0.8$ and $c \sim 3.5$ km) and upon taking the values $K_{Ic} \sim 100$ kPa m$^{1/2}$ (Chapter 9) and $\mu_i \sim 0.9$ (see below), we obtain the estimate that $\sigma_c \sim 36$ kPa. Moderate confinement would have increased this value somewhat. This estimate compares favorably with the measured value of ~30 kPa that Richter-Menge *et al.* (2002) obtained (see their Fig. 1a) from *in-situ* stress measurements (more below) within the SHEBA cover around the day that faults A–A′ and B–B′ were created.

15.2.3 Summary and further analysis

Thus, in answer to the first question posed in the Introduction, sliding lineaments appear to exhibit the character of Coulombic shear faults. While not all lineaments necessarily possess this character, the ones that do probably develop under a biaxial state of compressive stress that is marked by a moderate degree of confinement. The faults form, we propose, via the initiation, growth and interaction of secondary cracks that stem from appropriately oriented, pre-existing linear weaknesses or structural defects, such as thermal cracks, that serve as stress concentrators, as sketched in Figure 15.7. The process might be assisted by the creation of deformation damage ahead of the main front, in the manner of a process zone (Lockner *et al.*, 1991; Schulson, 2001). The speed with which the fault grows, once initiated, is not known, although judging from the crack velocity through sea ice of ~10 m s^{-1} at temperatures relevant to the ice cover (Petrenko and Gluschenkov, 1996), albeit obtained from measurements on a much smaller scale, we expect that the faults grew at a similar rate. The SHEBA faults, for instance, appear to have lengthened by more than 50 km in 3 days or >0.2 m s^{-1}, in keeping with this expectation. Faults form either conjugate sets, as sketched in Figure 15.7, where the members are oriented on either side of the direction of maximum principal stress, or single sets. The reason

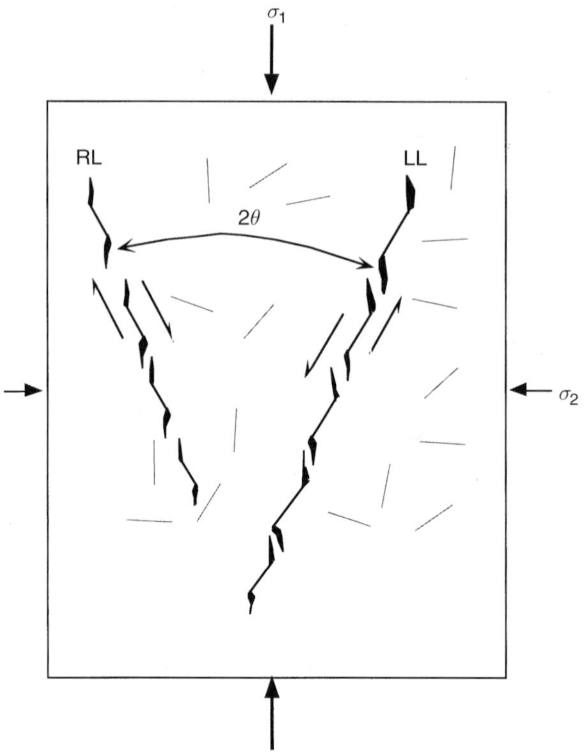

Figure 15.7. Schematic sketch of proposed wing-crack interaction mechanism for initiating intersecting faults. The straight lines denote pre-existing linear weaknesses of less than optimum orientation for frictional sliding under the imposed compressive stress state. (From Schulson, 2004.)

for one kind versus the other is not clear, although from the mechanics of strike-slip faults (Sylvester, 1988) the pattern may be related to the character of the deformation field and to whether it is one of pure shear (irrotational cum conjugate sets) or one of simple shear (rotational cum single set).

On the orientation of the faults about the principal stress direction, the theory of Coulombic failure (Jaeger and Cook, 1979, see Chapter 12) dictates that the internal coefficient of friction is the governing factor. Accordingly, the most favorably oriented weaknesses/structural defects from which the all-important secondary cracks develop are the ones on which the effective shear stress is a maximum. The effective shear stress is defined here in the same way as it is defined in Chapters 11 and 12; namely, as the difference between the applied shear stress resolved onto the plane of the defect (in reality, the ice sheet is a three-dimensional body, penetrated either partially or fully by the defect) and the product of the normal stress acting across that plane and the internal friction coefficient. From Chapter 12:

$$\tan 2\theta = 1/\mu_i. \tag{15.10}$$

where 2θ now denotes twice the acute angle between the strike of the defect and the principal stress direction or the included angle between conjugate features (Fig. 15.7). Equation (15.10) is the same as Equation (12.7). It approximates reasonably well the orientation of the fault that ultimately forms in laboratory specimens (Schulson *et al.*, 2006a). We assume this is true in the field as well. Variations in the internal friction coefficient, which could arise in the field through variations in temperature and sliding speed just as they do in the laboratory (Fortt and Schulson, 2007), could then account for the variations in orientation noted above, from the Landsat-7 value of $2\theta = 38 \pm 5°$ ($\mu_i = 1.3 \pm 0.2$) to the SHEBA value of $2\theta = 70°$ ($\mu_i = 0.36$) to the value of $2\theta \approx 20° - 50°$ ($\mu_i \approx 2.7$–0.84) for other lineaments noted in the Introduction to this chapter.

On the stress required to activate the defects and thus to begin the process of fault formation, an important dimension is defect length $2c$. This determines the factor by which the applied stress is concentrated, which from linear elastic fracture mechanics scales as $(2c)^{1/2}$. Given that the winter ice cover contains structural defects of lengths that span many orders of magnitude, from centimeters to over a kilometer, the stress concentration factor can vary by a factor of 10^3 or more. Of course, fracture toughness, K_c, is important as well, for that parameter determines the resistance to crack propagation. The defects, then, are expected to become activated under a far-field applied compressive stress σ_a that decreases as the size of the stress concentrator increases, scaling as $\sigma_a \propto [K_c/(2c)^{1/2}]f(\mu_i)$ where $f(\mu_i) = [(1+\mu_i^2)^{1/2} + \mu_i]/[(1+\mu_i^2)^{1/2} - \mu_i]$.

Within this scenario, confinement is important. It not only raises the stress to activate the defects, but, if too high, it suppresses frictional sliding altogether. Coulombic faults then cease to form. In that case, the cover is expected to deform not only in its plane, but also through its thickness, in the manner observed in a laboratory specimen (Schulson *et al.*, 2006b). The stress ratio $R = \sigma_2/\sigma_1$ that is required to completely suppress sliding in a direction within the plane of the cover is independent of size and, from the phenomenological analysis presented in Chapter 12, is given by the relationship:

$$R = \frac{\sigma_2}{\sigma_1} = \left[\frac{(1+\mu_i^2)^{1/2} - \mu_i}{(1+\mu_i^2)^{1/2} + \mu_i}\right]. \tag{15.11}$$

For instance, for the value $\mu_i = 1.3 \pm 0.2$, deduced from the Landsat-7 scene, we expect frictional sliding to be suppressed when $R = \sigma_2/\sigma_1 \approx 0.1$. In other words, the level of confinement under which in-plane sliding is expected to be suppressed is quite low.

A more complete understanding of fault formation requires information on the size distribution of structural defects, as well as to the final linking-up of the secondary cracks that stem from them. Additional insight must await satellite imagery of higher spatial and temporal resolution.

15.3 Rheological behavior of the winter ice cover: friction and fracture

We turn now to the second question posed in the Introduction and consider the rheological behavior of the winter sea ice cover. The situation is a complex one, where fractures and faults form intermittently and in that process contribute to the deformation of the cover, and then once formed contribute further to the deformation through intermittent opening/closing and sliding, complicated by "healing". In the language of the laboratory, the deformation might be viewed as a combination of terminal and post-terminal brittle compressive failure, exacerbated by the presence of linear stress concentrators of all sizes and accommodated by frictional sliding.

Current models of ice cover deformation incorporated within climate models are based not upon fracture and friction. Instead, they are based upon either elastic-plastic[3] (Coon *et al.*, 1974; Pritchard, 1975) or viscous-plastic (VP) (Hibler, 1979) mechanics. Elastic-viscous-plastic (Hunke and Dukowicz, 1997; Hunke, 2001) mechanics has also been incorporated, although in this case the elastic term is unphysical in that it was added to the viscous-plastic rheology for numerical reasons; the model is thus an approximation of viscous-plastic mechanics. A key component of the models is the failure envelope: stress states that end on the envelope lead to rate-independent plastic flow, while those that end inside the envelope lead to viscous flow/creep. Although the models offer reasonable predictions of ice-thickness distribution and attendant enthalpic effects, especially when they incorporate a Coulombic-shaped envelope that includes tensile strength (e.g., Zhang and Rothrock, 2005), they are challenged in their ability to predict oriented fractures/shear faults. An exception appears to be Hibler's VP model: when a random array of linear weaknesses of variable strength is incorporated, numerical modeling shows that the VP model leads to the formation of oriented shear faults of the kind described above (Hibler and Schulson, 2000; Hutchings and Hibler, 2003; Schulson and Hibler, 2004). Also, high-resolution calculations based upon an isotropic version of the VP model appear to capture some basic features of oriented faults (Overland and Ukita, 2000). The challenge, however, is to directly incorporate fracture mechanics and the distribution of flaw sizes.

[3] The plasticity invoked here is not the volume-conserving, pressure-insensitive von Mises plasticity of crystal physics, familiar to metallurgists and other materials scientists, but a broader kind of pressure-sensitive inelastic deformation.

Another concern with current models is the apparent assumption that the strength of the ice is independent of space and time. However, from an analysis of the stress record obtained during the SHEBA experiment, Weiss *et al.* (2007) found that the failure stress depends upon both time and space. We review this observation below.

The other concern is the assumption that under low stresses the ice cover deforms in a viscous manner. Implicit is the dictate that the strain rate increases with increasing stress. Years ago, Nye (1973) questioned that assumption. He postulated, instead, that the cover behaves in a brittle manner. As we noted above and show again below, SHEBA behavior sides with Nye. From an analysis of the SHEBA stress and displacement record, Weiss *et al.* (2007) found that the deformation rate of the ice cover does *not* scale with stress. Instead of behaving as a viscous-plastic or an elastic-viscous-plastic body, the cover appears to behave as an elastic-brittle one whose deformation is dominated by friction and fracture on a wide range of temporal and spatial scales.

15.3.1 Nested failure envelopes

Figure 15.8a shows the *in-situ* SHEBA stress states, plotted in principal stress space (Weiss *et al.*, 2007). The measurements cover the 1997–8 winter period, from mid October 1997 until the end of June 1998, and were obtained hourly from the "Baltimore" sensor[4] deployed within the camp (Richter-Menge *et al.*, 2002). In this depiction, compressive stresses have negative values, in keeping with the practice within the ice–ocean modeling community. The figure contains 6473 points, each point corresponding to a measurement. The data are plotted twice to reflect the (assumed) mechanical isotropy of the ice cover within the horizontal plane; hence the symmetry about the loading path $\sigma_1 = \sigma_2$. Plots from five other sensors, including one situated about 15 km away from the camp, give similar results (Weiss *et al.*, 2007).

Of particular interest is the boundary or the envelope of the stress states, shown by the dotted line. It exhibits Coulombic-like character and may be described by the same relationship that describes the Coulombic branch of the compressive failure envelope for specimens of first-year sea ice harvested from the winter cover and deformed in the laboratory (Chapter 12), namely:

$$\sigma_1 = \sigma_u + q\sigma_2 \tag{15.12}$$

where $\sigma_u = 250$ kPa denotes the unconfined compressive strength. (Equation (15.12) is the same as Equation (12.2).) For the ice sheet, σ_u is considerably lower than for a

[4] The sensor was a stiff, cylindrical device that housed wires loaded under tension. When frozen into a hole in the ice cover and then loaded biaxially by the *in-situ* stresses, the wall deformed. Correspondingly, the vibration frequency of the internal wires changed, leading to a measure of the stress state (Cox and Johnson, 1983).

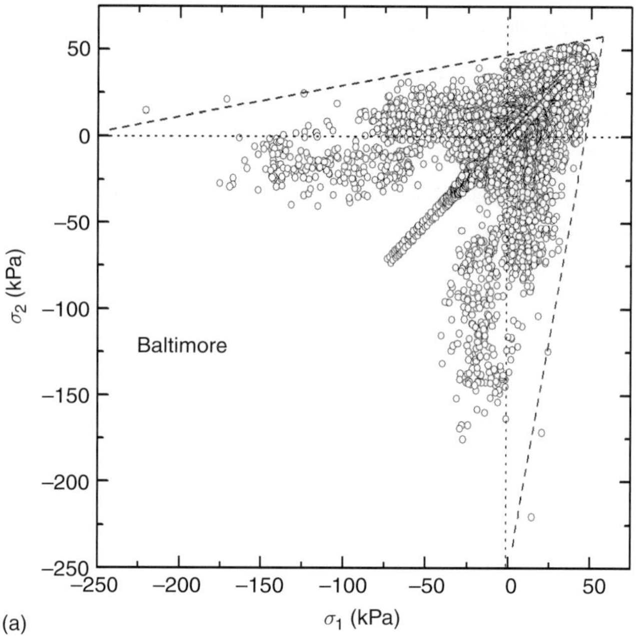

(a)

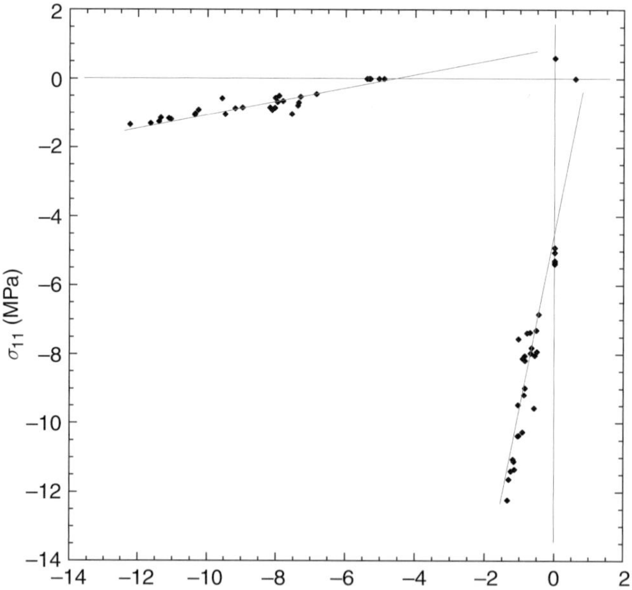

(b)

Figure 15.8. (a) Stress states within the winter sea ice cover on the Arctic Ocean during the SHEBA experiment, measured *in situ* at the "Baltimore" site by Richter-Menge *et al.* (2002) and plotted in principal stress space. (From Weiss *et al.*, 2007.) (b) Coulombic branch of the brittle compressive failure envelope (from Fig. 12.7) for columnar-grained S2 first-year sea ice. The material was harvested from the winter cover on the Arctic Ocean and then was loaded to terminal failure in the laboratory by compressing biaxially across the columns at $-10\,°C$ at $\dot{\varepsilon}_1 = 1.5 \times 10^{-2}\ s^{-1}$ along the loading path $\sigma_2 = 0.1\sigma_1$. Note that the slope of the laboratory envelope is the same as the slope of the field envelope (dotted line) that bounds the *in-situ* stresses shown in (a). (From Schulson *et al.*, 2006b.)

test specimen (for which $\sigma_u \sim 5$ MPa (Schulson *et al.*, 2006a)), presumably because the sheet contains larger stress concentrators. The slope, however, has the same value as the slope of the Coulombic branch of the envelope measured in the laboratory at $-10\,°C$, shown in Figure 15.8b (obtained from Figure 12.7), namely, $q = 5.2$. The implication of the common q-value is that the internal friction coefficient, which governs this parameter (as discussed in Section 12.6), is independent of scale. In other words, the field envelope appears to be a contracted version of the one obtained in the laboratory.

The degree of contraction depends not only upon the presence of larger defects, but also upon the period of time over which stresses are measured (Weiss *et al.*, 2007). For instance, Figure 15.9 compares the winter-long SHEBA envelope of Figure 15.8a with one obtained from stresses measured hourly over a 3-day period, from 10 January to 13 January 1998. (In this example the envelope is plotted not in terms of principal stresses, but in terms of shear stress, $\tau = (\sigma_1 - \sigma_2)/2$, and normal stress, $\sigma_n = (\sigma_1 + \sigma_2)/2$. To convert from $\tau{:}\sigma_n$ space to $\sigma_1{:}\sigma_2$ space, rotate by 45°.) The implication is that the ice cover is characterized not by a single failure envelope, independent of time and space, but by a set of closely nested envelopes, all characterized by the same slope but, owing to differences in the length of the attendant stress concentrators and to differences in the length of time that stress is allowed to build up, by different unconfined strengths. Thus, the concept of a single failure envelope, although meaningful in relation to the strength of a laboratory specimen, should be viewed with caution in relation to the ice cover. In other words, on the geophysical scale, "strength" depends upon space and time in a probabilistic manner.

To what processes do the SHEBA stresses relate? Weiss *et al.* (2007) attribute them principally to friction and fracture, in keeping with the interpretation of the laboratory failure stresses. Accordingly, stress states that lie along the Coulombic-like envelope appear to be related to smaller fractures located within the vicinity of the sensor, for the larger stresses that these states represent were not temporally correlated from sensor to sensor within the SHEBA array. On the other hand, stress states that lie below the boundary appear to be related to larger fractures more remote from the camp, for the smaller stresses that these states represent were temporally correlated to some degree. Friction and fracture also account for another characteristic of the SHEBA stresses; namely, their intermittency (Richter-Menge *et al.*, 2002). This activity relates to both the formation of new faults ($\sigma_u \neq 0$) and the sliding along pre-existing ones, either fresh ($\sigma_u = 0$) or partially healed ($\sigma_u > 0$): both kinds of event, judging from laboratory measurements (Schulson *et al.*, 2006a,b), are characterized by friction coefficients of similar magnitude. The other processes that contribute to the measured SHEBA stresses are thermal expansion and contraction, which lead to isotropic states

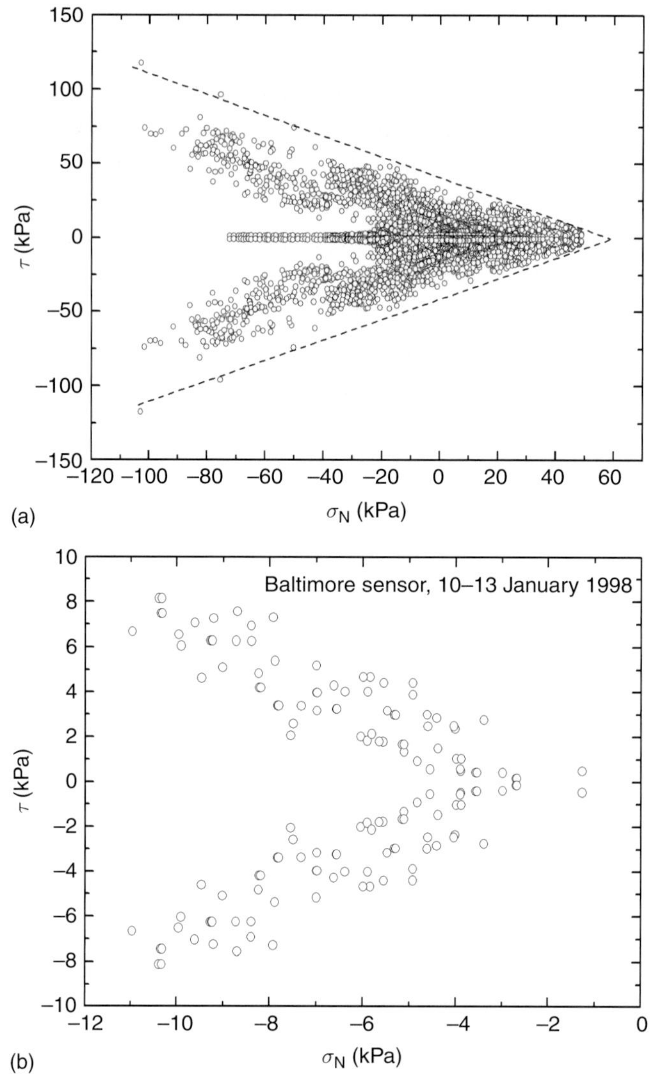

Figure 15.9. *In-situ* stress states within the SHEBA ice cover (a) measured over the SHEBA winter from mid October 1997 to end of June 1998, and (b) measured over the 3-day period from 10 to 13 January 1998. Note the different scales in (a) and (b). The comparison shows that the size of the envelope that bounds the stress states is temporally dependent and is smaller for the shorter-term, but that the slope of the envelope is temporally independent. (From Weiss *et al.*, 2007.)

(either tensile or compressive) superimposed upon those from frictional sliding and fracture.

The finding that the friction coefficient appears to be independent of scale may seem surprising. How, one might wonder, can the value of a temperature-dependent and a speed-dependent parameter (Chapter 12) (Fortt and Schulson, 2007) measured in the laboratory at a specific temperature and a specific sliding speed – namely, $-10\,°C$ and $0.4\times10^{-2}\,m\,s^{-1}$ – so closely match the value deduced for an ice sheet through which the temperature varies, say from around $-1.9\,°C$ to around $-20\,°C$, and across which sliding probably occurs at different speeds, say from around $3\,km\,d^{-1}$ or about $3\times10^{-2}\,m\,s^{-1}$ to lower? The answer is not clear. Perhaps the SHEBA coefficient represents a value integrated over the ranges of temperature and sliding speed encountered in the field.

On the same issue, one might also question why an arctic fault, which must contain much larger asperities than one created within a test specimen, appears not to exhibit much greater resistance to sliding. Perhaps the answer lies in the fractal character of the fracture surface. If we define its roughness in terms of the standard deviation of elevation (Russ, 1994), then the coordinate of the surface normal to the plane of the fault z is statistically related to the coordinates x and y in the plane, such that $\Delta z \sim (\Delta r)^H$ where $\Delta r^2 = \Delta x^2 + \Delta y^2$ and H is termed the Hurst exponent (Turcotte, 1992). It has the value $H<1$. Δz is the maximum difference in "elevation" over the distance Δr. From an analysis of the topography of Coulombic faults within both S2 fresh-water ice and S2 first-year arctic sea ice deformed in the laboratory, Fortt (2006) found that H(fresh-water ice)$=0.88$ and H(sea ice)$=0.77$, similar to the value $H=0.85$ obtained by Weiss (2001a) from the tensile fracture surface of S2 sea ice. He also found for a crack of uncertain origin within the first-year cover on the Beaufort Sea that H(crack)$=$ 0.90 (Fortt, 2006), similar to the value $H=0.86$ reported by Weiss (2001b) from an analysis of the profile presented by Parsons (1993) for a thermal crack within the sea ice cover on Allan Bay, Canada. Figure 15.10 summarizes these observations. Could the similarity in H-values on scales large and small account for the similarity in the friction coefficient? Again, as noted at the end of Chapter 12, friction is a property that must be better understood.

15.3.2 Stress versus strain rate

Returning to the question of viscous flow, Weiss *et al.* (2007) combined SHEBA stress data with RADARSAT-RGPS measurements of strain rate. More specifically, they compared RGPS-derived average shear strain rates $\dot{S}$ (defined above), obtained from displacements that occurred over periods of 3 days from winter-long observations of a $5\times5\,km^2$ cell NE of the SHEBA ship, against the maximum shear stress $|\tau|$ that corresponded to each RGPS observation. The maximum shear stress was obtained by averaging the hourly measured stresses over the same 3-day period, where $\tau=(\sigma_1-\sigma_2)/2$; the principal stresses σ_1 and σ_2 were averaged from five sensors within the cell. In so doing, they obtained a series of 78 values of average strain rate and average stress, for the period from early November 1997 to

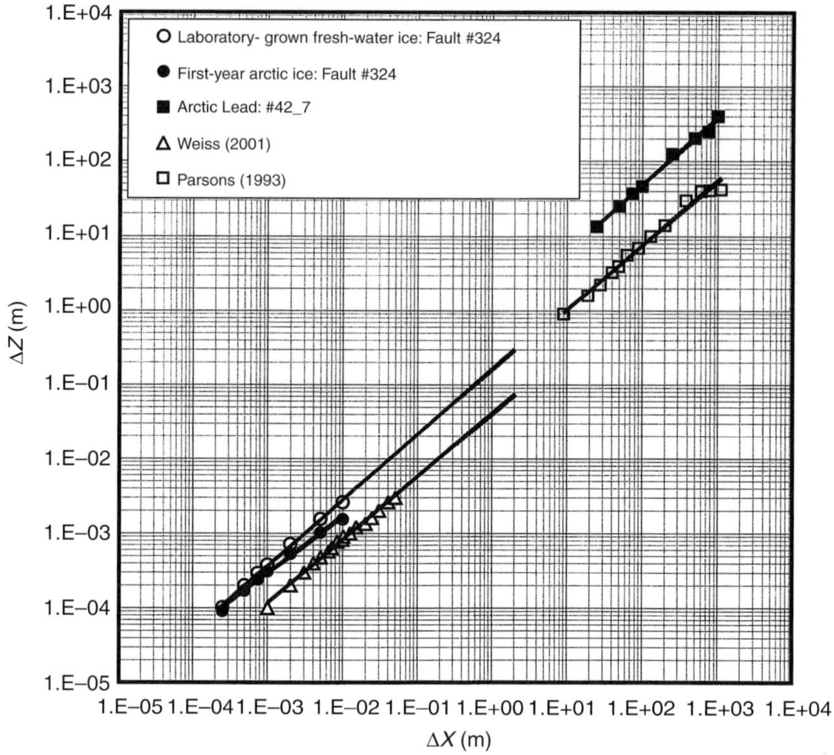

Figure 15.10. Hurst-type plots of the roughness of Coulombic shear faults within S2 fresh-water ice and S2 first-year sea ice deformed in the laboratory (from Fortt, 2006), of the roughness of the fracture surface of a tensile specimen of sea ice harvested from the Baltic Sea (from Weiss, 2001a), and of the roughness of cracks within the first-year ice cover on the Arctic Ocean (from Fortt, 2006) and on Allan Bay, Canada (from Parsons, 1993). The parameter Δz denotes the maximum difference in "elevation" over the distance Δx. (From Fortt, 2006.)

late April 1978. Figure 15.11 shows the result (a number of points overlap). Contrary to the dictates of viscous flow, no correlation is evident between stress and strain rate.

Had creep contributed significantly, the deformation rate would have been much lower than observed. That is to say, under a stress of $|\tau| = 50\,\text{kPa}$, which is about the highest SHEBA stress, the creep rate would have approached about $10^{-9}\,\text{s}^{-1}$ at $-2\,°\text{C}$. This is based upon measurements by Barnes *et al.* (1971) for fresh-water ice, multiplied by a factor of 10 to account for the accelerating effect of brine (de La Chapelle *et al.*, 1995; Cole *et al.*, 1998). However, the ice sheet deformed at rates about one hundred times greater than this (Fig. 15.11); i.e., around $\sim 10^{-7}\,\text{s}^{-1}$ or $10^{-2}\,\text{d}^{-1}$. Even when a correction is made for transmitting stress through thin ice using a two-level ice-thickness model, the difference between the rate expected

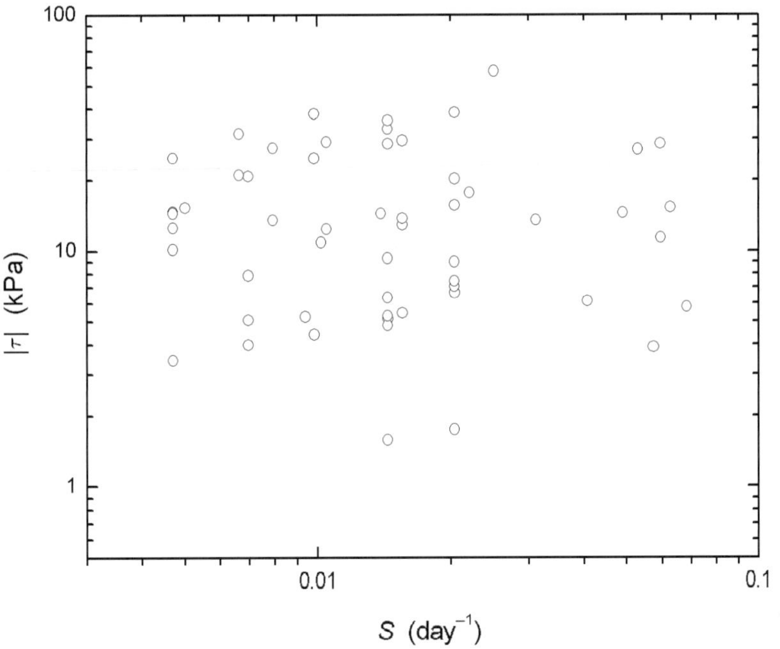

Figure 15.11. Average shear strain rate $\dot{S}$ measured over a $5 \times 5\ \mathrm{km}^2$ cell within the SHEBA camp, plotted against the maximum shear stress $|\tau|$ averaged over the time windows of the RGPS-displacement observations and over five stress sensors situated within the cell. No correlation is evident. (From Weiss *et al.*, 2007.)

from viscous flow and the observed rate is still large. Moreover, because the area fraction of thin ice is small, its contribution to global deformation is expected to be small as well. Furthermore, any additional increment that might result from twice-daily cyclic deformation (Kwok *et al.*, 2003) would be expected to increase the difference, as would a lower ice temperature.

Implicit in the preceding paragraph is the assumption that viscous flow originates from power-law or dislocation creep (described in Chapter 6). Hibler (1977), and Nye (1973) before him, demonstrated that flow with viscous-like character, in which deformation rate increases with increasing stress, could originate independently from the averaging of "plastic" or fracture events. In this regard, Nye's model is particularly instructive. He imagines a wind-loaded ice cover in which nothing happens most of the time, but in which storm-induced fractures occur once in a while that cause the cover to deform in the direction of the applied stress and by an amount proportional to the stress. The "viscosity" so observed would be inversely related to the frequency and intensity of the storms, rather than to any property of the ice. Whichever way one views viscosity, the SHEBA data point more to its absence than to its presence.

The skeptic might argue with this result. S/he might note that the actual cell within which RGPS-observed displacements occurred may have been smaller than 5 km, and that the actual time over which the displacements occurred may have been shorter than 3 days. The problem with that argument is that, for a given displacement, reductions in both cell size and time would have the effect of increasing the strain rate and thus of heightening the difference between the observed rate and the rate expected from viscous flow.

Thus, echoing Nye's (1973) sense of many years ago, the contribution from viscous flow/creep to the deformation of the SHEBA ice cover appears to have been very small indeed. The principal contribution, in keeping with the discussion in Section 15.2, probably originated from episodes of frictional sliding and fracture on a wide range of spatial and temporal scales, even under low stresses (Weiss *et al.*, 2007). The relative contribution from each scale is presumably a function of the size distribution of the sliding fracture features and of the relationship between differential displacement and the length of features. The values of these parameters are not yet known.

This interpretation, incidentally, does not preclude tensile fracturing (e.g., thermal cracking). Such events can lead to cracks of a variety of sizes that then can serve as sliding interfaces, as already noted. However, it does consider tensile fracturing itself to contribute only a small amount to the overall deformation of the cover, in keeping with the dominance of shear deformation versus total deformation that is generally observed in RGPS data (Lindsay, 2002; Stern and Moritz, 2002; Coon *et al.*, 2007).

15.3.3 Another case: scaling laws

The reader may wonder whether the SHEBA case alone is sufficient basis for taking the position that viscous flow does not contribute significantly to the deformation of the arctic sea ice cover. A more exhaustive study also supports this position.

In a recent analysis of data that have been obtained from the International Arctic Buoy Program (IABP) over a period of more than 20 years, on spatial scales from 300 m to 300 km and on temporal scales from 3 hours to 3 months, Rampal *et al.* (2008) found that the ice cover deforms at a rate, $\dot{\varepsilon}_{\mathrm{tot}} = \sqrt{\dot{D}^2 + \dot{S}^2}$, that depends upon both space L and time τ: the rate obeys scaling laws where the spatial scaling exponent is a function of time and the temporal scaling exponent is a function of space. Specifically, $\dot{\varepsilon}_{\mathrm{tot}} \sim L^{-\beta(\tau)}$ and $\dot{\varepsilon}_{\mathrm{tot}} \sim \tau^{-\alpha(L)}$. The exponent β decreases with increasing time scale, ranging in winter from to $\beta(1\,\mathrm{h}) = 0.85$ to $\beta(1\,\mathrm{month}) = 0.35$, Figure 15.12. Similarly, the exponent α decreases with increasing spatial scale, ranging from $\alpha(1\,\mathrm{km}) = 0.89$ to $\alpha(300\,\mathrm{km}) = 0.30$, Figure 15.13. Physically, β may

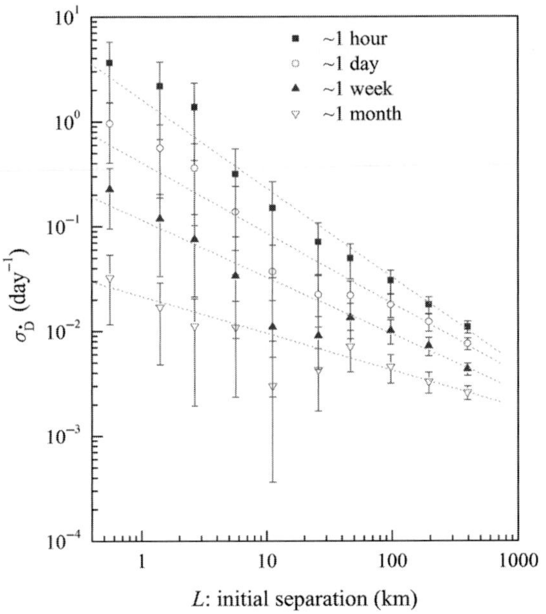

Figure 15.12. Strain rate vs. the initial separation distance for different temporal scales of arctic sea ice deformation during winter. (From Rampal *et al.*, 2008.)

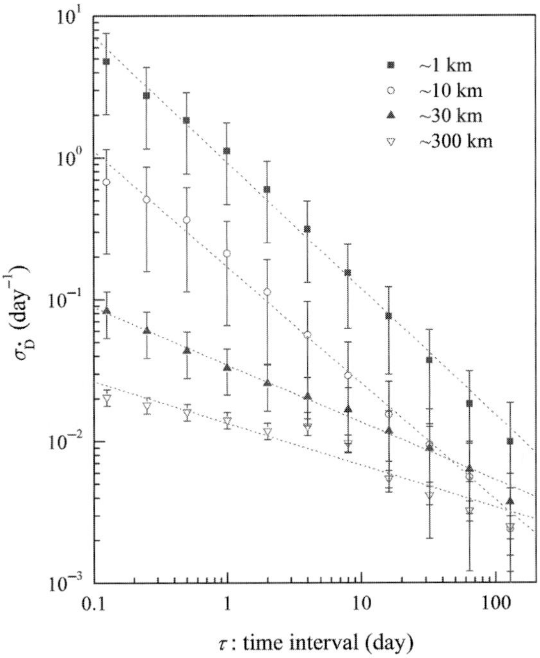

Figure 15.13. Strain rate vs. time for different spatial scales of arctic sea ice deformation during winter. (From Rampal *et al.*, 2008.)

be viewed as a measure of spatial heterogeneity of deformation and α, as a measure of the degree of temporal intermittency. This means that over the >20 year IABP period, the arctic sea ice cover deformed in a heterogeneous and intermittent manner, just like the SHEBA cover. Such behavior is inconsistent with viscous flow, even at the largest spatial and temporal scales, but is consistent with multiscale fracture and frictional sliding.

15.4 Ductile-to-brittle transition

We consider now the third question posed in the Introduction; namely, why the sea ice cover exhibits brittle behavior while laboratory specimens harvested from the cover exhibit ductile behavior when compressed at similar temperatures and deformation rates.

One explanation relates to the size of structural defects (Schulson, 2004). The theory of the ductile-to-brittle transition, described in Chapter 13, holds that for material that deforms homogeneously on a macroscopic scale the transition strain rate decreases with increasing defect size, scaling as $\dot{\varepsilon}_{tc} \propto 1/(2c)^{3/2}$. This means that upon increasing the size from the scale within test specimens, say 10 mm, to the scale within the ice cover, say >1 m, the transition strain rate is expected to decrease by a factor of $>10^3$. On this basis, and upon taking the value $\dot{\varepsilon}_{tc,lab} \sim 10^{-4}\,s^{-1}$ for test specimens (Chapter 13), the ice sheet would be expected to exhibit brittle behavior when deforming, as it generally does, at rates $\dot{\varepsilon}_{tc,field} < 10^{-7}\,s^{-1}$.

A problem with that explanation is that the ice cover does not deform homogeneously, as already noted. A better explanation relates to the heterogeneity or localization of the deformation (Marsan *et al.*, 2004; Rampal *et al.*, 2008). From a multifractal analysis Marsan *et al.* (2004) calculated the cumulative probability of observing a given deformation rate at different spatial scales. They found that at the spatial scale of 1 m on a time scale of 3 days about 15% of the deformation is accommodated in the brittle regime (i.e., $\dot{\varepsilon} > 10^{-4}\,s^{-1}$ or $> 10\,d^{-1}$), corresponding to about 0.2% of the surface area of the ice cover. A similar result was obtained by Rampal *et al.* (2008) from the analysis described above: on a spatial scale of 100–1000 m and a time scale of 3 hours about 20% of the deformation occurred at a strain rate $>10^{-4}\,s^{-1}$. In other words, the brittle behavior on the larger scale appears to originate in highly localized, high-strain-rate deformation on the smaller scale.

15.5 Is the physics of fracture independent of scale?

Finally, we return to the question posed in the Prologue. It is now apparent that certain deformation features (wing cracks, comb cracks, shear faults) and failure envelopes resemble each other on scales large and small; that similar physical mechanisms account for behavior on both the small and the large scales; and that on both scales the ductile-to-brittle transition can be understood in terms of a competition between creep and fracture. Large-scale deformation appears to be more complex, characterized as it is by intermittency and by spatial variability, but even those characteristics are probably scale independent as evident from the microstructural evolution in the development of a Coulombic fault (Chapter 12). These points seem to speak of scale-independent fracture physics.

Agreement, however, appears not to be unanimous. Overland *et al.* (1995) and McNutt and Overland (2003) argue against that view. Instead, they invoke hierarchy theory of complexity (for review, see O'Neill *et al.*, 1986) and hold that it is the degree of disconnectedness between scales that accounts for the organization of the system, not the similarity of physical mechanisms. The idea is that there are emergent properties that represent discontinuities in scale where the whole is greater than the sum of the parts. Transitions, for example, are specified as somewhere around the "floe scale" of 1 km, around the "multi-floe scale" of 2–10 km, and around the "aggregate scale" of 10–75 km. To support their view, Sanderson's (1988) observation (Chapter 14) is cited; namely, that the indentation pressure of arctic ice sheets decreases as the size of the sheet/indenter increases, from $\sim$10 MPa at $10^{-4}\,\mathrm{m}^2$ to $\sim$0.01 MPa at $10^5\,\mathrm{m}^2$.

A problem with the hierarchy argument is that the different scales seem to be defined rather arbitrarily. Also, hierarchy theory seems to defy the results of fractal analyses (Weiss, 2001, 2003), which reveal neither a characteristic length in the fragmentation pattern of ice in over eight orders of magnitude, from <1 mm to $\sim$100 km, which includes the proposed transition scales, nor a characteristic length in the deformation of the winter sea ice cover (Marsan *et al.*, 2004). As to the difference in strength, that observation could be explained, as already discussed, in terms of the difference in size of structural defects that weaken the ice.

Our sense at this juncture, therefore, is that the evidence for scale independence is stronger than the argument against.

References

Ashby, M. F. and S. D. Hallam (1986). The failure of brittle solids containing small cracks under compressive stress states. *Acta Metall.*, **34**, 497–510.

Barnes, P., D. Tabor, F. R. S. Walker and J. F. C. Walker (1971). The friction and creep of polycrystalline ice. *Proc. R. Soc. Lond. A*, **324**, 127–155.

Cole, D. M. R. A. Johnson and G. D. Durell (1998). Cyclic loading and creep response of aligned first-year sea ice. *J. Geophys. Res.*, **103**, 21,751–21,758.

Coon, M. D., G. A. Maykut, R. S. Pritchard, D. A. Rothrock and A. S. Thorndike (1974). Modeling the pack-ice as an elastic-plastic material. *AIDJEX Bull.*, **24**, 1–105.

Coon, M. D., G. S. Knoke, D. C. Echert and R. S. Pritchard (1998). The architecture of an anisotropic elastic-plastic sea ice mechanics constitutive law. *J. Geophys. Res.*, **103**, 21,915–21,925.

Coon, M., R. Kwok, G. Levy *et al.* (2007). Arctic ice dynamics joint experiment (AIDJEX) assumptions revisited and found inadequate. *J. Geophys. Res.*, **112**, doi: 10.1029/2005JC003393.

Cox, G. F. N. and J. B. Johnson (1983). Stress measurements in ice. *CRREL Report*, No. 83-23.

de La Chapelle, S., P. Duval and B. Baudelet (1995). Compressive creep of polycrystalline ice containing a liquid phase. *Scr. Metall. Mater.*, **33**, 447–450.

Erlingsson, B. (1988). Two-dimensional deformation patterns in sea ice. *J. Glaciol.*, **34**, 301–308.

Fortt, A. (2006). The resistance to sliding along coulombic shear faults in columnar S2 ice. Ph.D. thesis, Thayer School of Engineering, Dartmouth College.

Fortt, A. L. and E. M. Schulson (2007). The resistance to sliding along coulombic shear faults in ice. *Acta Mater.*, **55**, 2253–2264.

Heil, P. and W. D. Hibler (2002). Modelling the high-frequency component of arctic sea-ice drift and deformation. *J. Phys. Oceanogr.*, **32**, 3039–3057.

Hibler, W. D. (1977). A viscous sea ice law as a stochastic average of plasticity. *J. Geophys. Res.*, **82**, 3932–3938.

Hibler, W. D. (1979). Dynamic thermodynamic sea ice model. *J. Phys. Oceanogr.*, **9**, 815–846.

Hibler, W. D. (2003). Modeling the dynamic response of sea ice. In *Mass Balance of the Cryosphere*, ed. J. Bamber. Cambridge: Cambridge University Press.

Hibler, W. D. I. and E. M. Schulson (2000). On modeling the anisotropic failure and flow of flawed sea ice. *J. Geophys. Res.*, **105**, 17,105–17,120.

Hunke, E. C. (2001). Viscous-plastic sea ice dynamics with the EVP model: Linearization issues. *J. Comput. Phys.*, **170**, 18–38.

Hunke, E. C. and J. K. Dukowicz (1997). An elastic-viscous-plastic model for sea ice dynamics. *J. Phys. Oceanogr.*, **27**, 1849–1867.

Hutchings, J. K. and W. D. Hibler (2003). Modelling sea ice deformation with a viscous-plastic isotropic rheology. In *Ice in the Environment*, eds. V. Squire and P. Langhorne. Dunedin, New Zealand: University of Otago Press, pp. 358–366.

Jaeger, J. C. and N. G. W. Cook (1979). *Fundamentals of Rock Mechanics*, 3rd edn. London: Chapman and Hall.

Kerr, A. D. (1978). On the determination of horizontal forces a floating ice sheet exerts on a structure. *CRREL Report*, 78-15.

Kwok, R. E. (1998). The radarsat geophysical processor system. In *Analysis of SAR Data of the Polar Oceans*, eds. C. Tsatsoulis and R. Kwok. Berlin: Springer-Verlag, pp. 235–257.

Kwok, R. (2001). Deformation of the Arctic Ocean sea ice cover between November 1996 and April 1997: A qualitative survey. In *Scaling Laws in Ice Mechanics*, eds. J. P. Dempsey and H. H. Shen. Dordrecht: Kluwer Academic Publishing, pp. 315–322.

Kwok, R. (2006). Contrasts in Arctic Ocean sea ice deformation and production in the seasonal and perennial ice zones. *J. Geophy. Res.*, **111**, doi:10.1029/2005JC003246.

Kwok, R. and M. D. Coon (2006). Introduction to special section: Small-scale sea ice kinematics and dynamics. *J. Geophys. Res.*, **111**, doi:10.1029/2006JC003877.

Kwok, R., D. A. Rothrock, H. L. Stern and G. F. Cunningham (1995). Determination of ice age using lagrangian observations of ice motion. *IEEE Trans. Geosci. Remote Sens.*, **33**, 392–400.

Kwok, R., G. F. Cunningham and W. D. Hibler (2003). Sub-daily sea ice motion and deformation from radarsat observations. *Geophys. Res. Lett.*, **30**, 2218.
Laxon, S., N. Peacock and D. Smith (2003). High interannual variability of sea ice thickness in the arctic region. *Nature*, **425**, 947–950.
Lindsay, R. (2002). Ice deformation near SHEBA. *J. Geophys. Res.*, **107**, 8042.
Lindsay, R. W. and D. A. Rothrock (1995). Arctic sea-ice leads from advanced very high-resolution radiometer images. *J. Geophys. Res.*, **100**, 4533–4544.
Lockner, D. A., J. D. Byerlee, V. Kuksenko, A. Pnomarev and A. Sidorin (1991). Quasi-static fault growth and shear fracture energy in granite. *Nature*, **350**, 39–42.
Marko, J. R. and R. E. Thomson (1977). Rectilinear leads and internal motions in the ice pack of the western Arctic Ocean. *J. Geophys. Res.*, **82**, 979–987.
Marsan, D., H. Stern, R. Lindsay and J. Weiss (2004). Scale dependence and localization of the deformation of arctic sea ice. *Phys. Rev. Lett.*, **93**, 178501.
Maykut, G. A. (1982). Large-scale heat exchange and ice production in the central Arctic. *J. Geophys. Res.*, **87**, 7971–7984.
McNutt, S. L. and J. E. Overland (2003). Spatial hierarchy in arctic sea ice dynamics. *Tellus Ser. A*, **55**, 181–191.
McPhee, M. G., R. Kwok, R. Robins and M. Coon (2005). Upwelling of arctic pycnocline associated with shear motion of sea ice. *Geophys. Res. Lett.*, **32**, L10616.
Muehlberger, W. R. (1961). Conjugate joint sets of small dihedral angle. *J. Geol.*, **69**, 211–219.
Nye, J. F. (1973). *AIDJEX Bulletin* No. 21, July 1973, pp. 18–19, ed. A. Johnson. University of Washington.
O'Neill, R. V., D. L. DeAngelis, J. B. Waide and T. F. H. Allen (1986). *A Hierarchical Concept of Ecosystems*. Princeton: Princeton University Press.
Overland, J. E. and J. Ukita (2000). Dynamics of arctic sea ice discussed at workshop. *EOS*, **81**, July 11, 2000.
Overland, J. E., B. A. Walter, T. B. Curtin and P. Turet (1995). Hierachy and sea ice mechanics: A case study from the Beaufort Sea. *J. Geophys. Res.*, **100**, 4559–4571.
Overland, J. E., S. L. McNutt, S. Salo, J. Groves and S. S. Li (1998). Arctic sea ice as a granular plastic. *J. Geophys. Res.*, **103**, 21,845–21,867.
Parsons, B. L. (1993). *The Application of Fractal/Chaos Concepts to Ice Mechanics: A Review.* NRC Canada, Institute for Marine Dynamics.
Perovich, D. K., E. L. Andreas, J. A. Curry *et al.* (1999). Year on the ice gives climate insights. *EOS, Trans. Amer. Geophys. Union*, **80**, 481, 485–486.
Petrenko, V. F. and O. Gluschenkov (1996). Crack velocities in freshwater and saline ice. *J. Geophys. Res.*, **101** (B5), 11,541–11,551.
Pritchard, R. (1975). An elastic-plastic constitutive law for sea ice. *J. Appl. Mech. Trans. ASME*, **42**, 379–384.
Rampal, P., J. Weiss, D. Marsan, R. Lindsay and H. Stern (2008). Scaling properties of sea ice deformation from buoy dispersion analysis. *J. Geophys. Res.*, **113** (CO3002), doi: 10.1029/2007JC004143.
Richter-Menge, J. A. and B. C. Elder (1998). Characteristics of pack ice stress in the Alaskan Beaufort Sea. *J. Geophys. Res.*, **103**, 21,817–21,829.
Richter-Menge, J. A., S. L. McNutt, J. E. Overland and R. Kwok (2002). Relating arctic pack ice stress and deformation under winter conditions. *J. Geophys. Res.*, **107** (C10), 8040.
Russ, J. C. (1994). *Fractal Surfaces*. New York: Plenum Press.
Sanderson, T. J. O. (1988). *Ice Mechanics: Risks to Offshore Structures*. London: Graham & Trotman.

Schulson, E. M. (2001). Brittle failure of ice. *Eng. Fract. Mech.*, **68**, 1839–1887.
Schulson, E. M. (2004). Compressive shear faults within the arctic sea ice cover on scales large and small. *J. Geophys. Res.*, **109**, 1–23.
Schulson, E. M. and W. D. Hibler (1991). The fracture of ice on scales large and small: Arctic leads and wing cracks. *J. Glaciol.*, **37**, 319–323.
Schulson, E. M. and W. D. I. Hibler (2004). Fracture of the winter sea ice cover on the Arctic Ocean. *C. R. Physique*, **5**, 753–767.
Schulson, E. M., A. L. Fortt, D. Iliescu and C. E. Renshaw (2006a). On the role of frictional sliding in the compressive fracture of ice and granite: Terminal vs. post-terminal failure. *Acta Mater.*, **54**, 3923–3932.
Schulson, E. M., A. Fortt, D. Iliescu and C. E. Renshaw (2006b). Failure envelope of first-year arctic sea ice: The role of friction in compressive fracture. *J. Geophys. Res.*, **111**, doi:10.1029/2005JC003234186.
Stern, H. L. and R. E. Moritz (2002). Sea ice kinematics and surface properties from radarsat synthetic aperture radar during the SHEBA drift. *J. Geophys. Res.*, **107**, 8028–8038.
Stern, H. L. and D. A. Rothrock (1995). Open water production in arctic sea ice: Satellite measurements and model parameterizations. *J. Geophys. Res.*, **100**, 20,601–20,612.
Sylvester, A. G. (1988). Strike-slip faults. *Geol. Soc. Am. Bull.*, **100**, 1666–1703.
Turcotte, D. L. (1992). *Fractals and Chaos in Geology and Geophysics*. Cambridge: Cambridge University Press.
Vavrus, S. J. and S. P. Harrison (2003). The impact of sea ice dynamics on the arctic climate system. *Climate Dynamics*, **20**, 741–757.
Walter, B. A. and J. E. Overland (1993). The response of lead patterns in the Beaufort Sea to storm-scale wind forcing. *Ann. Glaciol.*, **17**, 219–226.
Walter, B. A., J. E. Overland and P. Turet (1995). A comparison of satellite-derived and aircraft-measured regional surface sensible heat fluxes over the Beaufort Sea. *J. Geophys. Res.*, **100**, 4584–4591.
Weiss, J. (2001a). Fracture and fragmentation of ice: A fractal analysis of scale invariance. *Eng. Fract. Mech.*, **68**, 1975–2012.
Weiss, J. (2001b). Scale invariance of fracture surfaces in ice. In *IUTAM Symposium on Scaling Laws in Ice Mechanics and Ice Dynamics*, eds. J. P. Dempsey and H. H. Shen. Dordrecht: Kluwer Academic Publishers, pp. 217–226.
Weiss, J. (2003). Scaling of fracture and faulting of ice on earth. *Surv. Geophys.*, **24** (2), 185–227.
Weiss, J., E. M. Schulson and H. L. Stern (2007). Sea ice rheology in-situ, satellite and laboratory observations: Fracture and friction. *Earth Planet. Sci. Lett.*, doi: 10.1016/j.epsl.2006.11.033.
Zhang, J. L. and D. A. Rothrock (2005). Effect of sea ice rheology in numerical investigations of climate. *J. Geophys. Res.*, **110** (C8), C08014.

Index